Lecture Notes in Electrical Engineering

Volume 208

For further volumes:
http://www.springer.com/series/7818

Wenjiang Du
Editor

# Informatics and Management Science V

Springer

*Editor*
Wenjiang Du
College of Elementary Education
Chongqing Normal University
Chongqing
People's Republic of China

ISSN 1876-1100 ISSN 1876-1119 (electronic)
ISBN 978-1-4471-4795-4 ISBN 978-1-4471-4796-1 (eBook)
DOI 10.1007/978-1-4471-4796-1
Springer London Heidelberg New York Dordrecht

Library of Congress Control Number: 2012952024



Printed on acid-free paper

Springer is part of Springer Science+Business Media (www.springer.com)

# Preface

Welcome to the proceedings of the International Conference on Informatics and Management Science (IMS) 2012, which will be held in December 21–23, 2012, in Kunming, China.

IMS 2012 will be a venue for leading academic and industrial researchers to exchange their views, ideas and research results on innovative technologies, and sustainable solutions leading to Informatics and Management Science. The conference will feature keynote speakers, a panel discussion, and paper presentations.

The objective of IMS 2012 is to facilitate an exchange of information on best practices for the latest research advances in the area of Informatics and Management Science. IMS 2012 will provide a forum for engineers and scientists in academia, industry, and government to address the most innovative research and development including technical challenges, social and economic issues, and to present and discuss their ideas, results, work in progress and experience on all aspects of Informatics and Management Science.

There was a very large number of paper submissions (2351). All submissions were reviewed by at least three Program or Technical Committee members or external reviewers. It was extremely difficult to select the presentations for the conference because there were so many excellent and interesting submissions. In order to allocate as many papers as possible and keep the high quality of the conference, we finally decided to accept 614 papers for presentations, reflecting a 26.1 % acceptance rate. We believe that all of these papers and topics not only provided novel ideas, new results, work in progress, and state-of-the-art techniques in this field, but also stimulated the future research activities in the area of Informatics and Management Science.

The exciting program for this conference was the result of the hard and excellent work of many others, such as Program and Technical Committee members, external reviewers, and Publication Chairs under a very tight schedule. We are also grateful to the members of the Local Organizing Committee for supporting us in handling so many organizational tasks, and to the keynote speakers for accepting to come to the conference with enthusiasm. Last but not least, we hope you enjoy the conference program, and the beautiful attractions of Kunming, China.

With our warmest regards.

December 2012

Wenjiang Du
Guomeng Dong
General and Program Chairs
IMS 2012

# Organization

IMS 2012 was organized by Electric Power Research Institute, YNPG, Yunnan Normal University, Wuhan Institute of Technology, Guizhou University, Chongqing Normal University, Chongqing University, Yanshan University, Xiangtan University, Hunan Institute of Engineering, Shanghai Jiao Tong University, Nanyang Technological University, and sponsored by National Natural Science Foundation of China (NSFC). It was held in cooperation with *Lecture Notes in Electrical Engineering* (LNEE) of Springer.

# Executive Committee

| | |
|---|---|
| General Chairs: | Maode Ma, Nanyang Technological University, Singapore |
| | Yuhang Yang, Shanghai Jiao Tong University, China |
| Program Chairs: | Yanchun Zhang, University of Victoria, Australia |
| | Rafa Kapelko, Wrocaw University of Technology, Poland |
| | Rongbo Zhu, Virginia Tech, USA |
| | Ming Fan, University of Washington, USA |
| Local Arrangement Chairs: | Qing Xiao, Chongqing University, China |
| | Wenjiang Du, Chongqing Normal University, China |
| Steering Committee: | Maode Ma, Nanyang Technological University, Singapore |
| | Nadia Nedjah, State University of Rio de Janeiro, Brazil |
| | Lorna Uden, Staffordshire University, UK |
| | Dechang Chen, Uniformed Services University of the Health Sciences, USA |
| | Mei-Ching Chen, Tatung University, Taiwan |
| | Rong-Chang Chen, National Taichung Institute of Technology, Taiwan |
| | Chi-Cheng Cheng, National Sun Yat-Sen University, Taiwan |
| | Naohiko Hanajima, Muroran Institute of Technology, Japan |
| | Shumin Fei, Southeast University, China |
| | Yingmin Jia, BeiHang University, China |
| | Weiguo Liu, Northwestern Polytechnic University, China |

# Program/Technical Committee

| | |
|---|---|
| Ragip Kur | Nokia Research, USA |
| Mina Gui | Texas State University, USA |
| Yanbin Sun | Beijing University of Posts and Telecommunications, China |
| Ven Prasad | Delft University of Technology, The Netherlands |
| Yajun Li | Shanghai Jiao Tong University, China |
| Mengze Liao | Cisco China R&D Center, China |
| Xiwen Hu | Wuhan University of Technology, China |
| Yajun Guo | Huazhong Normal University, China |
| Suresh Subra | George Washington University, USA |
| Mingyi Gao | National Institute of AIST, Japan |
| Yanliang Jin | Shanghai University, China |
| Haibing Yin | Peking University, China |
| Jianxin Chen | University of Vigo, Spain |
| Yuan Lin | Norwegian University of Science and Technology, Norwegian |
| Miche Rossi | University of Padova, Italy |
| Nils Asc | University of Bonn, Germany |
| Chunxiao Yu | Yanshan University, China |
| On Altintas | Toyota InfoTechnology Center, Japan |
| Guofu Gui | CMC Corporation, China |
| Dianxuan Gong | Hebei Polytechnic University, China |
| Haiyong Bao | NTT Co., Ltd, Japan |
| Shuang Cong | University of Science and Technology of China, China |
| Yangwen Zou | Apple China Co, Ltd, China |
| Tai-hon Kim | Defense Security Command, Korea |
| Sean McLoo | National University of Ireland, Ireland |
| Jian-Xin Peng | Queens University of Belfast, UK |
| Lui Piroddi | Technical University of Milan, Italy |
| Wi Richert | University of Paderborn, Germany |

| | |
|---|---|
| Sa Sharma | University of Plymouth, UK |
| Wang Bin | Chinese Academy of Sciences, China |
| Yongsheng Ding | Donghua University, China |
| Xiang Mei | The University of Leeds, UK |
| Yongning Tang | Illinois State University, USA |
| Wenbin Jiang | Huazhong University of Science and Technology, China |
| Jun Cai | University of Manitoba, Canada |
| Xingang Zhang | Nanyang Normal University, China |
| Veli Mumcu | Technical University of Yildiz, Turkey |
| Xiao Li | CINVESTAV-IPN, Mexico |
| Ruichun Tang | Ocean University of China, China |
| Michiharu Kurume | National College of Technology, Japan |
| Dong Yue | Huazhong University of Science and Technology, China |
| Qishi Wu | University of Memphis, USA |
| Zhichun Li | Northwestern University, China |
| Lisong Xu | University of Nebraska-Lincoln, USA |
| Mei Yu | Simula Research Laboratory, Norway |
| Gui-Rong Xue | Shanghai Jiao Tong University, China |
| Jalel Ben-Othman | University of Versailles, France |

# Contents

# Part I
# Control Engineering and Applications

# Chapter 1
# Research of Intelligent Uncertainty of Measurement Based on Computer-Aided Evaluation System

**Mingxiang Sui**

**Abstract** This paper discusses the relevant concepts, algorithms and the practical evaluation process on the intelligent evaluation of uncertainty of measurement with the use of computer technology. Through the optimization and integration of evaluation process, key points and regulative relationship, the author makes use of computer-aided calculation and database technology to achieve standardization of evaluation process, intelligent data analysis and automatic calculation of uncertainty.

**Keywords** Uncertainty of measurement • Regulative relationship • Fast evaluation • Nonlinear effects • Cause-effect graph

## 1.1 Introduction

With the implementation of GUM: 1993, the evaluation of uncertainty of measurement has become increasingly popular [1]. However, as a large number of terminologies, strong skills, complicated process and normal mathematical statistics foundation are required in the course of evaluation, which cause great difference on the evaluation level. Therefore, to set up a well-established computer-aided evaluation system has become an important task in the field of evaluation of uncertainty of measurement.

M. Sui (✉)
Zhongshan Polytechnic, Zhongshan, Guangdong Province, China
e-mail: hjlfoxes@163.com

W. Du (ed.), *Informatics and Management Science V*, Lecture Notes in Electrical Engineering 208, DOI: 10.1007/978-1-4471-4796-1_1,

## 1.2 Brief of System Design

Based on the in-depth study of relevant calibration regulations and guides such as "Guide to the Expression of Uncertainty in Measurement" (GUM: 1993), "Evaluation and Expression of Uncertainty in Measurement" (JJF1059-1999), "Detection and Calibration Laboratory Capacity Accreditation Criteria" (ISO/IEC 17025:2005), "General Requirements for Evaluating and Reporting Measurement Uncertainty" (CNAS-CL07), "International Vocabulary of Basic and General Terms in Metrology" (VIM) and "General Terms in Metrology and Their Definitions" (JJF1001-1998), and through the integration, optimization, summarizing and presupposition of evaluation process, evaluation technique, frequently asked questions, key points, correlation matching of elements and regulative relationship, intelligentized uncertainty of measurement computer aided evaluation system makes use of advanced theories such as monte carlo method, taguchi-method, rey system theory and bayesian theory and finally achieves standardization of evaluation process, intelligent statistical data analysis and automatic calculation of uncertainty with the help of computer-aided calculation and database technology.

## 1.3 System Design and Implementation

The main service management module of this system consists of basic information management, aided calculation system of uncertainty of measurement, intelligent analysis system and data import and export port. Users can log on the system via public portal.

### *1.3.1 Basic Information Management*

#### 1.3.1.1 Management of Name of Measuring Instruments

Based on "Norm of Designation for Working Measuring Instrument and its Classification Cades" (JJF1051-1996), and "Classification of Laboratory Accreditation Field" (CNAS-IL06), the author establishes the classified information base of name of measuring instruments by layers.

#### 1.3.1.2 Verification Regulation or Technical Specification Management

Preset information such as the implementation time, revocatory date, substitutional relation and receiving records of verification regulation or calibration specification, collect and organize the currently valid version of verification regulation or calibration specification (Word or PDF format), fetch and add the following information:

- Name of instruments and limitation in application scope.
- Measuring unit.
- Ambient condition.
- Verification or calibration items (excluding observation item, items that are unnecessary to be verified in using and do not affect the results of verification or calibration), clause number of corresponding method and maximum permissible errors in "list of verification or calibration items".
- Measuring range and technical requirements of equipments for verification or calibration.
- Verification result of measuring instruments.
- Uncertainty of result of a measurement.

#### 1.3.1.3 Management System of Measuring Unit

According to the "Legal Units of Measurement of the Peoples Republic of China" and other relevant documents, the system establishes units of measurement library based on classification of measurement by layers, and through the presupposition of modules in system, the system automatically generates units of measurement and mapping of their corresponding name of measuring instruments, verification regulation or technical specification, and the personnel in charge of measurement examination and determination. At the same time, the system also provides ordering functions for units of measurement, which can realize the automatic separation of suspected data and automatic correction of wrong units of measurement.

#### 1.3.1.4 Basic Information Management of Class B

Preset the source of uncertainty of measurement for class B and its corresponding degree of reliability.

#### 1.3.1.5 Management of Types of Probability Distribution

Management of input quantity distribution.

Establish types of probability distribution and mapping of corresponding applicable conditions, coverage factor $k$ value and the source of uncertainty.

Management of measured $Y$ distribution.

Establish the distribution type and judgment principle for probable value of measured y, so that the system can automatically compare and judge the values and distribution of input quantity $X$ according to the judgment principle, and automatically calculate the distribution of measured $Y$ and the corresponding coverage factor $k_p$ value.

Establish mapping of normal distribution, fiducial probability $p$ and coverage factor $k_p$.

Establish data set of T-distribution table for automatic calculation of effective degrees of freedom tp(v).

#### 1.3.1.6 Management of Source of Uncertainty in Measurement

According to authoritative document such as GUM 1993, define and classify the source of uncertainty in terms of personal error, integrated stability of the instrument, standard reference materials used, error of method and environmental error, and estimate the possible distribution.

#### 1.3.1.7 Management of Mathematical Model

Based collect and regulate various mathematic models, and preset linear, uncertain types (absolute uncertainty or relative uncertain) to be chosen, simplified processing principle of nonlinear model and variance expression of $Y$ to be measured.

#### 1.3.1.8 Correlation Management of Input Quantity

Preset the reason, judgment principle and processing method for the occurrence of input quantity correlation.

Set whether the automatic check has uncertainty components of measurement $u(\mathrm{q})$ resulted from resolution to judge $s(\mathrm{q}_k)$ value of experimental standard deviation to choose the bigger one.

Preset the suspected judgment principle of correlation between input quantities.

#### 1.3.1.9 Equipment Management

Users can preset the basic information of their experimental equipments and the basic information is classified into two categories. One is fixed information including metrological characteristics of equipment at the time of purchasing. The metrological characteristics include measurable parameters and corresponding resolution, minimum division value, measuring range, uncertainty of measurement, accuracy class, and maximum permissible error; the other is submittal information for inspection including metrological characteristics of equipments submitted for inspection each time.

## 1.3.2 Evaluation Management of Uncertainty of Measurement

The evaluation of uncertainty of measurement can be proceeded according to Fig. 1.1.

### 1.3.2.1 Formation of Cause-Effect Graph

Choose measuring instrument, and the system will automatically correlate the instrument's corresponding verification regulation or technical specification and its electronic copy. Users snatch principle of measurement in electronic copy to create

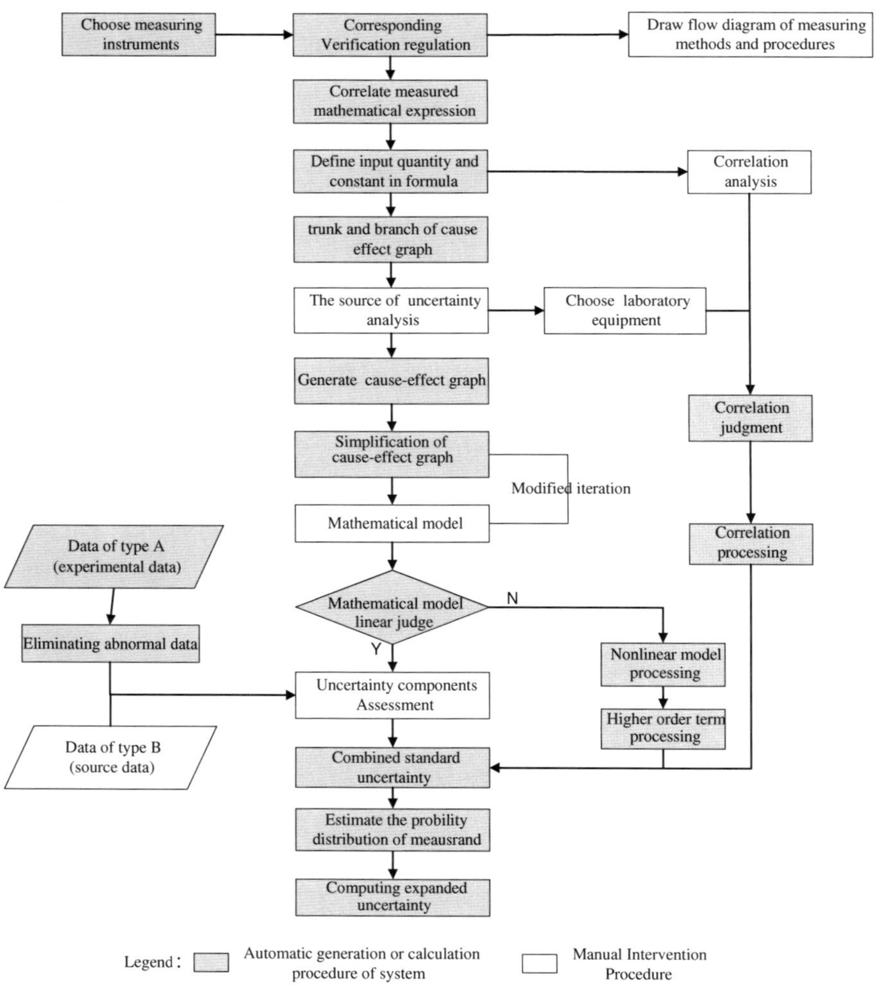


**Fig. 1.1** Flow diagram for evaluation of uncertainty of measurement

flow diagram for method and procedure of measurement by using Adobe Reader. At the same time, snatch the measured mathematic expression so that users can freely define the measurand, input quantity and constant. After being defined by users, the system will automatically generate measurand, input quantity and input quantity branch of cause-effect graph, namely the trunk and distribution of cause-effect graph [2].

#### 1.3.2.2 Analysis of Source of Uncertainty

Users analyze each distribution at the bottom in terms of the source of uncertainty and the system will provide possible list (Fig. 1.2) of the source of uncertainty for reference. Users can make analysis according to the following procedures:

- Define the quantity of branch based on the actual situation (e.g. define several experimental equipments).
- Re-expand the source of uncertainty provided by the system.
- Through the analysis of the source of uncertainty, choose the source of uncertainty provided by the system.
- Add other source of uncertainty excluded in the system freely.

#### 1.3.2.3 Simplification of Cause-Effect Graph

The system automatically simplifies the cause-effect graph and judges the repetition of source of uncertainty, and furthermore, marks the similar source. Users can freely delete and integrate the quits element according to comprehensive analysis of cause-effect graph.

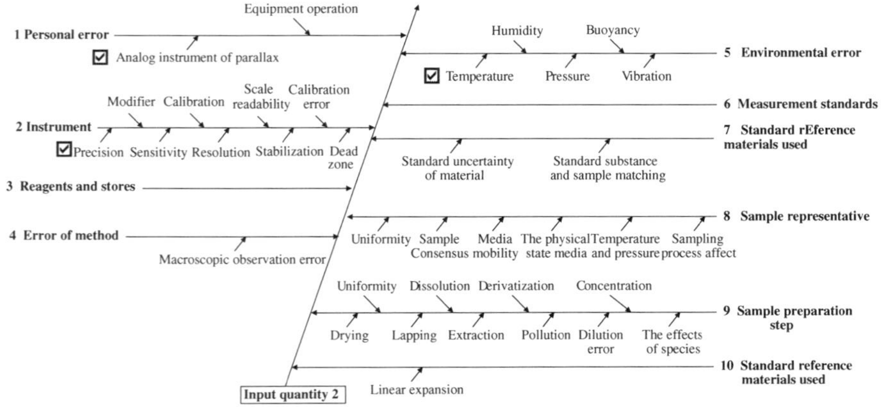


**Fig. 1.2** Aided analysis cause-effect graph of source of uncertainty

#### 1.3.2.4 Simplification of Source of Uncertainty

According to the following steps, users can delete source of uncertainty which has a negligible effect on the result.

If the same measuring instrument is used to measure a certain measurand, this measuring instrument may have systematic deviation which centralizes the source of uncertainty of measurement as main branches.

The source of uncertainty of measurement can be centralized as main branches. For example, in the increasing repeated branches, centralize the repeatability of each main branch to be a single component so that the effect element will not repeat and increase new items to formula.

#### 1.3.2.5 Establish Mathematical Model

Users can judge whether the expressions have made correction to all system error and random error according to measured math expressions. If not, users can increase correction term or correction factor based on real situation. In the course of increasing the correction term, system requires users to define the correction types. Correction from system error corresponds with the transparent box model in mathematical model, while correction from random error corresponds with the black box model in mathematical model [3].

As for the complex mathematical model, system will give users a hint to judge whether the demand of uncertainty component is too high. If not, whether users can use the black box model to estimate the size of uncertainty component according to the experience.

#### 1.3.2.6 Judgment of Nonlinear Model and Processing of High-Order Term

System judges linearity of mathematical model. As for nonlinear model, the system simplifies variance expression of measured $y$ with presupposition by use of mathematical model base, and defines type of uncertainty (absolute or relative). For these failed to be simplified, use propagation law of uncertainty and take processing of high-order term. Moreover, system helps to process nonlinear model by using tools such as Monte Carlo method, least squares method, etc.

#### 1.3.2.7 Evaluation of Uncertainty Component

After the mathematical model is built, the system will automatically supplement the cause-effect graph of source of uncertainty and produce a summary table of uncertainty component layer by layer with input quantity, input quantity branch and bottom layer branch in accordance with cause-effect graph.

In the production of sensitivity coefficient $c_i(x_i)$, if the mathematical model is nonlinear, the system will remind users to select mathematical expectation of input as input quantity. If there is no reliable sensitivity mathematical expression, the system will also remind users to use experimental method.

In the course of defining standard uncertainty, the system will take the following processing according to different evaluation methods that users have chosen.

Type A Evaluation of uncertainty flow, see the following graph (Fig. 1.3).

Type B Evaluation of uncertainty flow, shown below:

- If users can provide the expanded uncertainty $U(x)$ of measured $x$ and coverage factor $k$, the system will automatically calculate its standard uncertainty of according to formula $u(x) = \frac{U(x)}{k}$.
- If users can provide the expanded uncertainty U(x) of measured x and its corresponding fiducial probability p, calculate its standard uncertainty of according

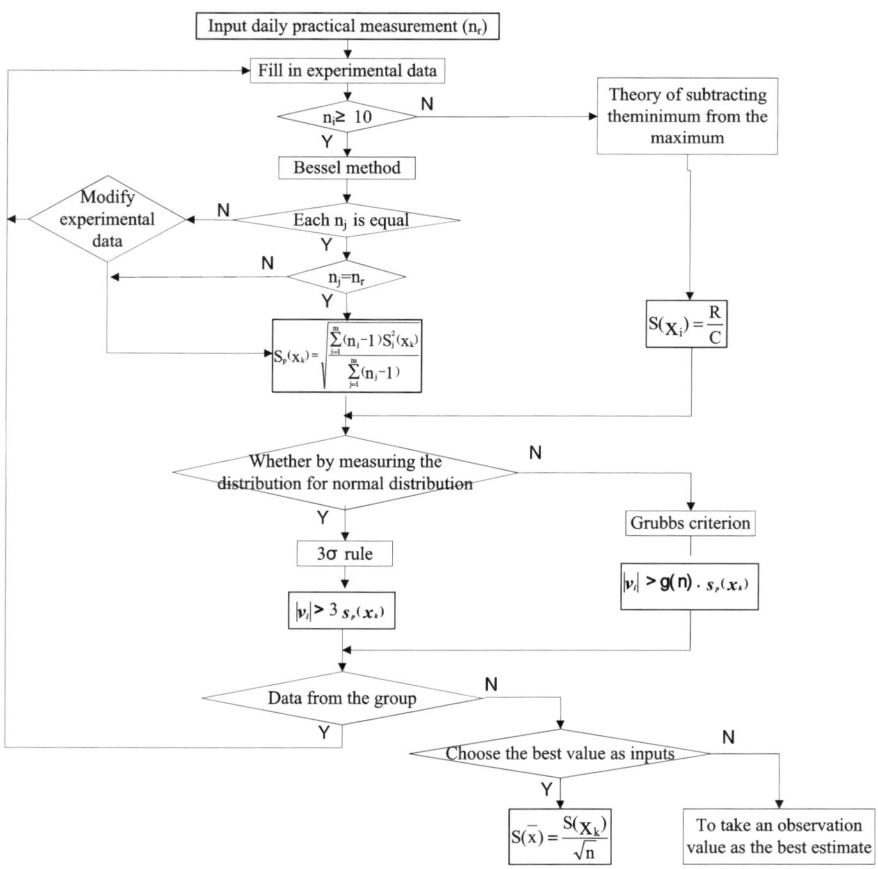


**Fig. 1.3** Type A evaluation flow diagram of uncertainty

to formula $u(x) = \frac{U(x)}{k_p}$ with the given kp. If there is no kp, look up kp according to the following table and calculate its standard uncertainty.
- As for $k_p$, it is possible to calculate it according to the estimation of the unreliability of data source made by evaluating personnel.

#### 1.3.2.8 Combined Standard Uncertainty

After users finish the evaluation of each component uncertainty, the system can automatically evaluate the measured distribution and $k$ value according to the presupposition information in the measured $Y$ distributing management module. It can also automatically calculate the measured combined uncertainty and expanded uncertainty and output uncertainty report that meet the requirement.

#### 1.3.2.9 Output of Evaluation Report of Uncertainty of Measurement

System can automatically output uncertainty of measurement report that meet the requirement.

## 1.4 Validation Management of Verification or Calibration Results

Add expanded uncertainty as a result of being measured by appraisal measurement criteria and senior measurement criteria and corresponding measuring result of measured value of the uncertainty for equipments marked as "verification standard" in the management of equipment. The system will automatically carry out calculation and have the validation of verification or calibration. When transmission comparison method is unable to be used, the system will provide the comparative measurement method for validation of measurement uncertainty [4].

## 1.5 Online Interactive Platform

Online interactive platform provides a space for users to communicate and study with each other and exchange information. With this platform, users can share, build and communicate information of uncertainty of measurement. In this way, users can achieve united and standardized management of evaluation flow of uncertainty of measurement to the great extent and avoid repetitive building.

## 1.6 Automatic Warning System

With the promulgation of new rules and standards, the system will remind users to update rules through real-time communication, short message, and confirm the measuring range, accuracy class, personnel training, operation guide, result of examine or calibration certificate, data model, evaluation of uncertainty through new methods.

## 1.7 Summary

Through the unified planning and building of basic information for evaluation of uncertainty of measurement, the system helps users free themselves from the previously complicated calculations and concentrate their attentions on the research of theory and evaluation method of uncertainty of measurement. Meanwhile, with the application of centralized data storage and online interactive platform, the system realizes the share, co-building and information interaction of uncertainty of measurement. Furthermore, it greatly guarantees the unification, normalization and standardization of evaluation flow of uncertainty of measurement and avoids the repetitive building.

## References

1. International Bureau of Weights and Measures, ISO (1995) Guide to the expression of uncertainty in measurements, 2nd edn. ISO, Geneva
2. EURACHEM/CITAC Guide (2001) Quantifying uncertainty in analytical measurement. 2nd edn. Laboratory of the Government Chemist, London
3. Ni Y (2007) Practical uncertainty of measurement assessment. 3rd edn, vol 3, China Metrology Publishing House, BeiJing, pp 29–36 (22)
4. International Standard of Testing and Calibration Laboratories, ISO, IEC ISO/IEC17025 (2005) General requirements for the competence of testing and calibration laboratories. Switzerland, ISO

# Chapter 2
# Sustainable and Dynamic Supervision Study on Safe Architectural Production Permit

**Jianguo Li**

**Abstract** Nowadays, safe architectural supervision in China is still in its primary stage and there is still much to be perfected and improved. From time to time, enactment of new laws and regulations play the role of a promoter and facilitator to the safe architectural supervision. This paper, mainly from the perspective of significance and purpose of construction safe production permit, offers some specific suggestions on sustainable and dynamic safety supervision to construction companies.

**Keywords** Construction unit · Security license · Dynamic security supervision

## 2.1 Introduction

According to Chinese laws and regulations, any construction unit that wants to run architectural business must have relevant construction safety license. Constructors who violate this provision will be subject to severe penalties [1]. Supervisions on construction enterprises and businesses are required to guarantee that contractor's production condition should meet the requirements of the construction safety license and dynamic safety supervision. The construction safety supervision and management in china is still in its infancy stage and there are many problems in its operation mechanism of supervision the management. Related scheme is still far from completion and corresponding personnel's quality is still low. All these factors become a stumbling block in sustainable and dynamic safety supervision.

J. Li (✉)
Constructional Engineering Institute, Binzhou Vocational College, Binzhou 256603, China
e-mail: bz_Ljg@yeah.net

W. Du (ed.), *Informatics and Management Science V*, Lecture Notes in Electrical Engineering 208, DOI: 10.1007/978-1-4471-4796-1_2,

For better work of this matter, related leaders of the architectural units have to pay more attention with more initiative [2]. Their work is to implement self-examination and actively promote safe construction work. Moreover, let the external organizations of supervision and management participates. And finally form a set of sustainable and dynamic supervision mechanism that ensures safe construction [3].

## 2.2 Meaning and Significance of Sustainable Dynamic Supervision on Safe Architectural Production Permit

### *2.2.1 Meaning of Sustainable Dynamic Supervision*

The so-called dynamic can be understood as the condition which is constantly developing and changing. To develop safe production dynamic supervision, we should have the goal and the rules to be formulated under the premise of definite rules and regulations according to the actual situation and environment which building company in. And finally we gain the corresponding evaluation results. In the supervision system, the definition of the relevant target and norms marks the purpose and implement of the supervision. In actual work, the dynamic supervision always changes according to the actual work needs. And then it generates supervision system corresponding with reality, it also reflects the principle of dynamic. In addition it is dynamic supervision makes evaluation at last that show the dynamic supervision has certain timeliness. In the process of the dynamic regulatory system it can well promote the safe production of the construction department. It provides the solid foundation guarantee for safe production of the project.

### *2.2.2 Scientific Dynamic Supervision on Safe Architectural Production Permit*

While carrying out dynamic supervision on safe production permit, it should be done on incentive basis. Fundamentally dynamic supervision on safe production permit is actually a kind of the supervision system which constantly encourages construction unit according to its actual condition. It makes use of every method to encourage people's safety motive, so that relevant personnel can take safety behaviour self-consciously and constantly make effort to meet this goal. Generally speaking, it aims at fully mobilizing the enthusiasm of related personnel. This supervision strategy usually begins from actual production needs, and constantly develops new motivation during the work. It can guide people's behavior through motive before reaching the goal eventually, the achievement of which, in turn, brings new needs and motive, thus forming a kind of sustainable state of periodic development, and finally realizing the purpose of dynamic supervision on safe production permit.

### 2.2.3 *Staff Participate in Dynamic Supervision on Safe Architectural Production Permit by Use of Systematic Theory*

At the beginning of dynamic supervision on safe production permit, the aim of using system theory is to let staff participate in and work as a master. The use of system theory is based on the basic characteristics of objective things, started from safe production of the whole construction. The changes of production and management should be timely diagnosed so that we can adopt scientific rational treatment to them according to the surroundings and internal enterprise characteristics. The launch of dynamic supervision on safe production permit is conditioned to safe production of the whole enterprise, take good care of the relationship between each employee and the enterprise and enable each employee to take part in the whole supervision process, constantly promote the employee's safety consciousness and the sense of responsibility, increase their positivity and initiative so as to create favorable conditions for successful process of the whole security supervision.

## 2.3 Implementation Strategy of Sustainable Dynamic Supervision for Construction Safety Production Permit

### 2.3.1 *Relevant Staff of Supervision Units Should Enhance the Communications of Safety Production Information with Enterprises*

The work content of current building operations is trivial and complicated. The corresponding responsible staff must face various works every day and the related safety production work is only a part of them. Therefore, if the relevant responsible staff are expected to pay more attention to safety production issue and put much more efforts on it as well as solve safety issue well from the bottom of their hearts, it is not only the responsible staff needs to have stronger safety responsibility, but also the correlative safety production supervision department needs to promote the communication with the responsible staff. For example, the responsible staff of every project should be gathered together periodically to hold safety production work meeting, learn various safety production rules and regulations, summarize and exchange based on the status in present industry, find the existing shortage and merits in working between enterprises, then learn from each other. All of the above acts are on the purpose of making every safety responsible staff have the concept that safety production is primary. In addition, as for the lacking phenomenon of safety management in some enterprises, the exact responsible staff should be aimed to proceed a necessary communication, say clearly the advantages and disadvantages so as to make safety production not only staying in word, but also penetrating into every corner of the works. The chord of safety production must be tensioned all the time in strict accordance with all of the rules and regulations. Only in this way, the supervision management work of construction safety production can be achieved indeed, all of the productions can be conducted on the premise of safety and the principle of safety first can be insisted all along. Through the way of communication between regulator authority and safety responsible staff in enterprise, it can not only make enterprise itself to conduct safety management work as better as it can to improve its own safety constraint strength, but also adequately play a function role of regulatory authority, thus, the real function of sustainable dynamic supervision and control for safety production can be played out better.

### *2.3.2 Related Supervision Department Should Play a Corresponding Role in Supervising and Urging Enterprises to Establish Safety Production Responsibility System*

In the process of safety production management work, the most important thing is to necessarily establish and effectively complete all kinds of safety rules and regulations, what's more, these rules and regulations should be strictly performed into safety production management work. Along with the continuous improved system on national construction safety production, some original system of enterprise cannot meet the current requirement and needs to do some necessary modification. But in reality, since the enterprise faces a variety of works, the insider frequently ignores the modification on these rules and regulations. From this point of view, it is necessary for the related safety supervision department to aperiodically precede inspection work on correlative contents to the enterprises according to the actual situation. In addition, the enterprises themselves should seriously carry on work in accordance with various rules and regulations, implement the responsibility system, carefully read the details in the system and attentively carry out in the work to make all of the safety production administrative staff authentically fulfill their duties and obligations. The purpose of the above measures is to achieve that the whole construction site can be smoothly proceeded on the premise of complying with the safety production regulations.

### *2.3.3 Focus on the Construction of Backbone Contingent, Promote Self-Development of Safe Production*

As for construction enterprises, safe production supervision cannot be carried out smoothly without the support of the related leaders. Only with the emphasis leaders put on safe production can the supervision be implemented smoothly. Besides the influence of related safety management personnel should not be ignored. With a good sense of safety responsibility and general knowledge of relevant laws and regulations, the management personnel can take proper safety measures to carry out their work, thus realizing the goal of safe construction. And in order to improve the sense of responsibility and knowledge of the personnel, we can take the following measures. Firstly, related safety supervision regulators should organize safety conferences in accordance with the company's actual situation, invite experts and scholars to deliver lectures, educate staff with cases to impart advanced safety management and enhance their safety consciousness. Secondly, organize the staff to visit, study and exchange experience of preeminent companies. Field study makes it easier to understand the merits of those experiences which can then be applied to the management of safety production. Finally, organize safety inspection activities among counterparts. Each company has its own particular organization method and successful experience. These activities provide them good chances to learn from others' strong points to offset the weakness, thus avoiding flaws in work. Only when the leaders improve their comprehensive qualities can the safety management of construction site be under full control and the enterprise move with favorable development.

### *2.3.4 Conduct Essential Safety Training for Employees, Enhance Their Safety Consciousness*

The fundamental problem of construction enterprise lies in the employees' safety consciousness because the basic safety production is carried out by them. In order to solve the problem for good and for all, companies should conduct essential safety training enabling them to obtain common safety knowledge and realize the importance of safety production. In this way they will gradually form safety consciousness from the inside.

Nowadays the majority of the employees in construction are peasants who have not taken any essential safety training, and most of them are just temporary workers. It is an important task to find a feasible solution to conduct effective safety training for them. Related safety supervision regulators can take the following methods: conduct essential 3-level safety training at construction site and make sure every trainee sign his name and then conduct random inspection periodically to see the effect; set up zones to propagandize safety production in the construction site and update information timely; related project directors should hold safety meetings regularly, each team keep a detailed record of the meetings and convey the meeting spirits to junior staff. Only with such various means can everyone establish safety consciousness in his mind.

### 2.3.5 *Enhance Project Management and Appraisal Strength of Construction Project*

Construction safe production dynamic management aims at guaranteeing that construction enterprises can conduct construction activities, safe production conditions meet the requirement of Construction Department. Administrative organizations of construction safety supervision should be in line with the implementation guideline proposed by Construction Department in the matter of carrying out safety quality standardization. When working out annual safe production plan, regulations should be made to the standard of construction enterprise production, corresponding liability statement of safe production should be signed. While supervising construction enterprise in the matter of drawing up safe production plan, efforts should be made to specify safe production subitems, qualified and good sub-project targets. Upon the commencement of the contracted project, the target drawn by construction enterprise engineering should be well understood. During the construction, project is subject to patrolled inspection. Further carry out sub-projects of the subsections and safe production conditions, realize integration of plan, implementation and result.

In order to ensure realization of safe production goal of the construction, administrative organizations of construction safety supervision should be set up. Certain measure should be taken to examine and appraise related personnel in enterprise departments and project department in line with the post responsibility system of the enterprise. Only in this way will related personnel fulfil their responsibilities in an effective way. Basic work should be well done, and hidden trouble at construction site should be discovered timely for rectifying and improving. Shortcomings concerning safety in daily work should be consciously completed.

## 2.4 Conclusion

In a word, it takes a long period of time to explore before carrying out safe production dynamic management against enterprises and establishing characteristic safe production permanent mechanism. Only thorough constant innovation and perfection for safe production dynamic management against enterprises will safe production conditions at construction site be suitable for human being when working. Provided that our safety supervision advances with the times in pursuit of safe development on the principle of people-oriented, a fresh new appearance will take on in safety supervision work.

## References

1. Rong X (2005) Research on safe management innovation of chinese architecture. Architectural Constr 6:78–80
2. Xiang Z, Yi B (2006) Safe input and economic benefit. Saf Health 1:4–8
3. Mingchen M (2007) Review on government regulation theory. Manage World 2:137–150

# Chapter 3
# Unloading Features Analysis of a New Combined Unloading System

**Kang Sun, Jianping Tan and Yuxiao Si**

**Abstract** A new combined unloading system of large-scale hydraulic press was designed and combined simulation was also executed. By simulation, unloading response characteristics in varying working conditions and the effect of component parameter to unloading response characteristics in unloading process were got. The research showed that three-order cartridge valve of drainage type can realize smooth and fast unloading in high pressure working condition. Unloading speed is proportional to opening degree of pilot throttle valve.

**Keywords** Unloading features analysis · Unloading system · Three-order cartridge valve

K. Sun (✉) · Y. Si
School of Mechanical and Electrical Engineering, Central South University, Changsha 410083, China
e-mail: tqskbest@yahoo.com.cn

Y. Si
e-mail: 709217681@qq.com

K. Sun · Y. Si
The State Key Lab of Fluid Power Transmission and Control, Zhejiang University, Hangzhou 310007, China

J. Tan
School of Mechanical and Electrical Engineering, Central South University, Changsha 410083, China
e-mail: jptan@mail.csu.edu.cn

W. Du (ed.), *Informatics and Management Science V*, Lecture Notes in Electrical Engineering 208, DOI: 10.1007/978-1-4471-4796-1_3,

## 3.1 Introduction

Large-scale hydraulic press has the feature of high working pressure, large flow rate, long travel distance, high frequency of action switching, large drive power and movement inertia, and complex hydraulic system. Because the large volume of its operating cylinder and high working pressure of its hydraulic medium (generally more than 25 Mpa) during the extrusion process, there will be large elastic deformation of its rack and relatively high elastic potential energy storing in the hydraulic medium. At the end of operation, if these energy was released unsteadily, it will generate huge vibration and noise, hydraulic press will work unsteadily, its mechanical components and hydraulic unit will be damaged or its service life will be shorten. In severe case, the whole device will be damaged. So, the unloading of main cylinder is usually a key link in designing hydraulic system of large-scale hydraulic press. And the unloading system which applied to high pressure and large flow situation is the key factor that decides the smoothness of the whole unloading process. Traditionally, single-stage cartridge valve is applied to unload. However, a huge vibration will be generated in high pressure [1].

The author designed a new combined unloading system which can choose difference unloading mode according to the pressure of hydraulic medium, created its combined simulation model and analyzed its unloading performances.

## 3.2 New Combined Unloading System and Its Principle

Figure 3.1 showed the new combined unloading system. In the hydraulic control system of large-scale hydraulic press, the main impact and vibration occurred at the instant of main-cylinder's unloading. But in different operating condition, pressure beard by hydraulic system is different. In medium and low pressure situation, pressure-controlled and flow-controlled type valve can meet the need of unloading. This also can save time and improve efficiency. However, in high and ultra-high pressure, in order to smooth unloading of the main cylinder, and decrease shocking vibration, the unloading valve of the main cylinder had better open a small orifice first to release pressure of hydraulic oil, then, completely open orifice to make oil come back to tank. Applying the new combined unloading can ensure the unloading process steady and fast.

The new combined unloading system is actually a three-order cartridge valve block. Hydraulic-control pilot two-way cartridge valve is inserts between control cavity of main valve and inlet of two-stage cartridge valve. As pilot control element, it was used to control main valve, make its orifice open slightly to relieve pressure, then completely open to discharge oil. Actually, the opening speed of this two-way valve block is from slow to fast. When pilot valve is electrified, the two-stage valve opens quickly, but just slightly. Then, the control cavity of three-order valve discharge a small quantity of oil by flow path of throttle check valve, which

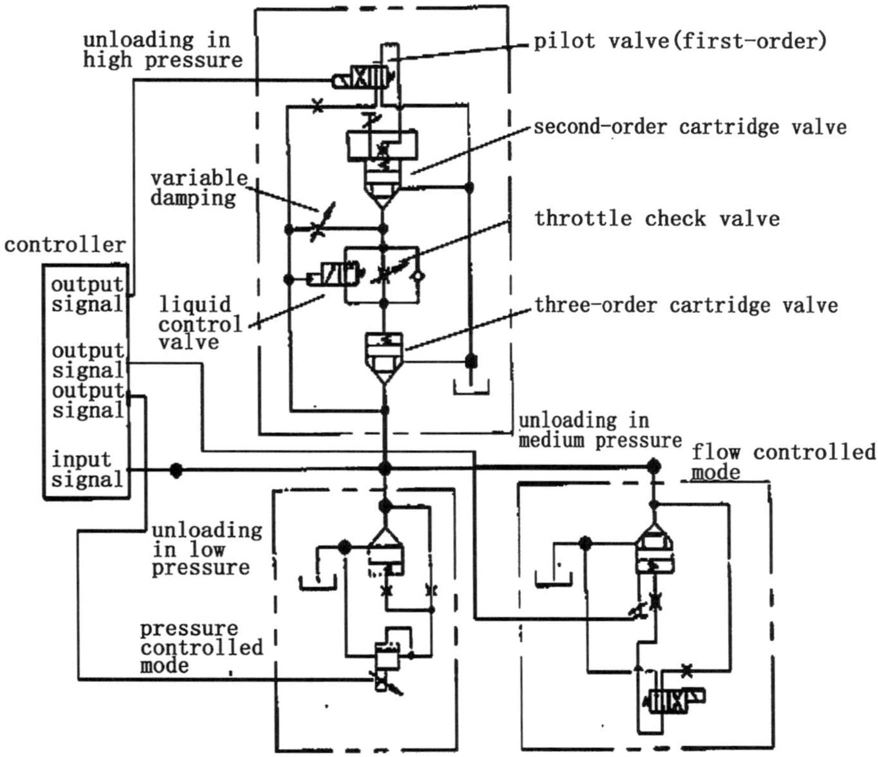


**Fig. 3.1** Schematic diagram of new combined unloading system

makes the three-order cartridge valve open slowly under differential pressure. Consequently, the orifice of it begins to unload. When the pressure of orifice is lower than the spring force of hydraulic control valve, three-order cartridge valve resets, there upon, oil in its control cavity discharge from hydraulic control valve and two-stage cartridge valve, and that can make three-order cartridge valve open rapidly. When electricity is cut off, control oil from orifice of three-order cartridge valve enters control orifice of two-stage valve by pilot valve and damping hole, turns it off. When the pressure of control oil is higher than the spring force of oil controlled valve, left position of hydraulic control valve works, that makes the pressure in control cavity of three-order cartridge valve goes up, the orifice gradually turns off.

In Fig. 3.1, throttle check valve is used to adjust opening speed of three-order cartridge valve while depressurizing, thus opening of orifice port is relatively small; hydraulic control valve is used to control the three-order cartridge valve depressurizing slowly at first, then open quickly. Spring force of hydraulic control valve is used to set the time that valve will be open again. Generally, it is 10 Mpa. Pressure in oil tank is gained by pressure detect component. Control system controls the actions of pre-relief valve, proportional relief valve, proportional dynamic valve, cartridge valve of drainage type and so on by pre-set programs, shortens dead cycle time of large-scale hydraulic press caused by unloading, makes unloading in steady and fast way, improves its production efficiency [2].

## 3.3 Combined Simulation Model

According to the principle, new combined unloading system is composed by three-order cartridge valve block of drainage type, cartridge valve of pressure-controlled type, and cartridge valve of flow-controlled type, which can realize unloading and depressurization in different pressure level.

Applying the powerful hydraulic modeling function of AMEsim, choosing suitable parts, combining with the design of Matlab control system, combined simulation model is built which was in Fig. 3.2. Some important parameters are set in Table 3.1 [3]. The input of this system is variable pressure source encapsulated.

Main In simulink model set by Matlab, the pressure in AMEsim model is used as control input variables to control unloading process in the range of different pressure. Its output is used as input signal of pressure control valve, flow control valve, and three-order pilot valve. In simulink model, "if-in-out" subfunction is called as pressure chosen signal of executive control system of the unloading system. Signals of pressure valve and flow valve in AMEsim adopt "ramp" signal as input. Pilot valve and electro-hydraulic control valve adopt "step" signal as input, all of these are showed in Fig. 3.3.

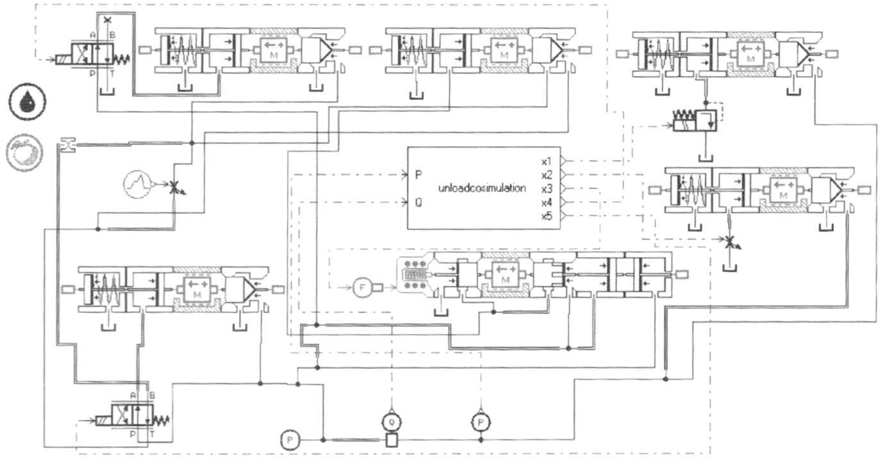


**Fig. 3.2** AMEsim/Matlab combined simulation model

**Table 3.1** Parts parameter of unloading system

| Parameters | Value |
|---|---|
| Path of cartridge valve (mm) | 63 |
| Valve spool maximum displacement of cartridge valve (mm) | 130 |
| Valve spool quality of cartridge valve (kg) | 0.493 |
| Maximum flow of hydraulic pump (L/min) | 465 |
| Maximum load pressure (MPa) | 36 |
| Valve spool quality of pilot valve (kg) | 0.09 |
| Valve spool maximum displacement of pilot valve (mm) | 2 |

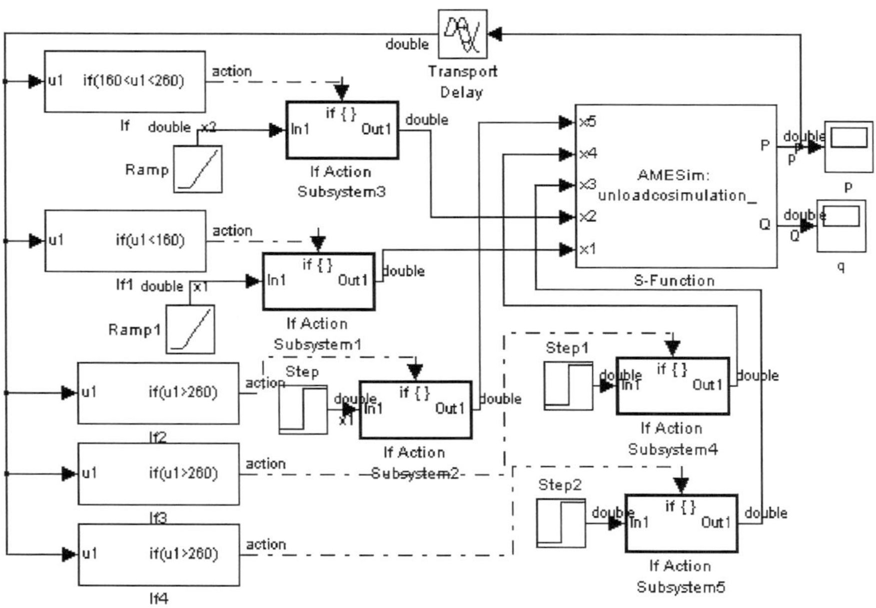


**Fig. 3.3** Simulink control system model of unloading system

### 3.3.1 Result of Combined Simulation

In the range of high pressure, unloading process of large-scale hydraulic press divided into two steps: first depressurization, then discharge oil. Just as it is showed in Fig. 3.4, drainage cartridge valve is controlled by controllers. By changing the opening degree of back-pressure orifice port in cartridge valve blocks, different unloading response curve can be got. Curves 1–6 in Fig. 3.4 are unloading curves when flow coefficient of throttle valve is 0.1–0.6. And depressurization speed is proportional to opening degree of orifice port. According to different working conditions, depressurization speed can be adjusted with opening degree of orifice port.

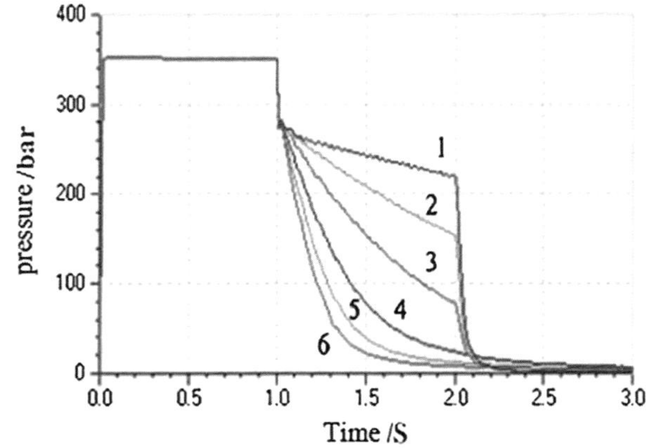


**Fig. 3.4** Unloading response characteristics in different flow coefficient

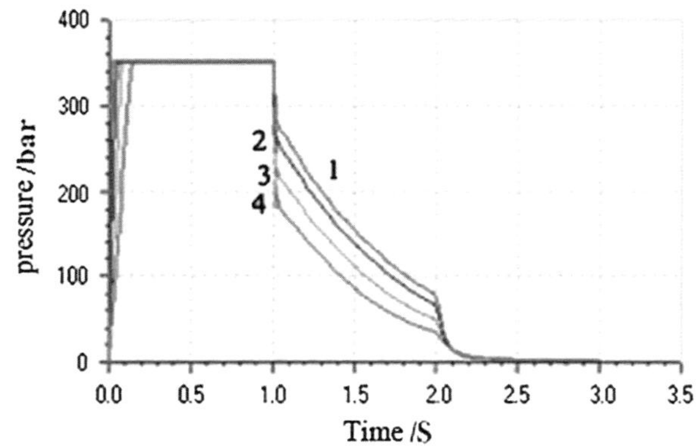


**Fig. 3.5** Unloading response characteristics at different flow rate

Flow of the hydraulic system is needed to change with extruding speed of large metal extruder. At varying flow rate, unloading curves of this system are showed in Fig. 3.5. A curve 1–4 is unloading curves when flow rate is 465, 365, 265 and 125/ min. In else same condition, the larger flow rate of the metal extruder, the larger pressure of initial depressurization pressure, but depressurization time is the same.

## 3.4 Conclusion and Discussion

This paper built combined simulation model of unloading process of large-scale hydraulic press by applying AMEsim and Matlab software. AMEsim is applied to build hydraulic system model, and Matlab/simulink is applied to build controller model. The simulation results showed: a three-order drainage cartridge valve block is adopted to unload in high pressure. Different unloading process can be controlled by different opening degree of orifice port. Depressurization speed is proportional to opening degree of orifice port.

In summary, according to simulation, unloading process simulation characteristics and parameter conditions of steady unloading can be got. In the future study, this conclusion can be used to optimize control model of electro-hydraulic, and realize unloading in a steady and fast way.

**Acknowledgments** The project was supported by national important scientific special (No.2009ZX04005-031), open fund project from the state key lab of fluid power transmission and control (No.GZKF-201023) and scientific planning project from the education department of Jiangxi province (No.GJJ1146).

## References

1. Wei J (2005) Metal extruder. Chemical Industry press, vol 1 pp:123–126
2. Zhang LP (2005) Using and maintenance of hydraulic valves. Mechanical Industry Press vol 1 pp 345–347
3. Xu Y (2009) Modern manufacturing engineering. Trans Mechatron 28:15–19

# Chapter 4
# An Acousto-Optic Tunable Filter and Digital Micromirror Device Based Projection Display System

**Qingli Li, Yiqing Liu, Yinghong Tian, Xiaojin Li and Shuxian Wang**

**Abstract** The digital light processing (DLP) technology has made significant inroads in the projection display devices to create high quality images. In this paper, an acousto-optic tunable filter (AOTF) based DLP projection display system is introduced. This new DLP projection system uses an AOTF to replace the rotating color wheel or prism in the traditional DLP systems. As the AOTF can filter out any monochromatic light at a certain wavelength from 400 to 800 nm, more colors can be modulated by only one digital micromirror device (DMD) chip and more color gamut can be got. The AOTF based DLP projector offers the advantage of having no rotating parts and can switch the color at very high rates, which can avoid the "rainbow effect" and makes the new system more simple, reliable, and cost effective.

**Keywords** Digital light processing · Acousto-optic tunable filter · Digital micromirror device · Projection displays

## 4.1 Introduction

Projection display system is a specialized computer display that projects an enlarged image on a movie screen. The projection display industry has undergone a period of explosive growth in recent years. There are several kinds of projection display devices such as the cathode-ray tube (CRT), the active-matrix liquid

---

Q. Li (✉) · Y. Liu · Y. Tian · X. Li · S. Wang
Key Laboratory of Polor Materials and Devices, East China Normal University, Shanghai 200241, China
e-mail: qlli@cs.ecnu.edu.cn

W. Du (ed.), *Informatics and Management Science V*, Lecture Notes in Electrical Engineering 208, DOI: 10.1007/978-1-4471-4796-1_4,

crystal display (LCD), and the electron beam-addressed oil films and liquid crystal light valves (LCLV's). These kinds of systems have limitations that compromise their performance or the spectrum of their applicability [1]. A newer projection display scheme is known as digital light processing (DLP), a proprietary technology developed by Texas Instruments (TI). The DLP technology has made significant inroads in the projection display devices enabling the world's smallest data and video projectors, HDTV's, and digital cinema [2]. This technology was originally developed in 1987 by Dr. Larry Hornbeck of Texas Instruments. In a DLP display system, tiny mirrors are used instead of transparent panels [3]. Each mirror represents one pixel. The light, rather than passing through the panel, is reflected from it. The mirrors move back and forth, varying the amount of light that reaches the projection lens from each pixel [4]. This kind of tiny mirror are called digital micromirror device (DMD), a light switch that uses electrostatically controlled MEMS mirror structures, producing stable, high-quality imagery on screen [5]. DMD is the heart device of the DLP projection display system. The DMD covers each memory cell of a CMOS static RAM with a movable micromirror. Electrostatic forces based on the data in the cell tilt the mirror either +15 or −15°, modulating the light incident on its surface. Light reflected from any on-mirrors passes through a projection lens and creates images on a screen [6]. Then this device can be used to modulate light digitally in a DLP system [7].

With the development of DMD technology, different kinds of DLP projection display systems have been proposed. For example, there are single DMD-based, double DMD-based, and three DMD-based DLP projection display systems. The advantages of DLP technology include light weight, high contrast, and lack of pixelation. Figure 4.1 shows a typical scheme of single DMD-based and three DMD-based DLP system, respectively. From the figure we can see that the color is obtained by passing the light from the lamp through a rotating wheel with red, green, and blue filters. This subjects the mirrors to light at each of the primary colors in a rapid rotating sequence. Each micromirror will reflect the incident colored light in varying proportions and for varying lengths of time to provide the appropriate color in the pixel which it projects. The result is a color-modulated image that the human eye sees as natural color. Therefore, the color filter (rotating wheel or prism) is very important for the DLP system as color can be added to the gray scale capability of the DMD by using the filter, in combination with one, two, or three DMD chips and a single projection lens [1, 2]. However, there are some limitations using this kind of color filters in the DLP projection display systems [8, 9]. First, the rotating color wheel need to rotate fast to reconstruct a color image, which will make the time schedule control become complex. Second, the switches between the red, green, and blue colors will induce the "rainbow effect", which will influence the quality of the reconstructed image. This is best described as brief flashes of perceived red, blue, and green "shadows" observed most often when the projected content features high contrast areas of moving bright/white objects on a mostly dark/black background. This effect is caused by the way the eye follows a moving object on the projection. Finally, when a color filtering prism is used in a DLP system, it will need at least three DMD chips to modulate the red, green, and blue light respectively. This will make the DLP projection display systems more expensive.

In this paper, we developed a new kind of DLP projection display system. This system uses an acousto-optic tunable filter (AOTF) instead of a rotating wheel or a prism to get different color and one DMD chip to modulate all color lights. This new system offers the advantage of having no rotating parts and can switch the color at very high rates. This will make the DLP projection display system more simple, reliable, and cost effective.

## 4.2 AOTF Based Color Reconstruction

The acousto-optic tunable filter was first proposed in 1969 and was subsequently demonstrated in the visible and then the infrared wavelength [10, 11]. AOTF is a rapid wavelength-scanning solid-state device that operates as a tunable optical band pass filter. The acoustic wave is generated by radio-frequency signals, which are applied to the crystal via an attached piezoelectric transducer. The basic idea is to utilize collinear acousto-optic diffraction in an optically anisotropic media. Changing the driving acoustic frequency changes the band of optical frequencies that the filter passes. When such a filter is placed in the optical train, different wavelengths can pass through it as a function of time. AOTF has been designed and incorporated into some optical systems such as laser wavelength tuning, spectroscopy, hyper spectral imaging, etc., [12, 13]. However, AOTF has little been used as a color filter in a DLP projection display system [14, 15].

Studies show that the color of objects may be communicated in many forms. Color may be described by physical samples, by color terms or by numerical parameters that encode its appearance [16]. Human color vision is based on three light-sensitive pigments called red, green, and blue [17]. Common color images are captured and displayed as a set of three black and white images collected with

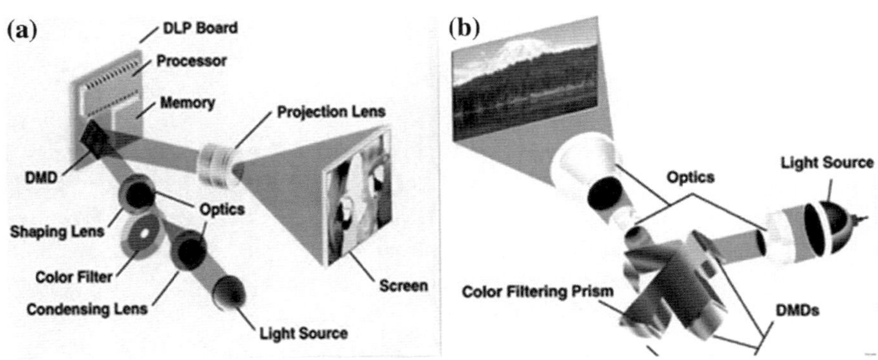


**Fig. 4.1** Typical scheme of **a** single DMD-based and **b** three DMD-based DLP projection display systems

red (R), green (G), and blue (B) light, i.e., at wavelengths of approximately 630, 545 and 435 nm, respectively. So these color images are often called RGB images. To reconstruct the color, a color wheel is used in a single DMD chip DLP projector to produce full-color for projector. The color wheel is made up of red, green, and blue filter components. If the color wheel runs with 60 Hz rotation frequency, 180 images (60 frames) per second can be got [18]. The color wheel filter efficiency ($\tau$CF) and the system colorimetry are determined by a number of parameters, including the characteristics of the dielectric filters on the color wheel and the lamp spectral content, as can be seen in the following Eq. 4.1:

$$\tau_{CF} = \frac{\int [R(\lambda) + G(\lambda) + B(\lambda)]D(\lambda)L(\lambda)O(\lambda)V(\lambda)d\lambda}{\int D(\lambda)L(\lambda)O(\lambda)V(\lambda)d\lambda} \tag{4.1}$$

where D($\lambda$) is the spectral efficiency of the DMD, R($\lambda$), G($\lambda$), B($\lambda$) are the filter spectral responses, L($\lambda$) is the spectral response of the lamp, O($\lambda$) is the spectral response of the system optics exclusive of the color filters, V($\lambda$) is the photopic response of the eye. From the equation we can see that how to get the red, green, and blue light exactly is very important for the DLP projector to display the real color.

System colorimetry is often quantified by using the CIE 1931 chromaticity coordinate system [19]. Figure 4.2a shows the color gamut of a typical color wheel based DLP projector and the REC. 709 of high definition television (HDTV). The CIE standard observer color-matching functions are shown in Fig. 4.2b. From the figure we can see that more color gamut can be got if an AOTF is used to replace the color wheel in the DLP system. The reason is that the AOTF not only can filter out the R($\lambda$), G($\lambda$), and B($\lambda$) light at 630, 545, and 435 nm, respectively, but also can filter out any other light at a certain wavelength from 400 to 800 nm. That is, the wavelength $\lambda$ in R($\lambda$), G($\lambda$), and B($\lambda$) can be changed as demanded in the AOTF based DLP system while it is constant in the traditional DLP system. Therefore, more colors can be modulated by using an AOTF and a single DMD chip compared with the traditional one.

## 4.3 The New DLP Projector Based on AOTF

The optical path and functional layout of the AOTF based DLP projector is illustrated in Fig. 4.3. The system consists of light source, convergent lens, AOTF (Brimrose, CAV-200), shaping lens, one DMD chip, projection lens, AOTF driver (radio-frequency drive unit), micro control unit (MCU), and some interface units. From the figure we can see that the instrument has similar configuration with the traditional one DMD based DLP projector except that an AOTF adapter was used to replace the rotating color wheel in the new system. The AOTF adapter with Brimrose synthesizer electronics can provide narrow bandwidth, rapid wavelength selection, and intensity control. The designed wavelength ranges is 400–800 nm

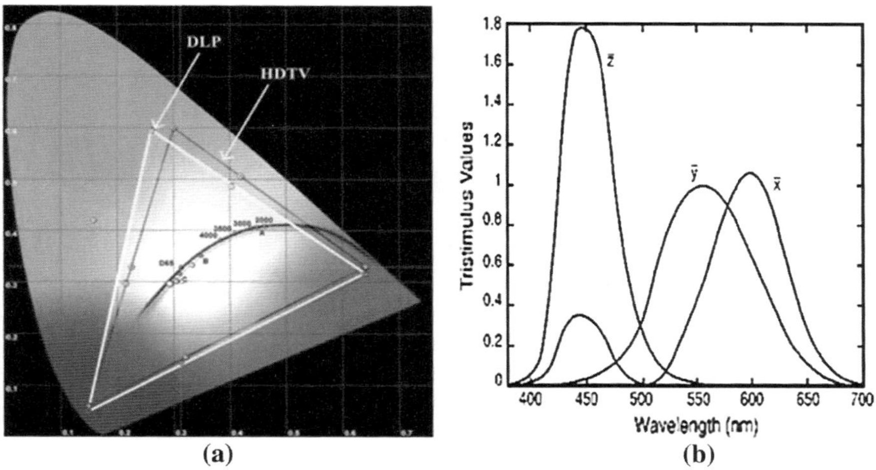


**Fig. 4.2** CIE 1931 chromaticity coordinate and CIE standard observer color-matching functions

and the spectral resolution is 2–6 nm (2 nm @ 543 nm; 5 nm @ 792 nm). The minimum wavelength selection sweep interval of the AOTF is 20 ns, which is faster than the rotating speed of the color wheel and can meet the color modulate speed demand of DLP projector.

The new projector uses an ultra high pressure (UHP) lamp as a light source. This is a common light source for digital data/video front projectors and rear projection televisions because it has very high light output available for the small environments of micro display projectors. The white light focused down by

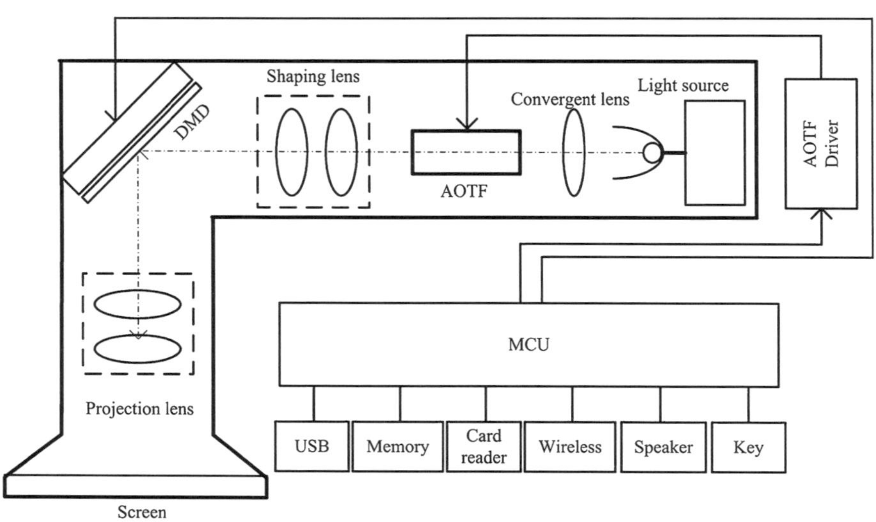


**Fig. 4.3** Scheme of AOTF-based DLP projection display system

convergent lens onto the input port of AOTF. The output light from the AOTF illuminates the DMD sequentially with red, green, blue, or other monochromatic light through the shaping lens under the control of AOTF driver. At the same time RGB video signal is being sent to the DMD mirror-pixels. The mirrors are turned on depending on how much of each color is needed and project the light on the screen through the projection lens. Then the human eye integrates the sequential images and a full color image is seen.

As a new DLP projector, it also contains MCU to control the AOTF adapter, the DMD chip, and other interface units. The interface units include more than one USB interface to connect with other mobile equipments or computers. It also can read multimedia data from some memory cards by using the card reader. Some wireless devices can access the new DLP projector through the wireless unit. There are also some keys to accept the setup instructions of the user.

## 4.4 Conclusion

Digital light processing projection technology based on the digital micromirror device light modulator is currently used in a wide variety of display applications to create high quality images [20]. The fundamental operating principles of the DMD modulator and the basic processing and control methodologies of DLP have been studied in recent years. In this paper, a new DLP projection display system is introduced. In this DLP projector, an AOTF instead of a rotating color wheel is used to filter out different monochromatic light and modulate color image by using one DMD chip. The AOTF based DLP projector offers the advantage of having no rotating parts and can switch the color at very high rates, which can avoid the "rainbow effect". This also makes the DLP projection display system more simple, reliable, and cost effective. Another advantage of this DLP system is that the color gamut can be larger than the traditional one because of the rich monochromatic light the AOTF filtered out. The high level of integration in this new system has enabled a rapid migration from early projectors to very compact, high performance projection systems.

**Acknowledgments** This work is supported in part by the National Natural Science Foundation of China (Grant No. 61177011, 60976004), the Project supported by the Shanghai Committee of Science and Technology, China (Grant No. 11JC1403800), and the Project supported by the State Key Development Program for Basic Research of China (Grant No. 2011CB932903).

## References

1. Van Kessel PF, Hornbeck LJ, Meier RE, Douglass MR (1998) A MEMS-based projection display. Proc IEEE 86:1687–1704
2. Tousain R, van Casteren D (2007) Iterative learning control in a mass product: light on demand in DLP projection systems. American Control Conference 07, pp 5478–5483

3. Monk DW (1997) Digital light processing: a new image technology for the television of the future. Broadcast Convention Int 1:581–586
4. Gergelyi D, Foldesy P (2010) Digital micromirror device (DMD) projector based test bench for vision chips. Cell Nanoscale Netw Appl (CNNA) 12:1–4
5. Gale R (1999) Principles and applications of the digital micromirror device in projection displays. Lasers and electro-optics society. In: 12th annual meeting LEOS 99 IEEE, vol 211. pp 212–213
6. Younse JM (1993) Mirrors on a chip. Spectrum. IEEE 30:27–31
7. Hornbeck LJ (1996) Digital light processing and MEMS an overview. Advanced applications of lasers in materials processing. Broadband optical networks/smart pix-els/optical MEMs and their applications IEEE/LEOS summer topical meetings, vol 1. pp 7–8
8. Feather GA, Monk DW (1995) The digital micromirror device for projection display. Wafer scale integration. In: Proceedings of 7th annual IEEE international conference, vol 1. pp 43–51
9. Hung C-C, Fang Y-C, Tsai C-M, Lin C-C, Yeh K-M, Wu J-H (2009) Optical design of high performance con-focal microscopy with digital micro-mirror and stray light filters. Opt Int J Light Electron Opt 121:2073–2079
10. Harris SE, Wallace RW (1969) Acousto-optic tunable filter. J Opt Soc Am 59:744
11. Harris SE, Nieh STK, Winslow DK (1969) Electronically tunable acousto-optic filter. Appl Phys Lett 15:325
12. Coquin GA, Cheung KW (1988) Electronically tunable external cavity semiconductor laser. Electron Lett 24:599–600
13. Tran CD, Furlan RJ (1992) Acousto optic tunable filter as a poly chromator and its application in multidimensional fluorescence spectrometry. Anal Chem 64:2775–2782
14. Inoue Y, Penuelas J (2001) An AOTF-based hyper spectral imaging system for field use in ecophysiological and agricultural applications. Int J Remote Sens 22:3883–3888
15. Martin ME, Wabuyele M, Panjehpour M (2006) An AOTF-based dual-modality hyper spectral imaging system (DMHSI) capable of simultaneous fluorescence and reflectance imaging. Med Eng Phys 28:149–155
16. Zuffi S, Santini S, Schettini R (2008) From color sensor space to feasible reflectance spectra. IEEE Trans Signal Process 56:518–531
17. Nathans J, Thomas D, Hogness DS (1986) Molecular genetics of human color vision: the genes encoding blue, green, and red pigments. Science 232:193–202
18. Feng T, Xiang W, Jingao L (2008) Research and realization of innovative LED illumination system for DLP projector. Audio, language and image processing, ICALIP International Conference on 1:194–199
19. Hunt RWG, Pointer MR (1985) A color-appearance transform for the CIE 1931 standard colorimetric observer. Color Res App 10:165–179
20. Van Kessel PF (2001) Electronics for DLP < sup > TM </sup > technology based projection systems. VLSI circuits, Digest of technical papers symposium on pp: 91–94

# Chapter 5
# Port Intelligence of Storage Management System

**Yingsun Sun**

**Abstract** Through the computer network technology port construction intelligent storage management system, get the data RFID reader, identify and collection of items with RFID tag information and real-time monitoring items state. Through to the warehouse of the items in the recognition and monitoring, and to realize the real-time warehouse, automatic, smart management, the main application have inventory operation, outbound operation, goods move, inventory operation, since the quantity of goods adjust.

**Keywords** Storage management · RFID · Ports · Intelligence

## 5.1 Introduction

The goods warehouse management as a port a powerful, intelligent warehouse management system of intelligent building for port logistics has important influence on. RFID technology is a non-contact automatic identification technology [1]. The basic principle is to use the rf signal transmission characteristics, and its space coupling, and to realize the stationary or moving to identify items automatic machines recognition. The RFID system is composed of at least include 2 parts: electronic Tag (Tag) and Reader (the Reader). Electronic tag installed in the items to identify, when electronic tag of goods into reader range have, the reader can through wireless way will the information to read out the tag, and to realize the automatic identification of the item. In intelligent warehouse management system

Y. Sun (✉)
Anhui Business College of Vocational Technology, Anhui wuhu 241003, China
e-mail: sunyingsun@sohu.com

W. Du (ed.), *Informatics and Management Science V*, Lecture Notes in Electrical Engineering 208, DOI: 10.1007/978-1-4471-4796-1_5,

using RFID technology, can finish all kinds of efficient operation, such as automatic identification of the goods [2].

## 5.2 Intelligent Warehouse Management in Port Logistics the Application Fields

For the intelligence of port logistics warehouse management main application fields include the following two aspects:

### *5.2.1 Port Container Management (Containers in and Out of the Yard, in and Out of Port)*

At this time, the container port can be seen as a warehouse, in each container with RFID tags, effectively identify container of relevant information. Through some of the place in yard installed RFID tags are able to identify the reader, can real time monitoring in the container yard information management in the container yard pass in and out and scheduling management (Fig. 5.1) [3].

### *5.2.2 Port Warehouse Management*

In the warehouse, intelligent warehouse management systems for cargo can automatic and intelligent identification and management. Intelligent warehouse management system according to the different needs of RFID level, for ZhengTuo

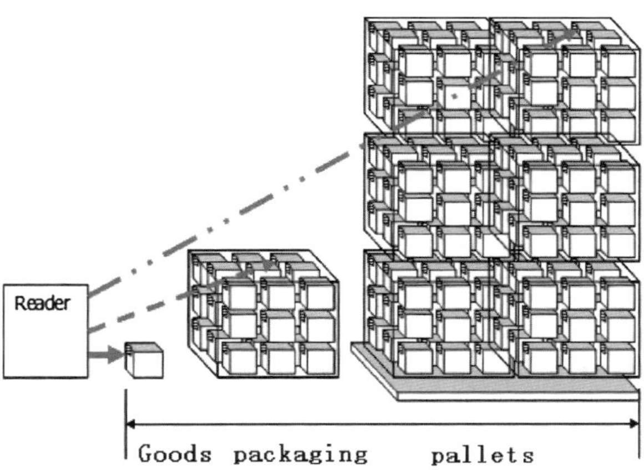


**Fig. 5.1** RFID application of different levels

in and out of items, with tray level appropriate; In the case of goods for the unit operation, tray or box level application level [4]. The goods are placed in the warehouse RFID tag, can use warehouse in fixed RFID reader and armed with RFID reader, automatic identification of goods, and automation of the inventory, take orders, inventory, and mobile operation. Intelligent warehouse management system using RFID tags, reader, can effectively finish all kinds of business operation, such as the designated area, shelves stacked take orders, etc. so as to the intelligence of the warehouse management.

Whether for the port container management or the warehouse management, warehouse management system of intelligent building is unified; RFID tags level is different, so no longer separately.

## 5.3 The Whole Building Intelligent Storage Management System

Intelligent warehouse management system using RFID to be automatic identification, computer network technology and wireless communication technology, integration of advanced automatic identification equipment and improve warehouse management software system in a body, set up wireless real-time warehouse management system automation, automation, intelligent warehouse management.

Port on the storage items RFID tags, RFID reader through the wireless communication technology to read the label information, automatic identification tag information. Intelligent warehouse management, identify and collection of items with RFID tag information and real-time monitoring items state. Through to the warehouse of the items in the recognition and monitoring, and to realize the real-time warehouse, automatic, smart management.

The database is a face of the theme, integrated, stable, contains the historical data collection of data, used to support management decision-making. Data warehouse completed the collection of data, integration, storage, management, intelligent face is processed data, make intelligence can focus more on information extraction and knowledge discovery, data warehouse for rip or load of raw information, merge various data sources of data to support management and decision making.

In warehouse management, the main use portable hand ChiJi and fixed the reading and writing to complete realization to warehouse material (labelled) storage management. In practical applications, besides the warehouse management outside, still can involve materials and materials on the real-time information query, and follow up on materials.

If fixed for reading and writing of goods into the library, outbound, etc. for the management and the basic operation is: material handling, through reading and writing with fixed the RF coverage area, read the reading and writing material label information. Fixed the reading and writing through the COM mouth/the etheric WangKou and other communications, data to the warehouse workstation PC, terminal PC through the network and backend database finish data interaction, PC and finish all kinds of terminal of the service request.

## 5.4 Intelligent Storage Management System in Port Logistics of the Principal Applications are

### *5.4.1 Warehousing Operation*

After the arrival of the storage: each before storing goods are housed a RFID tags, tag contains information related to the goods. The goods to the repository, the operating personnel in management terminal according to delivery of the goods, inquires the documents to the related information (such as the name of commodity, quantity, suppliers code, etc.), and then the scanning to the goods coding, while testing the packaging of goods have damaged, then check the delivery system of the documents and the project is the actual delivery tally. If there is not accord with the data on the document delivery, the system will direct warning, warehouse operators will be rejected the goods; If Treasury consistent with documents, system is given the corresponding information. At the same time will need the related information after inspection and acceptance by the RFID reading and writing automatic identification equipment sent to system.

The goods after the Treasury: incoming goods, goods in warehouses, pile up by the system to determine the position, the position of the systems are piling up, System to verify rack, will need to update the inventory information into the system, and in the system of incoming confirm finish, and print storage of documents.

### *5.4.2 The Outbound Operation*

#### 5.4.2.1 Outbound Goods Before

The operating personnel in intelligent warehouse management system according to the outbound vouchers input relevant information, the system to meet the outbound certificates of a query goods, to meet the requirements of generating the picking orders, including single number, has chosen as material, code, destination, the whole package number, number of packages, the selection of rack in pre-set rules to be automatic instructions generate graphics interface hint, allow operators in the system manual specified. Workers in designated as the goods have chosen, of the goods, the quantity, as automatic champions league recognition and calibration, begins to prepare.

#### 5.4.2.2 Outbound Goods

Transport vehicles delivery to the repository, operation personnel for outbound inspection, confirm the material chosen with single or delivery notice is consistent, complete the confirmation of the outbound work; Here will eventually have to request the delivery of goods information synchronization to system, and in the system to complete the outbound confirmation, update the information system database and warehouse, and print outbound order.

### *5.4.3 The Goods Storehouse Mobile*

When a group of outbound goods distribution work near the end, inventory a little while, and the next batch or receive a bulk cargo warehousing notification, and need to move library storehouse, make the goods to meet new location coming. System identification information related to the goods and according to the information and the transfer of goods to allocation. The goods move out, move up as and the check, make sure the goods transfer target is correct, and synchronization system of information.

### *5.4.4 Inventory Operation*

The system will cycle or when goods are removed to move around in the warehouse inventory of the goods. The system through the items RFID tags in the warehouse of the items to be automatic recognition, and the amount of goods, such as position identify calibration. Figure 3.4 warehouse management system for intelligent inventory operation flow chart.

### *5.4.5 The Quantity of Goods Adjustment*

System of inventory goods regularly counted, or system real time monitoring items in the warehouse state, when found some goods inventories less than alert value will remind management personnel to supplement the goods; when you find that some goods more than alert value remind management personnel avoid excessive storage of the goods.

## References

1. Shuang L (2010) The enterprise warehouse management system the research and implementation. Inf Comput 1:170–171
2. Song L (2010) Puli the design and realization of the system. Comput Fujian Province 2:134–135
3. Liang WS (2007) Warehouse management system design and implementation. Tianjin Univ 12:111–114
4. XiaoMing D, Ge SL (2010) Searched RFID and bar code of small and medium-sized enterprise warehouse management system. Combination Mach Tools Autom Process Technol 2:107–112

# Chapter 6
# Research on Dynamic Job Shop Scheduling

**Jingmin Zhang and Xia Li**

**Abstract** In the paper, the development of dynamic job shop scheduling problem were summarized comprehensively. It discusses the conception of dynamic job shop scheduling, dynamic events, evaluation indicator, dynamic scheduling strategy, and dynamic scheduling methods. The research methods are divided into two classes: the precise methods and the approximate methods. Characters of each method are analyzed. At last, problems which need further investigation and possible research directions are pointed out.

**Keywords** Job shop scheduling · Dynamic optimization · Dynamic job shop scheduling · Method

## 6.1 Introduction

Production scheduling is the core of production management. Many job shop scheduling problems (JSSP) is already proved as typical NP problem [1]. During the real production system, environment is mostly dynamic, dynamic events like the breakdown maintenance of machine etc., are unavoidable, which exactly need

---

J. Zhang (✉)
College of Information and Technology, Shijiazhuang University of Economics, Shijiazhuang 050031, China
e-mail: zhangjingmin5254@163.com

X. Li
College of Information and Technology, Shijiazhuang University of Economics, Shijiazhuang 050031, China
e-mail: 53871486@qq.com

W. Du (ed.), *Informatics and Management Science V*, Lecture Notes in Electrical Engineering 208, DOI: 10.1007/978-1-4471-4796-1_6,

the dynamic scheduling. The dynamic scheduling is one of hotspots in the production scheduling research.

At present, there are few summarized documents about the dynamic scheduling research methods, and the document makes comprehensive summary and analysis about the different research methods of previous dynamic scheduling before 2001. In recent years, the fast development of computer technique and the generation and development of artificial intelligence, neural network, genetic algorithm, emulation technique, particle swarm algorithm, ant colony algorithm, immune algorithm, differential evolution algorithm, quantum algorithm etc. new methods open up new train of thought for the dynamic JSSP, therefore, it's necessary to make comprehensive and systematic statements about the dynamic JSSP [2].

## 6.2 Description of Dynamic JSSP

### *6.2.1 Definition of Dynamic JSSP*

Jackson first distinguished the static scheduling and dynamic scheduling in the year of 1957. Static scheduling is to show all the jobs for processing in processing condition, Once scheduling plan is identified, the plan don't need another schedule [3, 4].

In the actual production process, because the production operation environment continue to appear disturbance factors, such as processing overtime or ahead, the emergency order to join, and the processing time inaccurate, etc., scheduled plan is destroyed. So according to the changes of the actual conditions, we must constantly adjust the scheduling plan, which is called dynamic scheduling [5]. Obviously the dynamic scheduling is more complex than static scheduling, more difficult to solve.

### *6.2.2 Dynamic Events*

The random events which cause the dynamic scheduling are called dynamic events. The dynamic events were divided into the following four classes [6]:

The work piece related events: the processing time is not sure, the work piece arrived randomly, delivery date is changed, and dynamic priority and order are changed.
The machine related events: the machine damage load limited, the conflict between machine blocked and production capacity.
The process related events: process delayed, quality rejected and production unstable.
Other events: the absence of operating personnel, the late arrival of raw materials etc.

### 6.2.3 Dynamic Scheduling Strategy

At present, most of the dynamic JSSP strategies can be classified as totally reactive and predict reactive [7].

#### 6.2.3.1 Totally Reactive Scheduling

No pre-scheduling scheme, only use scheduling strategy to re-schedule according to dynamic scheduling time information. Because the totally reactive scheduling makes scheduling plan without considering the possible dynamic interruption in the future, it cannot supply a plan benchmark to the relevant activities of the job shop. So it will result in modifying scheduling plan frequently and then enlarge scheduling cost when the interruption happens.

#### 6.2.3.2 Predict Reactive Scheduling

Predict reactive scheduling is common strategy and it includes two basic steps: (a) To generate a pre-scheduling scheme without considering the dynamic events of job shop in the future, (b) Update the pre-scheduling plan which is triggered by dynamic events and dynamic scheduling can keep the feasibility of scheduling scheme or improve the scheduling performance.

#### 6.2.3.3 Evaluation Indicator of Dynamic JSSP

The indicator of dynamic JSSP is mainly divided time indicator, economic indicator and system indicator [5]. Time indicator includes job delivery time, job completion time, average circulating time of work piece etc. Economic indicator includes cost of production, penalty fees of production delay, storage cost of earlier completion etc. System indicator includes device availability, productivity etc. The usual scheduling cannot take all indexes into account but make choice according specific requirement.

## 6.3 Research Methods of Dynamic JSSP

By analysis of lots of documents, this paper divides the research methods of dynamic JSSP into two classes: the precise methods and the approximate methods.

## 6.3.1 Precise Methods

Precise methods build one or many optimization models of target function to get optimal solution by transform the production scheduling problems to equality constraint or inequality constraints. Most scheduling problems are, which are already proved, NP problems. As the enlargement of scheduling problems scale, its feasible solution rises as exponential order and the precise methods cannot get optimal solution within valid time easily so that it cannot be used to solve practical problems. For example, Song Ye designs one real-time scheduling algorithm based on the branch and bound algorithm and artificial neural network to solve some scheduling problems which need high precision and instantaneity [8]. Klein-man makes the scheduling arrangements of work pieces which arrive at random by adopting dynamic programming and proves it as feasible by test [9].

## 6.3.2 Approximate Methods

Approximate Methods include many kinds.

### 6.3.2.1 Rule of Priority Assignment

The most traditional method to schedule the production and processing tasks is the schedule rule. The earliest assignment rule is posed by Jackson [10] etc, it could be used in the dynamic real-time scheduling system because of its simplicity, easy to come true, low computational complexity and etc. In recent years many new rules are created. For example, the conclusion of document is: the online scheduling based on the assignment rule is stronger than offline scheduling [11]. A document carries out dynamic scheduling with the method combined with the rule and genetic algorithm. Simulation result shows that the algorithm is effective and corrects [12].

### 6.3.2.2 Genetic Algorithm

The advantages of genetic algorithm are mature coding technology and simple operation. But when solving large combinatorial optimization problem, the algorithm often exist in the search space is large, search time is longer, easy to premature convergence, etc.

How to use the GA to solve scheduling problem efficiently, has been considered as a challenging problem and a research hotspot. Wang Ling conducts comprehensive research on the application of genetic algorithms in the field of scheduling [13]. Document puts forward generalization model of JSSP which can apply to static, dynamic and uncertain production environment, based on which the solution method based on genetic algorithm is designed to minimize the average flowing time, and makes experiment [14]. The results show that it's more superior to the optimization method which is based on classic rule of priority assignment.

#### 6.3.2.3 Expert System

Scheduling expert system can generate complex heuristic rules and is intelligent, but the development period is long, the cost is very high and it is difficult to obtain the necessary experience and knowledge. ISIS is the first expert system for solving JSSP. Against the continuous casting production scheduling problem in steel enterprise, document adopts IBM's knowledge tool and collaboration technology to solve continuous casting production scheduling problem [15].

#### 6.3.2.4 Simulated Annealing Algorithm

The simulated annealing algorithm is a kind of serial optimization algorithm with low calculation efficiency, which requires a high enough initial temperature and slow enough cooling speed so as to restrain to overall optimum, so it usually needs to be improved [16]. The biggest advantage of simulated annealing algorithm is that it can be used in combination with other algorithm, and combined algorithm can achieve the effect of fostering strengths and circumventing weaknesses. Document puts forward the parallel machine tool replaceable scheduling algorithm towards the alterable process route, the Job Shop dynamic scheduling problem which aims at both production period and key work piece delivery time under the condition of limited resource, and adopt the combination of both way scheduling strategy and GASA hybrid genetic algorithm for solving problems, which not only satisfy the on-time delivery of key work piece, but also shorten the production cycle [17].

#### 6.3.2.5 Ant Colony Optimization Algorithm

Ant colony optimization algorithm finds the optimal solution of the problem with the help of group information, and the advantages are simple calculation, fast convergence, but there is the defect of falling into local optimum. Yu et al. [18] proposed to use ant colony optimization algorithm after treatment of mutant so as not to fall into local optimum easily, and to solve this urgent order inserting in dynamic scheduling problem. Document [19] proposed a multi-agent dynamic scheduling method of negotiation strategy based on intelligence and reinforced study and it is applied to coloration shop scheduling problems.

#### 6.3.2.6 Particle Swarm Optimization

Particle swarm optimization is widely used in solving various optimization problems for its high efficiency, less parameter to adjust, simple algorithm and strong robustness and other characteristics. Its main flaw is that it is easy to emerge premature convergence and poor local optimization ability etc. Document [20] solved job shop dynamic scheduling problem using the particle swarm algorithm combined with genetic algorithm, and take simulation experiments of a number of dynamic events in the actual production. Simulation results show that, it is feasible and effective to solve static scheduling and dynamic scheduling problems by GAPSO.

#### 6.3.2.7 Immune Algorithm

Artificial immune algorithm is a global search intelligent algorithm of simulating function of the biological immune system, with its fast convergence speed and strong global search ability, and immune algorithm has opened up new ideas for the scheduling problem [21]. For example, Wu Xiuli proposed a dynamic scheduling optimization algorithm based on multi-objective immune genetic algorithm when quality problem occurs and adopting the scheduling strategy of combining event-driven and cycle-driven [22]. Proposed an immune algorithm for dynamic scheduling of flexible manufacturing cell and brought in sliding window technique and scheduling task pool concept [23].

#### 6.3.2.8 Hybrid Optimization Scheduling Method

Recently it has become a hotspot that combination of the various algorithms research to make up for their shortcomings, and achieve a high degree of sub-optimal goals. In order to arrange quick and efficient equipment adjustment to ensure the continuous machine casting after an incident in the steel mill, rule-based dynamic scheduling method is proposed, using the method of linear programming model solution [24]. Liu Min et al. minimize the completion time of parallel machine scheduling problem using genetic algorithms to optimize the scheduling strategy [25].

## 6.4 Summary

Above all, there is still not a series of systematic method and theory for dynamic job shop scheduling research till now. To reduce the difficulty of research, the production scheduling researches are all simplified so it is far from the practical production. The achievements of dynamic production scheduling research are not applied in practical production ideally and it has many limitations which are mainly showed as the singleness of method and narrow scope of application. Therefore, how to study more effective scheduling methods based on the classic scheduling theory is the hot issue which is highly concerned by academic circles and business circles for so many years. Therefore, the writer thinks it can go from the following aspects to expand the research:

By adopt new optimization method, such as: artificial fish-swarm algorithm, quantum method, Imitation plant growing method etc. to solve the dynamic JSSP.

To closely concern the development of the research in fields of computer science, operational research life science, artificial neural network and anthropology etc., and accept various thoughts and inspiration of theory widely to put forward new practical scheduling algorithm.

To fully concern the inside and outside resources and different restrictions of enterprise to close to the practical production,

To make scheduling process networked.

In a word, as the further development of mathematics theory, the study for the dynamic JSSP with NP-hard characteristic must go deeper towards the directions of integration, practical, multi-target and high sub optimization.

**Acknowledgments** This research was supported by the science and technology project of Hebei Province, China, contract no. 11213525D and no. 09213547.

## References

1. Johnson SM (1954) Optimal two-and three-stage production schedules with set-up times included. Naval Res Log Quart 1:61–68
2. Qian X, Tang L, Liu W (2001) Dynamic scheduling: a survey of research methods. Control Decis 16(2):41–145
3. Jackson JR (1957) Simulation research on job shop production. Naval Res Log Quart 4(3):287–295
4. Graves S (1981) A review of production scheduling. Oper Res 29(4):646–675
5. Wang W, Wu Q (2007) Intelligent production scheduling algorithm and its application. Science Press, Beijing
6. Suresh V, Chandhuri D (1993) Dynamic scheduling–a survey of research. Int J Prod Econ 32(1):53–63
7. Bao L (2010) The research on dynamic scheduling based on the event-driven strategy, vol 92. Shandong University Press, Shandong, pp 9–11
8. Song Y, Yang G (2008) Real time scheduling using branch-bound algorithm and artificial neural network. Microcomput Appl 24(004):10–12
9. Miao X, Kleinman DL (1990) Dynamic job scheduling with strict deadline. In: Proceedings of the 29th IEEE conference on decision and control (Cat.No.90CH2917-3), vol 1. pp 16–121
10. Jackson JR (1955) Scheduling a production line to minimize maximum tardiness, research report 43, management science research projects, vol 23., Los Angeles University of California, California, pp 124–125
11. Sabuncuoglu I, Kizilisik OB (2003) Reactive scheduling in a dynamic and stochastic FMS environment. Int J Prod Res 41(17):4211–4231

12. Fox MS, Smith SF (1984) ISIS: a knowledge-based system for factory scheduling. Expert Syst 1(1):25–49
13. Wang L (2003) Shop scheduling with genetic algorithms, vol 230. Tsinghua University Press, Beijing, pp 39–43
14. Bierwirth C, Mattfeld DC (1999) Production scheduling and rescheduling with genetic algorithms. Evol Comput 7(1):1–17
15. Shah MJ, Damian R, Silverman J (1990) Knowledge based dynamic scheduling in a steel plant. In: Proceedings of the conference on artificial intelligence applications, vol 5. pp 108–113
16. Kolonko M (1999) Some new results on simulated annealing applied to the job shop scheduling problem. European J Oper Res 113(1):123–136
17. Dou D, Sun S, Zhuge J et al (2010) Research on resource-constrained dynamic job-shop scheduling. Mech Sci Technol Aerosp Eng 29(2):159–163
18. Yu YH, Yan JQ (2005) Flow shop rescheduling problem under rush orders. Zhejiang Univ Sci 6A(54):1040–1046
19. Xu X, Hao P, Wang W (2010) Multi-agent dynamic scheduling method and its application to dyeing shops scheduling. Comput Integr Manuf Syst 16(3):611–620
20. Wang C, Wang S, Feng D et al (2010) Application of hybrid PSO in job-shop dynamic scheduling problem. Comput Eng Appl 46(260):219–222
21. Mo H (2002) Theory and application of artificial immune system. Harbin Institute of Technology Press, Haerbing
22. Wu X (2008) Research on flexible job shop dynamic scheduling problem. J Syst Simul 20(14):3828–3832
23. Yu J, Sun S, Wang J et al (2007) Flexible manufacturing cell dynamic scheduling by immune algorithm. Acta Aeronautica et Astronautica Sinica 28(2):464–469
24. Yu G, Tian N, Xu A et al (2010) Research on robustness and dynamic scheduling of equipment adjustment in steel plant. Control Eng China 17(6):861–865
25. Liu M, Wu C, Yang Y (2000) Genetic algorithm method based on combinatorial rules in identical parallel machine scheduling problem. Acta Electronica Sinica 28(5):52–54

# Chapter 7
# Overall Difficulty Control Scheme of Examination Paper Based on Balanced Strategy Group Volume Algorithm

**Xianye Zhang**

**Abstract** The paper put forward a kind of test bank group volume algorithm which is based on the balance of the strategy. This algorithm is based on topics, to grasp the requirements as constraint conditions of small set of item bank division, and adjust the difficulty of all kinds of questions to the corresponding balance coefficient approximation, In order to achieve overall control of the difficulty of examination paper. Through The Higher Algebra group roll verification. This algorithm is based on the volume efficiency of higher to achieve control of the examination paper difficulty.

**Keywords** The difficulty coefficient of test paper · Balanced algorithm · Balanced point

## 7.1 Introduction

Along with the computer in the application and development of teaching field, The establishment and application of item bank is more and more popular. And Automatic group volume database system developed key points. It require the user to enter little parameter values (Such as the total score, the difficulty coefficient of test paper, each topic quantity, each item in the score). A test mode could be generated by computer. And then take problem sets test paper, Forms one both to meet the user requirement finally and to conform to the teaching request

X. Zhang (✉)
Department of Mathematics and Information Science, Weinan Teachers University, Weinan 714000, China
e-mail: zhangxy71327@163.com

W. Du (ed.), *Informatics and Management Science V*, Lecture Notes in Electrical Engineering 208, DOI: 10.1007/978-1-4471-4796-1_7,

examination paper. And must design in reasonably the examination paper is the test question difficulty distribution to achieve the basic or to the requirements of the customers [1, 2].

The examination paper automatic generating system's examination paper difficulty control lies in the group volume algorithm on realization. The common group volume algorithm has along with the random group volume law; the recollection probe group volume law and the genetic algorithm group volume law and so on [3, 4]. Based on the test bank group volume algorithm, is tries the question bank set to divide first according to the topic set, then carries on each small set according to the group volume constraints the pretreatment, carries on the group volume each kind of topic request's difficulty coefficient as the balance point, if in the group examination paper this kind of topic average difficulty coefficient is consistent with this kind of topic expectation difficulty coefficient, then this kind of topic group volume is successful, carries on the next kind of topic group volume, otherwise on the continual readjustment difficulty coefficient, causes it to this kind of topic balance point stem for stem, realizes the group volume [5, 6].

## 7.2 Examination Paper Difficulty Evaluation Index System

Papers difficulty coefficient is the problem with difficulty coefficient for weighted average score is calculated, value range for (0, 1), namely

$$H = \sum_{i=1}^{i=n} K_i \times H_i / \sum_{i=1}^{i=n} K_i$$

Among them, $i$ is frage number, n is the question total number, $K_i$ and is the full score of the $i$th question. $H_i$ Is the difficulty coefficient of the ith question? Obviously $\sum_{i=1}^{n} K_i$ is the test score with a full, $\sum_{i=1}^{n} K_i \times H_i$ is the all examinee the entire volume to average lose point.

The greater of $H$, the more difficult, Generally speaking, when Difficulty coefficient value is 0.5, Test difficulty is Medium difficulty, when Difficulty coefficient is less than 0.3, Test is too easy. When Difficulty coefficient is greater than 0.7, Test is too difficult.

Test range have also influence about Difficulty coefficient.

Paper structure such as questions proportion, content ratio, and difficult problem of the proportion of the examination paper will also affect the difficulty.

## 7.3 Balance Strategy Group Volume Algorithm

Assume that test bank has already built; each test question's difficulty coefficient already gave. Has the volume before to publishing the volume to carry on the definition to the test scope, through inputs the chapter which each kind of test question needs to inspect, user input examination paper perfect score A, and each kind of topic total score $A_i$, thus calculates each kind of topic the topic quantity $N_i$, before publishing the volume to have an expected value to the examination paper difficulty coefficient, and estimates each type test question by this expected value the average difficulty coefficient, further input each average difficulty coefficient $B_i$.

Step 1: Carries on the pretreatment Question bank set {L} in ith kind of topic's trial question bank $\{L_i\}$ according to the group volume constraints, according to the investigation of the exam that range, choose the corresponding sections of the $C_{KL}$ for all 1 topics, get sub base $\{L_i\}$.

Step 2: According to the set $\{L_{Si}\}$ difficulty coefficient by ascending order, To the requirements of the type of topic test paper difficulty coefficient $B_i$ as the set $\{L_{Si}\}$ point of balance, the $\{L_{Si}\}$ is divided into $\{L_{Si}\}_{low}$ and $\{L_{Si}\}_{high}$ two orderly set. And in $\{L_{Si}\}_{low}$ deposits the difficulty coefficient to be lower than the balance point all test questions, in $\{L_{Si}\}_{high}$ deposits the difficulty coefficient to be greater than or equal to the balance point all test questions.

Step 3: Put the questions quantity $N_i$ according to the requirements of the proportion of the $\{L_{Si}\}_{low}$ and $\{L_{Si}\}_{high}$ divided into two parts $N_{i1}$ and $N_{i2}$, and $N_i = N_{i1} + N_{i2}$.

Step 4: In $\{L_{Si}\}_{low}$ draw an item at random of $N_{i1}$ topic, in $\{L_{Si}\}_{high}$ draw an item at random of $N_{i2}$ topic, $N_i = N_{i1} + N_{i2}$, the $N_i$ topic has composed the foundation examination paper $\{S_i\}$.

Step 5: In $\{L_{Si}\}$ all the remaining problem consisting of set $\{R_i\}$, $R_{\max}$ is the difficulty of the questions for maximum coefficient, $R_{\min}$ is the difficulty of the questions for minimum coefficient, find out the difficulty of the questions for maximum coefficient $S_{\max}$ of the foundation examination paper $\{S_i\}$ and minimum coefficient $S_{\min}$, extracts in foundation examination paper $\{S_i\}$ that all topic average difficulty coefficient $B_L$ to take the temporary balance point, calculate the difference between balance $B_i$ and temporary balance $B_i$ that $\Delta B = |B_i - B_L|$.

Step 6: When temporary balance point and balance point error in permission error range, that is $\Delta B < \varepsilon$, ($\varepsilon$ for allowable error value), then this topic group volume is successful, continues to carry on the next type test question the group volume, that is running Step 1, otherwise Step 7 operation.

Step 7: $\Delta B * N_i$ less than the difficulty of the coefficient $R_{\min}$ of remaining question bank $\{R_i\}$, if $B_L > B_i$, carries on difficulty coefficient biggest topic $S_{\max}$ with the difficulty coefficient closest topic ($S_{\min}$ difficulty coefficient $-\Delta B * N_i$) the replace, if $S_{\max}$ difficulty coefficient uses difficulty coefficient smallest topic $R_{\min}$ to replace $S_{\max}$.

Step 8: If $B_L < B_i$, then difficulty coefficient smallest topic $S_{\min}$ the difficulty coefficient closest difficulty coefficient's topic (difficulty coefficient $S_{\min} + \Delta B * N_i$) carries on foundation examination paper $\{S_i\}$ in with surplus test question set $\{R_i\}$ in the replace, If ($S_{\min}$ difficulty coefficient $+\Delta B * N_i$) is bigger than in surplus question bank $\{R_i\}$ the $R_{\max}$ difficulty coefficient, then carries on the replace with difficulty coefficient greatest topic $R_{\max}$.

Step 9: Now obtains new foundation examination paper $\{S_i\}$ and recent surplus topic set $\{R_i\}$, re-calculated $R_{\max}$, $R_{\min}$, $S_{\min}$, $S_{\max}$ and temporary balance point $B_L$, calculate differential value between balance point $B_{\mathrm{i}}$ and temporary balance point $B_L$, this is $\Delta B = |B_i - B_L|$. Carries on the Step 6.

Step 10: If all topic group volume completes, then withdrawal movement, otherwise, moves Step 1.

Algorithm characteristic:

Carry on the comparison through the balance point and the temporary balance point, makes the adjustment to the examination paper, Causes the temporary balance point to restrain unceasingly in the balance point, and may complete the group volume process fast.

Through group volume constraints, to tries the question bank to carry on the pretreatment, tries the question bank to divide meets the condition slightly to try the question bank, the group volume process slightly is trying in the question bank to carry on, has saved the time for the group volume, raised the group volume efficiency.

Because pulls out the topic process is stochastically carries on the balance point both sides, therefore had guaranteed pulls out the topic difficulty approaches in the balance point both sides in the equal distributions, has avoided difficult or the simple topic has appeared.

## 7.4 Control Method in Automatic Group Volume Process Realization

This algorithm is in the existence tries the question bank in the foundation to propose, must therefore confirm this algorithm to probably establish one to conform with the actual trial question bank, has "Advanced algebra" maturely in our institute tries the question bank. Among them is radio question bank kind, choose more question bank kind, and fills up the question bank kind, and judgment of problem, math class and Jane a question bank kind. First defers to the programmed control request user input examination paper name, various topics total score as well as various topics average difficulty coefficient, procedure automatic computation entire examination paper difficulty, if this difficulty coefficient is consistent with the user to the entire examination paper's difficulty coefficient expected value, then further chooses in the test scope in each type's question bank all test questions, otherwise the user must adjust each kind of topic average difficulty coefficient, causes the entire examination paper the difficulty coefficient to satisfy the user request.

**Table 7.1** Test after completion of the judgment problem

| Topic serial number | Difficulty coefficient | Chapter | Section |
|---|---|---|---|
| Question 1 | 0.23 | 1 | 3 |
| Question 2 | 0.23 | 1 | 5 |
| Question 3 | 0.35 | 1 | 2 |
| Question 4 | 0.35 | 1 | 2 |
| Question 5 | 0.36 | 3 | 3 |
| Question 6 | 0.34 | 3 | 2 |
| Question 7 | 0.42 | 3 | 3 |
| Question 8 | 0.42 | 3 | 4 |
| Question 9 | 0.58 | 2 | 8 |
| Question 10 | 0.58 | 2 | 7 |
| Average difficulty | 0.386 | | |
| Expatiated average difficulty | 0.380 | | |

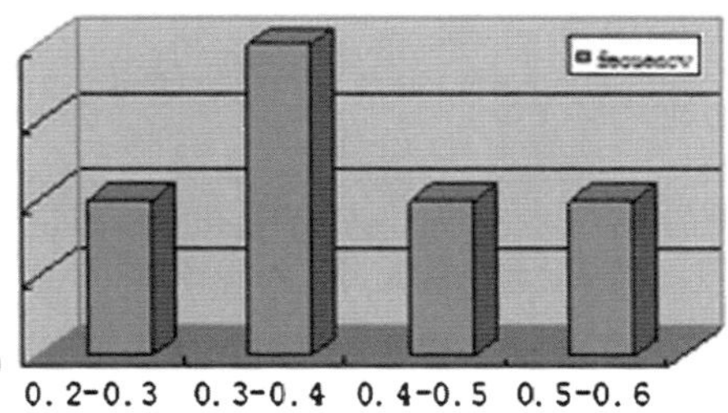


**Fig. 7.1** Difficulty coefficient frequency destitution of the judgment problem

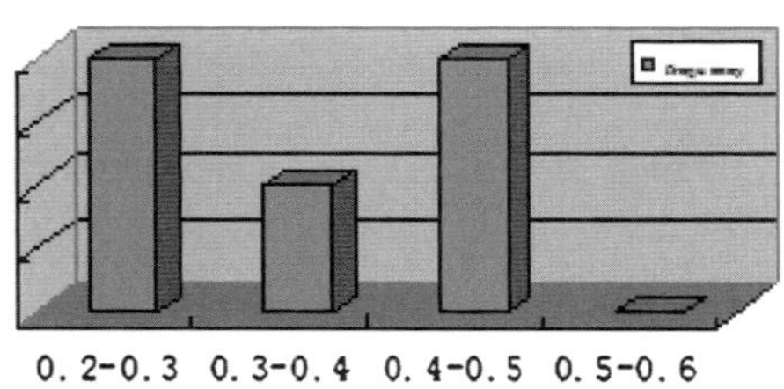


**Fig. 7.2** Expectation judgment topic difficulty algorithm

Example application: Take the Weinan Teachers university, Department of Mathematics and Information Science, 10 level of mathematics education specialty, 10–11 school year first semester advanced algebra test question as an example, by VF6.0 for the tool for analysis: Examination scope: The first semester of the first three chapters study content. The 1st chapter altogether 5, 2nd chapter of altogether 8, 3rd chapter of altogether 5,

Take judges the topic and fills up the topic to use this control method as the example to realize the automatic group volume. Each 10 small topics, carries out the result (Table 7.1).

The expectation judgment topic difficulty is 0.380, the actual production's judgment topic's average difficulty is 0.386, expects fills up the topic difficulty is

**Table 7.2** Test after completion of the problem filling in the blanks

| Topic serial number | Difficulty coefficient | Chapter | Section |
|---|---|---|---|
| Question 1 | 0.23 | 1 | 1 |
| Question 2 | 0.23 | 1 | 2 |
| Question 3 | 0.23 | 3 | 1 |
| Question 4 | 0.23 | 2 | 6 |
| Question 5 | 0.35 | 3 | 5 |
| Question 6 | 0.35 | 2 | 5 |
| Question 7 | 0.47 | 2 | 4 |
| Question 8 | 0.47 | 3 | 1 |
| Question 9 | 0.47 | 1 | 3 |
| Question 10 | 0.47 | 2 | 4 |
| Average difficulty | 0.350 | | |
| Expatiated average difficulty | 0.340 | | |

0.340, the actual production fills up the topic the average difficulty is 0.350, explained that this algorithm to the examination paper overall difficulty's control is very precise (Figs. 7.1, 7.2, Table 7.2).

## 7.5 Concluding Remark

The design and the control examination paper's difficulty is establishes the examination paper the important aspect, is also one of examination paper quality rating important targets. After the procedure confirmation, this article proposed applies based on the balanced strategy examination paper overall difficulty control method in the automatic group volume process has made the good progress, explained that this control examination paper difficulty the method was successful, also has built the good foundation for this control method in the later practical application.

**Acknowledgments** Shaanxi education projects (11JK0478).

## References

1. Li X (2003) Based on VF6.0 school test automatic group volume system. Comput Proj Des 11:66–70
2. Wen H (2005) In intelligence examination paper automatic generating system examination paper difficulty control technology research. Hunan Sci Technol Inst J 5:153–157
3. Ren Z, Shan R (2011) Based on MATLAB examination paper quality analysis research. Guangxi Nationality Norm Sch J 6:30–32
4. Jiao R, Li X (2005) The examination paper automatic generating system's design with realizes. Shanxi Med coll J (Found Med Educ Vers) 8:446–448

5. Ren Z (2011) MathType in the advanced algebra tries the question bank to realize the applied research. Inf Technol 3(9):45–47
6. Liu B (2004) Visual basic.NET database development specialized course vol 23(4). Tsinghua University publishing house, Beijing, pp 398–404

# Chapter 8
# Automation Structural Analysis Based on the VB and ANSYS

**Feng Liu, Xiankuan Qi and Mingqing Sun**

**Abstract** Taking automation mechanical analysis of automobile connecting rod structure in VB as an example, this paper discusses the automation analysis basic principle of mechanical mechanism in VB and ANSYS. Through the use of Visual Basic 6.0 friendly user interface and ANSYS APDL parameter analysis function, it in detail discusses the building process of analytical file, the starting and ending process of ANSYS analysis and the automatic display method of ANSYS results. The developed application program displays more efficient, fast and easy to use by using Visual Basic 6.0 to package ANSYS functions. Furthermore these functions make the application program more intelligent.

**Keywords** VB · APDL · Parametric calculation · Parametric analysis · Automatic analysis

## 8.1 Introduction

ANSYS software is the large-scale finite element analysis software which may analyze structure, electric field, magnetic field, fluid, sound field and so on. It is developed by ANSYS companies which is the world's largest finite element

F. Liu (✉) · X. Qi · M. Sun
Department of Engineering Structures and Mechanics, School of Science,
Wuhan University of Technology, Wuhan 430070, China
e-mail: whlfok@126.com

X. Qi
e-mail: qxk002@126.com

M. Sun
e-mail: sunmqing@whut.edu.cn

W. Du (ed.), *Informatics and Management Science V*, Lecture Notes in Electrical Engineering 208, DOI: 10.1007/978-1-4471-4796-1_8,

analysis software companies. So it can be widely used in civil, mechanical, aerospace, automotive industry, biomedical, bridges, construction, electronics and other fields with its strong function. It usually consists of pretreatment module, analysis module and post processing module. But because ANSYS software involves deep mechanical and electronic theory, ordinary users can't in a short time master the use method of ANSYS software. In this paper, VB is used to establish the analysis interface of mechanical mechanism. In the interface, we may call the APDL program in ANSYS to automatically analysis mechanical structure through the importation of certain parameters. Finally, through importing the data of ANSYS result database into the database of ACCESS, the user can easily query the analysis results of mechanical structure.

## 8.2 ANSYS Parametric Analysis and VB Program Interface Design

### *8.2.1 APDL Parametric Design Language*

The APDL language in ANSYS is a kind of programming design languages which can conduct two times development of ANSYS, namely it is the ANSYS parametric design language. In the mechanical structure mechanics analysis, there are generally two kinds of analysis methods: (1) GUI mode. (2) Command flow mode. In two kinds of modes, the command flow mode is provided by ANSYS APDL analysis method. Usually, in the ANSYS for the mechanical parameter analysis of the mechanical structure, we organize and management ANSYS finite element analysis command with the APDL program language and macro technology, thereby realizing the automatic analysis of the mechanical structure [1, 2].

The use of APDL can do mechanism parametric modeling, parametric material definition, parametric grid generation, parametric load and boundary conditions, parametric solution, parameter display results and so on. But because of the complexity of ANSYS APDL programming language as well as the optimal design of mechanical structure, it is necessary to develop a set of parameters software based on automatic analysis.

### *8.2.2 VB Program Design Language*

VB is an object-oriented programming language and has strong interactive function. VB is a kind of visual structured program design language which is developed by Microsoft Corporation. It is easy to learn BASIC language development. The VB uses event-driven way to run the application program, and it can decomposes a very complicated problem into a few even more small development module and has the strong function of database management, so this is for the exchange of data between application programs to provide convenient and favorable conditions. In the APDL program design language, graphical interface development and the interaction energy is not strong, and it is very difficult to control the flow of the program, but the VB in interactivity, program control and integration of various program has a strong advantage, therefore the VB is used in the development of program interface and the program flow's control, while the APDL is used in the mechanical analysis of mechanical structure.

## 8.3 Basic Principle of Automation Analysis Program of Mechanical Structure

### *8.3.1 ANSYS Macro File Creation and Call*

ANSYS macro is usually composed of several ANSYS command, and its file extension name is mac or MAC. Under normal circumstances, a complete analysis procedure is divided into different parts, and different parts are organized by different macros. For example when automatically analysing the mechanical properties of mechanical structure, we uses macro file to organize different parts. The macro file will greatly improve the readability and clarity of entire program. Different macro files are composed of the macro file of Modeling part, the macro file of material part, the macro file of solving and analysis part, the macro file of results processing part and so forth.

### *8.3.2 ANSYS Macro File Build*

Usually the macro files of ANSYS are edited in the WINDOWS Notepad. After the macro files are edited, it will be saved under ANSYS working directory as a mac or MAC file format. But some macro files of mechanical properties analysis of parametric mechanical structure are built by manual input parameters, and VB developed application program is automatically established.

### *8.3.3 Call of ANSYS Macro File in VB*

The call modes of ANSYS macro file are divided into internal call mode and external call mode. The internal call mode in the ANSYS directly uses a macro file, but the external call mode uses a macro file through the parameter transfer in VB. The internal call mode of macro files basically has the following kinds: (1) in the ANSYS command window input * USE, macro file name; (2) in the ANSYS command window directly input the macro file name; (3) in the ANSYS command window input/INPUT, macro file name. The external call mode uses a batch command of ANSYS to call macro files.

### 8.3.4 ANSYS Program Package and Call

Usually when using VB to package ANSYS program, at first we build the application program interface with VB in which the model parameters and material parameters of ANSYS analysis are included. After the application program interface is built, the various parameters are determined that are in need in the operating of analysis procedures. Thereafter, these parameters as ANSYS command flows together form a macro file and wait for the operation through the VB own internal mechanism. The "SHELL" command is used when VB calls ANSYS programs. If it is successful that the VB called ANSYS program, the application program will return a program number or it will return 0. Therefore, we need judge it that whether it is successful that VB calls ANSYS programs with judgment sentence. The command is as follows [3]:

Result = shell ("C:\ProgramFiles\Ansys Inc\v81\ANSYS\bin\intel\ansys81–b–i input_file–o output_file");
Among them: the input_file is the ANSYS input file of APDL language (i.e.: txt file format);
The output_file is the ANSYS output file (i.e.: dat file format).

In VB, we use the timer to determine whether the running ANSYS program is an end. From time to time you may check the *. err file, if the file is empty, the application program can judge that the analysis is the end. If the analysis result is not correct, the ANSYS program will write an error to the file. So the analysis programs need to reset and analysis.

### 8.3.5 Conversion of ANSYS Results Database and ACCESS Database

The data of ANSYS database are usually in the form of a table stored. Sometimes it is difficult to find the inner contact of the data such the form of a table, and the data analysis will also bring great difficulties, and the ACCESS database can be very convenient with EXCEL data processing software to exchange data, thereby it can improve data display and analysis function of application program. Figure 8.1 shows that the results of ANSYS database and ACCESS database conversion method flow chart.

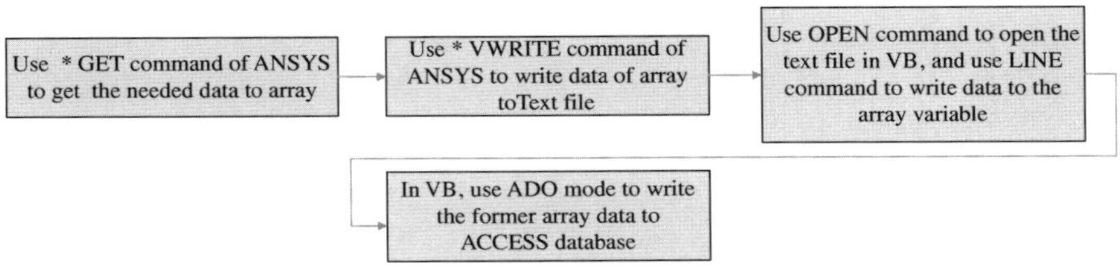


**Fig. 8.1** Conversion method of ANSYS results database and ACCESS database

## 8.4 Example

Taking the automobile connecting rod as an example, this part describes automatic analysis process of the mechanical structure (simplified). The automobile connecting rod dimension is shown in Fig. 8.2. The small right half ring of the connecting rod withstands the pressure of 25 MPa load. The left big hole coordinates with the fixed shaft. The shaft can be considered to be rigid. The elastic modulus of the connecting rod is 210 GPa, its Poisson's ratio is 0.3. Stress analysis doesn't consider the weight of parts [4, 5].

### *8.4.1 The Design of the Main Program*

The design of the main program usually includes the establishment of the ANSYS analysis file (i.e.: command flow file creation) and the ANSYS promoter analysis. (1) Use "Print # file number" command to establish the ANSYS analysis command flow. For example the meaning of Print # 1, "CLEAR, START" command is that the ANSYS analysis command flow/CLEAR, START in the macro file is

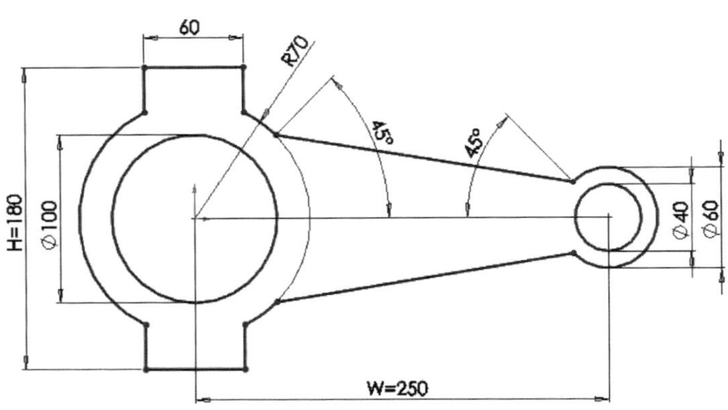


**Fig. 8.2** Automobile connecting rod size

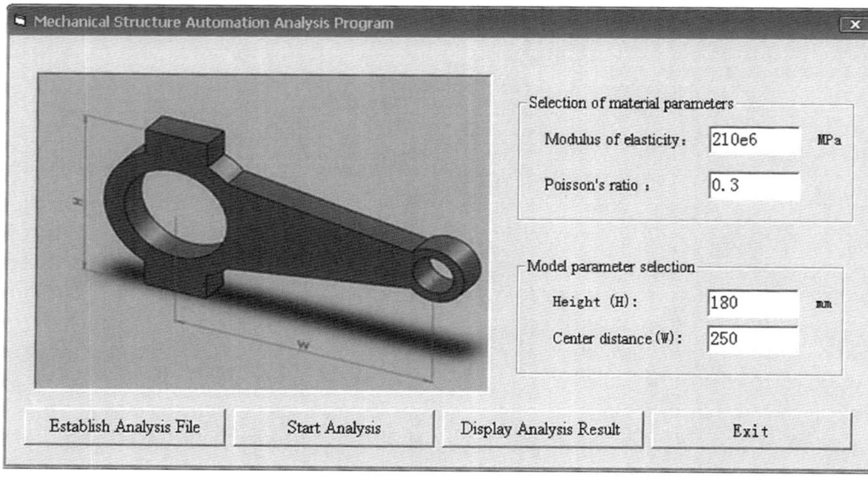


**Fig. 8.3** The main program interface

established (i.e.: clear memory, ready to begin the analysis). (2) we usually use the above two methods to start ANSYS analysis with the first step establishment file. When the *. err file is empty, the analysis is correct, and it will be finished. Figure 8.3 shows the main program operation interface.

### *8.4.2 The Design of the Results Display Part*

The analysis results show part contains graphic display and data display. Usually, the analytical results of the developed application program is displayed in the ANSYS, but because the common user can not skillfully use ANSYS to deal with the analysis results, this section will directly display the results of ANSYS in VB.

### *8.4.3 Graphic Display Part*

Graphic display part usually is composed of deformation nephogram, node stress nephogram, node stress nephogram and so on. We use the following method to automatically display the various nephogram of ANSYS analysis results in VB. Figure 8.4 shows the X direction displacement nephogram.

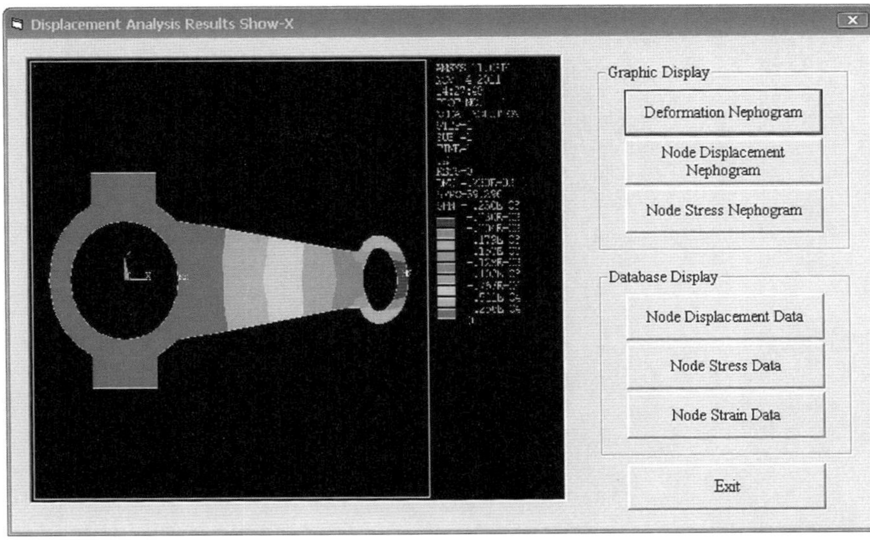


**Fig. 8.4** X direction displacement nephogram

To display the step of the nephogram in VB is as follows:

```
!Recovery the database of ANSYS analysis results
RESUME,'fenxiwenjian','db','E:\chengxu\fenxiwenjian\',0,0
......
!Output X direction displacement results file
/SHOW, JPEG, 0
  PLNSOL, U, X, 0, 1.0
/SHOW, CLOSE
```

### *8.4.4 Data Display Part*

After we convert ANSYS database into ACCESS database, the analysis results of ANSYS will be displayed in VB. The data display section contains the node displacement, the node stress and so forth. This section will realize the conversion of ANSYS database and ACCESS database. Node X, Y and Z direction displacement data in VB are shown in Fig. 8.5. The Fig. 8.6 shows the flow chart of data display method in VB.

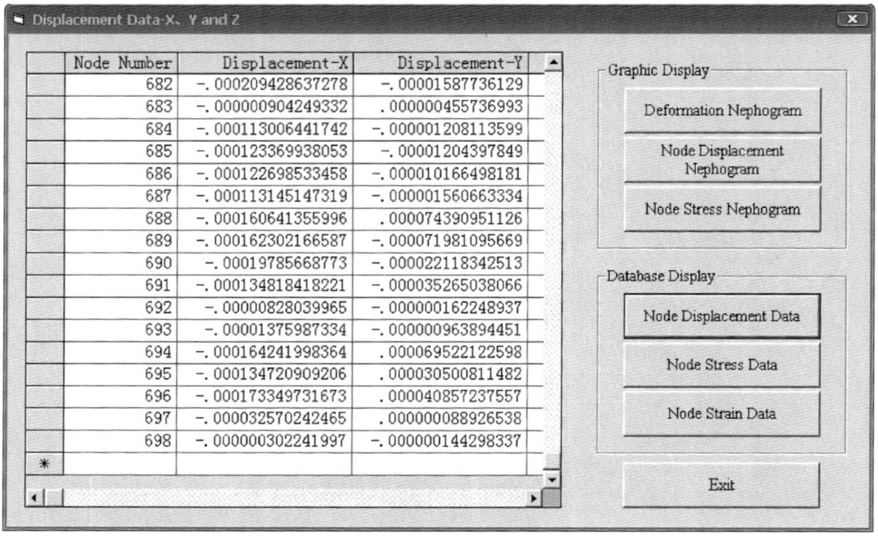


| Node Number | Displacement-X | Displacement-Y |
|---|---|---|
| 682 | -.000209428637278 | -.00001587736129 |
| 683 | -.000000904249332 | .000000455736993 |
| 684 | -.000113006441742 | -.000001208113599 |
| 685 | -.000123369938053 | -.00001204397849 |
| 686 | -.000122698533458 | -.000010166498181 |
| 687 | -.000113145147319 | -.000001560663334 |
| 688 | -.000160641355996 | .000074390951126 |
| 689 | -.000162302166587 | -.000071981095669 |
| 690 | -.00019785668773 | -.000022118342513 |
| 691 | -.000134818418221 | -.000035265038066 |
| 692 | -.00000828039965 | -.000000162248937 |
| 693 | -.00001375987334 | -.000000963894451 |
| 694 | -.000164241998364 | .000069522122598 |
| 695 | -.000134720909206 | .000030500811482 |
| 696 | -.000173349731673 | .000040857237557 |
| 697 | -.000032570242465 | .000000088926538 |
| 698 | -.000000302241997 | -.000000144298337 |

**Fig. 8.5** Displacement data-X, Y and Z

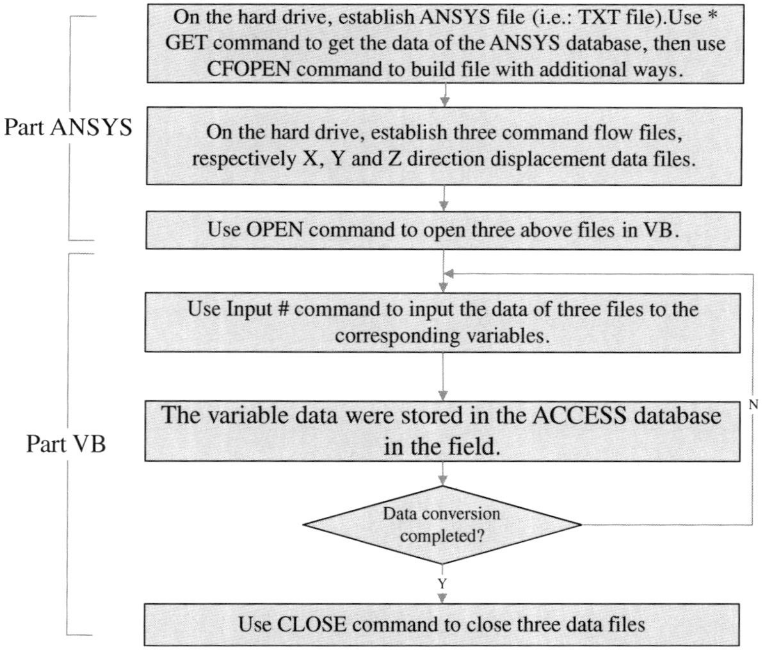


**Fig. 8.6** Flow chart of data display method in VB

## 8.5 Conclusions

Use VB visual program design language to package ANSYS analysis function, the mode forms of so-called “black box” function. When users only input some parameters, the automatic analysis of the mechanical structure can be realized, and achieve automatic display of results. At the same time, it also provides a basis optimization design of automatic analysis. This paper expounds the basic principle of automatic analysis of mechanical structure, and puts forward the thought of packaging ANSYS result data into the VB and the thought of transforming the data of ANSYS database into the data of ACCESS database. After the database is transformed, the foundation is established in which we use the EXCEL data analysis software to conduct the twice analysis of the data.

**Acknowledgments** This work was supported by the National Natural Science Foundation of China (NO. 50878170).

## References

1. Zhang HJ (2008) The technical research on secondary development of ANSYS graphical user interface. Dev Innovation Mach Electr Prod 21(3):116–125
2. Shao J (2006) The parametric design analysis of ANSYS based on visual basic. J Chongqing Univ Sci Technol 8(3):98–100
3. Peng GF (2004) A method of invoking ANSYS based on VB control developing. J Wuhan Univ Technol 28(1):148–150
4. Liu X (2006) The ANSYS foundation and application guide vol 73. Science Press, Beijing, pp 163–165
5. Sun X (2009) Visual basic development of technology, 2nd edn, vol 36. People’s Posts and Telecommunications Press, Beijing, pp 255–258

# Chapter 9
# Risk Early Warning System Preventing Natural Disaster in Yunnan Power Grid

**Zhengzhi Li, Tong Han, Yumei Li and Zhigang Liu**

**Abstract** In order to prevent natural disaster on Yunnan power grid caused harm, research team in collecting and processing a lot of data, built natural disaster prevention model, and according to the model, risk assessment model, build a database, combined with GIS software to design the risk early warning system, this paper focuses on major geological disaster risk point of monitoring and warning to solve scheme, based on Meteorological rainfall early warning of geological disasters solutions, software system function modules, system features and other aspects of analysis and design.

**Keywords** Risk early warning · Geological disaster · Disaster model · System design · YN power grid

## 9.1 Introduction

Risk early warning system is the future state warning process. According to the basic theory study proposes a theoretical model; disaster risk early warning is in disaster on disaster bearing system under the action of the disaster system status, future changes to the warning [1]. The natural disaster risk early warning depends on monitoring data and risk management of collected information [2]. Due to

Z. Li · T. Han · Y. Li
Safety Technique Research Institute, Electric Power Research Institute, Yunnan Power Gird, Kunming 650217, China

Z. Liu (✉)
School of Computer and Information, Southwest Forestry University, Kunming 650224, China
e-mail: fiberhome@126.com

W. Du (ed.), *Informatics and Management Science V*, Lecture Notes in Electrical Engineering 208, DOI: 10.1007/978-1-4471-4796-1_9,

various professional fields are provided with corresponding monitoring network, as the grid system to guard against natural disasters risk early warning, set monitoring system should be considered first directly from the relevant departments to obtain professional monitoring data (such as the meteorological disasters, geological disasters), then considering the network set up its own monitoring system [3].

## 9.2 The Solutions of Monitoring and Warning on the Risk Point of Major Geological Disaster

For a direct threat to the security of power grid the major geological disasters risk point, grid can consider setting up special monitoring and early warning system. Early warning of geological disasters in the overall framework of the system is shown in Fig. 9.1; the monitoring of hardware system is shown in Fig. 9.2.

Monitoring system using displacement sensor, gauge, video network monitoring and other relevant professional equipment, combined with GIS, with supplementary professional geological disasters, early warning, and decision-making

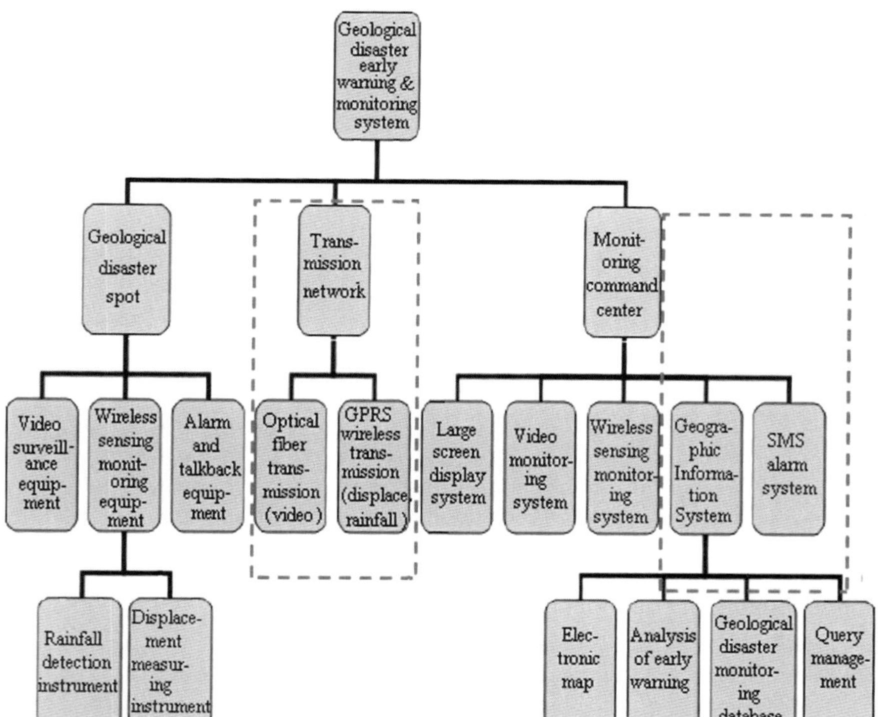


**Fig. 9.1** The frame of early warning system of major geological disasters risk

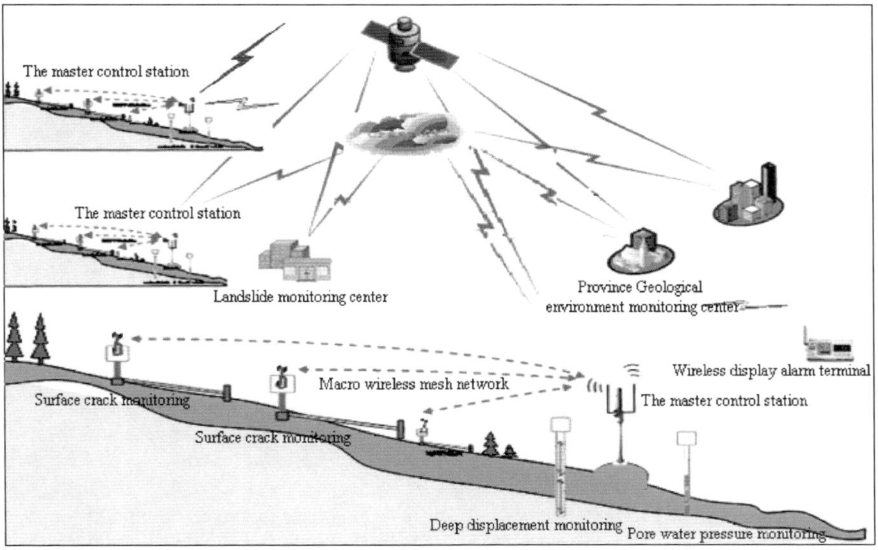


**Fig. 9.2** The hardware diagram of risk monitoring system of major geological disaster

system to build geological disaster prevention and measurement system a new method of geological disasters, continuous, practical, dynamic monitoring, timely access to comprehensive and accurate data automation requirement, satisfy, coordination efficiency is high, thereby preventing the occurrence of geological hazards, reduce the loss of life and property [4].

According to the actual situation of disaster points, corresponding to the selected high-tech detection equipment. Video surveillance system, if the distance is appropriate and with construction conditions, by laying an optical fiber; [5] also can be used for mobile GPRS wireless transmission channel, the range is wide, environment, technology, high quality cheap, convenient, the region from the time space constraints; long on the potential dangers of geological disasters the implementation of on-line monitoring. In situ monitoring specific methods may refer to Table 9.1.

## 9.3 The Solution of Early Warning of Geological Disasters Based on the Meteorological Rainfall

According to the study on the relationship between landslide and debris flow and rainfall, landslide and debris flow occurrence not only with the excitation of rainfall, rainfall and early process closely, as shown in Fig. 9.3. In the picture,

**Table 9.1** The method selection of reference table for the geological disaster monitoring

| Kinds | | Practical |
|---|---|---|
| Deformation monitoring | Macro geological monitoring | All kinds of geological fire field macro geological bowls |
| | Surface displacement monitoring | Collapse, landslide and debris flow and ground subsidence of geological hazards, such as surface integral and fracture displacement deformation monitoring |
| | Deep displacement monitoring | For monitoring has obvious characteristics of the deep sliding slope disasters deep the displacement monitoring |
| Physics and physicochemical monitoring | Stress field monitoring | Used to collapse, landslide and debris flow geological disasters body the special site or the whole stress field change monitoring |
| | To sound monitoring | Apply to the rocky collapse, landslides and mudslides and geological disasters in the process of the acoustic emission characteristics of events |
| | Electromagnetic field monitoring | Apply to monitoring the electric field in the evolution process of the disaster, the change of the electromagnetic field information |
| | Disaster body temperature monitoring | Apply to monitor landslides and debris flows of the geological hazards in the process of the body temperature change disaster information |
| | Radioactive survey | For monitoring crack, subsidence disaster body special parts of the radon anomalies |
| | Mercury gas measurement | For monitoring crack, subsidence disaster body special parts of the mercury gas abnormal |
| Inducing factors monitoring | Weather monitoring | Apply to clear the influence of atmospheric precipitation by geological hazard factors to trigger monitoring. Such as: collapse, landslide and debris flow, ground subsidence, geological hazards, such as to crack |
| | The earthquake monitoring | Apply to clear the affected geological disaster caused by the monitoring factors, such as: collapse, landslide and debris flow, ground subsidence, etc |
| | Human activity | For monitoring human activity of geological disasters in the process of the formation, development impact |
| Groundwater monitoring | Groundwater dynamic monitoring | Apply to monitoring of landslide and debris flow, ground subsidence of geological hazards, such as the dynamic change of groundwater |
| | The pore water pressure monitoring | Apply to landslide and debris flow geological disasters in pore water pressure monitoring |
| | Groundwater quality monitoring | Apply to monitoring landslide and debris flow, ground subsidence, seawater intrusion of geological hazards, such as dynamic changes of groundwater |

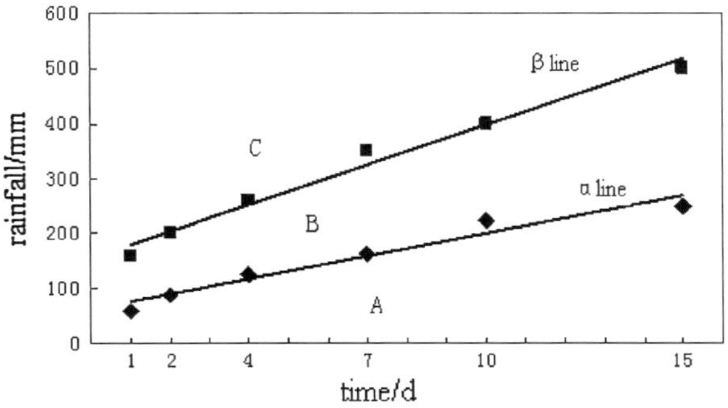


**Fig. 9.3** Weather warning criterion model modules of the geological disaster

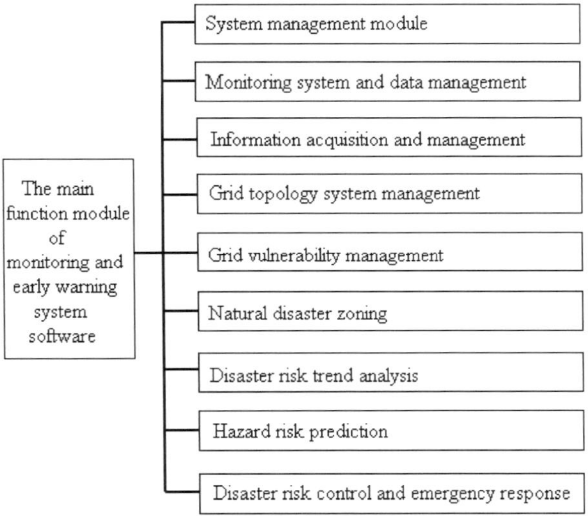


**Fig. 9.4** The main function of the monitoring system

alpha and beta line two landslide and debris flow occurrence of critical rainfall line, a line below the A zone to forecast area (1, 2, the possibility of smaller), alpha to beta line between B area for geological disaster forecast area (3, 4, the possibility of a larger) beta, above the line of the C area for geological disaster alert zone (Level 5, the possibility of a large). Actual warning solutions: criterion models of Fig. 9.4 is transformed into warning software system calculation, according to the meteorological observatory issued the rainfall, artificial rainfall value input, automatic calculation and judgment, and according to the results of published corresponding early warning of geological disasters. The warning for regional geological disasters forecast [6].

## 9.4 The Software System Design

### *9.4.1 The Function Module of Software System*

Software system main function module is shown in Fig. 9.4.

The system management module

Responsible for the management of software system, such as basic data, system parameter, the user information, such as user authority management [7].

Monitoring system and data management

Responsible for access to real time monitoring system display, monitoring data, monitoring video display and monitoring data processing operation.

Information collection and management

Responsible for non real-time monitoring information (such as meteorological, geological disaster, power line inspection and etc.) the collection, processing, display, query.

The network topology system management

Responsible for the power network topology system generation, modify, coloring, zoom, zoom and other functions. These functions should be based on geography, such as local amplification function not only enlarge power itself, its geographical should also with amplification (equivalent to the functions of electronic map). The module is the foundation of the whole system (equivalent to "stage"), the other modules (including network vulnerability zoning, natural disaster risk zoning, disaster risk trend analysis, disaster warning) uses this to finish. The request of two-dimensional, three-dimensional long-term consideration due function [8].

The grid vulnerability management

According to the power system vulnerability assessment classification, different sections, on the grid position, equipment, coloring, marking, analysis, query, save display operation, and according to the inspection, a new evaluation results for dynamic modification.

Natural disaster zoning

According to the regional natural disaster (geological disaster, earthquake, volcano, the development mainly to geological disaster risk analysis results) is responsible for the regional natural disaster risk zoning, according to area in different position in different risk levels with different colors (such as red, orange, yellow, blue, green, etc.) for coloring, and mark the grade and legend. According to the condition change, the evaluation model of regionalization, modified. With input, evaluation, storage, query, modify, coloring, marking, display and other functions.

Disaster risk trend analysis

According to the real-time monitoring data or information collection data, according to the analysis model, grid vulnerability and disaster risk overlay analysis, determination of regional power grid by natural disasters risk undermining the size and change trend. And the dynamic demonstration. According to needs but also for typical disaster simulation demo.

Disaster risk early warning and forecast

According to the analysis of the forewarning model, warning degree standard, Web issued warning.

### *9.4.2 The Characteristic of Software System*

It includes specifically:

1. Based on the GIS or MAPGIS development "Yunnan power grid to prevent natural disaster monitoring and early warning system" software system;
2. The system has a natural disaster monitoring system data access interface and monitoring data management function;
3. Information collection and management functions (including grid basic information system, natural disaster information, related geographic information)
4. The network topology system management function;
5. Power system vulnerability zoning management function;
6. The natural disaster risk zoning management functions (including geological disasters, earthquake disaster, fire disaster);
7. The dynamic risk analysis and warning function;
8. Risk management, risk management module mainly includes risk zoning, survey and control in group, annual plan, emergency response module.

## 9.5 Conclusions

This article from the thought and technology of network of natural disaster risk early warning system design, mainly from the geological disaster risk point rainfall monitoring, early warning of geological disasters aspects, analyzed the software system function modules, system features and other aspects of the analysis and design for Yunnan power grid to prevent natural risk, provide decision-making support.

**Acknowledgments** This work was supported by the Science and technology Program of Yunnan Power Grid (NO. 20110034) and the National Natural Science Foundation of China (NO. 50804038). In the process of doing experiments, I have got some help from others. Especially, I would love to thank Xiaozhi Wang and PiaoPiao Bie (Jiangcheng College, China University of Geosciences) for photo-processing.

## References

1. Qing D, Zhang M, Jianshe L et al (2006) "9/26" in Hainan power grid blackout accidents analysis and summary. Autom Electr Power Syst 1:1–7
2. Pengfei Z (2008) Influence of natural calamity on power grid and Enlightenment in power network planning. Jiangxi Electr Power 8(2):20–28

3. Guoxin X, Qing X, Chongqing K (2010) The anti-disaster power system planning model and model. Autom Electr Power Syst 2:45–48
4. Liu Y, Cai B, Wu S (2008) Power grid ice disaster accident emergency treatment and reflection. Autom Electr Power Syst 32(8):124
5. Laura B, Gerson CO, Mario P (2001) A mixed integer disjunctive model for transmission network expansion. IEEE Trans Power Syst 16(3):560–565
6. Sun H (1996) Electric power network planning. Chongqing University Press, Chongqing
7. Sharifnia A, Aashtiani H (1985) Transmission network planning: a method for synthesis of minimum 2 cost secure networks. IEEE Trans Power Apparatus Syst 104(8):2026–2034
8. Xiaohui B, Hui H (2008) Based on ice disaster of Yunnan power grid risk and disaster countermeasures. J Wuhan Univ Technol (Information and Management Engineering Edition) 3(4):17–19

# Chapter 10
# Study of Velocity Control of Hydraulic Servo System Based on Server/Client Architecture

**Cheng-Yi Chen**

**Abstract** In this paper, server/client control architecture, based on simple proportional-integral-derivative (PID) controller, is proposed to achieve a velocity control of the electro-hydraulic system with system uncertainties, and disturbances. In the server control system, a embedded controller is applied to support the real-time control for general testing with PID controller and data latched functionality. However, in client control system, a personal computer is adopted to provide the human–machine interface for monitoring and transferring the control command with expected parameters from and to the embedded controller, respectively. According to the concept of the PID Ziegler-Nichols tuning rules, the proposed firmware design in embedded controller provides an easy testing and quick setting for the PID's parameters by the experimental approach through the Ethernet communication between server and client system. The experimental results verified that the tracking control of the designated velocity profile cannot be achieved by one set of the PID's parameters. Therefore, the control strategy for combining the five sets of the different PID's parameters is proposed through directly switching function. The experimental results show that the proposed method can efficiently improve the control performance for the velocity trajectory of the cylinder in wide range speed.

**Keywords** Embedded controller · PID control · Remote control

C.-Y. Chen (✉)
Cheng Shiu University, Kaohsiung, 83347 Taiwan
e-mail: albert@csu.edu.tw

W. Du (ed.), *Informatics and Management Science V*, Lecture Notes in Electrical Engineering 208, DOI: 10.1007/978-1-4471-4796-1_10,

## 10.1 Introduction

Hydraulic system have been widely used in automotive, manufacturing, military industry and civil engineering, because it possess the advantages of small size-to-power ratio, variable speed operation, smooth behavior, force control and so on. The hydraulic system conventionally adopts a constant-flow type of vane pump to supply the required hydraulic pressure and flow so that it need to additionally consider flow control valve in the hydraulic oil circuit to achieve the piston velocity control. Basically, its hydraulic circuit is more complex and the components are cheaper but not accuracy. With advanced electronic flow-control elements, such as proportional valves and servo valves, the feedback control approach has become an attractive option for electro-hydraulic systems, as it permits accuracy and bandwidth requirements to be maintained [1]. Servo control valve provides a faster frequency response as well as better dynamic performance for precision control, compared with conventional valve. However, servo valve is manufactured by multi-layer approach so that the demand of its oil quality is very high. As long as the oil is dirty, it easily leads to mistake behavior because of its high precision requirement. For example, an experimental optimization formulation is proposed to determine the effective control gains for an electro-hydraulic position control system, such that the variations of the system dynamic caused by various factors can be properly compensated [2]. A nonlinear control algorithm is developed to address motion synchronization of a dual-cylinder electro-hydraulic lifting system [3]. In contrast, the operation of proportional valve is controlled by the solenoid force. It almost possesses the same performance with servo valve. However, its control performance is better than the conventional flow control valve does, and its price is also cheaper than servo valve [4]. Besides, nonlinearities and parameter uncertainties are the two main challenges associated with developing control algorithms for electro-hydraulic systems. The self-tuning of PI controllers using fuzzy logic is studied to shorten start-up time and reduce the initial cost of building an automatic tuning system [5]. Although many control methods can be used, hydraulic control system generally adopt the controller design based on the PID tuning rules because its control structure is simple and easily implemented in real-time controller [6]. In this paper, server/client control architecture, based on simple proportional-integral-derivative (PID) controller, will proposed to achieve a velocity control of the electro-hydraulic system using the proportional control valve.

## 10.2 Hydraulic System Description

An electro-hydraulic system shown [4] is considered as experiment setup. In order to obtain a system model the controller design. Considering the velocity response of a step input voltage of 6 V applied to the hydraulic servo system, second order under damped system is generally adopted to model the hydraulic servo dynamic [1].

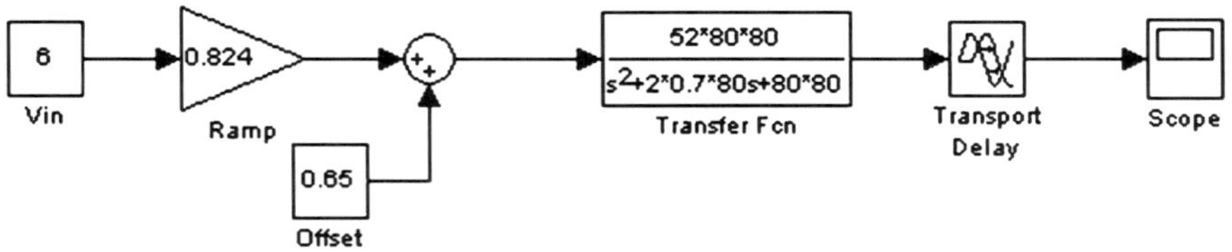


**Fig. 10.1** Block diagram of Simulink model

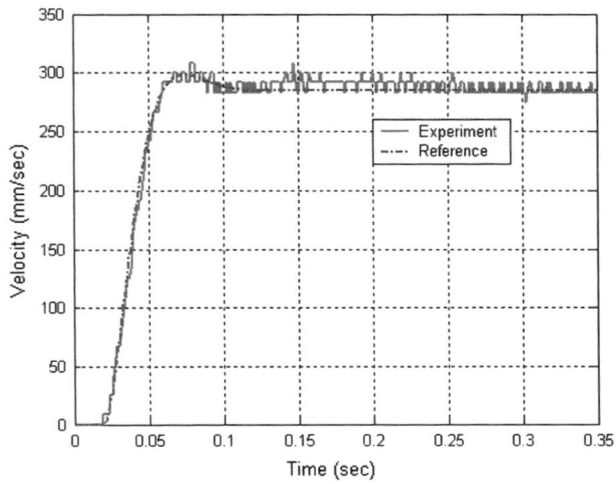


**Fig. 10.2** Compared velocity response of simulation model and experiment system

$$G(s) = \frac{a \cdot \omega^2}{s^2 + 2 \cdot \xi \cdot \omega \cdot s + \omega^2} \tag{10.1}$$

where $a$ is the gain, $\omega$ is the natural resonance frequency, and $\xi$ is the damping Ratio. Here, we simply implements a block diagram of Simulink model as shown Fig. 10.1, where offset gain of 0.65 and ramp gain of 0.824 are respectively applied to capture dead zone operation and ramp response of proportional valve; transport delay block is for response delay phenomenon. Note that $a$ is adjusting for steady response, $\omega$ is adjusting with respect to response speed, and $\xi$ is adjusting for oscillation response. Based on the above tuning rules, Fig. 10.2 presents the velocity responses of the cylinder for both the experiment and its corresponding simulation system when $a = 51$, $\omega = 80$, $\xi = 0.7$, and transport delay of 0.018 s. It shows that the system response is almost matched compared Simulink model with experimental response. In this paper, PID controller design is simply based on the first method of Ziegler and Nichols tuning rule shown in Table 10.1, where $R$ is the slope and $L$ is the time delay. For the input voltage of 6 V applied to the cylinder hydraulic servo system, $K_P = 0.0128$ and $T_i = 0.02$ for PI controller; $K_P = 0.01714$, $T_i = 0.12$, and $T_d = 0.03$ for PID controller are obtained. For fine tuning criteria, it basically has the following features: (1) increasing $K_P$ and $1/T_i$ tend to reduce system error but may not be capable of also producing adequate stability, and (2) increasing $T_d$ tends to improve stability [7].

**Table 10.1** Ziegler-Nichols tuning for the regulator, $D(s) = K_P\,(1 + 1/T_i s + T_D s)$

| Type of controller | $K_P$ | $T_i$ | $T_d$ |
|---|---|---|---|
| P | $1/RL$ | | |
| PI | $0.9/RL$ | $L/0.3$ | |
| PID | $1.2/RL$ | $2L$ | $0.5L$ |

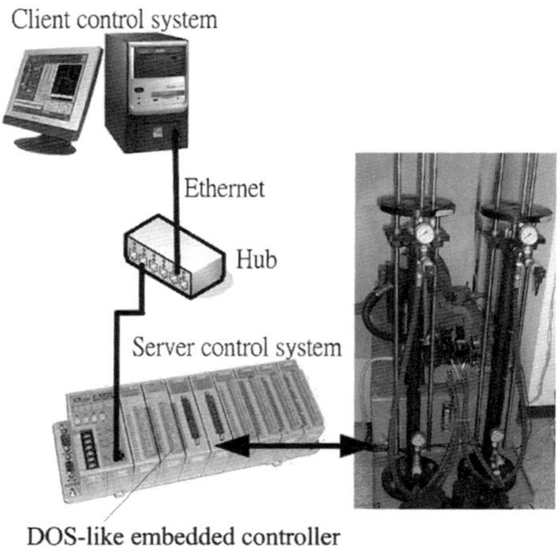


**Fig. 10.3** Experimental hydraulic system for server/client control architecture

## 10.3 The Hydraulic Server/Client Control Architecture

The server and client control architectures for hydraulic system are implemented as Fig. 10.3. Based on PID control law, it provides a easy testing platform for tuning its parameters in embedded real-time system. As to client system, it can transfer the controller's parameters to server system, and then trigger the experiment and log the data as well as analysis it in Mat lab tool. The essential composition of both systems will be investigated in below.

### *10.3.1 The Structure of Controller Server*

An embedded controller I-8000, produced by ICPDAS Corporation, is adopted as server controller since it provides a DOS-like operating system and supports Ethernet communication, through which user-designed program can be uploaded from the client system into the target system for testing. Figure 10.4 presents a Ethernet server control flowchart named Xserver for easily implementing a customized real-time controller. Basically, this flowchart is composed by two parts of VxComm.lib and user. VxComm.lib is library calls and supports the well-developed fundamental functions, which includes hardware configuration, Ethernet and modbus server, especially the first entry of main module. As to User.c, it allows the

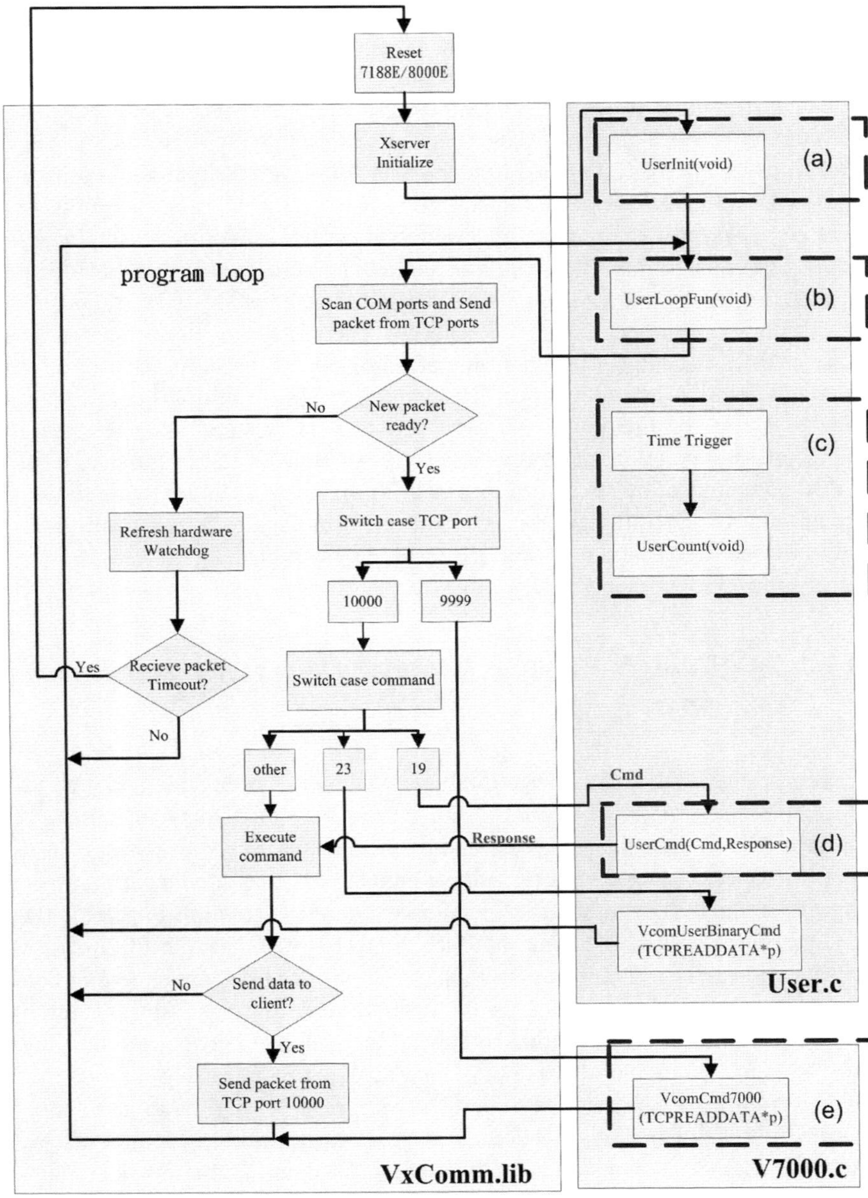


**Fig. 10.4** Xserver control flowchart

designer to embed the customized program through the six function calls (UserInit, UserLoopFun, UserCount, UserCmd, VcomCmd7000 and VcomUserBinaryCmd). UserInit function enables designer to initialize I/O devices and declares control-oriented variables. As shown in Block (b), Xserver will periodically execute the function of UserLoopFun(). According to this feature, the mechanisms of warm-up hydraulic system, searching home, manual operation, enable/disable hydraulic pump etc., will be programmed in this portion, and they can be trigged by the command string which will be activated by UserCmd function. As shown in Block (c), Xserver provides a interrupt function of UserCount, which can be executed by timer software interrupt of embedded controller. However, the periodic trigger time can be set up by AddUserTimerFunction (UserCount, time) in UserInit() function. Since digital controller is generally designed in the way of constant sampling rate, the PID controller in this article will be discretely implemented during this function. For Block (d), UserCmd function allows user to send messages from the client to the server with two arguments of (cmd, Response). Therefore, programmer can use cmd in the character argument to determine and implement the different user-defined command string. The argument of "Response" is used to return the string to the client system. Furthermore, Xserver provides the other channel through VcomCmd7000 function for large amounts of data transferring by the port 9999 of TCP/IP communication, as shown in Block (e). In single-cylinder hydraulic servo system, we expect to transfer the experimental data back to host system for data analysis when experimental testing is immediately completed.

### 10.3.2 The Human–Machine Interface Design in Client Personal Computer

After constructing the server programming in the target system, client system need to use the TCP socket of Visual Basic 6.0 development tools to build a the graphical human–machine interface and connections with server system. In this paper, we use VB 6.0 to design a graphical human–machine interface for prototype controller testing of hydraulic servo control system, as shown in Fig. 10.5. Basically, graphical human–machine interface control system has the following two main functions: (1) communication: toolbar buttons, PID parameter setting fields, the communication speed setting frame, manual mode settings frame, IP settings frame, communication port setting frame and the message window response from server system; (2) graphics display: as long as experiment testing is done in the server system, human–machine interface can receive the large amounts of experimental data from server side, and then show up in the predefined tab window. This client system can also save the graphical figure into a raw data files for further analysis in Mat lab software.

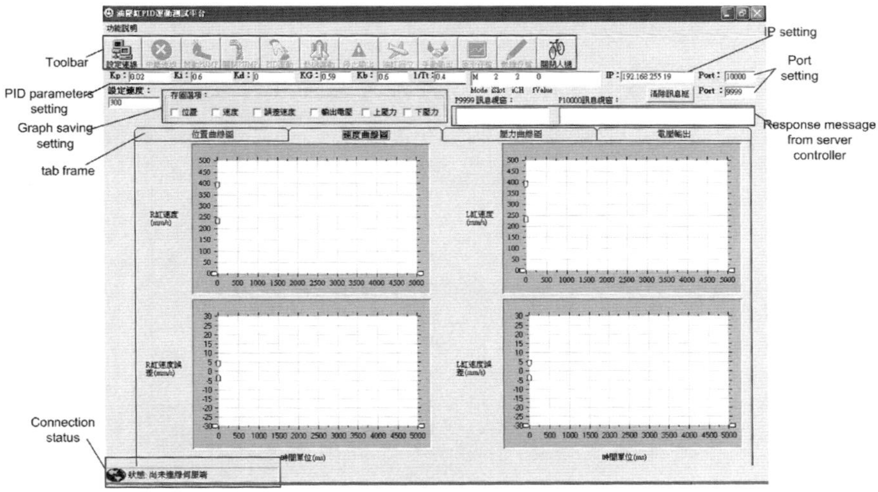


**Fig. 10.5** Human-machine interface at client personal computer

## 10.4 Experimental Results

In this article, a single-cylinder hydraulic servo system is considered for the validation of the proposed control approach, when hydraulic working pressure is 50 kg/cm$^2$. Figure 10.6 shows the simulation and experimental results of step response of 300 mm/sec by apply the PID's parameters ($T_b = 0.1$, $K_P = 0.0128$, $1/T_i = 19$, $T_t = 0.4$), which is designed from simulation system. It is observed that the PID's parameters obtained from the simulation model provide an effective speed control in the experiment system, but produce a sluggish response. Compared experimental and simulation results, its response rise time the maximum

**Fig. 10.6** Simulation and experimental velocity control by $K_b = 0.1$, $K_P = 0.0128$, $1/T_i = 19$, $T_t = 0.4$

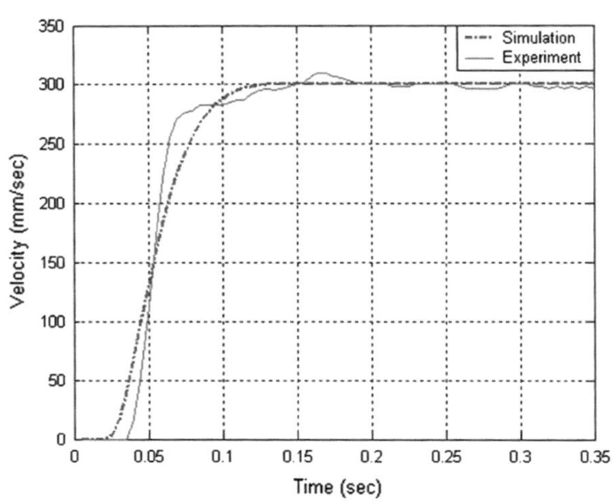

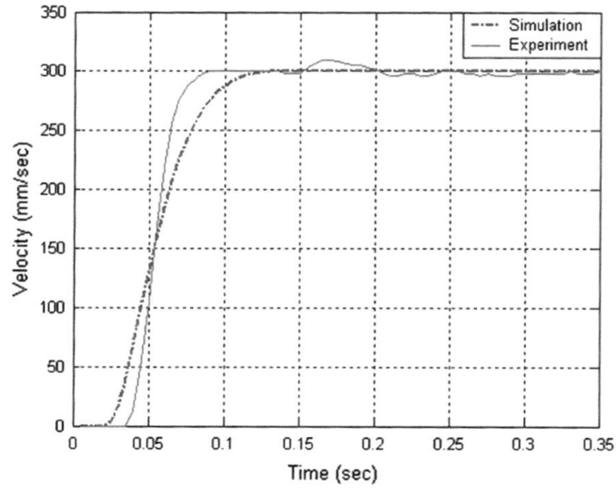


**Fig. 10.7** Simulation and experimental velocity control by $K_b = 0.1$, $K_P = 0.0128$, $1/T_i = 20.2$, $T_t = 0.4$

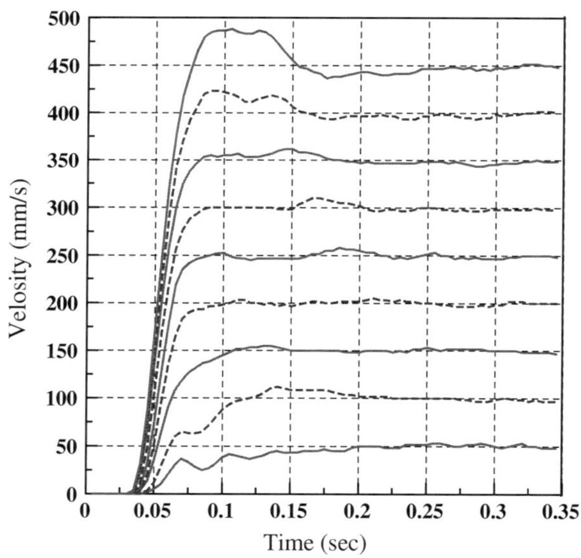


**Fig. 10.8** Step response of 50–450 mm/s as PID parameters is $T_b = 0.1$, $K_P = 0.0128$, $1/T_i = 20.2$, $T_t = 0.4$

overshoot can, respectively, be further increased and reduced. Therefore, when the integral parameters ($1/T_i$) are considered to be 20.2, Fig. 10.7 shows that speed control performance is improved with significantly reducing the maximum overshoot. Figure 10.8 shows the velocity responses from 50 to 450 mm/s, when PID's parameters of $T_b = 0.1$, $K_P = 0.0128$, $1/T_i = 20.2$, and $T_t = 0.4$ is used. It can be observed that these control parameters is only suitable for velocity control of 200–300 mm/s. As to velocity control of 50–150 mm/s, it depicts over damped response and its settling time is more than 0.1 s. As to velocity control of 350–450 mm/s, it significantly produces larger overshoot response, which also exceeded the maximum amount as set speed increases. It is conducted that one

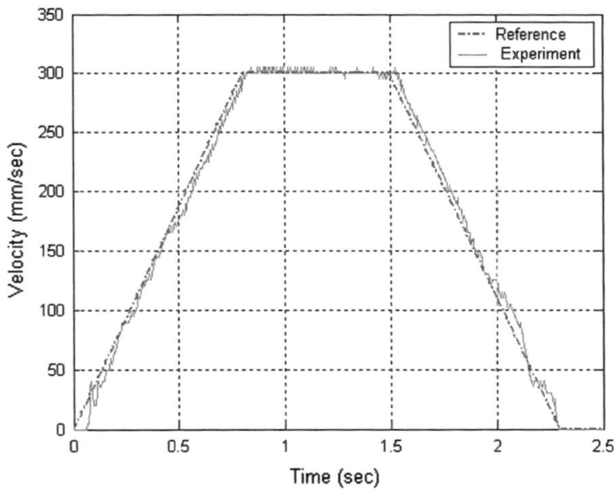


**Fig. 10.9** Velocity response for trapezoidal velocity profile by directly switching the five control parameters

controller's parameters is not suitable for wide range of speed control. Figure 10.9 show the experimental results of tracking a trapezoidal velocity profile by directly switching the control parameters among the five rules, which is designed from individual speed control. It shows that velocity response is very smooth and speed tracking error is also very small. It is suggested that the results of the combined five controllers is better than single controller does. Although the switching rules are intuitively for each velocity region, it provides control results for hydraulic servo system because of hydraulic damper phenomenon.

## 10.5 Conclusions

In this paper, server/client control architecture has proposed to achieve the velocity control of hydraulic servo system by simply applying a PID controller with adjusting parameters in the embedded controller. Through the help of client control system, we can quickly set the PID's parameters into sever control system, immediately execute the experimental testing, and get the experimental results from server control system as well as analysis the velocity, positioning and control input in graphical view. Therefore, qualified PID's parameters can be easily found for the velocity control of hydraulic servo system. The experimental results verified that the tracking control of the designate velocity profile cannot be achieved by one set of the PID's parameters since hydraulic system possess strong nonlinear phenomenon in different moving speed. Furthermore, a control strategy of directly combined five sets of the PID's parameters is verified successfully in the velocity control for a trapezoid curve profiles.

## References

1. Jelali M, Kroll A (2003) Hydraulic Servo-system modeling, identification and control. Springer, London
2. Kim MY, Lee C-O (2006) an experimental study on the optimization of controller gains for an electro-hydraulic servo system using evolution strategies. Control Eng Pract 14(2):127–147
3. Sun H, Chiu GT-C (2002) Motion synchronization for dual-cylinder electro hydraulic lift systems. IEEE/ASME Trans Mechatron 7(2):171–181
4. Chen CY, Liu LQ, Cheng CC, Chiu GTC (2008) Fuzzy controller designs for synchronous motion in a dual-cylinder electro-hydraulic system. Control Eng Pract 16(6):658–673
5. Berger M (1996) Self-tuning of a PI controller using fuzzy logic for a construction unit testing apparatus. Control Eng Pract 4(6):785–790
6. Cheng CC, Chen C-Y (1998) A PID approach to suppressing stick-slip in the positioning of transmission mechanisms. Control Eng Pract 6(4):471–479
7. Franklin GF, Powell JD, Emanmi-Naeini A (1991) Feedback control of dynamic system. Addison-Wesley publishing company, New York

# Chapter 11
# Robust Stability of Discrete Interval Systems with a State Time Delay

**Chien-Hua Lee and Chien-Chih Chang**

**Abstract** This paper studies robust stability analysis for discrete interval time delay systems. By selecting properly a positive definite matrix $Q$, we derive a simple upper bound of the solution of the discrete Lyapunov equation. Then, by using Lyapunov equation approach associated with this bound, several concise criteria are developed to guarantee the robust stability of the mentioned systems. The feature of these present criteria is that they are independent of any Lyapunov equation although the Lyapunov equation approach is adopted.

**Keywords** Lyapunov equation approach · Solution bound · Robust stability · Time-delay · Interval system

## 11.1 Introduction

Interval matrices are caused by unavoidable parametric variations, changes in operating conditions, aging, etc., and are real matrices in which all entries are known only to the extent that each belongs to a special closed interval. A system that contains an interval matrix (or interval matrices) is called an interval system and therefore can be considered as a system with parametric perturbations. In the recent literature, there are considerable interests in the stability analysis and

C.-H. Lee (✉) · C.-C. Chang
Department of Electrical Engineering, Cheng Shiu University, 833 Kaohsiung, Taiwan, People's Republic of China
e-mail: chienhua@csu.edu.tw

C.-C. Chang
e-mail: chang968@ms13.hinet.net

W. Du (ed.), *Informatics and Management Science V*, Lecture Notes in Electrical Engineering 208, DOI: 10.1007/978-1-4471-4796-1_11,

stabilization design of the mentioned systems [1]. In practice, due to information transmission between elements or systems, natural property of system elements, computation of variables, etc., time delay(s) exist(s) in real-life systems and should be integrated into system model. It is known that time delay is always a source of instability of controlled systems. Therefore, the research of the impact of time delay(s) for linear or nonlinear systems has become more and more attractive during the past decades. In the recent years, a number of research results have been devoted to stability analysis and/or stabilization controller design for interval time-delay systems [2]. Therefore, this paper addresses the robust stability for discrete interval time-delay systems. By using the Lyapunov equation approach associated with a very simple upper bound of the solution of the discrete Lyapunov equation, a delay-independent stability criterion is derived for the mentioned systems. An interesting consequence is that this obtained criterion does not involve any Lyapunov equation. Comparison between the present result and a well-known result is also made [3]. It will be shown that the obtained result is better. Finally, we also give a numerical example to verify the correctness of the presented schemes.

The following symbol conventions are used in this paper. Symbol $\lambda_1(A)$ means the maximal eigenvalue of a symmetric matrix $A$. Furthermore, $|A| = [|a_{ij}|]$ for matrix $A = \{a_{ij}\}$, and $|A| \le B$ means $|A| = [|a_{ij}|]$ and $|a_{ij}| \le b_{ij}$ for $B = [b_{ij}]$.

## 11.2 Main Results

Consider the discrete time-delay system

$$x(k+1) = A_I x(k) + B_I x(k-d) \tag{11.1}$$

$$x(k) = \Phi(k), \quad k \in [-d, 0] \tag{11.2}$$

where $x(\cdot) \in R^n$ represents the state, integer $d > 0$ denotes the delay, $\Phi(k)$ is a known time function, $A_I = [a_{Ipq}]$ and $B_I = [b_{Ipq}]$ are interval matrices with appropriate dimensions and have the properties:

$$A_I \in N[U, V] \text{ with } U = [u_{pq}], \; V = [v_{pq}] \tag{11.3}$$

$$B_I \in N[E, F] \text{ with } E = [e_{pq}], \; F = [f_{pq}]. \tag{11.4}$$

Functions $N[U, V]$ and $N[E, F]$ present the set of all matrices $A_I$ and $B_I$ satisfying

$$u_{pq} \le a_{Ipq} \le v_{pq}, \quad e_{pq} \le b_{Ipq} \le f_{pq}, \quad p, q = 1, 2, \ldots, n. \tag{11.5}$$

**Lemma 1** *Matrices* $A, B \in R^{n\times n}$. *Then for any given positive* $\alpha$, *the following inequality is satisfied* [4].

$$A^T B + B^T A \leq \alpha A^T A + B^T B/\alpha. \tag{11.6}$$

*Utilizing the Lyapunov equation approach and Lemma 1, we derive the following criteria.*

Define

$$A = [a_{pq}] \equiv 0.5(U + V) \text{ and } G = [g_{pq}] \equiv 0.5(V - U), \tag{11.7}$$

$$B = [b_{Ipq}] \equiv 0.5(E + F) \text{ and } H = [h_{pq}] \equiv 0.5(F - E), \tag{11.8}$$

where $p, q = 1, 2, \ldots, n$. *Then (11.1) can be represented as*

$$\dot{x}(t) = (A + \Delta A)x(t) + (B + \Delta B)x(t - d) \tag{11.9}$$

where $\Delta A$ *and* $\Delta B$ *denote parametric uncertainties with properties:*

$$|\Delta A| \leq G \text{ and } |\Delta B| \leq H \tag{11.10}$$

which means $|\Delta A| = [|\Delta a_{pq}|]$, $|\Delta B| = [|\Delta b_{pq}|]$ *and* $|\Delta a_{pq}| \leq g_{pq}$, $|\Delta b_{pq}| \leq h_{pq}$ *for* $p, q = 1, 2, \ldots, n$.

**Theorem 1** *If*

$$(\|A\| + \|G\| + \|B\| + \|H\|)\left(\frac{A^T A}{\|A\|} + \|G\|I + \frac{B^T B}{\|B\|} + \|H\|I\right) < I, \tag{11.11}$$

*Then the interval time-delay system* (11.1) *is robustly stable.*

*Proof* For system (11.9), we choose a Lyapunov function as

$$V = x^T P x + \sum_{j=1}^{d} (\|A\| + \|G\| + \|B\| + \|H\|)x^T(k - j) \times \left(\frac{B^T P B}{\|B\|} + \frac{\Delta B^T P \Delta B}{\|H\|}\right)x(k - j) \tag{11.12}$$

where the positive definite matrix $P$ satisfies the following Lyapunov equation

$$(A^T + \Delta A^T)P(A + \Delta A) - P = -Q. \tag{11.13}$$

with

$$Q = q\left[(\|B\| + \|H\|)\left(\frac{A^T A}{\|A\|} + \|G\|I\right) + (\|A\| + \|G\| + \|B\| + \|H\|)\left(\frac{B^T B}{\|B\|} + \|H\|I\right)\right] \tag{11.14}$$

where $q > 0$ is an arbitrary constant. Now, we rewrite the Lyapunov Equation (27) as

$$\begin{aligned}&(A+\Delta A)^T(qI-P)(A+\Delta A)-(qI-P)\\&\quad=Q+q(A+\Delta A)^T(A+\Delta A)-qI\\&\quad=Q+q(A^TA+A^T\Delta A+\Delta A^TA+\Delta A^T\Delta A)-qI\end{aligned} \tag{11.15}$$

In light of Lemma 1, we have

$$\begin{aligned}A^T\Delta A+\Delta A^TA\le\frac{\|G\|}{\|A\|}A^TA+\frac{\|A\|}{\|G\|}\Delta A^T\Delta A\le\frac{\|G\|}{\|A\|}A^TA\\+\|A\|\|G\|I\le 2\|A\|\|G\|I\end{aligned}, \tag{11.16}$$

$$\Delta A^T\Delta A\le\|\Delta A\|^2I\le\||\Delta A\||^2I\le\|G\|^2I. \tag{11.17}$$

Then substituting (11.16), (11.17), and (11.14) into (11.15), it is seen that condition (11.11) can assure the positive definiteness of the solution $(qI-P)$ of Eq. (11.15). This infers $P<qI$. Taking the forward difference for $V(x(k))$ along the trajectories of (11.9) yields

$$\begin{aligned}\Delta V=x^T[(A+\Delta A)^TP(A+\Delta A)-P]x+x^T(A+\Delta A)^TPBx_d\\+x^T(A+\Delta A)^TP\Delta Bx_d+x_d^TB^TP(A+\Delta A)x+x_d^T\Delta B^TP(A+\Delta A)x\\+x_d^T(B^TPB+B^TP\Delta B+\Delta B^TPB+\Delta B^TP\Delta B)x_d\end{aligned}$$

$$\begin{aligned}+(\|A\|+\|G\|+\|B\|+\|H\|)\Bigg[x^T\left(\frac{B^TPB}{\|B\|}+\frac{\Delta B^TP\Delta B}{\|H\|}\right)x\\-x_d^T\left(\frac{B^TPB}{\|B\|}+\frac{\Delta B^TP\Delta B}{\|H\|}\right)x_d\Bigg]\end{aligned} \tag{11.18}$$

Using $P<qI$ and the facts

$$\begin{aligned}&x^T(A+\Delta A)^TPBx_d+x_d^TB^TP(A+\Delta A)x\\&\le\frac{\|B\|}{\|A\|+\|G\|}x^T(A+\Delta A)^TP(A+\Delta A)x+\frac{\|A\|+\|G\|}{\|B\|}x_d^TB^TPBx_d\end{aligned} \tag{11.19}$$

$$\begin{aligned}&x^T(A+\Delta A)^TP\Delta Bx_d+x_d^T\Delta B^TP(A+\Delta A)x\\&\le\tfrac{\|H\|}{\|A\|+\|G\|}x^T(A+\Delta A)^TP(A+\Delta A)x+\tfrac{\|A\|+\|G\|}{\|H\|}x_d^T\Delta B^TP\Delta Bx_d\end{aligned} \tag{11.20}$$

$$B^TP\Delta B+\Delta B^TPB\le\frac{\|H\|}{\|B\|}B^TPB+\frac{\|B\|}{\|H\|}\Delta B^TP\Delta B, \tag{11.21}$$

Equation (11.18) becomes

$$\begin{aligned}\varDelta V \le & x^T\left[-Q + \frac{\|B\| + \|H\|}{\|A\| + \|G\|}(A+\varDelta A)^T P (A+\varDelta A)\right. \\ & \left.+(\|A\|+\|G\|+\|B\|+\|H\|)\left(\frac{B^T PB}{\|B\|}+\frac{\varDelta B^T P \varDelta B}{\|H\|}\right)\right]x \\ < & x^T\left[-Q + q\left[(\|B\|+\|F\|)\left(\frac{A^T A}{\|A\|}+\|G\|I\right)\right.\right. \\ & \left.\left.+(\|A\|+\|G\|+\|B\|+\|H\|)\left(\frac{B^T B}{\|B\|}+\|H\|I\right)\right]\right]x = 0\end{aligned} \tag{11.22}$$

It shows if (11.11) holds then $\Delta V < 0$. This guarantees the robust stability of system (11.9). Thus, the proof is completed.

*Remark 1* If $A_I = A$ and $B_I = B$, then the system (11.1) becomes the following time-delay system

$$x(k+1) = Ax(k) + Bx(k-d). \tag{11.23}$$

For system (11.23), a concise stability condition was given in [5] as

$$\|A\| + \|B\| < 1. \tag{11.24}$$

Furthermore, the condition (11.11) now becomes

$$(\|A\| + \|B\|)\left(\frac{A^T A}{\|A\|} + \frac{B^T B}{\|B\|}\right) < I. \tag{11.25}$$

For this condition, we have

$$\begin{aligned}\frac{A^T A}{\|A\|} + \frac{B^T B}{\|B\|} & \le \lambda_1\left(\frac{A^T A}{\|A\|} + \frac{B^T B}{\|B\|}\right)I \\ \le \lambda_1\left(\frac{A^T A}{\|A\|}\right)I + \lambda_1\left(\frac{B^T B}{\|B\|}\right)I & = (\|A\| + \|B\|)I\end{aligned}. \tag{11.26}$$

This means that the condition (11.25) is better than that proposed in [5].

*Remark 2* Note that Lyapunov equation (11.13) is unsolvable. However, an interesting consequence of the obtained result is that it is not needed to solve these equations for stability condition (11.11). Furthermore, this stability condition is also independent of the constant $q$.

## 11.3 A Numerical Example

Consider the discrete time-delay system (11.1) with.

$$A_I = \begin{bmatrix} [-0.1 \quad -0.06] & [-0.1 \quad 0] & 0 \\ 0 & [-0.4 \quad -0.3] & [-0.06 \quad 0] \\ 0 & [0 \quad 0.1] & [0 \quad 0.1] \end{bmatrix}$$

And

$$B_I = \begin{bmatrix} [-0.1 \quad 0] & [0 \quad 0.06] & 0 \\ [-0.35 \quad -0.25] & 0 & 0 \\ 0 & 0 & [0 \quad 0.1] \end{bmatrix}.$$

Then, from definitions (11.7) and (11.8), we have

$$A = \begin{bmatrix} -0.08 & -0.05 & 0 \\ 0 & -0.35 & -0.03 \\ 0 & 0.05 & 0.05 \end{bmatrix}, B = \begin{bmatrix} -0.05 & 0.03 & 0 \\ -0.25 & 0 & 0 \\ 0 & 0 & 0.05 \end{bmatrix},$$

$$G = \begin{bmatrix} 0.04 & 0.1 & 0 \\ 0 & 0.1 & 0.06 \\ 0 & 0.1 & 0.1 \end{bmatrix}, H = \begin{bmatrix} 0.1 & 0.06 & 0 \\ 0.1 & 0 & 0 \\ 0 & 0 & 0.1 \end{bmatrix}.$$

We now check the condition of Theorem 1 and obtain

$$(\|A\| + \|G\| + \|B\| + \|H\|)\lambda_1\left(\frac{A^TA}{\|A\|} + \frac{B^TB}{\|B\|} + \|G\|I + \|H\|I\right) = 0.6844 < 1.$$

It shows that the proposed stability condition can assure the robustly stability of this system. For this case, if $A_I = A$ and $B_I = B$, this system becomes system (11.23). Since

$$(\|A\| + \|B\|)\lambda_1\left(\frac{A^TA}{\|A\|} + \frac{B^TB}{\|B\|}\right) = 0.2227 < 1.$$

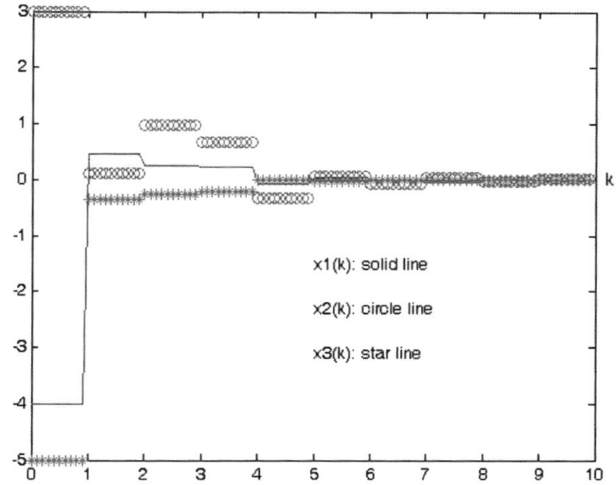


**Fig. 11.1** The trajectories of state $x(k)$ of example 1

Let $d = 2$ and $x(k) = [-4 \quad 3 \quad -5]^T$ for $k \in [-2,\ 0]$. The simulation results for the states are shown in Fig. 11.1. It is seen that all states are regulated to zero irrespective of the time-delay.

## 11.4 Conclusions

In this paper, a simple upper solution bound of different Lyapunov equations can be obtained by choosing properly the positive matrix $Q$. Then, by utilizing the Lyapunov equation approach associated with this upper bound, the robust stability for discrete time-delay interval systems has been studied. We have derived a robust stability condition. It is easy to be checked. The feature of this condition is that it does not involve any Lyapunov equation although the Lyapunov approach is utilized. Finally, an illustrative example has been given to verify the correctness of the presented schemes.

**Acknowledgments** The authors would like to thank the National Science Council, the Republic of China, for financial support of this research under the grant NSC 100-2221-E-230-009.

## References

1. Biglarbegian M, Melek WW, Mendel JM (2010) On the stability of interval type-2 TSK fuzzy logic control systems. IEEE Trans Syst Man Cybernet Part B 40:798–818
2. Chang YC, Chen SS, Su SF, Lee TT (2004) Simultaneous static output-feedback stabilization for discrete-time interval systems with time-delay. IEEE Proc Control Theory Appl 151:445–452
3. Chang YC, Su SF, Chen SS (2004) LMI approach to static output-feedback simultaneous stabilization of discrete-time interval systems with time-delay. In: Proceedings of the third international conference on machine learning and cybernetics, Shanghai, vol 43(3). pp 4144–4149
4. Zhou K, Khargonekar PP (1988) Robust stabilization of linear systems with norm-bounded time-varying uncertainty. Syst Control Lett 10:17–20
5. Lu JG, Chen G (2009) Robust stability and stabilization of fractional-order interval systems: an LMI approach. IEEE Trans Autom Control 54:1294–1299

# Chapter 12
# Design of Visual Embedded Mooring Control System

**Yuliang Liu, Zuoyu Zhou and Xiaomin Shi**

**Abstract** A design scheme of visual embedded mooring control system is proposed in this paper, including hardware scheme based on an embedded processor and software scheme referring to clipping of Linux operation system and coding of mooring control tasks. Main function of the visual system refers to controlling casting and collecting mooring, i.e. to controlling the direction, speed and rhythm of anchor chain movement, with an aim of completing mooring control only on the shore. This paper focuses on improving safety, flexibility and extension of mooring control, especially in poor weathers such as big wave or typhoon. The total scheme, hardware design, software design are narrated respectively. Finally, the safety and flexibility of the visual system has been confirmed from some experimental results. The extension can be guaranteed by ideal equipment cost in that the equipment is programmable.

**Keywords** Mooring system · Visual control · Embedded processor

Y. Liu (✉) · Z. Zhou · X. Shi
School of Electromechanical Engineering, Zhejiang Ocean University,
Zhoushan 316004, China
e-mail: liu.yuliang@hotmail.com

Z. Zhou
e-mail: dianxin2011@126.com

X. Shi
e-mail: sxm@126.com

W. Du (ed.), *Informatics and Management Science V*, Lecture Notes
in Electrical Engineering 208, DOI: 10.1007/978-1-4471-4796-1_12,

## 12.1 Introduction

Nowadays there are a large number of anchorages where mooring systems are still manipulated artificially in China. Anchor equipments often move out of control due to poor performance of monitoring and controlling, which leads to some collision accidents between ships or between ship and bridge in the past years, so visual control of mooring system becomes a novel research hot point in ship domain [1, 2]. We are aware of progress of this field, where barge controller's successful development is representative, however, almost all the hardware designed in point is based on the technology of application special integrated circuit (ASIC) with well-known higher price and lower flexibility. In addition, maximum communication range of barge controller can only reach dozens of meters, i.e. it is often used in the inland river, so its application and extension are limited. In view of the present situation, in this paper we propose a visual mooring control scheme [3, 4].

As shown in Fig. 12.1, necessary mooring information, including distance between ships, image of mooring equipments, water speed and direction, wind speed and direction, and mooring force, can be sampled by the corresponding sensor and can be collected and saved by an information processor, then can be analyzed by hypogynous machine. Next the information is submitted to upper machine by a wireless receiving and transmitting device [5]. On the contrary, some order information such as casting mooring or collecting mooring can be transmitted from upper machine down to hypogynous machine, then to drive anchor machine by actuating mechanism. Wireless receiving and transmitting device in Fig. 12.1 is similar to a couple of wireless network interface. One part of the wireless receiving and transmitting device is built-in information processor adjacent to hypogynous machine in ship, and the other part is built-in upper machine onshore.

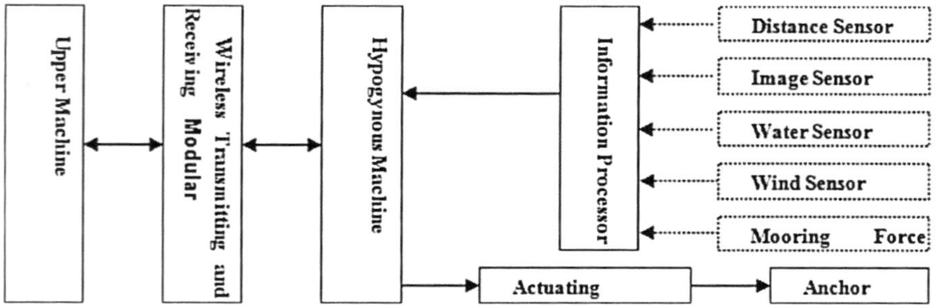


**Fig. 12.1** Scheme chart of visual mooring control system

## 12.2 Working Principle

The hardware design interferes with upper machine, wireless receiving and transmitting device, hypogynous machine, sensor group and actuating mechanism.

The sensor group includes ship distance sensor, image sampling sensor that is a digital camera, wind speed and wind direction sensor, water speed and water direction sensor and mooring force sensor. The ship distance sensor is fixed on bow and it is used to detect distance between ships in case of collision with the front ship. Wind speed and wind direction sensor is also equipped on bow to detect wind. Water speed and water direction sensor is equipped underwater to obtain the wave information. Mooring force sensor is fixed on mooring chain to detection force from shape change. The function of sensor group is to provide environmental data and force data for safety assessment of mooring system.

In the process of casting anchor, first the order of opening hypogynous machine is transmitted by wireless channel. Next the hypogynous machine's power supply is turned on and each sensor's state can be scanned. Then the hypogynous machine start to receive, calculate and judges the safety level of mooring system. If it finds that some dangers exit, then it will emit some alarming signals to upper machine. In the same time, the hypogynous machine will sample and submit image data to main machine. If the data is sent to upper machine successfully, the upper machine will submit confirm information and then submit order information to drop an alternate mooring or to collecting some mooring on basis of the ship loading and the sensor data.

## 12.3 Hardware Design

As shown in Fig. 12.2, in this paper the upper machine is design based on embedded processor chip s3c2440a developed by Korea Samsung corporation, interfaced with LCD displayer, clock chip, keyboard control and reset circuit, etc. LCD screen displays information such as time, operation menu, keyboard state, mooring image and alarming information. The information can also be displayed alternately by keyboard operation. The standard time of mooring operation is provided by clock chip S12887. The function of upper machine includes opening the power of submachine, detecting ship's position, wind state, wave state and mooring image, sending order, recognizing information from mooring system, etc.

The function of hypogynous machine includes obtaining mooring state data, calculating the ship stress based on experienced formula, giving the probability of mooring movement by comparing mooring force with other total stress, then making a feedback to upper machine. The hardware design of hypogynous machine about image sampling is shown in Fig. 12.3, including an embedded processor chip S3C2440A, CMOS digital camera, FLASH memory, USB driver CY7C68013A, built-in wireless receiving and wireless emitting chip nRF2401 [5].

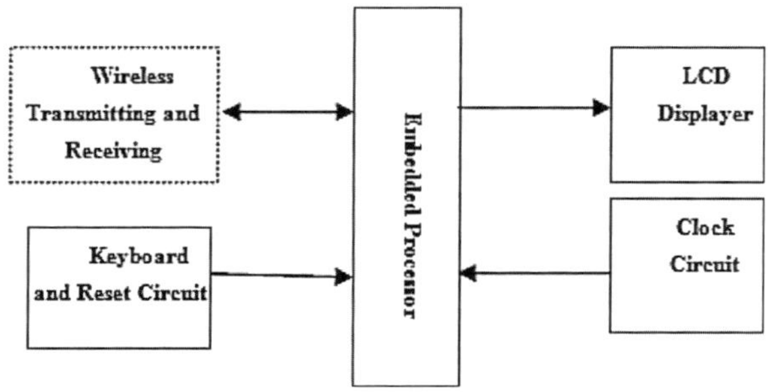


**Fig. 12.2** Hardware construction of upper machine

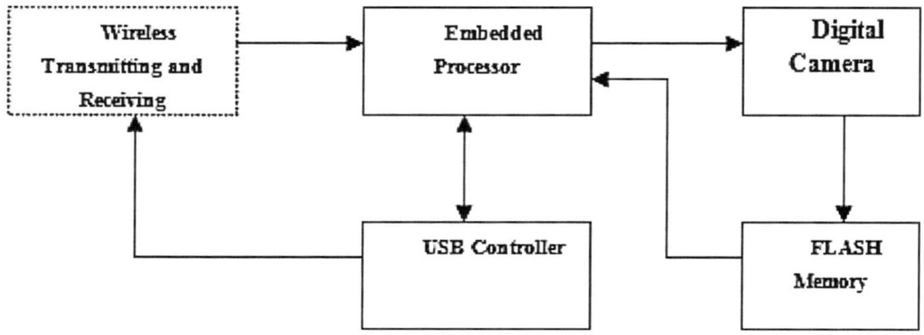


**Fig. 12.3** Hardware construction of hypogynous machine about image sampling

## 12.4 Software Design

Software design in point includes constructing an embedded platform and realizing the concrete control order, where the transmitting speed of image is a key issue. Software flow chart of image sampling is shown in Fig. 12.4, whose function steps include initialization containing initialing of USB controller, wireless receiving and emitting, and Linux operation system, utilizing digital camera to form image with PPM style and to save it in FLASH memory, compressing image with PPM style to JPG style by JMPG program, then transmitting JPG image to USB2.0 controller terminal.

## 12.5 Conclusion

In this paper a novel visual control scheme of mooring system is presented. The scheme includes upper machine, wireless receiving and transmitting, hypogynous machine, execution mechanism, etc. The upper machine is fixed onshore with

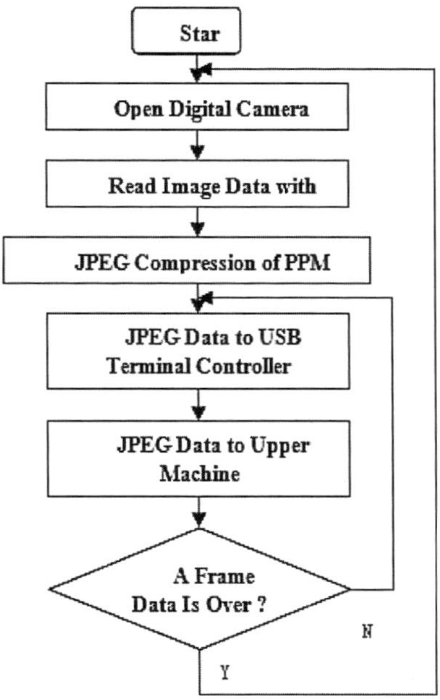


**Fig. 12.4** Software flow chart of image collection

hardware design based on a programmable embedded chip. Such a scheme makes operator work without entering a marine, so the safety can be improved obviously. In addition, the flexibility is better than that of other schemes based on ASIC technology in that the control function can be planted and changed by software coding. Both the safety and flexibility are confirmed from affirmative experiment results made at Shuanghe wharf of Zhejiang province of China in 2011. This paper tries to use information technology to improve mooring control, and we will concentrate on researching application and extension of the visual embedded mooring control system in the next.

**Acknowledgments** This work is supported by non-profit scientific project of science and technology department of Zhejiang province of China (2010C33178), project of science and technology department of Zhoushan city of China (10249) and project of Zhejiang Ocean University of China (200913, 201023, 11182100710).

## References

1. Ma NQ (2002) Auto control research of anchor machine based on a fuzzy algorithm. J Wuhan Univ Technol (Sci Eng) 26(3):361–364
2. Wang CX (2010) Anchorage and pre-test of mooring ship's safety management countermeasure. Transp Sci Technol 241(4):108–110

3. Yang QT (2008) PLC application in three speed windlass's control system. China Sci Inform 2(24):112–114
4. Jiang W, Liu XH, Zhang H et al (2008) Remote controller of casting and collecting mooring system of unmanned barge. Patent Bureau of the People's Republic of China (G08C17/02; G05B19/05; B63B21/22)
5. Tan H (2009) Principal and advanced application of NRF wireless SOC single machine, vol 2. Publishing House of Beihang, Beijing, pp 39–47

# Chapter 13
# Efficient Online Examination System Based on Java

**Yongliang Li and Caixia Wen**

**Abstract** The system with the writing of the Java language, function of the realization of students' test, results inquiries, the teacher inquires the, monitor examination process, management and maintenance of the database information. The system is composed of four is module: test the client system module, the test server system module, the client inquires the system module, database information management and maintenance module.

**Keywords** Java language · The test the client · Database · The system module

## 13.1 Introduction

This system as a student on online examination system the requirements of the network examination system the basic function from simple and easy operation into consideration, integrated the above two kinds of realization of advantage, take the client/server/database (c/s/d) three layer system structure, the system structure of the selection, which is based on the characteristics of the local area network and design school [1]. Simple said is made a system c/s mode expansion, users in the

Y. Li (✉) · C. Wen
Shandong Transport Vocational College, Weifang, 261206 Shandong, China
e-mail: Sdjtlyl@163.com

C. Wen
e-mail: jyjcbw@163.com

W. Du (ed.), *Informatics and Management Science V*, Lecture Notes in Electrical Engineering 208, DOI: 10.1007/978-1-4471-4796-1_13,

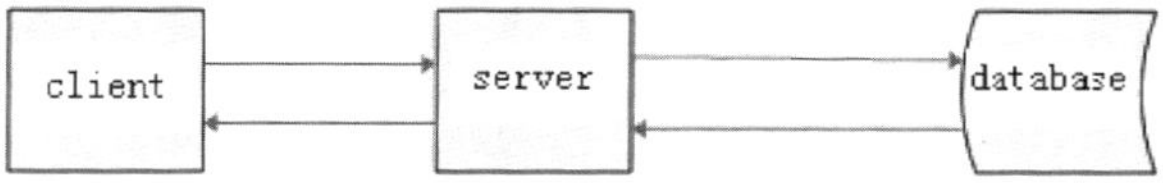


**Fig. 13.1** The user logs in data flow graph

machine the pre-installed system, can be connected directly to the server, the server of the receipt of the analysis of the data processing, the database of the corresponding operation, the results back to the client [2].

From security speak, using c/s mode has its own defects, will be set on the client database stores increase the security of data. And this structure is simple compared web model the configuration server and cumbersome web structure maintenance, also make use of the crowd is more targeted [3].

Second, system composition and functions (a) the composition of the system, the flow chart of the system is as follows:

System flow charts (Fig. 13.1):

The system is composed of four is module: test the client system module, the test server system module, the client inquires the system module, database information management and maintenance module. Their respective functions as described below [4].

## 13.2 The Exam Client System Module

The system is mainly student use. Examination system is the core part of the network exam system, is also the focus of the design. In order to unified management, students must be specified in the exam room system installed in the exam, eliminate the will in no man under the possibility of cheating, maintain the exam of justice [5, 6]. The examinee to land the server is set in advance user name and password to login the system, if the user name or password and the database of the stored data not, will not be the test. Success after landing, the examinee to answer questions from a selection, at the same time, and the thread of control by the timer starts to run in the required time to finish the test, because the exam time control by thread, and machine time is irrelevant, also maintained the test of fairness [7]. After the exam, users click the submitted, is the answer to the user submitted to the server is driven by the server so as to conduct the performance evaluation. And will result in a database. The user to finish the test, which in the next test before the start of not into the examination room, avoid the repeated login server raw scores to make changes.

## 13.3 The Client Scores Query System

In order to let the examinee timely and easy to understand their test scores, this system also provides the examinee inquires the function, the examinee in system interface input your information, and the server can keep the performance information transmitted to the examinee machine [8].

In addition, the system also provides the function of teacher inquires. The teacher according to the prompt input information may inquire any or all student performance information.

## 13.4 Server Terminal System

The server by the user graphic interface design, database administrator with the administrator password to login, set the test information, server, the administrator can use the system to provide the information query functions, real-time monitoring in the process of test server information, customer information network connection information, make the administrator can convenient understand the examination process the information, so as to deal with the problems. The server connection for clients and database, receiving user information and the database corresponding operation the server using multiple threads realization mechanism set up multiple threads, complete different operation. Has the following function module:

Students landing test module: receiving user login information and information in the database and students compare if the same, then landed successfully. In order to maintain the server performance, and maintain exam discipline this module joined the control function of people online, if the number of online at maximum amount will not be receiving system new users [9].

The selected topic module: receiving user selected topic request, and to select the corresponding title database returns to the client, at the same time, the number of questions to return to the test time as the client end the sign of the exam.

Scores were module: receive client from the examination results, and with the database of test answers compare the information, make corresponding scores judgment. Will result to return to the client and the fraction in a database of students' information table, the convenient user inquires query and administrators to score after statistics.

Students inquire the module: receiving query information will be stored in the database user information (class, achievement, etc) to return to the client.

Teacher login module: receiving teachers' login information, and teachers' information in the database is if the same success do login query, different cannot enter the system.

Teacher inquires the module: teacher's success after login, receiving teachers' information, and will the database of the students' information returns to the corresponding client.

## 13.5 Database Connection and Maintenance Module

This module to the information in the database (students' information, exam) for maintenance, the main use of object is the system administrator (only run on a server, not for remote maintenance. The aim is to increase the security of the system). This system database to be part of myself 1.4/Microsoft access 2000 to achieve.

Connect to database using java Bean tools and the configuration files of the system of combining the mode, with a single class for reading configuration files of database connection information, connection with database, convenient database maintenance, if database change simply changing the configuration file can be achieved new database connection. And not in every need to connect to the database function module change the database connection information.

Maintain the visualization of the database graphical interface and realize the query, update, increase, delete operation.

### *13.5.1 System Flow Chart*

Figures 13.2, 13.3 and 13.4.

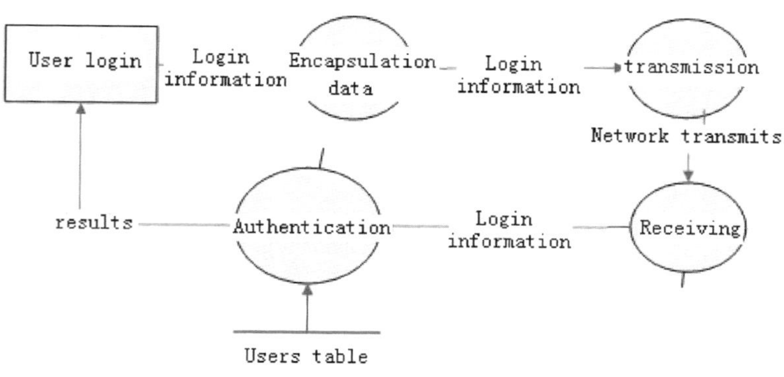


**Fig. 13.2** The user logs in data flow graph

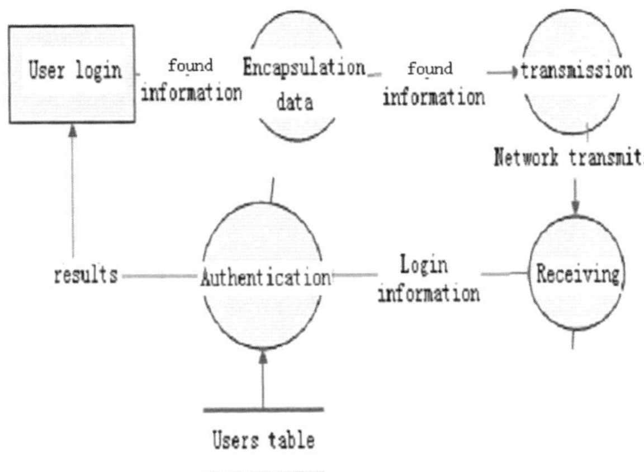


Fig. 13.3 The user found data flow graph

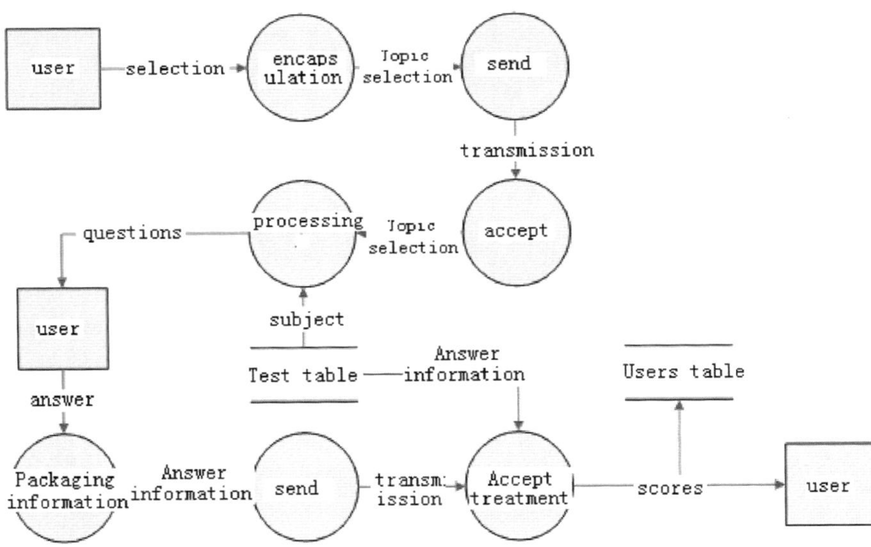


Fig. 13.4 Students' test data flow graph

## *13.5.2 The Realization of the Function of the System*

Interface aspects: the strong Java GUI design, the system has "students' test", "teacher inquires the", "the server management" three interface. Each corresponding to the operation of the system is different interface state (specific login system by the identity of the person and decide).

"Students' test interface", complete student login, student information query, students' test etc. Function

"The teacher inquires interface", complete the teacher login, student information query etc. Function

"The server management interface", complete the view, set the initial test information, server, monitor examination process, management and maintenance of the database etc. Function.

Content:

Try aspects: due to the characteristics of the online examination restriction, exam content in an objective topic primarily. By multiple choice and fills up the topic composition, are stored in the database of the examination.

Database aspects: initial system to use the database for Oracle and I, in view of the particularity of graduation design, to switch to Mascot access 2000.

Safety aspects: setting up the user name and password to verify way, at the same time, the user has finished the operation of the step to the operation to prevent illegal user login and excessive operation. At the same time will be important information to the maintenance of server, avoid the safety hidden danger of remote maintenance. The development environment, Windows server (sp2) + 1.41 + 3.0 + j2dk Eclipse. Oracle and I/Microsoft access 2000.

## References


1. Araon M (2005) From the JDKTM 5.0 documentation 83:23–26
2. John (2003) Murkowski the master java2 J2SE java 1.4, vol 71. Electronic Industry Press, China, pp 340–373
3. Scheldt H (2006) The java: the complete reference jews 5 edn. vol 84. Tsinghai University Press, Beijing, pp 250–298
4. Veneers B (1993) The inside the java virtual machine, 2nd edn., vol 23. Machinery Industry Press, Beijing, pp 54–130
5. Ekes J (2005) The datebase orally company with JDBC and java, 2nd edn. O 'Reilly Associates Inc 923:8132–8134
6. Deckle B (2005) The thinking in java, 3rd edn. In: Chen WP, Rao RN et al. (eds) Mechanical Industry Press, Beijing, pp 162–166
7. Flanagan D (2000) The java examples in a nutshell, 2nd edn. O 'Reilly and Associates Inc 91:128–131
8. Horstmann C, Cornell G (2004) Core java (TM) 2, volume I-fundamentals (7th edn), vol 39. Prentice Hall, PTR, pp 123–131
9. Slle K, Cadenhead R, Lemay L (2003) To teach yourself java2 in 21 days, 3rd edn., 23:112–115

# Chapter 14
# Logistics Enterprise Warehouse Management System Optimization

**Hengwei Wang**

**Abstract** The warehouse management system of the enterprise is analyzed, the loading and unloading warehouse work flow optimization, by building a learning enterprise, rational planning storage sites, promote warehousing information management, establishing effective inventory control model and so on a number of measures, in the management system of the improvement to the warehouse management system is more and more perfect.

**Keywords** Storage · Management system · Optimization design

## 14.1 Introduction

Storage in logistics system plays an important role for the enterprise that, the discretion of the logistic cost on enterprise performance also play a crucial role. Warehousing logistics activity is the important pillar. Storage is through the warehouse storage and preservation of materials [1]. In the logistics system, reasonable to the warehouse management, can reduce supplies in the process of material loss in storage and labor consumption, speed up the circulation and capital goods flow, which can save cost and improve the economic efficiency of enterprises. In the warehouse of the warehouse management system in the management of a subsidiary role, perfect management system to improve storage management level [2].

H. Wang (✉)
Shandong Transport Vocational College, Weifang, 261206 Shandong, China
e-mail: whw737@yahoo.com.cn

W. Du (ed.), *Informatics and Management Science V*, Lecture Notes in Electrical Engineering 208, DOI: 10.1007/978-1-4471-4796-1_14,

## 14.2 The Loading and Unloading Warehouse Operation Process Optimization

The stand or fall of loading and unloading process warehouse often determines the entire warehouse management level have reached a certain level, the company warehouse in the process for the many shortcomings, too simple operation, the process is not clear, detailed the content is not enough [3].

Figure 14.1 is I according to your idea of improved flow chart, mainly is the increased a lot of details of the content. So after the design improvement, the whole process is much more perfect outbound warehousing, such as inventory custody, the original process just chine to the preservation of the storage products, improve the optional custody after not, but according to the characteristics of different products to keep, thus not only separate in order to access, also avoided for products such as bad affect the other reason product etc.

## 14.3 Establish a Learning Enterprise, and Improve the Quality of Employees

Man is the biggest power enterprise progress, personnel quality promotion, enterprise some place also is corresponding improvement. Create a learning enterprise is more widely in the sense of the training work, he then we previously engaged in training more profound more comprehensive is more targeted, pay more attention to individual subjective people move the play of sex [4]. From the start, to adhere to the people-oriented, promote the all-round improvement of the enterprise and the comprehensive development of people, deepen the learning enterprise concept of understanding and the understanding, make enterprise gradually improved [5].

## 14.4 Reasonable Planning Storage Sites, Perfect Warehouse Facilities and Equipment

Reasonable planning storage site is vital, and a reasonable plan for the surplus space arrangement can be other things, achieve maximize the use of space. Storage facilities and equipment not only save the perfect cost and time, also greatly improve the quality of the products, so as to ensure that the products of name recognition.

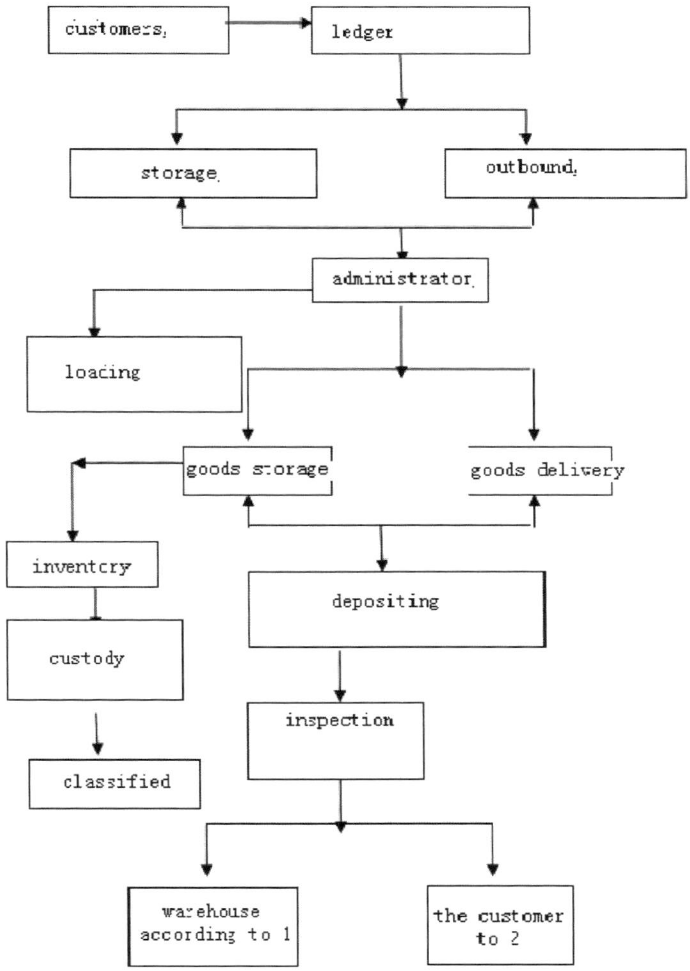


**Fig. 14.1** The improved flow chart

## 14.5 Promote Warehousing Information Management

Do the unity of the logistics information of modern logistics industry planning construction is very important. I think that provide the information planning to pay attention to the following problems: (1) It is to seriously study and establish industry in 3–5 years of informatization overall development planning, in order to adapt to the industry on the basic management level requirement [6]. (2) it is to continue to improve the industry cooperative marketing share information platform of industry and commerce of the construction, enhance the digital storage management system level. (3) it is to want to in all the industry promotion product logistics in road information system, strengthen the logistics information control in transit. 4 is to actively to department as a unit, establish unified GPS vehicle tracking and warehouse monitoring system, a province of within the scope of the vehicles, warehouse on the safety operation of the unified monitoring, centralized management, and unified evaluation [7].

## 14.6 To Establish an Effective Inventory Control Model

In guarantee enterprise production, management requirements, under the premise of make inventories often keep in reasonable level, Master inventory dynamic, timely, and put forward the order right amount, avoid Chao Chu or out of stock, Reduce inventory space occupied, reduce inventory total cost, Control inventory capital takes up, speed up the capital turnover [8]. The purpose of the inventory management according to the classification and can be divided into two types of economy and safety. Economic model's main purpose is to save money, improve economic benefit, the main purpose of safety model is the guarantee of the normal supply, regardless of the increasing safety stock and security reserve period, make the possibility of stock to a minimum. Although many of the inventory management model, but comprehensive considering all conflicting factors for good economic result is a common principles of inventory management [9].

## 14.7 Designated Unified Code

Material once coding, material correctly and record quickly, material storage sleek, can reduce the occurrence of events fraud. Inventory materials are well the correct uniform name and specification to code [10]. The recipient of materials department of materials development and material warehouse is very convenient. The supplies coding, can for some performance similar or the same as the materials are unified, merger and simplified, and compression of the material varieties and specifications.

## 14.8 Plan Implementation Security Measures

No plan is one hundred percent sure of success, we must not only ensure the perfection of the scheme will also make sure that scheme failed after the damage, so security measures is necessary, he can maximize safeguard the interests of the company, even if the failure would also damage to a minimum, this is essential for every successful enterprise link.

## 14.9 Emergency Management Measures Such as Warehouse

Fire emergency plan: when there is a fire, according to the first place and the surrounding terrain size to judge, if the fire small the surrounding terrain and convenient for fire fighting, can call and report on level 1 department leaders, and according to the cause of the fire extinguishing method to determine the characteristics of the material.

High altitude falling objects emergency plan: the accident report on the first level leadership, according to the damage extent determine treatment methods, such as minor injury to the clinic treatment, such as seriously, and immediately call 120 first aid, and at the same time the clinic for corresponding processing the arrival of waiting for an ambulance.

Get an electric shock the emergency response plan: if there are people in electric shock accidents, the first to cut off the power supply, then take emergency measures. It is strictly prohibited to not cut off power supply in rescue them down, in case the serial electric shock. Find out reason to get an electric shock, report on level 1 leadership.

The dangerous product leakage emergency plan: the dangerous goods such as a small leak happened, first report security officer and the related leadership, because each of the dangerous product processing method is different, should be based on the corresponding method for processing. The dangerous goods such as the outbreak of leakage, should be based on the chemical properties of leakage product and MSDS have dug a pit to collect, diluted the or burning and other methods treatment, the researchers must according to leak the characteristics of our product for the corresponding preventive measures (such as a gas mask, the acid and alkali clothing gloves, etc.).

Mechanical damage emergency measures: high altitude falling objects with the same processing.

## 14.10 Conclusion

Taking a company warehouse management system as the object, the system of this formula make the analysis and research, and then according to these system optimization and improve the, warehouse management system with more perfect and reasonable. Establish a learning enterprise, and improve the quality of the personnel. Reasonable planning storage sites, perfect the strengths of facilities and equipment. Promote warehousing information management. Establish an effective inventory control model. Designated unified code plan implementation security measures.

## 14.11 The Company Warehouse Management System Present Situation and Problem Analysis

Enterprise by formulating warehouse operation regulation and rewards and punishment system, guide and regulate the store personnel daily homework behavior, through the rewards and punishments measures have incentives and examination personnel role. But in actual implementation process management system also has a lot of drawbacks. This is my work in found in the warehouse management system of the existing problems.

### *14.11.1 Personnel Allocation is Not Reasonable*

Storage management personnel allocation is very important, personnel arrangement inappropriate can lead to some areas in the warehouse management, to do the right people in the right post, to maximum warehouse management system to the more perfect reasonable.

### *14.11.2 Portage Infrastructure Backwardness*

Storage infrastructure is the foundation of the logistics activity, infrastructure is critical to the future of the storage. Increase the storage infrastructure investment, warehousing and infrastructure to improve conditions, only infrastructure to improve conditions can maximum security enterprise warehouse the improvement of management.

### *14.11.3 Warehousing Information Management Behind*

The development of the logistics industry is an important feature is to pursue "zero inventories" for the supreme goal. "Zero inventory" meaning is stored in the form of a warehouse or some item storage quantity is low a concept, can even for "zero", i.e. not keep inventory. Not to inventory the form can free warehouse inventory of a series of problems, such as warehouse construction, management fees, inventory maintenance, storage, loading and unloading, handling costs and inventory takes up liquidity and inventory of content, losses, such as aging metamorphism. Although at present the enterprise logistics from the current situation, in the whole in the whole process of the social reproduction, "zero inventories" is a kind of ideal model, we cannot become reality, but through the lower inventory, the enterprise can save more cost. In the process, a storage aspect of information is very important.

### 14.11.4 Inventory Control is not Reasonable

At present, most of the problems with the enterprise are high inventory. In many enterprise boss eyes, inventory is a kind of assets, is "this", have stock have sales. But, from economics point, not a reasonable inventory itself is a huge waste, it'll tie up enterprise of liquidity, reduce the capital turnover, and eats of the profits of the business. At the same time inventory will also cover the problems of enterprise management, cause enterprise "false prosperity". So only through the establishment of effective warehouse management system, do the tedious work and to further strengthen the inventory to the application of information technology and inventory management measures to perfect and help enterprises to achieve the purpose of inventory control.

## References

1. Zhi-Tai W (1995) Modern logistics learn. vol 93, China Supplies Press, Beijing, pp 23–25
2. You S (2005) Storage practice. vol 912, Foreign Economic and Trade University Press, Beijing, pp 83–89
3. Sith J (2004) The latest XuJie purchasing and warehouse management. vol 62, Tsinghai University Press of Beijing Jiaotong University Press, Beijing, pp 128–131
4. Yongsheng L, Wenling Z (2004) Warehousing and distribution management. vol 87, Mechanical Industry Press, Beijing, pp 38–41
5. Beibin W (2003) Modern warehousing management. vol 28, People's Traffic Press, Beijing, pp 81–87
6. Beams army (2003) Storage management practice. vol 39, Higher Education Press, Beijing, pp 138–142
7. Kejun Z, Zhongwen Y, Dai-fen C (2005) Warehousing and distribution management. vol 91, Science Press, Beijing, pp 47–52
8. Wan-sen Z (2005) Storage and distribution management. vol 42, Peking University Press, Beijing, pp 18–23
9. Zhong-wen Y, Dai-fen C (2003) Logistics technology and practice. vol 39, People's Traffic Press, Beijing, pp 13–15
10. Xiong W, Jia H (2006) Zhen purchasing and warehouse management. vol 02, Higher Education Press, Beijing, pp 48–52

# Chapter 15
# Activity-Based Costing Calculation Model of Milk-Run Inbound Logistics for Auto Parts

**Bengang Gong, Wei Huang, Yunmiao Gui and Tinglong Zhang**

**Abstract** In order to achieve precise cost accounting and inbound logistics management of auto parts, the paper presents Activity-Based Costing (ABC) calculation models for Milk-Run (MR) inbound logistics. Based on the analysis of its operation processes, MR logistics operation processes of auto parts are classified into four sections, which are largely the logistics activity of the supplier, the third party logistics (3PL), a hub of auto parts, and one auto manufacturing enterprise. Along with the process of MR inbound logistics and the characters of cost composition, the paper proposes a modified ABC accounting model of MR inbound logistics, which is finally illustrated.

**Keywords** Auto parts · Milk-run · Logistics costing accounting · Activity-based costing · Inbound logistics

B. Gong (✉) · W. Huang · Y. Gui · T. Zhang
Anhui Polytechnic University, Wuhu 241000, China
e-mail: bggong@ahpu.edu.cn

W. Huang
e-mail: hwycyt@sohu.com

Y. Gui
e-mail: ymgui789@163.com

T. Zhang
e-mail: saztl@mail.ustc.edu.cn

W. Du (ed.), *Informatics and Management Science V*, Lecture Notes in Electrical Engineering 208, DOI: 10.1007/978-1-4471-4796-1_15,

## 15.1 Introduction

How to seek an effective way to decrease logistics operation cost based on the analysis of the manufacturing enterprise logistics system operation mode becomes the focus of the experts and scholars at home and abroad. The traditional volume-based cost accounting method results in cost distortion and lack of relevant cost information, which makes cost accounting and management subject to certain constraints. Therefore, work-based cost accounting method, namely Activity-based Costing (ABC) is introduced into the manufacturing enterprise logistics cost accounting [1–3]. ABC has some research and application in the logistics field, but up to now the research and application for ABC problem in inbound logistics of auto parts is poor. As a new auto parts logistics operation mode, MR has been spread and applied in the Toyota, General and other famous auto manufacturing company [4]. But at present the traditional cost accounting methods based on volume/product can not properly reflect MR inbound logistics process of the key activities, activity motivations, as well as resources consumption quantity in implementation of activities so as to make the cost accounting and management subject to certain constraints.

ABC method can eliminate the activities of no value activities, increasing the accuracy of cost accounting and decision scientific data [5]. Combing with the characteristics of MR inbound logistics operation, the paper puts forward an improved ABC accounting model.

## 15.2 Auto Parts MR Operation Cost Analysis

According to MR operating process, MR logistics operation processes largely consists of logistics activity of suppliers, of 3PL, of auto parts hub, of automobile manufacturing enterprise [6]. The four operation processes are as four sub-operation centers of logistics activities, which are consisting of a complete MR logistics center. In addition, this assumes that the system operating in the implementation process is not related to the procurement, quality inspection etc., and according to the order information to each supplier/auto parts hub, all operations are to pick up service. The resources and operation motive for MR operation process is shown in Table 15.1 [1, 6].

**Table 15.1** Motive analysis of auto parts MR operation cost

| Activity centre | Resource | Resource cost motive | Operation | Operation cost motive |
|---|---|---|---|---|
| Supplier | Wage | Employee number | Sorting | Working time |
| | Equipment depreciation | Depreciation rate | | Service time |
| | Wage | Employee number | Order processing | Handle number |
| | Equipment depreciation | Depreciation rate | | Service time |
| | Wage | Employee number | Handling | Working time |
| | Equipment depreciation | Depreciation rate | | Service time |
| | Fuel cost | Fuel consumption | | Service time |
| Manufacturing enterprises | Wage | Employee number | Checking | Inspection time |
| | Wage | Employee number | Order processing | Handle number |
| | Equipment depreciation | Depreciation rate | | Service time |
| | Wage | Employee number | Handling | Working time |
| | Equipment depreciation | Depreciation rate | | Service time |
| | Fuel cost | Fuel consumption | | Service time |
| Hub | Wage | Employee number | Sorting | Working time |
| | Equipment depreciation | Depreciation rate | | Service time |
| | Wage | Employee number | Order processing | Handle number |
| | Equipment depreciation | Depreciation rate | | Service time |
| | Wage | Employee number | Handling | Working time |
| | Equipment depreciation | Depreciation rate | | Service time |
| | Fuel cost | Fuel consumption | | Service time |
| 3PL | Wage | Employee number | Order processing | Handle number |
| | Equipment depreciation | Depreciation rate | | Service time |
| | Wage | Employee number | Transport operation | Working time |
| | Vehicle depreciation | Depreciation rate | | Service time |
| | Fuel cost | Fuel consumption | | Distance |

## 15.3 Auto Parts MR Model of ABC

### 15.3.1 Collect Resources to Operation, Establish Operation Cost Pool

Count and summarize various types of indirect resources $i$ that the workers involved in participate in auto parts MR operations. The resources $i$ being consumed values attribute to resource pool of the total activity center, and then according to the known conditions, the different motivational rate of resources $i$ is calculated, $r_i = \frac{c_i}{a_i} = \frac{c_i}{\sum_{j=1}^{m} q_{ijx}} \quad (i = 1, 2, \ldots, n;\ j = 1, 2, \ldots, m; x = 1, 2, \ldots, s)$. The distribution of total activity center resource cost is distributed to each specific activity in each operations center, and total cost for operations is calculated, $c_{jx} = \sum_{i=1}^{n} r_i \times q_{ijx} \quad (i = 1, 2, \ldots, n;\ j = 1, 2, \ldots, m; x = 1, 2, \ldots s)$. According to $C_k = \sum_{j=1}^{m} c_{jx}(k = 1, 2, \ldots, t)$ the consumption cost of different operation center with same operation is distributed to the total operation cost pool in MR process. Where $i$ is the different resource type, $j$ is the different operating activities, $r_i$ is the resources reason rate, $c_i$ is the resource cost for $i$, $a_i$, is the resource driver quantity of $i$, $c_{jx}$, is the operating cost of operation $j$ in operating centre $x$, $q_{ijx}$ is the resource $c_{jx}$ driver quantity of consumption of $i$ for operation $j$, $C_k$ is the total operating costs of all operation, and $x$ is the every operating centre.

### 15.3.2 Determining Operation Motivation and Allocating Operation Cost

Based on the previous analysis, the activity motivation rate of operation $j$ is calculated, i.e., $R_k = \frac{C_k}{A_k} \quad (k = 1, 2, \ldots, t)$, then the actual cost occurred during the operating process of cost objects $p$ is separated by the formulation $C_p = \sum_{k=1}^{t} R_k \times Q_{kp}$. The summation of operating cost is got in MR operating process [6]. Where $A_k$ is the operating driver quantity of operation $j$, $R_k$ is the operating driver rate of operation $j$, $Q_{kp}$ is the operating driver quantity consumed by operation $j$ of operating centre cost pool of cost object $p$, and $C_p$ is the total cost assigned for the object $p$.

### 15.3.3 Calculating the Total Cost and Unit Cost of the Cost Objects

Because of no involving the cost of auto parts themselves, the cost object is just the summation of all cost consumed in the MR operating process, i.e., the total cost $T_p$ and unit cost $T_p'$ of MR $p$ is as follows:

$$T_p = Z_p + C_p = Z_p + \sum_{k=1}^{t}\sum_{j=1}^{m}\sum_{i=1}^{n} \frac{c_i \times q_{ijx}}{A_k a_i} \times Q_{kp}, T_p' = \frac{T_p}{Q_p} = \frac{(Z_p + C_p)}{Q_p} \quad (15.1)$$

where $Z_p$ is the MR special cost which is directly charged to cost accounting, and $Q_p$ is the freight transport total volume of MR.

## 15.4 Application Example

An auto manufacturing enterprise takes MR to the local auto parts logistics. The related information of auto manufacturing enterprise is as follows: the enterprise owns three auto parts suppliers $S_1, S_2, S_3$, one auto parts hub in this locality. Taking-stock-vehicles sent by the 3PLs among the supplier, hub and manufacturing enterprises run a loop in a total distance of 20 km, every 2 h a loop, the 2 vehicles alternately picking up, a cycle of 12 times a day by the specified time. In addition, employee's working time is 8 h in three-shift. According to the accounting, in the process of MR operations the cost for a single month (1 month = 30 days calculating) includes in labor costs 1,15,200 Yuan, depreciation of equipment 3726.64 Yuan (equipment and vehicle depreciation calculated according to 10 years), and fuel cost 45,828 Yuan (just indirect cost of MR operating process).

The operating workers are about 81 people in the process of MR operating process (specific allocation is shown in Table 15.2). The order processing employees are 18 people, monthly wages for a total 36,000 Yuan. The employees for sorting cargoes are 30 employees, monthly wages a total of 60,000 Yuan. The handling employees (i.e., drivers) are 24 people, monthly wages for a total of 60,000 Yuan. The transport operators (i.e., drivers) are 6 people, monthly wages for a total of 16,800 Yuan. The inspection personals are 3 people, monthly wage for a total of 5,400 Yuan. In MR operating process, the order processing equipment (computer) has 6 units and a total of 25,200 Yuan. The sorting cargo equipment (trolley) has 7 units and a total of 14,000 Yuan, and the handling equipment (forklift) has 8 units. A total of 400,000 Yuan is amount to consuming 6 L oil/h; there are 2 transporting vehicles; a total of 112.000 Yuan is amount to consuming 6 L oil /100 km, there is oil for the vehicle price 6.7 Yuan/L. It is assumed that the service life of the equipment shall last 10 years, and depreciation charges according to the purchase value by annual average apportion.

**Table 15.2** Employee and equipment allocation table

| | | $S_1$ | $S_2$ | $S_3$ | Manufacturing enterprises | Hubs | 3PLs |
|---|---|---|---|---|---|---|---|
| Order processing | Employee number | 3 | 3 | 3 | 3 | 3 | 3 |
| | Equipment number | 1 | 1 | 1 | 1 | 1 | 1 |
| Sorting cargo | Employee number | 9 | 3 | 6 | | 2 | |
| | Equipment number | 2 | 1 | 1 | | 3 | |
| Loading and unloading cargo | Employee number | 3 | 3 | 3 | 9 | 6 | |
| | Equipment number | 1 | 1 | 1 | 3 | 2 | |
| Checking cargo | Employee number | | | 3 | | | |
| Transporting cargo | Employee number | | | | | | 6 |
| | Vehicle number | | | | | | 2 |

In this month, the number of order processing is 2,160 copies during MR operating process, and a total of 1,800 h on sorting cargoes consumed, 288 h shared on loading and unloading, a total consumption of 720 h on transportation operating, 3,240 cubic meters of the actual total packing volume for every month MR operation, where during MR operating process, the same operation that suppliers, manufacturers, hubs and 3PL have the same resource cost motive rate and operation motive quantity.

## 15.4.1 Calculating the Total Cost and Unit Cost of the Cost Objects

### 15.4.1.1 Determining the Allocation of Auto Parts MR Operating Resources

According to MR operating activities, allocating the operating resources is shown in Table 15.3.

### 15.4.1.2 Collecting Resource Cost to the Operating Center Cost Pool

According to the above example information and in Table 15.1, we obtain collecting resource cost to the operating center cost pool shown in Table 15.4, where fuel costs are not as operation cost accounting, included in the accounting cost during the MR operating process.

**Table 15.3** Resource allocation table unit: Yuan

| Resource project | Wage | | | | Depreciation | | | Fuel costs |
|---|---|---|---|---|---|---|---|---|
| Operation | Order processing | Sorting | Handing | Checking | Order processing | Sorting | Handing | |
| value/yuan | 36,000 | 60,000 | 60,000 | 5,400 | 210 | 116.67 | 3333.33 | 23,472 |

**Table 15.4** Resource consumption of operating centre

| Operating centre | Operation | Resource | $r_i = {}^{c_i}/_{a_i}$ | $q_{ijx}$ | $c_{jx}$ |
|---|---|---|---|---|---|
| $S_1$ | Order processing | Wage | 2,000 | 3 | 6,000 |
| | | Depreciation | 35 | 1 | 35 |
| | Sorting | Wage | 2,000 | 9 | 18,000 |
| | | Depreciation | 16.67 | 2 | 33.34 |
| | Handing | Wage | 2,500 | 3 | 7,500 |
| | | Depreciation | 416.67 | 1 | 416.67 |
| $S_2$ | Order processing | Wage | 2,000 | 3 | 6,000 |
| | | Depreciation | 35 | 1 | 35 |
| | Sorting | Wage | 2,000 | 3 | 6,000 |
| | | Depreciation | 16.67 | 1 | 16.67 |
| | Handing | Wage | 2,500 | 3 | 7,500 |
| | | Depreciation | 416.67 | 1 | 416.67 |
| $S_3$ | Order processing | Wage | 2,000 | 3 | 6,000 |
| | | Depreciation | 35 | 1 | 35 |
| | Sorting | Wage | 2,000 | 6 | 12,000 |
| | | Depreciation | 16.67 | 1 | 16.67 |
| | Handing | Wage | 2,500 | 3 | 7,500 |
| | | Depreciation | 416.67 | 1 | 416.67 |
| Manufacturing enterprise | Order processing | Wage | 2,000 | 3 | 6,000 |
| | | Depreciation | 35 | 1 | 35 |
| | Handing | Wage | 2,500 | 9 | 22,500 |
| | | Depreciation | 416.67 | 3 | 1250.01 |
| | Checking | Wage | 1,800 | 3 | 5,400 |
| Auto parts hubs | Order processing | Wage | 2,000 | 3 | 6,000 |
| | | Depreciation | 35 | 1 | 35 |
| | Sorting | Wage | 2,000 | 12 | 24,000 |
| | | Depreciation | 16.66 | 3 | 49.98 |
| | Handing | Wage | 2,500 | 6 | 15,000 |
| | | Depreciation | 416.67 | 2 | 833.34 |
| 3PL | Order processing | Wage | 2,000 | 3 | 6,000 |
| | | Depreciation | 35 | 1 | 35 |
| | Transporting | Wage | 2,800 | 6 | 16,800 |
| | | Depreciation | 466.67 | 2 | 933.34 |

### *15.4.2 Allocating the Cost Pool of Operating Center to Cost Accounting Object*

As the operating mode of MR is a strict closed-loop logistics system, the request of logistics operations for the operations center is standardized, and the processing power for operation employees also has a uniform standard. So the same operation of the logistics operations center has the same motive quantity. According to the established model, the total operating resources value consumed for different operation can be accounted in Table 15.5.

**Table 15.5** The actual resources cost of different operations during the MR process

| Operation | | Motive | $C_k$ | Permonth ($A_k$) | $R_k = \frac{C_k}{A_k}$ | $Q_{kp}$ | $C_p$ |
|---|---|---|---|---|---|---|---|
| Order processing | Wage | Handing copies/hour | 36,000 | 4,600 | 7.830 | 2,160 | 1,6912.8 |
| | Depreciating cost | Using time/hour | 210 | 4,320 | 0.049 | 2,160 | 105.84 |
| Sorting | Wage | Working time/hour | 60,000 | 5,400 | 11.110 | 1,800 | 19,998 |
| | Depreciating cost | Using time/hour | 116.67 | 5,400 | 0.022 | 1,800 | 39.60 |
| Handing | Wage | Working time/hour | 60,000 | 1,728 | 34.722 | 288 | 9,999.94 |
| | Depreciating cost | Using time/hour | 3,333.33 | 1,728 | 1.929 | 288 | 555.55 |
| Checking | Wage | Checking time/hour | 5,400 | 150 | 36 | 60 | 2,160 |
| Transporting | Wage | Working time/hour | 16,800 | 1,440 | 11.667 | 720 | 8,400.24 |
| | Depreciating cost | Using time/hour | 933.34 | 1,440 | 0.648 | 720 | 466.56 |
| Total | | | | | | 58638.53 | |

### *15.4.3 Accounting Total Operating Costs and Unit Costs of the MR Operation*

According to total cost accounting formulation $T_p = Z_p + C_p$, we calculate the total operation cost of auto Parts MR operation $T_p$, as follows:

$$T_p = Z_p + C_p = 0 + (58638.53 + 23472) = 82110.53\ (yuan) \tag{15.2}$$

Similarly, the unit cost of auto Parts MR operation $T_p'$, as follows:

$$T_p' = \frac{T_p}{Q_p} = \frac{(Z_p + C_p)}{Q_p} = \frac{82110.53}{3240} = 25.34\ (yuan) \tag{15.3}$$

The calculating results show that we can not only enable enterprises to truly understand how to happen and transfer the MR cost of operations, but also make accounting cost data more scientific and reasonable via applying the idea of ABC to analyze the cost motive in the process of MR, and to collect the cost.

## 15.5 Conclusions

The MR inbound logistics with some characters like low batch, high frequency, timing, which are very suitable to auto parts supply, has many advantages including a high transport efficiency, low inventory levels, quick response and good service level. But the traditional cost accounting methods are hard to meet the needs of the MR logistics cost accounting. Therefore, this paper puts forward the ABC of MR logistics based on the analysis of the ABC principle and the typical model in combination with the characters in the process and cost composition of auto parts MR logistics. This model cannot only provide a relatively accurate cost accounting for auto parts MR logistics, but also provide precise and timely cost information to achieve precise management of operating cost.

**Acknowledgments** This work was financially supported by the NSFC (71171002, 70901001), Natural Science Foundation of Anhui Province (11040606M24) and Natural Science Foundation of Anhui Provincial Higher Education (KJ2010A039).

## References

1. Pohlen TL, Lalonde BJ (1994) Implementing activity-based costing in logistics. J Bus Logistics 15(2):1–23
2. Baykasog A, Kaplanog V (2008) Application of activity-based costing to a land transportation company: a case study. Int J Prod Econ 116(2):308–324
3. Askarany D, Yazdifar H, Askary S (2010) Supply chain management, activity-based costing and organizational factors. Int J Prod Econ 127(2):238–248

4. Nemotoa T, Hayashib K, Hashimotoc M (2010) Milk-Run logistics by Japanese automobile manufacturers in Thailand. Procedia Soc Behav Sci 2(3):5980–5989
5. Tian J, Liu ZC, Zhao ZP, Yang YF (2009) ABC-based logistic cost calculation model and its application. Chin J Manag 6(4):447–452
6. Huang W, Gong BG, Wang AJ, Gui YM (2010) Constitution and evaluation on milk run logistics capability of manufacturing enterprises. Comput Integr Manuf Syst 16(11):2454–2459

# Chapter 16
# Economic-Oriented Efficient Practical Ability Training Scheme

**Wei Cao**

**Abstract** As a branch of the social charitable and cultural career, college students' art team shoulders the mission of transmitting the music culture. The author investigates and analyzes the present situation and influencing factors of the art team; meanwhile ponders about how to serve the local economy, so as to better serve the social economic and cultural development of Qinhuangdao.

**Keywords** University · College students' art team · Practical ability · Local economy

## 16.1 Introduction

Practical teaching is an important part of college teaching system, and closely associates with the service of the local economy, which is on the basis of social needs and the development of local economy and culture [1]. Based on this, the author did a survey on the present situation of college students' art team of Qinhuangdao. The objects of this questionnaire survey are college students in Qinhuangdao city. The process includes designing, issuing, recovering, counting and analyzing [2]. The questionnaire is delivered 1,200 and took back 1,134 of which 1,061 is effective.

W. Cao (✉)
Ocean College of Hebei Agricultural University, Qinhuangdao, 066003 Hebei, China
e-mail: caowei8603383@126.com

W. Du (ed.), *Informatics and Management Science V*, Lecture Notes in Electrical Engineering 208, DOI: 10.1007/978-1-4471-4796-1_16,

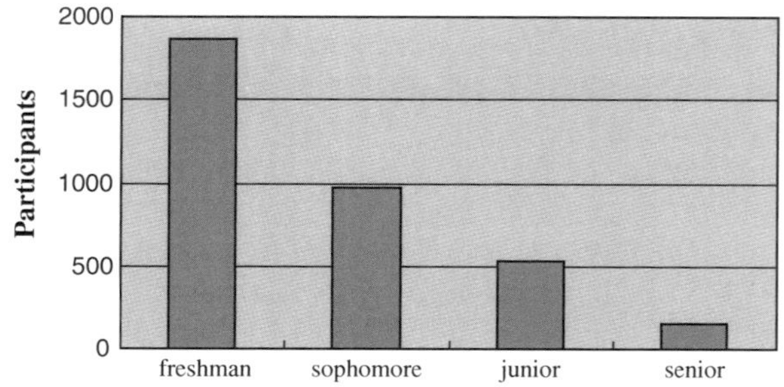


**Fig. 16.1** Participate in the number of art changes

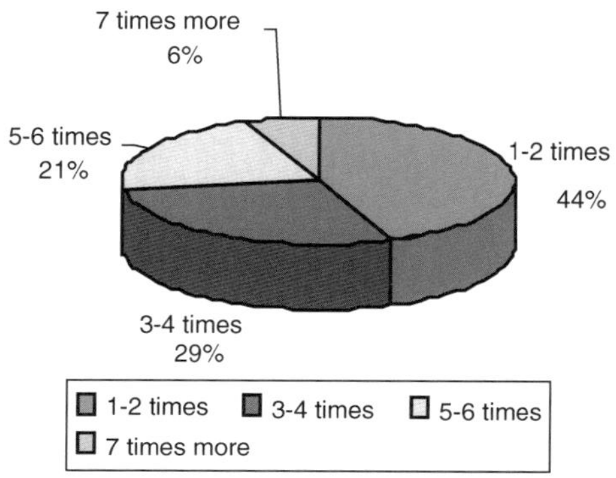


**Fig. 16.2** Art activities in an average number

## 16.2 The Status Quo of the College Students' Art Team

### *16.2.1 Participate in the Number of Art Change*

From the distribution in different grade, the number decreases from grade on to grade four [3]. The time that college students join in the art team generally is the first semester. Freshmen are the main objects of each college social team [4]. One year later, the number that joins in each team is decreases sharply, and the junior and senior seldom take part in such activities (Fig. 16.1).

### *16.2.2 The Monthly Activity Number of the College Students' Art Team*

Investigation shows, a proportion of the total is 73 %, whose weekly routine training activities is once or less than once. Of which over 44 % has a routine practice every two week. From the results of the survey came to see, group activities and members' demand is still have a large gap. Low frequency and low level is the main cause that leads to the withdrawal of the team members (Fig. 16.2).

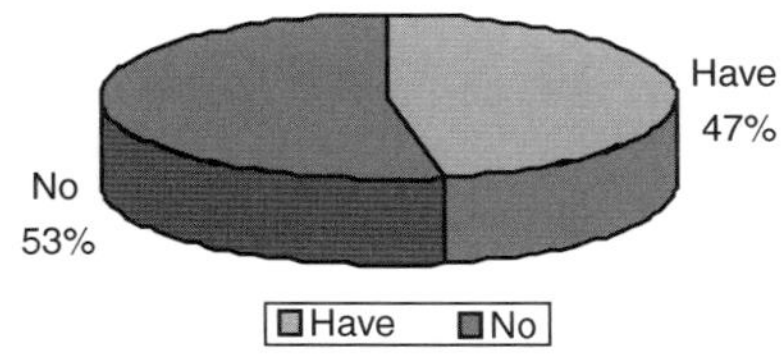


**Fig. 16.3** The cooperation with other colleges and the society

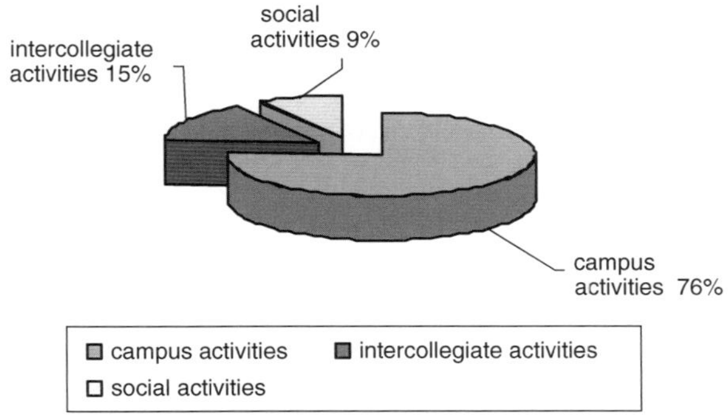


**Fig. 16.4** Range of the activities

### *16.2.3 The Cooperation with Other Colleges and the Society*

The survey shows that 47 % of the group has established cooperation with other colleges or other organizations, 53 % has not such cooperation (Fig. 16.3).

### *16.2.4 Range of the Activities*

Most activities of the college students' art team is inside the college. Only a small part is between the colleges and in the society. The results show that the campus activities take up 76 %, intercollegiate activities 15 %, social activities 9 % (Fig. 16.4).

## 16.3 Influencing Factors of the Practical Effectiveness

### *16.3.1 College and Students Diverge Greatly About the Course Education and Practical Education*

The data show that college students believe, among the factors individual influence the practice mostly. But college practice should have shown more influence, its mean is more than 2. That is, in reality it does not reach the expected standards.

**Table 16.1** Practice and factors

| Practice and factors | N | Minimum | Maximum | Mean | Standard deviation |
|---|---|---|---|---|---|
| Own ability and quality | 601 | 1 | 5 | 1.91 | 0.890 |
| Own character | 603 | 1 | 5 | 1.94 | 0.903 |
| Necessary funds for practice | 603 | 1 | 5 | 1.96 | 0.966 |
| Economic reasons | 605 | 1 | 5 | 2.00 | 0.959 |
| Related information | 605 | 1 | 5 | 2.14 | 0.976 |
| Social relations | 604 | 1 | 5 | 2.16 | 1.005 |
| The current employment situation | 605 | 1 | 5 | 2.23 | 1.018 |
| Family factors | 603 | 1 | 5 | 2.31 | 0.907 |
| College practical education | 597 | 1 | 5 | 2.53 | 1.009 |
| The opportunity of course selection | 601 | 1 | 5 | 2.73 | 0.997 |
| Practice in college | 535 | 1 | 5 | 2.79 | 1.119 |
| Effective N (list state) | 522 | | | | |

*Note* 1 very big, 5 very small, the more the mean tends to be 1, the bigger the factor is

This shows that the current students practice depends more on its capacity and resources, rather than the school support and help. This is not only against the original intention of practice education, also greatly limits students' practical activities. When supply can't keep up with demand, the students and college show great differences or conflict. The great disparity between reality and the theory must cause the attention of education workers (Table 16.1).

## *16.3.2 Low Practical Effectiveness of the College Entrepreneurship Education*

Practice education of the effectiveness of the influence factors depends mainly on two aspects: one is whether students with; Second is the school's efforts are put in position. Can be seen from the table, the students' own factors on the practice of the larger role, students for practice high enthusiasm, and also has the ability to relate to and conditions. And the problem is mainly come from school, students think that school education practice or related practice theory education, also or school the practice of internal and external environment, are not realize the proper role (Table 16.2).

This group of data also on another level explained above doubts, college students thought that practice of college can improve their abilities but not willing to take part in it. The only reasonable explanation is that, although some colleges and universities in certain scope are developing practice activities, but it quite ineffective, and cannot help the students and mobilize their enthusiasm.

**Table 16.2** Factors influencing college students' practical training

| Factors influencing college students' practical training | N | Minimum | Maximum | Mean | Standard deviation |
|---|---|---|---|---|---|
| Enthusiasm | 599 | 1 | 5 | 1.68 | 0.765 |
| Ability | 568 | 1 | 5 | 1.71 | 0.763 |
| Character and personality | 593 | 1 | 5 | 1.76 | 0.777 |
| Outside college environment | 591 | 1 | 5 | 2.10 | 0.920 |
| College's attitude and regulation | 599 | 1 | 5 | 2.19 | 0.842 |
| Classmates' and friends' practice | 597 | 1 | 5 | 2.27 | 0.891 |
| College's education and guidance | 597 | 1 | 5 | 2.32 | 0.813 |
| Others | 47 | 1 | 3 | 2.06 | 0.734 |
| Effective N (list state) | 36 | | | | |

*Note* 1 very big, 5 very small, the more the mean tends to be 1, the bigger the factor is

## 16.4 The Main Form that College Students' Art Team Serve the Local Economy

Combine the college students' practice with their service, meet the need of society, and cultivate music talent for the urban and region. Colleges play an important role in this way. The forms are as follow:

### *16.4.1 Scientific Researches with the Local Cultural Department*

A main form is to cooperate with the local cultural department, and establish music culture research center or institute. The college students' practice should develop classical music culture and combine national culture, which is quite important to inherit and develop the local culture.

### *16.4.2 Service Performance Under the Cooperation of Local Enterprises*

Foreign performance is one of the most important ways to exercise the students' ability of, and it is an effective way for the students' to practice and to serve the society. Such performance must be managed and organized. Such cooperation performance, can not only enrich the local people's cultural life, improve their cultural quality, promote local spiritual civilization, but serve the development of local economy, achieve high social reputation.

### 16.4.3 Social Education Service with the Cooperation of Educational System

Music education practice is a comprehensive application of the college students' professional knowledge. In this practice the students learn the basic ability of teaching organization, and conducting the class. Additional, students should be encouraged to go to the community, rural schools to practice. Through this, they can counsel the primary and middle school students' extracurricular music learning consolidate the professional knowledge, accumulate their service experience.

The above kinds of practice guarantee college students' "normal" comprehensive qualities, laids a solid foundation for their later service.

### 16.4.4 Spirit Civilization Service of the Rural Area and Community with the Cooperation of Local Government

High education is an important position for the social spiritual civilization. College and university play a powerful role with its special condition and function. To serve the local spiritual civilization construction, the students' practical ability should be certain features. College can form college students' performance groups or teams. With the cooperation of the local government, carry out "literature and art to fields" activities, carry out "college students, farmers and citizens are heart with heart". Through bringing the local characteristics and times melody to the community and countryside, serving "agriculture, countryside and farmers", college students can contribute their strength to the construction of spiritual civilization.

## References

1. Wang B, Li ZT, Luo JF (2006) College students' art community in campus culture in the role. Shandong Inst Technol Prof Bus J 33(2):83–85
2. Yang MJ, Cui XD (2006) Full play of local university service local economic and social development of the important role. Lang Fang Norm Coll J 18(2):105
3. Su HQ (2008) Enhance university students' social practice some ideas about the pertinence and effectiveness. Henan Edu 20(8):67–68
4. Chen DS (2005) About the scientific research as the local economic construction service thinking. The Southwest National University Press vol 67(2), p 3

# Chapter 17
# Bus Arrive Time Prediction Based on Weighted Distance Feedback

**YaQin Luo, Cheng Sun and MuTian Cheng**

**Abstract** Traffic Information release is an important part of ITS system. Bus arrive time prediction (BATP) is one of key and difficult site in that system which can provide comprehensive bus information service as well as improve traffic environment. In this article, we take historical data as a measurement method in estimating operation stability, coupling with real time GPS data (which will be handled in KFM), to release ultimate error weighted result generated from these two kinds of predictions. Simulated results prove the max error time for passenger waiting time prediction is limited within 1 min. This method has a better veracity as well as efficient which makes it easier for realization and promotion.

**Keywords** Bus arrive time prediction · Historical data · KFM · Error weighted

## 17.1 Introduction

Huge traffic system which required for accommodating with society and economic development, coordinating with urban industrial layout, building sustainable urban foundation, must provide smooth, safe and comfortable service. Intelligent

Y. Luo (✉) · C. Sun · M. Cheng
School of Electrical Engineering and Information, Anhui University of Technology, Maanshan 243002, People's Republic of China
e-mail: lyq6699@163.com

C. Sun
e-mail: skcncspeaker@163.com

M. Cheng
e-mail: mutcheng@gmail.com

W. Du (ed.), *Informatics and Management Science V*, Lecture Notes in Electrical Engineering 208, DOI: 10.1007/978-1-4471-4796-1_17,

Transportation System (ITS) play an increasingly important role in urban development.

Traffic release system, as inseparable component of ITS, has achieved rapid promotion these years. But compare with the same use in rail transportation, traffic release system in bus arrive time prediction still have some way to go and attract more negative appraise.

To that resident who live in metropolis with properties of long average travel distance and more bus transfer times, a better performance bus information service can, to a great extent, reduce uncertain factors in trip, which may draw more attention on choosing bus as vehicle and reduce the pressure of rail transportation, the frequency of accounting traffic jam. A comprehensive bus information service contains at least two sections: scheduled bus lines service, bus arrive time prediction and release system. Scheduled bus lines service may gain better performance in the situation of less bus stations with stable operation environment, such as inter-city bus lines [1]. Bus arrive time prediction and release system can provide passengers dynamic operation information which may work well at city center with more bus stations and unstable environment. Because of the properties of unobvious regular pattern that seems better in rail transportation, Bus arrive time prediction needs other useful method.

Prediction method using nowadays can be classified from different data source (historical data [2, 3] and real-time data), from the difference of accuracy (station, time), from the specific calculate method [4–8] (Multiple Linear Regression, Time series analysis, K-filtering [4], Artificial neural network [5], Stochastic Queuing System, etc.).

## 17.2 Data

Condition of running environment has a great influence in bus operation. Any kind of prediction algorithm base on real-time GPS data need to take into consideration of the accuracy and speed of algorithm convergence. Through statistical analysis of historical data, we can get the operation mode in the same environment (time and distance) and provide a historical reference for real-time data. The combination of error weighted historical and real-time data will take into consideration while release ultimate information.

### *17.2.1 Data Category*

While choosing historical data, considering city development and road reconstruction, we'd better extract data cover a week or a month in the same year.

With help of ArcGIS software, we store date, specific time and distance of every return time in each return site position, delete data return from return position between stations. These remaining data is received while bus arrive stations.

We classify historical data as following:

Direction: Up going, Down going
Date: Workday Monday, workday Friday; workday Tuesday, workday Wednesday, workday Thursday; weekend
Period in one day: morning peak, evening peak, noon
Collect time between stations

Distance between two stations can be got from the subtraction between the next station's odometer record and former one. Assume the distance between station i and station j is $D_{ij}$. Then we can get the average running time and distance. In order to reduce error, the biggest and smallest value of record will be deleted directly.

The average running time between two stations is as function Eq. 17.1:

$$\overline{t_{i,j}} = \frac{1}{N}\sum_{ID=1}^{N} t_{i,j}(ID) \tag{17.1}$$

ID is identifier of bus. N is total bus number of a bus line.

For the case of simplification, we assume the speed is (but actually not) a constant. With this assumption, we design the speed of bus is combination of a constant and a Gaussian error:

$$\bar{v}_{i,j} = \frac{D_{i,j}}{\overline{t_{i,j}}} + w_v, w_v|N(0, \sigma_{i,j}^2) \tag{17.2}$$

## 17.2.2 Analytical Standard

With the foundation of former works, we still need a standard deviation of statistic operation time. Take the patient degree of passenger into consideration; we set the standard deviation one minute. As to say, if the prediction time has gap over than one minute, result of that period is not stable and has a big degree of dispersion, which does not have reference value.

Caused by the GPS drift, the error in stations distance is $\sigma_{Di,j}^2$. Standard deviation of average running time is shown in function Eq. 17.3:

$$\sigma_{\overline{t_{i,j}}}^2 = \frac{1}{N}\sum_{ID=1}^{N} (t_{i,j}(ID) - \bar{t}_{i,j})^2 \tag{17.3}$$

The changing rate of average running speed which can be obtained from derivation of average running speed in function Eq. 17.4:

$$\delta_{\bar{v}_{i,j}} = \frac{\delta_{D_{i,j}} \cdot \bar{t}_{i,j} - D_{i,j} \cdot \delta_{\bar{t}_{i,j}}}{\bar{t}_{i,j}^2} = \frac{1}{\bar{t}_{i,j}} \cdot \delta_{D_{i,j}} - \frac{D_{i,j}}{\bar{t}_{i,j}^2} \cdot \delta_{\bar{t}_{i,j}} \tag{17.4}$$

The "quality" of average speed can be measured with deviation. If we assume the separation between error in distance between stations and average running time, deviation of average speed can be obtained from function Eq. 17.5:

$$\sigma_{\bar{v}_{i,j}}^2 = Var(v|D_{i,j}) \approx \frac{1}{\bar{t}_{i,j}^2} \sigma_{Di,j}^2 + \left(\frac{D_{i,j}}{\bar{t}_{i,j}^2}\right)^2 \cdot \sigma_{\bar{t}_{i,j}}^2 \tag{17.5}$$

## 17.3 Algorithm

### 17.3.1 Kalman Filtering

Operation of Kalman filtering includes two aspects: prediction and update. In prediction period, we use the estimate value of last status to predict current status; in update period, we use current observed value to optimize the prediction value.

The status of slot k comes from slot k−1, as in function Eq. 17.6:

$$x_k = F \cdot x_{k-1} + w_k \qquad w_k \sim F \cdot x_{k-1} + w_k \tag{17.6}$$

Which $w_k$ is Gaussian white noise, $Q_k$ is covariance matrix. The observation of real status is in function Eq. 17.7:

$$Z_k = H \cdot x_k + V_k \qquad V_k \sim N(0, R_k) \tag{17.7}$$

### 17.3.2 Algorithm Model

Which $V_k$ is Gaussian white noise, $R_k$ is covariance matrix. It is usual of occasional delay while bus running, such as waiting red light or traffic jam. The feedback based on time will have big error, so we use feedback based on distance in this article. Specific algorithm as follow (Fig. 17.1):

$x_{d|d} = \begin{pmatrix} t(d) \\ v^{-1}(d) \end{pmatrix}$: Estimated value comes from dth reporting site. $t(d)$ is the time of dth reporting point. $v_{-1}(d)$ is reciprocal of the specific reporting site. $F = \begin{pmatrix} 1 & \Delta d \\ 0 & 1 \end{pmatrix}$: $\Delta d$ is the interval time between GPS data updating. $P = \begin{pmatrix} \sigma_t^2 & \mathrm{cov}(t, v^{-1}) \\ \mathrm{cov}(v^{-1}, t) & \sigma_{v^{-1}}^2 \end{pmatrix}$: Error related matrix comes from dth reporting

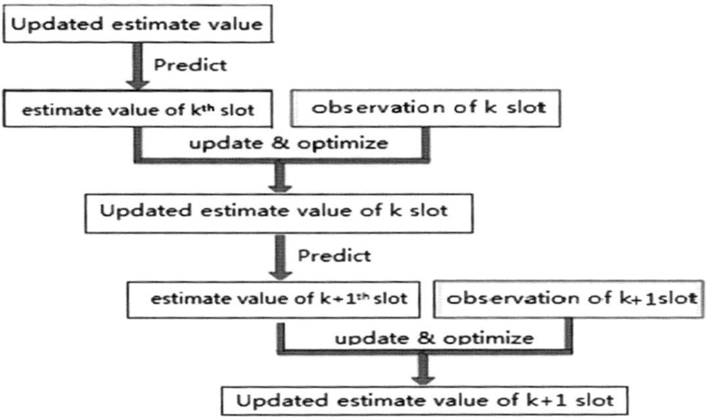


**Fig. 17.1** Workflow of kalman iterative algorithm

site, which represent the accuracy of estimated value. According to the function $\bar{v}_{i,j}^{-1} = \frac{\bar{t}_{i,j}}{D_{i,j}}$, we can calculate $\sigma^2_{\bar{v}_{i,j}^{-1}}$ as the follow function $\bar{v}_{i,j}^{1} = \frac{\bar{t}_{i,j}}{D_{i,j}}$

$$\sigma^2_{\bar{v}_{i,j}} \approx \frac{1}{D_{i,j}} \cdot \sigma^2_{\bar{t}_{i,j}} + \left(\frac{\bar{t}_{i,j}}{D^2_{i,j}}\right)^2 \cdot \sigma^2_{D_{i,j}} \tag{17.8}$$

$x_{d|d-1}$: The estimated status value comes from the dth reporting site, which calculated by the value from d−1th reporting site. $P_{d|d-1}$: The error related matrix comes from the dth reporting site, which calculated by the value from d−1th reporting site. $y_d$: Represent the subtract value between actual and estimated value. $Z_d$: Actual status observations (got from dth reporting site) $S_d$: Covariance of measurement remaining amount; kg(d) Optimal Kalman gain.

### *17.3.3 Steps of Kalman Iterative Algorithm*

Forecast period:

$$x_{d|d-1} = \mathrm{F} \times x_{d-1|d-1} \tag{17.9}$$

$$P_{d|d-1} = F \cdot P_{d-1|d-1} \cdot F^T + Q_d \tag{17.10}$$

Update period:

$$y_d = Z_d - H \cdot x_{d|d-1} \tag{17.11}$$

$$S_d = H \cdot P_{d|d-1} \cdot H^T + R_d \tag{17.12}$$

$$x_{d|d} = x_{d|d-1} + kg(d) \cdot y_d \tag{17.13}$$

$$P_{d|d} = (I - kg(d) \cdot H) \cdot P_{d|d-1} \tag{17.14}$$

$$kg(d) = P_{d|d-1} \cdot H^T \cdot S_d^{-1} \tag{17.15}$$

## 17.4 Algorithm Analysis

### *17.4.1 Algorithm Foundation*

In order to predict the arrival time of next dth reporting site, we build the model simulating bus operation between ith station and jth station.

The interval distance between GPS data updating is 100 m. In which station ith is considered as 0th reporting site and there are n−1 reporting sites between these two stations. jth station represent nth reporting site, as in Fig. 17.2:

1. Distance between two stat ions is $D_{i,j}$.
2. Prediction value of time when bus arrive the dth reporting site is $t_d$.
3. Measurement value of time when bus arrive the d−1th reporting site is $Z_{d-1}$.
4. Average speed between d−1th reporting site and jth station (if assuming the actual speed) is:

$$\bar{v} = \frac{100}{t_d - Z_{d-1}} \tag{17.16}$$

5. According with real-time GPS data, the formula of calculating the actual arrive time is as function Eq. 17.17:

$$T_r = \frac{D_{i,j} - 100 \times (d-1)}{\bar{v}} \tag{17.17}$$

**Fig. 17.2** Reporting topology based on distance

### 17.4.2 Error on Real GPS

Error in prediction procedure can be got from error related matrix $P_{d|d}$. Cause the adoption of feedback based on distance in this article, we can use the variance of arriving next reporting site as the generated error, also means this element in 2 × 2 dimensions' matrix:

$$\sigma_r^2 = P_{d|d}(1,1) \tag{17.18}$$

Error weighted and arrival time prediction.

Each procedure of calculation still has its error and while release ultimate information; more accuracy prediction value can be obtained by function Eq. 17.19:

$$T_{pred} = \frac{\sigma_h^2}{\sigma_h^2 + \sigma_r^2} T_r + \frac{\sigma_r^2}{\sigma_h^2 + \sigma_r^2} T_h \tag{17.19}$$

### 17.4.3 Experimental Result and Analysis

We choose No. 1 line of Nanjing bus system as experimental line, for the reason that its covers downtown area as well as stations distance in the line are various, and use method in this article to predict bus arrive time. To detail, we limit area to operation period between Central-Gate to Xin-Jie-Kou as scenarios. Here, we do analysis within GPS data collected from 2010-5-30 to 2010-6-6.

To definition the degree of accuracy, we develop average related error in function Eq. 17.20:

$$\varepsilon_{\text{mean}} = \frac{1}{N} \sum_t \left| \left\{ \frac{T_{pred(t)} - T_{real(t)}}{T_{real(t)}} \right\} \right| \tag{17.20}$$

We also define the max error in function Eq. 17.21:

$$\varepsilon_{\max} = \left| \left\{ \frac{T_{pred(t)} - T_{real(t)}}{T_{real(t)}} \right\} \right| \tag{17.21}$$

The following figure shows the performance comparison between method in this article and BPnet method (Figs. 17.3 and 17.4).

Though the restriction of experiment condition (such as less of reporting site and training sample is limited), which may cause the churning, result using BPnet still keep error within 30 s. But error of each minute will become huge if has a long time operation. Method provided in this article is more accuracy as well as smaller error. The example bus line has the longest operation time between two neighbor stations in 6 min, while will keep the maximum cumulative errors in 1 min (Table 17.1).

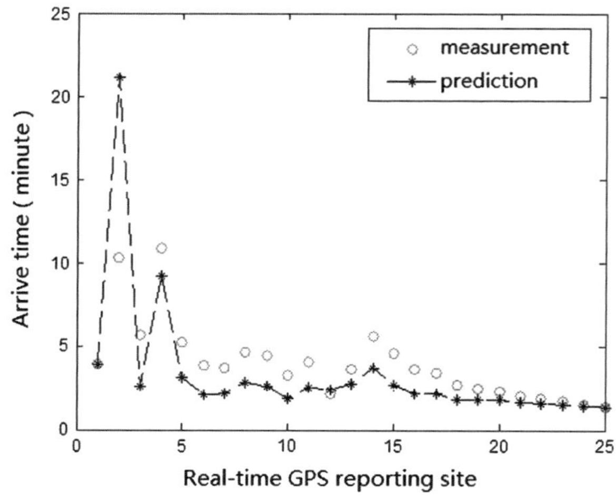


**Fig. 17.3** Comparation between measurement and prediction

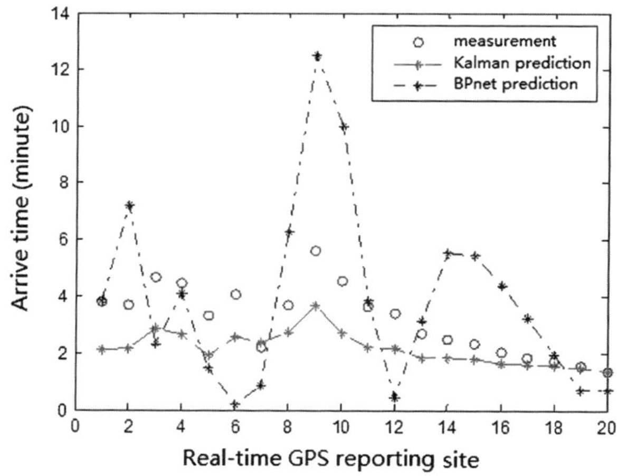


**Fig. 17.4** KPI in different methods

**Table 17.1** Error prediction

| | (ε mean) | (ε max) |
|---|---|---|
| (BPnet) | 0.68 | 1.329 |
| (Kalman) | 0.31 | 1.049 |

## 17.5 Conclusions

Bus arrive time prediction as an important technology in information release system, has gained significant progress these years. Classic Kalman filtering method with good performance has become a mainstream tool in prediction. In real environment, the amount of reporting sites need select a proper quantity.

Because of too many sites may reduce accuracy and not enough sites will not guarantee the astringency, some other prediction methods are needed to assist the method.

## References


1. Tan CW, Park S, Zhou K, Liu H (2006) IEEE intelligent transportation systems conference, vol 3. Toronto, Canada, pp 1478–1482
2. Chu H, Cai Y, Yang X (2007) Telecommunications, 7th international conference on ITS vol 21, pp 1–5
3. Wei CH, Lee Y (2007) IEEE transactions on vehicular technology conference, VTC Spring vol 56, 3682–3694
4. Zhu T, Kong X, Lv W (2009) Research on transportation. Comput Intell Softw Eng 1:1–5
5. Jeong R, Rilett LR (2004) IEEE intelligent transportation systems conference, vol 4. Washington, D.C., USA, pp 4:988–993
6. Zhang F, Shen Q, Clifton KJ (2009) Transportation research record. J Transp Res Board vol 2082 (2008). Transportation Research Board of the National Academies vol 1, pp 107–115
7. Coffey C, Pozdnoukhov A, Calabrese F (2011) Proceeding CTS'11 Proceedings of the 4th ACM SIGSPATIAL international workshop on computation transportation science vol 19, pp 377–384
8. Padmanaban RPS, Vanajakshi L, Subramanian SC (2009) IEEE, intelligent vehicles symposium, vol 12. Xi'an, pp 955–959

# Chapter 18
# ZigBee-Based Intelligent Home System

**Guopeng Song and Yunfeng Zhou**

**Abstract** Based on the application of the Internet of things, we analyze the intelligent home system, bring forward ZigBee-based intelligent home system and develop new software and hardware. The entire network has a star topological structure, wherein the central node works as data processing node to accept data, to transmit data to the data server, and to communicate with the Internet and GSM/CDMA network. Therefore, the environmental parameters and body parameters communicate with the remote terminal successfully and the remote terminal controls the indoor environment successfully. Its low-power, short distance and low transmission rate make it more suitable for an intelligent home.

**Keywords** ZigBee · Internet of things · Intelligent home · Remote monitoring

## 18.1 Introduction

With the rapid social development, people require higher quality of life and begin to think about how to apply intelligent control technology into daily life, such as intelligent control of home appliances, remote monitoring of the elderly and the children at home, automatic warning and suggesting of possible risks; at the technical level, intelligent home system requires a large amount of data collection

G. Song (✉) · Y. Zhou
School of Information System and Management, National University of Defence Technology, Changsha, China
e-mail: rocsgp@yahoo.com.cn

Y. Zhou
e-mail: zyf298515649@gmail.com

W. Du (ed.), *Informatics and Management Science V*, Lecture Notes in Electrical Engineering 208, DOI: 10.1007/978-1-4471-4796-1_18,

nodes and a variety of effective interconnection between networks, and restricts the use of cables.

In recent years, the rapid development of wireless communication technology solves these problems [1, 2]. Driven by actual needs and technological development, intelligent home systems with different applications appear. Further research finds that system endurance, scalability, security, reliability and other actual functions require higher of open data interface system, real-time data transmission and data link security.

## 18.2 Comparison of Several Wireless Transmission Technologies

### *18.2.1 Common Wireless Transmission Technologies*

At present, major wireless communications technologies include Bluetooth, Wi-Fi, infrared and ZigBee technology, which have their own characteristics in network of components. The following table compares the major parameters for several wireless technologies (Table 18.1).

We may find that, ZigBee technology is a communication technology with relatively low power, low transmission rate and large communication range. According to the estimates, ZigBee devices can support up to 6 months to 2 years relying on two AA batteries, which is unmatched by other wireless communication technologies. The large network capacity of ZigBee technology also lays a foundation for the large amount of data collection nodes of the intelligent home system. At the same time, each ZigBee module has a low cost of 6 dollars, applicable for large-scale network organization [3].

**Table 18.1** Comparison of wireless transmission modes [5]

| | Blue Tooth (802.15.1) | Wi-Fi (802.11b) | IrDA | ZigBee (802.15.4) |
|---|---|---|---|---|
| System overhead | High | High | Low | Low |
| Battery life | Shorter | Short | Long | Longest |
| Number of network nodes | 7 | 30 | 2 | 255/65,535 |
| Physical range | 10 m | 100 m | Directional 1 m | 100 m and above |
| Transmission rate | 1 Mbps | 11 Mbps | 16 Mbps | 250 kbps |
| Transmission medium | 2.4 GHz radio frequency | 2.4 GHz radio frequency | 980 nm infrared | 2.4 GHz radio frequency |

### *18.2.2 Advantages and Application of ZigBee Technology*

From the beginning, ZigBee was designed to construct the wireless network composing of thermostats, safety devices, gas meters and other devices, which is determined by its main technical features [4].

- Low data transmission rate: Only 10 k bytes/s to 250 k bytes/s, focusing on low transmission effect in daily home life.
- Low consumption: In low-power standby mode, two ordinary AA batteries can support for 6 months to 2 years, eliminating the need of charging or frequent replacement. It is also the unique advantage that the developer of ZigBee has been proud of.
- Low cost: ZigBee is of low data transmission rate, simple protocol and no royalty, greatly reducing costs.
- Large network capacity: Each ZigBee network can support 255 devices.
- Short time delay: Typically between 15 ms and 30 ms.
- Safety: ZigBee provides data integrity checking and authentication functions, adopting AES-128 encryption algorithm.
- Effective range: Covering 10–1,000 m, determined by the actual transmission power and different application modes, basically covering the average home or office environment.
- Flexible work frequency band: 2.4 GHz, 868 MHz (Europe) and 915 MHz (USA), all license-free.

On the contrary, other wireless communication technologies are basically designed as substitutes for cable transmission, used by mobile phones, headsets or PDA network. These technologies can work for several days or weeks without charging, but cannot meet the requirements of longer durability.

Generally, Bluetooth devices need to manually configure and maintain network connections; it can be used to effectively deal with data transmission between eight devices (one master and seven slaves); the communication rate is decreased significantly in case of more devices. 802.11 (Wi-Fi) has a similar problem; although it is a good solution for laptop and desktop access, its power consumption may be very high. The above-mentioned features of ZigBee wireless sensor network give it a good application prospect in individual physiological signal monitoring and remote home monitoring [6].

## 18.3 Structure of ZigBee-Based Intelligent Home System

### *18.3.1 Basic Composition of Intelligent Home System*

Intelligent home system needs to consider energy consumption, coverage, transmission rate, multiple network connection and other factors. In this study, we adopt ZigBee-based wireless network to monitor home appliances, to collect physiological signals and combines Internet and GSM/CDMA network communication technology to achieve remote intelligent control of daily home system.

Wherein, the physiological signal monitoring system uses the portable programmable sensor module with a sensor as its collection terminal. The home appliance monitoring system uses the embedded control module as its control and data collection terminal, and both data transmitting and data processing are through ZigBee network. The data processing terminal communicates with ordinary mobile terminals (such as mobile phones) via the Internet or GSM/CDMA network. Through SMS and other manners, users can monitor the system devices remotely, know the physiological parameters of the elderly and the children at home at any time, receive emergency alarm on time and take a series of response actions using the pre-set threshold functions to achieve intelligent control.

### *18.3.2 Network Topological Structure*

The network topological structure of ZigBee protocol has three types: Star structure, network structure and tree structure, as shown in Fig. 18.1.

In the intelligent home system, data collection nodes may be distributed in many rooms, but they need to transmit the data collected to the data processing nodes, so that our system adopts the star topological structure. This adopted structure has its exceptional advantages in that it facilitates data transmitting. The central node can inquire each data collection node, when the data collection node is inquired, the data will be transmitted to the central node.

### *18.3.3 Structure of Intelligent Home System*

Based on the above analysis, the intelligent home system designed in the thesis has the following structure shown in Fig. 18.2.

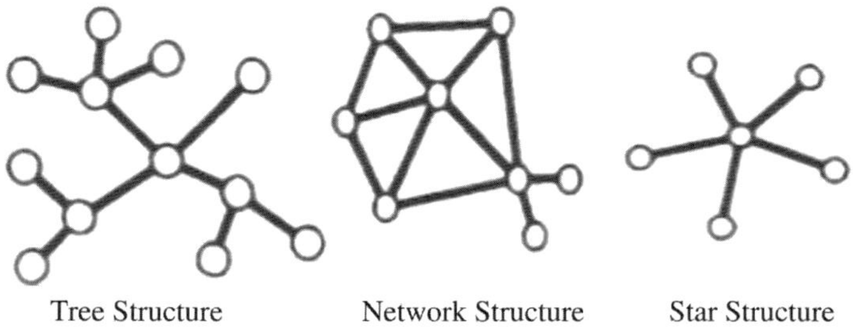


**Fig. 18.1** Three types of ZigBee network

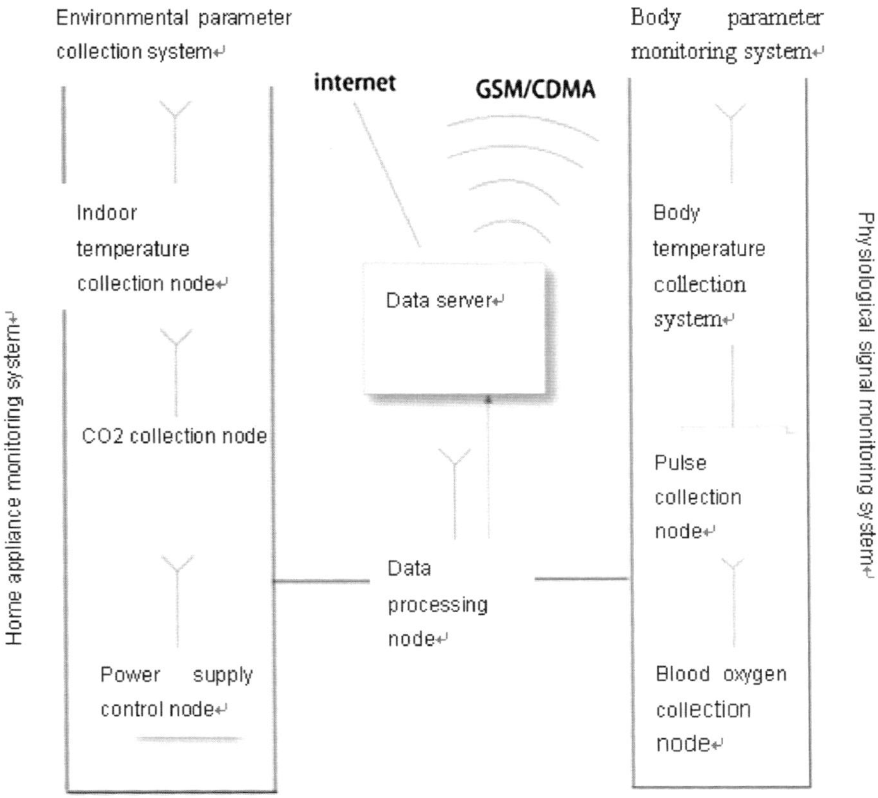


**Fig. 18.2** Structure of intelligent home system

The whole system is divided into three parts, wherein, home appliance monitoring system is used to collect environmental parameters and control home appliances, it contains the following functions: collection of room temperature and carbon dioxide concentration, intelligent cut off of general power supply. Physiological signal monitoring system is used to collect body parameters, its functions are as: collection of body temperature, pulse and blood oxygen concentration. Data collected by the two parts are transmitted to the data processing node via ZigBee network and then sent to the data processing server; after calculation and comparison with the threshold values, effective information will be extracted.

The Internet or GSM/CDMA network then will be connected to communicate with remote users. After receiving data at home, remote user terminals can send feedback command to the data processing server. After receiving the feedback command, data processing server sends the user command to corresponding control circuits through communication between data processing nodes and data collection nodes, realizing remote control of home appliances and environmental regulation.

**Fig. 18.3** CC2430 Node

## *18.3.4 Realization of Intelligent Home System*

Figure 18.3 shows ZigBee CC2430 node. As a major component of intelligent home system, it is the carrier of ZigBee module and various sensors.

The core of CC2430 node is an industrial 8051 singlechip packaged with ZigBee module, which has 21 programmable I/O pins and a serial port. As forepart node, it can be equipped with abundant data collection sensors; as central node, it can communicate with data processing server through the serial port. It has low power consumption in sleep mode and can be waken up by external interruption.

Figure 18.4 shows $CO_2$ sensor (S-100) and temperature sensor (DS18B20) for environmental parameter collection.

Various data collection sensors packaged can be directly inserted in the forefront collection node according to the user requirements to adapt to various environments. The sensors can be fitted flexibly according to different external environments and data to be collected to complete various tasks.

**Fig. 18.4** Environmental parameter sensor

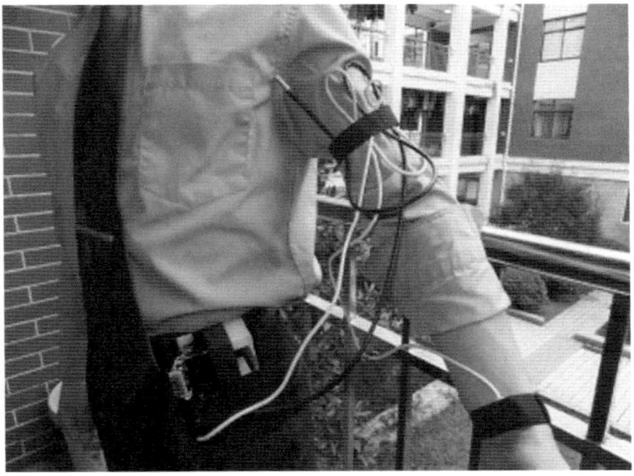

**Fig. 18.5** Packaged body parameter collection module

Figure 18.5 is a packaged physiological signal monitoring system. It integrates pulse and body temperature data collection and is arranged with buttons for the convenience of users to send emergency signals.

The module integrates body temperature and pulse collection functions and is made into a wearable device, which has little impact on daily life and facilitates to record physiological parameters in a timely manner and to deal with emergency situations.

## 18.4 Conclusion

The advantages of ZigBee technology are suitable for intelligent home applications. Its low energy consumption and short-distance low-rate transmission make it better fit the home environment. Users can control home appliances remotely through connection with the Internet and GSM/CDMA network. With our system imagine, users only need to send a message on the way home to warm up the air conditioner, to turn on water heater, or let themself and the hospital receive signals at the first time when the physical conditions of the elderly or the children worsen suddenly. The system we developed can bring people a lot of benefit and give life much convenience in the near future.

## References

1. Akyildiz IF, Cayirci E (2002) Wireless sensor networks: a survey. Comput Netw 38(4):393–422
2. Malan D, Jones TF (2004) An ad hoc sensor network infrastructure for emergency medicate international workshop on wearable and implantable body. Sens Netw 42(4):202–207

3. Yi Y, Honggen S (2009) The Status of Wireless Sensor Networks. Comput Appl Softw 12(03):11–14
4. Ruihong G, Zhang H (2000) ZigBee-based wireless network technology and its application. Appl Electron Tech 23(4):18–25
5. Yiwu Z, Zhihao L, Qinqin W (2003) Wireless communication technology and investigation on its application. Process Autom Instrum 123(2):48–53
6. Thaddeus RF, Fulford J, Gu YW (2004) A portable low-power wireless two-lead EKG system. In: Proceedings of the 26th annual international conference of the IEEE EMBS, San Francisco, CA, USA, vol 23(43). pp 289–295

# Chapter 19
# UAV Simulator Speediness Designing Based on Man in Loop Simulation Platform

**Chao Yun, Xiao-Min Li and Zong-Gui Zheng**

**Abstract** With UAV is widely applied in military action, in order to make the UAV operator know well the UAV equipment, we should train the UAV operator by high efficiency, and high fidelity training system could do the training work in stead of the real UAV aircraft, which can make sure the aircraft for safety flying, heighten the training efficiency and economize the outlay. Paper bring forward the design project of the man in loop UAV flying simulator base on Matlab/Simulink, which contrapose flying control system's characteristic and the requesting of the man in loop simulation. Compared with tradition program in manual, the modeling design base on Simulink have flexible, speediness, high efficiency and few cost. Which can afford great value to design the UAV training system.

**Keywords** Unmanned aerial vehicle · Simulation training · Man in loop; simulation platform · Applied value

## 19.1 Introduction

With the development of aviation and electron technology. The function of UAV is increasing, thus made the flying control system is getting more complicated (include hardware and software), which can bring malfunction risk for the UAV equipment is getting higher, in order to make sure that the UAV operator can in

---

C. Yun (✉) · X.-M. Li
Department of Optics and Electronics Engineering, Ordnance Engineering College, Shijiazhuang, China
e-mail: jeffyunchao@163.com

Z.-G. Zheng
Academe of the second artillerist, Beijing, China

W. Du (ed.), *Informatics and Management Science V*, Lecture Notes in Electrical Engineering 208, DOI: 10.1007/978-1-4471-4796-1_19,

charge of the UAV's flying safely and efficiency, we should do much job for the simulate training to increase the training fidelity and immersion. Therefore, simulate training play an important role in daily training [1]. For these requirements, the fast establishing for the simulation system of UAV is a problem should be put forward, Matlab/Simulink tool can afford technology and platform assist to this problem. This method can reduce the empolder training periods and avoid the faults by manual programming.

## 19.2 Hardware in Loop Simulation

System simulation technology is base on resembling principle, information technology and system technology. It rely on computer and other physics equipments, which using simulating modeling to research the system and do the experiment. It becomes a new integration subject. The simulation technology provides us an advance method to do the scientific research, problem analysis, decision making, and designing for the professional researcher. It improve the cognize ability for people and could promote the ration analysis in science instead of determine the nature analysis in past [2].

Hardware-in-loop simulation (Half-object in loop simulation) technology is developing by the development of army equipment research, training system design and computer technology. The actual military experiment always to be so valuableness. Hard-in-loop simulation technology offer great valuableness measure to weapon system experiment and simulate training. In other words, we can test the weapon performance in entire lifecycle, so it can reduce the research periods and the training cost. Therefore, the key researching technology of hardware-in-loop simulation system make great sense for the army.

Compare with other style simulations, hardware-in-loop simulation have great authenticity, it is a reliability simulation method, in the view of system, the hardware-in-loop simulation system permit the actual object in the loop, so it means that we can check the object performance in the loop, in this way the whole capability of the system could be detected in the simulation loop, therefore, hardware-in-loop simulation is perfecting simulation for system experiment.

## 19.3 Man in Loop Simulation Platform

### *19.3.1 Man in Loop Simulation Platform Analysis and Presentation*

UAV training system usually adopt mathematics modeling in computer to replace the real aircraft, in this way we can control the mathematics modeling instead of the real airplane, the control system and dynamic modeling is designing in mathematics simulation environment, then we can contrast the flying parameter in mathematics model with the parameter in flying environment, if the dynamic flying data generate by the model have much variance from the real flying data, it make out that the flying modeling is not true or flying modeling is imprecise, and then we should renewedly design the dynamic UAV simulation model and compare the simulation output data again, repeat the process while the simulation data is satisfied for the guide line. Matlab also put out RTW, xPC toolbox, it can generate the embedded code in time. We can using speediness design idea for UAV rapid prototyping.

Because there have being objects in the simulation loop, so the time measure of the simulation should as length as real time, in other words, hardware-in-loop simulation must work in real time [3], for simulation computer is connect with real object, so the simulation system must acquire dynamic parameters and export dynamic response in real time.

Simulation step must insure two aspects: first is insure that simulation modeling must work in real time; second is insure that simulation step must incarnate the real signal's characteristic, and not too long [4].

## *19.3.2 UAV Modeling Compared and Select Base on Matlab Platform*

Matlab is an scientific computational software, which contact the numerical value account with the videotext, What offer abundant internal function, therefore, matlab is widely used in scientific computation, control system and information management. With developing more than 10 years, it has been recognized that popularity and computational standard software in the world.

Combine the control system modeling with UAV flying dynamic modeling base on Matlab/simulink hardware-in-loop simulation system, which reduce the periods system of design and arithmetic researching. This simulation technology have opening and general. It can make different equipment working in a same platform. In this way we can provide a simulation environment for empoldering more function for the system.

Matlab environment platform have 3 styles modelings for object characteristic: M-file, Simulink-block and S-function [5].

### 19.3.2.1 M-File

M-file using user-defined expression structure by itself, which is readability for C programmer and it comprise abundant internal functions, however, every step calculation could employ Matlab solver, thus play down implement efficiency of the modeling, such mode should not accept the simulation for the system which require by high real time.

#### 19.3.2.2 Simulink-Block

Simulink-block is a dynamic interaction environment which afford us about dynamic modeling, simulation and analysis. This method make the researcher do not consider much of the non-linear factors and stochastic factors. In the environment of simulink, the modeling can use differential eguation to describe the dynamic characteristic, the user can operate mouse by hand in stead of write programe in text, which can establish visual simulation modeling [6].

Simulink contain much basic modeling such as Aerospace module and so on, Aerosim toolbox is aim at UAV simulation researching, we can make use this toolbox for the simulation analysis.

#### 19.3.2.3 S-Function

S-function come from the C program according to mex-file style, what can compile and link by Matlab compiler. At last, it can form.dll file which can transfer in Matlab environment, which could heighten the efficiency of the programme. S-function have the characteristic of readability and easily modification.

Compare with 3 modeling methods in Matlab and the simulation system designing we requirement. Speediness designing of the dynamic simulator is mainly used for daily training. According to performance of each method, the simulink modeling have high efficiency, function distinguish prominence, module modify and implement simpleness, facility maintenance and expand easily. Which adapt to empolder simulation modeling in UAV training system, so paper adopt simulink-block to plan the dynamic modeling of the training system.

## 19.4 Dynamic Simulation Modeling for UAV

### *19.4.1 Flying Simulation Block File Configuration Base on Aerosim Tool Box*

Simulink have much function about dynamic modeling, relative to other high level programming language (such as C language), it has mightiness calculating ability, operate easily and maturity module, for more the modeling could translate to real time code. Simulink have many toolbox for engineering application, include aerospace for aircraft flying simulation, adopt these toolboxes could to save manpower and time. Because Matlab is exoteric, there are many plugins could do help for our own project, We could using these basic module to set up the complete 6 degree of freedom aircraft model, some models are not modify or do a little modify for our project using, such as atmosphere model and earth model (Fig. 19.1).

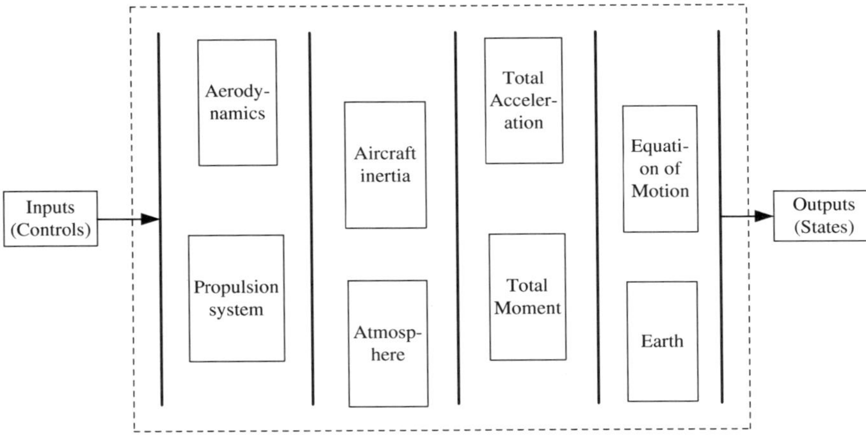


**Fig. 19.1** Complete 6-degree-freedom aircraft modeling and internal structure

The Aerosim aeronautical simulation blockset provides a complete set of tools for the rapid development of nonlinear 6 degree of freedom aircraft dynamic models. In addition to the basic aircraft dynamics blocks, the library also includes complete aircraft models which can be customized through parameter files [7].

A new aircraft parameter file can be generated from a custom Matlab script. To create this script, we need to open the template aircraft configuration script config_template.m, which specify the aircraft aerodynamic, propulsion and inertia parameters. By running this script at the Matlab command prompt, a new aircraft parameter file of the form filename.mat will be created. The file name then can be used in any of the complete aircraft blocks available in the Aerosim library. The first variable that should be specified is the name of the aircraft parameter file that will be generated. Type the chosen name of our aircraft parameter file, as a string, without the.mat extension [7].

% Insert the name of the MAT-file that will be generated

% without.mat extension

cfgmatfile='myairplanecfg'

Next we double-click the block to open the block parameters dialog. We can specify the aircraft parameter file—Myairplane.mat, the initial conditions (position, velocity, attitude, an gularrates, fuel, enginespeed), the ground altitude with respect to the sea-level, the simulation sample time and so on.

### *19.4.2 UAV Simulation Modeling Design and the Results Analysis*

We construct the modeling by Aerosim toolbox, before starting the simulation, we can also add a constant source for the controls input to the aircraft model. This will provide the model with a set of constant actuator commands. Finally, go to the

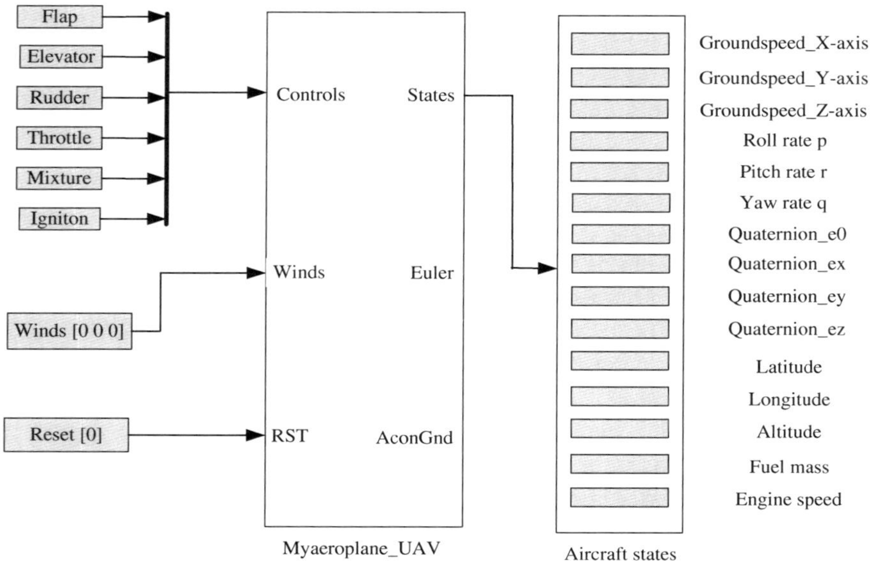


**Fig. 19.2** Dynamic simulation of UAV block diagram

simulink model pull-down menu, under simulation and open the simulation parameters. Set the solver type to Fixed-step, the integration scheme to ode-4 or ode-5, and set the fixed-step size to match the aircraft model sample time. The simulink model diagram is shown in Fig. 19.2.

After the commands is given to the intergrated UAV modeling, the simulation modeling start compute the UAV state according to the simulation time span and simulation step. The airspeed settle to a value which depends on the elevator

**Fig. 19.3** UAV airspeed outputs

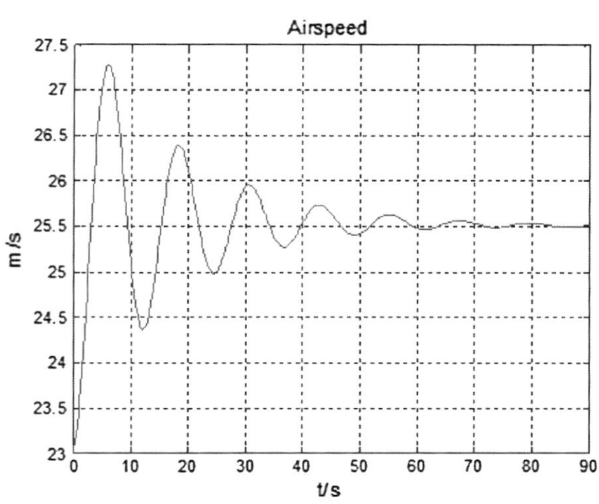

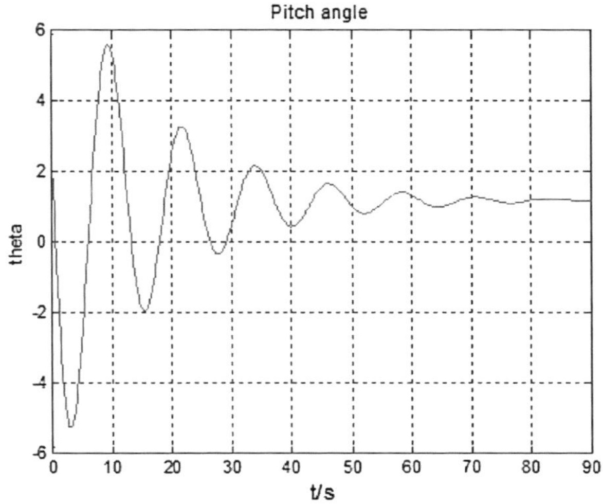


**Fig. 19.4** UAV pitch angle outputs

setting. The model adds an airspeed control loop using proportional, integral, derivative (PID) control laws.

By running the simulation, we could see a plot of the air speed and pitch angle similar to Figs. 19.3 and 19.4. The airspeed does indeed settle to the commanded value about 25.5 m/s, while the oscillations are virtually eliminated and the pitch angle settles to avalue of approximately 1.7°.

## 19.5 Conclusion

Paper introduce an project designing for man-in-loop simulator of UAV training system. Compare 3 methods in Matlab platform, we adopt Aerosim toolbox to make up the whole dynamic simulation model. Emphasis the designing process and flying parameters file configuration. We explain that there are feasibility and advantage in the designing course base on Simulink/Aerosim toolbox. The specifically type UAV modeling could be speediness upbuild which we should configure parameters file and make.mat extension file. Since the integrated modeling is setting up, we can use real time-workshop to automatically generate source code from our aircraft models. Thus the source code from models could do the foundation work for our fidelity training simulator of UAV.

## References

1. Anonymous (1999) Current and future UAV military users and applications. Air space Eur 1:5–6
2. Jiang Y-X, Zhu E (1998) Control system and simulation, vol 12(3). Beijing university of aeronautics and astronautics publishing company, Beijing, pp 480–487

3. Xiao T-Y, Zhang Y-Y (2000) System simulation introduction, vol 12(4). Qing hua university publishing company, Beijing, pp 28–34
4. Yang C, Du G-N (2003) Real time simulation system design base on Matlab. Appl Micro Comput vol 19(3):843–847
5. Jing Z-Y, Zhang Z-B (2008) Banausic manual of MATLAB7.0, vol 12., pp 03–09
6. Bi K-B, Wang X-D (2009) Flying navigation, controls and MATLAB simulation technology, vol 21(32)., pp 30–37
7. Information on aeroSim block user's guide. http://www.u-dynamics.com

# Chapter 20
# Global Optimization of UAV Heading Controller Parameters Based on Improved Genetic Algorithm

**Weiping Zhao, Zhanshuang Hu and Ming Yang**

**Abstract** In this paper, the genetic algorithm is introduced into the small UAVs heading control problem, using genetic algorithm to global optimization of PID parameters. On the basis of stability of the Roll Angle control inner loop, Get feedback signal of course Angle and Constitute the controller parameters of flying heading stable outer loop. According to the diversity of the individuals attenuate too fast, easy to fall into local optimum's characteristics in the genetic algorithm, to make sure the diversity of the individual and the quickness of optimization. Necessary improvements have been done to the genetic algorithm. The simulation results: The improved genetic algorithm design the PID controller has better adaptability, flexibility, stability and can ensure the control effect of the system, improve the system performance. The simulation results verify that the effectiveness of the proposed method.

**Keywords** UAV · Heading · Improved genetic algorithm · PID

W. Zhao (✉) · Z. Hu · M. Yang
Shenyang Aerospace University, General aviation key laboratory, ShenYang 110136, China
e-mail: 1621927296@qq.com

Z. Hu
e-mail: huzhanshuang@126.com

M. Yang
e-mail: ym1642@163.com

W. Du (ed.), *Informatics and Management Science V*, Lecture Notes in Electrical Engineering 208, DOI: 10.1007/978-1-4471-4796-1_20,

## 20.1 Introduction

The inner loop of The flight control system which is the base of the outer loop control such as flying height, heading, track circuit and so on [1]. Among them, based on the pitch control inner loop, the ALT HOLD of the UAV introducing high pressure feedback signal constitute a flying altitude stable outer loop; Heading control and stability's achieve is through heading signal feedback to the Roll control channels and Constitute flying heading stable outer loop; based on the flight navigation control circuit, introducing Cornering feedback from constitute track control outer loop achieve autonomous navigation fly [2].

In order to obtain satisfactory system performance, in this paper, the genetic algorithm is introduced into the small UAV heading control problem, using genetic algorithm to global optimization of PID parameters. On the basis of stability of the Roll Angle control inner loop, Get feedback signal of course angle and constitute the controller parameters of flying heading stable outer loop [3]. According to the diversity of the individuals attenuate too fast, easy to fall into local optimum's characteristics in the genetic algorithm, to make sure the diversity of the individual and the quickness of optimization. Necessary improvement has been done to the genetic algorithm. The simulation results: The improved genetic algorithm design the PID controller has better adaptability, flexibility, stability and can ensure the control effect of the system, improve the system performance. The simulation results verify that the effectiveness of the proposed method.

## 20.2 The Mathematical Model and Control Diagram of the System

The aircraft itself is a nonlinear system with multiple input and multiple output. To facilitate the analysis and design of the controller, With its fly straight and level for benchmark movement, Linearization of the state equation, get the UAV's linear differential equation, can choose the proper state vector, establish the plane's small disturbance state space equation. When a certain unmanned aerial vehicle at H = 12127 m, V = 0.787 Ma, the system state equation is [4],

$$\begin{bmatrix} \dot{\beta}_x \\ \dot{\omega}_x \\ \dot{\omega}_y \\ \dot{\gamma} \end{bmatrix} = \begin{bmatrix} -0.2331 & 0.0937 & 1 & 0.0422 \\ -34.773 & -1.21 & -1.1550 & 0 \\ -17.515 & 0.0067 & -0.2792 & 0 \\ 0 & 1 & 0 & 0 \end{bmatrix} \begin{bmatrix} \beta_x \\ \omega_x \\ \omega_y \\ \gamma \end{bmatrix} + \begin{bmatrix} 0 \\ -64.24 \\ 0 \\ 0 \end{bmatrix} A_t \tag{20.1}$$

Among them, $\beta_x$ is the lateral spreads Angle, $\omega_x$, $\omega_y$ is the helicopter angular velocity component in the body of the coordinate system, $\gamma$ is the slant angle for helicopters.

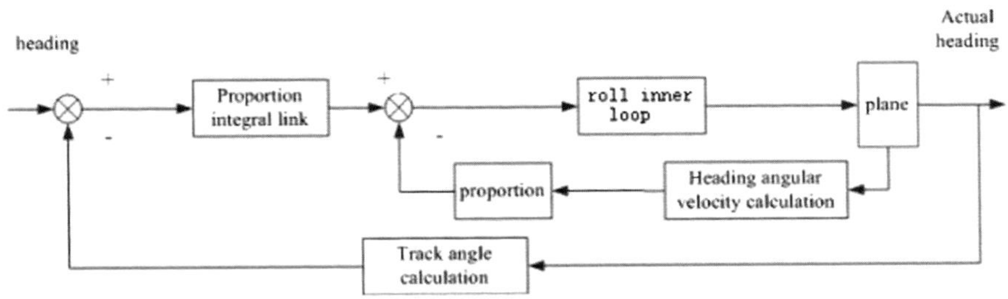


**Fig. 20.1** Course angle control diagram

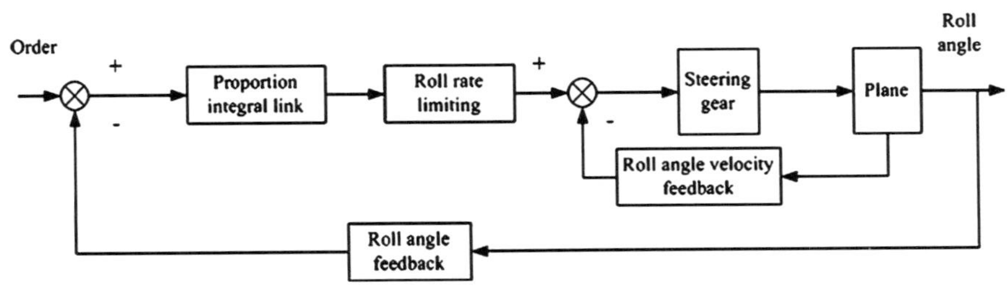


**Fig. 20.2** Roll angle control diagram

Heading to keep/control circuit can realize the function of heading to keep, heading given. Heading to keep is let heading Angle keep on that moment heading of value when this function take effect; Heading given is that flight control computer select a flight which can take plane to a given course and make it remain in the plane heading [5]. Heading control diagram shown as shown in Fig. 20.1.

Generally speaking, almost all the heading control circuit use tilt attitude control loop as its inner loop. Because, between the plane's roll Angle and the course Angle, there has a certain relation, the plane's heading change, mainly by the plane to tilt [6]. Roll attitude control use the angular rate and angular position's double closed loop control structure to make sure the circuit has good damping characteristics. Roll attitude control principle as shown in Fig. 20.2. The choice of rolling corner feedback coefficient should be in a certain range, need to coordinate choose parameters, increase damping, at the same time, reduced open-loop gain, otherwise it will cause system adjustment too slow, and even make the system instability.

In the flight control system design, the commonly method is to use small perturbation analysis make pitch, roll, heading and highly four channel's approximate linear model, then select the PID parameters of this model, which not only guarantee good dynamic quality characteristics (such as efficiency and weak super tonal, etc.), and should take the performance robustness and stability robustness [7]. In order to obtain satisfactory system performance, this paper uses the genetic algorithm to optimize the PID parameters.

## 20.3 The Improved Genetic Algorithm

The controller design goal is reasonable chose controller parameters Kp, Ki, Kd made the system to meet the design requirements. But, between PID controller parameters and the system does not exist direct, obvious causal relationship [8]. To solve this problem, we based on genetic algorithm optimal choice controller parameters, and through the simulation of MATLAB, and make the design and simulation's process automation.

Genetic algorithm is a random search algorithm, which uses genetic operation such as copy, crossover and mutation to simulate natural evolution, complete problem optimal method [9]. Essential genetic algorithm block diagram is shown in Fig. 20.3.

But genetic algorithm itself has shortage, the diversity of individual attenuate too fast, easy to fall into local optimum. In addition, because this chapter the parameters of the search are more, this article use the improved algorithm to guarantee the diversity in the individual and optimal efficiency. The improved genetic algorithm diagram is shown in Fig. 20.4.

From the flow chart of genetic algorithm we can see: the key of the genetic algorithm is choosing, crossover and mutation process. In the genetic algorithm's parameters, the choice of the crossover probability Pc and mutation probability Pm are the key factor to influence behavior and properties of the genetic algorithm, it direct impact algorithm convergence [10]. Pc greater, the speed of the creation of the new individual is faster. However, when Pc too big the chance of genetic pattern destroyed is greater, and the high fitness individual structure was quickly destruction; but, if Pc is smaller, can make the search process slow. For mutation

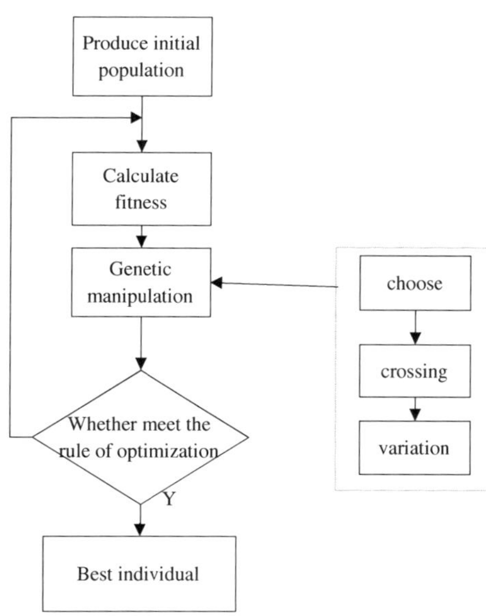


**Fig. 20.3** Flow chart of genetic algorithm

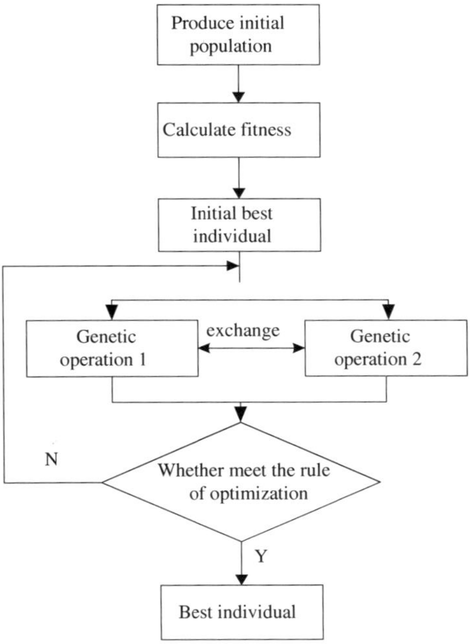


**Fig. 20.4** The improved genetic algorithm flow chart

probability Pm, if mutation probability Pm is too small, should not be create new individual; if mutation probability Pm is too large, genetic algorithm becomes a kind of pure random search method.

The improved genetic algorithm in inherited genetic algorithm has the above advantages, at the same time, make some improvement, that the algorithm convergence and diversity become unity. The first half of the flow chart, first of all, calculated fitness, the purpose is produce a initial best individuals. Use this individual procedure of parallel to go on the genetic operation 1 and 2, genetic population is divided into two child population 1 and 2 of genetic operations. Genetic operation 1 and 2 has the same steps with the genetic algorithm, including selection, crossover and mutation steps. And the difference is that the selection of Pc and Pm parameters, in this paper, in the genetic operation 1 the value of Pc and Pm parameters are smaller, the purpose is to consider the convergence and accuracy of the algorithm; In the genetic operation 2 Pc and Pm parameter values are larger, the purpose is to consider the diversity and the algorithm's search capability of population. Genetic operation 1 and 2 has communication of information and comparison of the results, easy to inherit and precise outstanding individual. Finally, according to the iterative conditions to judge whether the termination procedure.

This paper the biggest difference with genetic algorithm is that the parallel genetic operation steps to achieve the unity of algorithm convergence and population diversity, and use the comparison of the two results between 1 and 2 genetic operation make parallel two steps organic combine.

According to the model features of the small UAV stability control, and choose the rise time, steady-state error and overshoots's proportion combination as the optimization of the objective function. In order to guarantee the controller control effect, to reduce the oscillation of the system, introduce the oscillation frequency $\tau$, and multiplied by the penalty function $\lambda_3$. So the objective function eventually becomes [11]:

$$J = \int_0^\infty \lambda_1 |e(t)| dt + \lambda_2 t + \lambda_3 \tau \tag{20.2}$$

Among them, $\lambda_1$, $\lambda_2$–weighting coefficients; e(t)–the system error.

When searching should first optimization for roll Angle control inner loop control parameters, and then for flying heading stable outer loop controller parameters which constituted by the course Angle feedback signal. So, global optimal objective function into:

$$J = K_1 \int_0^\infty \lambda_1 |e_1(t)| + \lambda_2 t_1 + \lambda_3 \tau_1 + K_2 \int_0^\infty \sigma_1 |e_2(t)| + \sigma_2 t_2 + \sigma_3 \tau_2 \tag{20.3}$$

Among them, K1, K2–weighting coefficients of roll Angle control inner circuit and heading stability outer loop.

## 20.4 The Simulation Results

In the genetic algorithm optimization, in this paper the number of samples is 50, in the genetic operation 1, Pc and Pm parameter values are 0.25 and 0.1, the purpose is to consider the convergence and accuracy of the algorithm; In the genetic operation 2 Pc and Pm parameter values are 0.45 and 0.7, the purpose is to think about the diversity of population and search capability of the algorithm. The numeric area of control parameters values Ki are [0, 5], the weighting value step is 0.01 s. Using the binary code, go through 50 generation optimization.

Transfer function of Aileron rudder loop use inertia link, its transfer function as follows

$$W_f(s) = \frac{\alpha(s)}{u(s)} = \frac{-10}{s + 10} \tag{20.4}$$

Considering the actual system, introduce output restrictions of steering engine, the rudder amplitude limit is $\pm 20°$. In the above conditions finally get heading PID controller parameters for [1.4 4.3 1.45], the simulation results as shown in Fig. 20.5. The simulation results show that: up time of the system is quickly, the time of stabilization is short, the jitter small, achieve system requirements.

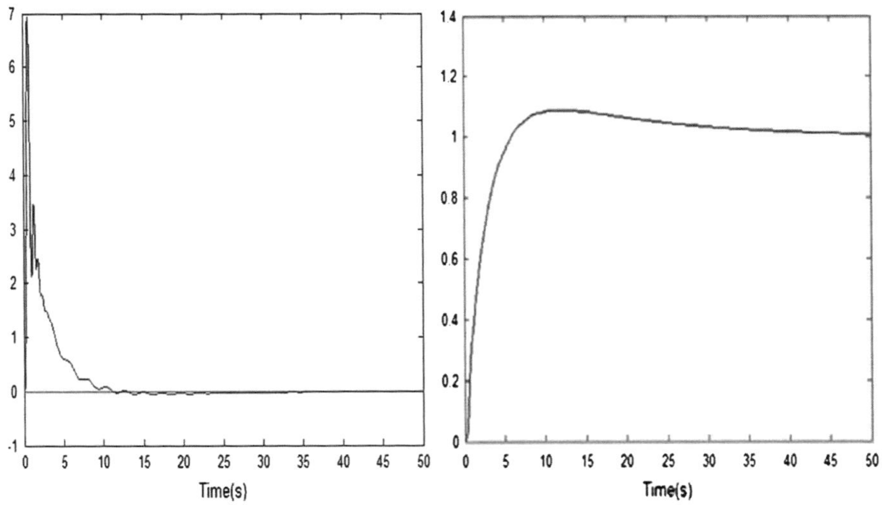


**Fig. 20.5** Genetic algorithm optimal PID control simulation results

## 20.5 Conclusions

Classic setting method not only process cockamamie, and can't ensure that you get the controller is optimal [12]. Therefore, in order to obtain satisfactory system performance, in this paper, the genetic algorithm is introduced into the small UAV heading control problem, a genetic algorithm is adopted in the PID parameters of the global optimization. On the basis of stability of the Roll Angle control inner loop, Get feedback signal of course Angle and constitute the controller parameters of flying heading stable outer loop. According to the diversity of the individuals attenuate too fast, easy to fall into local optimum's characteristics in the genetic algorithm, to make sure the diversity of the individual and the quickness of optimization. Necessary improvement has been done to the genetic algorithm. The simulation results: The improved genetic algorithm design the PID controller has better adaptability, flexibility, stability and can ensure the control effect of the system, improve the system performance. The simulation results verify that the effectiveness of the proposed method.

## References

1. Li L (2000) Unmanned helicopter flight control method and GPS application research. China's agricultural mechanization research institute 55(2):551–560
2. Franklin GF, David Powell J, Emami-Naeini A (2004) Feedback control of dynamic systems, vol 12, 4th edn. Publishing House of Electronics Industry, pp 7–13

3. Ge Z, Zhou Z, Zhi R (2007) The application of Robust control in the flying winger UAV control law design. Comput measur control 23(4):975–982
4. Ogata K (2003) Modern control engineering, vol 4(1), 4th edn. Publishing House of Electronics Industry in USA, pp 47–54
5. Zhou Z (2009) The key technology research of unmanned aerial vehicle flight control, vol 3(22), Northwestern polytechnical university, pp 291–296
6. Duan Z (2008) The UAV control law design. Northwestern polytechnical university
7. Xin G (2008) Simulation and software realization of UAV flight control, vol 19(23), Northwestern polytechnical university, pp 71–76
8. Liu J (2003) Advanced PID control and the simulation of MATLAB, vol 10(4), Electronic industry press, BeiJing, pp 581–588
9. Zhou M, Sun S (1999) Genetic algorithm theory and application, vol 34(4), Defense industry press, BeiJing, pp 192–195
10. Chen T, Guo P (2007) Data chain and flight control system of independent design in small UAVs. Microcomput comput inf 10(4):91–95
11. Li M (2002) Genetic algorithm's basic theory and application, vol 12(3), Science press, Beijing, pp 487–492
12. Hu S (1994) Automatic control principle, vol 9(1), 3rd edn. Defense industry press, BeiJing, pp 28–34

# Chapter 21
# Design of Unmanned Aerial Vehicle Coordinate Turn Controller Based on INA

**Weiping Zhao, Jun Yang, Zhanshuang Hu and Ming Yang**

**Abstract** In the process of unmanned aerial vehicle coordinated turn the transverse and Lateral Path coupling problem, This paper using inverse Nyquist method to decouple the system, In order to avoid tedious manual try gather together, using the pseudo-diagonalization method to obtain the pre-compensation matrix. On this basis, this paper introduced the genetic algorithm to the UAV coordinated turn control system, using the lateral spreads Angle feedback realize coordination control parameters of turning to set multivariable systems PID controller parameters. The simulation results show that: The proposed method to achieve good control effect.

**Keywords** UVA · Inverse Nyquist array · Genetic algorithm

## 21.1 Introduction

With the rapid development of the UAV system, its mobility greatly improved due to non-physiological limits, But it is also following coupling enhancement between its channel [1]. Furthermore, the increase of the comprehensive function

---

W. Zhao (✉) · M. Yang
General aviation key laboratory, Shenyang Aerospace University, ShenYang 110136, China
e-mail: 1621927296@qq.com

M. Yang
e-mail: ym1642@163.com

J. Yang · Z. Hu
Electronics and Information Engineering, Shenyang Aerospace University, Shen Yang 110136, China
e-mail: yangjun19880520@163.com

Z. Hu
e-mail: huzhanshuang@126.com

W. Du (ed.), *Informatics and Management Science V*, Lecture Notes in Electrical Engineering 208, DOI: 10.1007/978-1-4471-4796-1_21,

system also makes the coupling between the channel increase and the increase of control variables. Especially when there is a strong interaction between the various States, the design of single-variable is no longer appropriate. The multivariable control technology has provided an effective method for the complex systems control design [2].

Turning flight is a common flying mode of the unmanned helicopter. In the actual flight, the rolling and the yawing rotation are not independent, often producing large sideslip angle when turning. The existence of sideslip angle will cause the lateral acceleration, making the resistance increases and poor flying qualities, this is not conducive to the maneuvers. Therefore, to achieve coordinated turning needs to eliminate the sideslip angle, that is, minimum the coupling between the roll and yaw movement, realizing the coordinated turn. Coordinated turn is refers to the unmanned aerial vehicle to change the flight direction continuously in the horizontal plane, guaranteed the angle of slide is a zero, and can maintain the highly invariable one kind of curve condition.

There are three ways to achieve coordinated turning: The first method is use angle of slide feedback realization synergetic turn. Second, using curved lateral acceleration feedback to achieve synergy. The second is the use of the yaw angular velocity calculated value feedback turns together.

Due to the existence of the coupling between the channels, this paper using inverse Nyquist method to decouple the system, In order to avoid tedious manual try gather together, using the pseudo-diagonalization method to obtain the pre-compensation matrix. On this basis, this paper introduced the genetic algorithm to the UAV coordinated turn control system, using the lateral spreads Angle feedback realize coordination control parameters of turning to set multivariable systems PID controller parameters. The simulation results show that: The proposed method to achieve good control effect.

## 21.2 The Introduction of Inverse Nyquist Array

The inverse Nyquist array design method is used widely as a multi-variable frequency domain design method. The key to this approach is to pre-compensation control object, make its transfer function matrices become diagonal dominance of nature. On this basis, we can use a single variable system Nyquist diagram design method in each diagonal element. The method is applied to the inverse of the system transfer function matrix are particularly effective. Therefore, become the inverse Nyquist array design method. Control system diagram as shown in Fig. 21.1.

Where G($s$) represent the control object, $K_p(s)$ represent the advance compensator, they are generally added prior to the controlled object, Used to weaken the cross-linked between each loop of the control object. This makes the system of the transfer function matrix becomes a diagonal matrix or matrix of corner advantage. F represents the sensor's feedback gain matrix. It's typically a constant

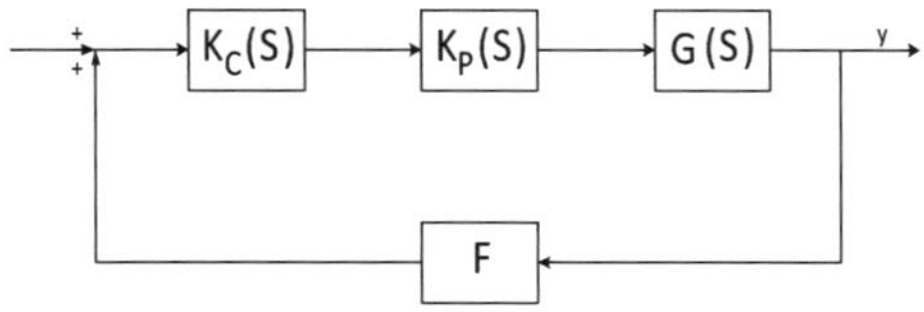


**Fig. 21.1** Control system decoupling chart

diagonal matrix. $K_c(s)$ is a dynamic correction of the compensator for each loop, it's also a diagonal matrix.

$$K_p(s) = G^{-1}(s)Q(s) \tag{21.1}$$

The compensation of the goal is to make G($s$) $K_p(s)$ turn into diagonal matrix or pseudo-diagonal matrix Q($s$). In this way we can decompose the multivariable system into a series of single-variable subsystem, the compensator transfer function matrix should be designed to be: $K_p(s) = G^{-1}(s)Q(s)$.

If you do not limit the complexity of the compensation, the diagonal dominance matrix can always realized. But for a complex array of pre-compensation will result in the engineering difficulties. In the project, weal ways want the pre-compensation array as simple as possible, preferably to be a constant matrix, or even a sparse matrix. For that reason and taking into account the characteristics of the process, we use the pseudo-diagonalization method for obtaining compensation array. That is, for a specific frequency $\omega$, getting a real constant matrix $K_p$, Make the squares of $K_p$ G($s$) non-diagonal elements as small as possible and the square of the modulus of the diagonal elements are equal to a constant. To achieve the goal of diagonal dominance matrix of $K_p$ G($s$), If cannot reach the diagonal dominance, and then change the frequency calculation again until satisfaction.

The Pseudo Diagonalization was first proposed by Method Hawkins in 1972, latter after Rosenbrock, Johnson et al. improved and the consummation, the approach has been widely used. Suppose in the frequencies of $jw_0$, the system transfer function matrix inverse Nyquist arrays are represented as:

$$\hat{g}_{ik}(jw_0) = \alpha_{ik} + j\beta_{ik},\ i,\ k = 1, \ldots, \mathrm{m} \tag{21.2}$$

Here m is the number of output variables, and assume that the number of input and output of the system is same. If you want to get an optimal compensation matrix kp, you can use the following steps:

1. Select a function of frequency point as jw0 and find the system of inverse Nyquist arrays $\hat{g}_{ik}(jw_0)$.
2. For each q value of (q = 1,..., M), constitute a matrix Aq which:

$$\mathrm{a}_{il,q} = \sum_{k=1,k\neq q}^{m} [\alpha_{ik}\alpha_{lk} + \beta_{ik}\beta_{lk}],\ i,\ l = 1, \ldots, m \tag{21.3}$$

3. Seeks the Aq matrix characteristic value and the characteristic vector, And the eigenvector corresponding to the minimum eigen value denoted by Kq.

4. From the above all q worth some of the smallest feature vector can make compensation matrix Kp.

$$Kp^{-1} = [k1, k2, \ldots, km]^T \tag{21.4}$$

This method is based on a frequency type, which frequency should aim at to design also need through the method gather together to finish.

Furthermore, we may carry on the weighting to some frequency band to realize the false diagonalization method in the basis of this method, Select N frequency point and assume that the R frequency weighting coefficient is $\psi_r$, Constructed Aq matrix as follows:

$$A_{il,q} = \sum_{r=1}^{N} \psi_r \Big( \sum_{k=1, k \neq q}^{m} [\alpha_{ik,r}\alpha_{lk,r} + \beta_{ik,r}\beta_{lk,r}] \Big) \tag{21.5}$$

So that you can enter the previous algorithm 3. Kp to calculating the pseudo-diagonal matrix Kp.

Kp is the dynamic compensator of each loop to meet the quality requirements of the dynamic. As a result of the constant matrix compensation, the mutual coupling of the channel is not completely eliminated. Therefore, this article uses the genetic algorithm to carry on the trundle and the direction channel PID parameter optimization.

## 21.3 Genetic Algorithm

Genetic algorithm is to adopt a "build and test" model, its basic operations including encoding, to generate initial population, fitness calculation, to judge whether meet the optimum conditions or not [3], return if not satisfied with the optimum conditions and via genetic operators (selection, crossover and mutation), and then return to calculate fitness of the initial population, followed by cycle. The following picture shows the flowchart of genetic algorithm. The steps are as follows:

- A certain amount of randomly generated initial population, calculate the fitness of each individual.
- According to certain rules to select the species, we usually use the roulette wheel method, select the population.
- Then, the population of the selected can generate new individuals through the pair wise crossover operation.
- In the new population's individual has the variation according to certain probability.
- Determining whether it meets the end optimized conditions, not met, go to the second Step.

## 21.4 Simulation Conditions and Results

A lateral movement of UAV linear model is:

$$\begin{bmatrix} \dot{\beta} \\ \dot{p} \\ \dot{\varphi} \\ \dot{r} \end{bmatrix} = \begin{bmatrix} -0.135 & 0.084 & 0.041 & 1 \\ -28.57 & -1.414 & 0 & -0.854 \\ 0 & 1 & 0 & 0 \\ 4.61 & -0.056 & 0 & -0.258 \end{bmatrix} \begin{bmatrix} \beta \\ p \\ \varphi \\ r \end{bmatrix} + \begin{bmatrix} 0 & -0.022 \\ -19.66 & -7.62 \\ 0 & 0 \\ -0.34 & -2.51 \end{bmatrix} \begin{bmatrix} \delta_a \\ \delta_r \end{bmatrix} \tag{21.6}$$

Of which: $\beta$—angle of side slip. $\varphi$—roll angle. $p$—roll velocity. $r$—yaw rate. The lateral spreads Angle feedbacks realize coordination control parameters of turning the setting [4, 5].

The above equation after the essential reorganization, withdraws two inputs and two output systems. Figure 21.2 is the decoupling control system Simulink simulation chart. By adopting the pseudo diagonal dominance compensated method obtain the compensation matrix.

$$K_p = \begin{bmatrix} 0.0054 & -1.0 \\ -0.064 & -0.99 \end{bmatrix} \tag{21.7}$$

Figures 21.3, 21.4 is on compensation system respectively before and after the Gershgor in compensation with inverse Nyquist figure. It can be seen from the

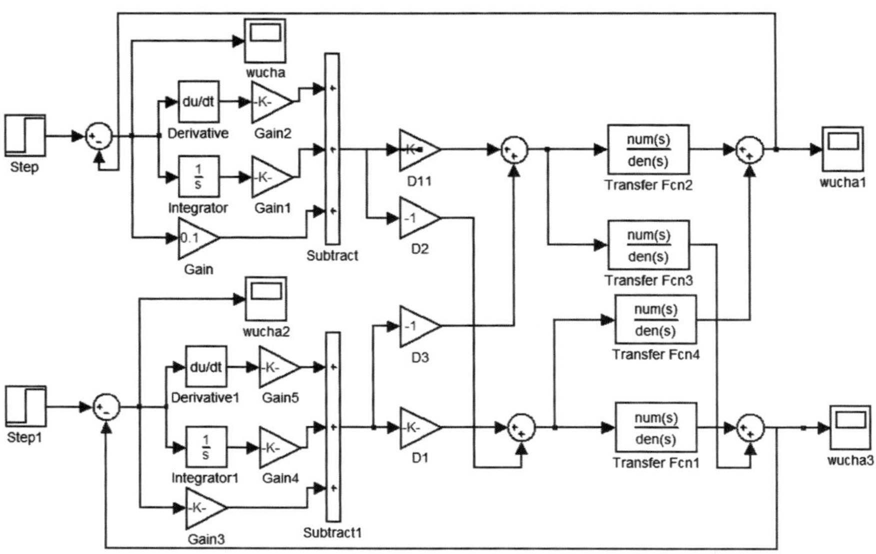


**Fig. 21.2** The decoupling control system Simulink simulation

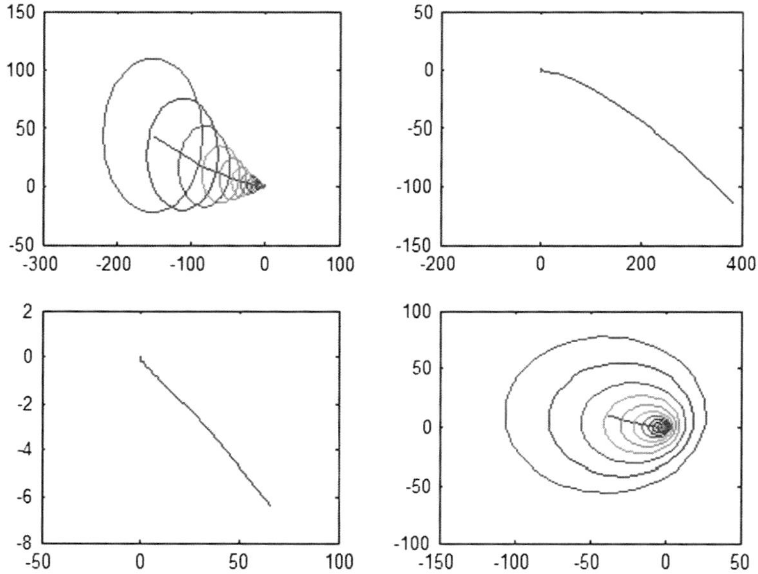


**Fig. 21.3** Compensation before the turn with a Gershgorin Nyquist figure

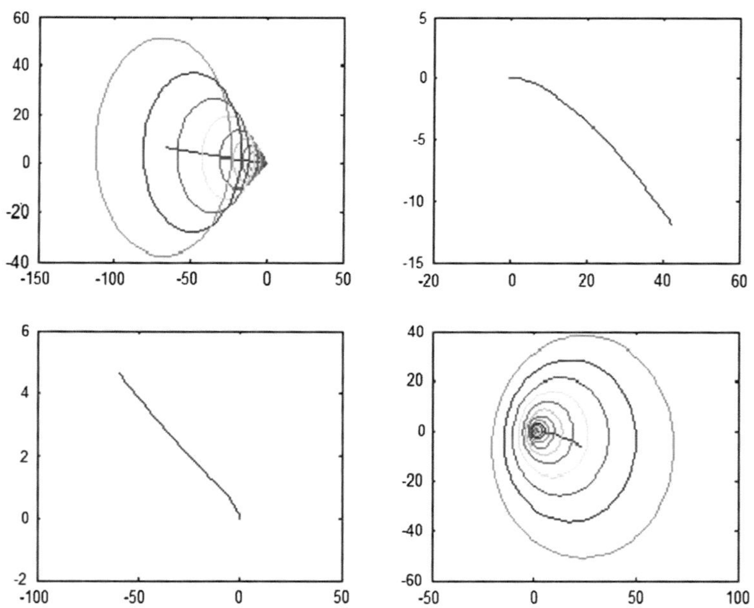


**Fig. 21.4** Compensation after the turn with a Gershgorin Nyquist figure

system transfer function matrix, the system has certain diagonal dominance, but there is also a strong coupling for UAV coordinated turn, decoupling needs on the system. In UAV coordinated turn control process require without position overshoot, for this reason, we must be coupled to the rear of the control system, make sure the systems have no overshoot and as far as possible in the shortest amount of time to reach steady state.

According to the characteristics of Dc motor position tracking system, this paper choose the rise time of system, combination of proportion of the steady-state error and overshoot and as a optimum target function. At the same time, in order to ensure the controller control effects, to reduce the oscillation of the system, introduce the oscillation frequency $\tau$ of the system, and multiply the penalty function $\lambda_3$. So the objective function eventually becomes:

$$J = \int_0^{\infty} \lambda_1 |e(t)| + \lambda_2 t + \lambda_3 \tau \tag{21.8}$$

Here, $\lambda_1$, $\lambda_2$–Weighting coefficients. e(t)—System error.

When optimizing, according to the control diagram and Constant coefficient matrix compensation Kp, Using genetic algorithm Setting the rolling angle control circuit and slide the corner circuit PID controller parameters. In this way, the Global Search optimization of the objective function becomes

$$J = K_1 \int_0^{\infty} \lambda_1 |e_1(t)| + \lambda_2 t_1 + \lambda_3 \tau_1 + K_2 \int_0^{\infty} \sigma_1 |e_2(t)| + \sigma_2 t_2 + \sigma_3 \tau_2 \tag{21.9}$$

In the formula, $K_1, K_2$-Roll angle for control loop weighting factor and slide angle control circuit of the weighting factor.

Selecting the parameter number of the species is 30, the iterative algebra is 500, overlapping and the variation probability respectively is pc = 0.8 and pm = 0.02, Obtaining the objective function optimal solution. Namely Fig. 21.3 is the genetic algorithm objective function most superior restraining curve. Under these conditions, the angle of slide and the angle of roll PID controller parameter respectively is: [4.5214 1.7621 1.4575], [4.5214 1.7621 1.4575]. Figures 21.5, 21.6, respectively be angle of roll and angle of slide response.

**Fig. 21.5** Roll angle of response

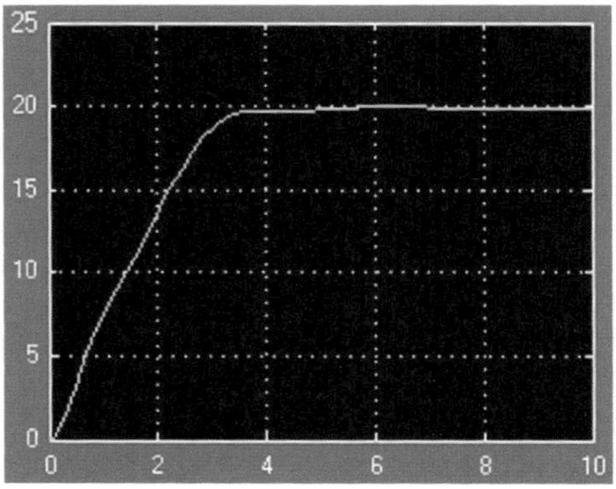

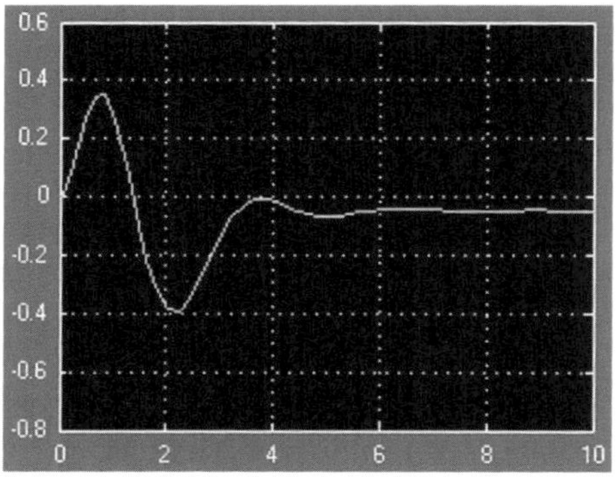


**Fig. 21.6** Sideslip angle of response

It can be seen: The angle of slide is finally controlled in the very small scope and the roll process within about 4 s. The system does not have overshoot and meet the requirements of the system.

## 21.5 Conclusion

In the process of unmanned aerial vehicle coordinated turn the transverse and Lateral Path coupling problem, This paper using inverse Nyquist method to decouple the system, In order to avoid tedious manual try gather together, using the pseudo-diagonaizing method to obtain the pre-compensation matrix. On this basis, this paper introduced the genetic algorithm to the UAV coordinated turn control system, using the lateral spreads Angle feedback realize coordination control parameters of turning to set multivariable systems PID controller parameters. The simulation results show that: The proposed method to achieve good control effect.

## References

1. Xue D (2009) Control system simulation and computer aided design, vol 2, Mechanic Industry Press, pp 235–245
2. Jinkun L (2003) Advanced PID control and MATLAB simulation. 12(44):89–109
3. Haupt RL, Haupt SE (2004) Practical genetic algorithms. 2nd edn. vol 12(4), Wiley, Pennsylvania, pp 102–110 (vol 2)
4. Li L (2009) Unmanned helicopter flight control method and GPS application research. China Agric Mechanization Res Inst 12(11):382–388
5. Tang Y (2000) Helicopter control system design, vol 3(54), National Defence Industry Press, pp 137–145

# Chapter 22
# Illegal Invasion of Computer Information Systems

**Kai Zhang**

**Abstract** This article describes the illegal invasion of computer systems, mainstream means of hidden Trojan horse property, describe the characteristics of several types of malicious code, and stressed the importance of computer information system security. Describe the basic idea of the invasion of the system means, mostly utilize the loopholes of the software vulnerability or system loophole, or inject malicious code to the target system, and then elevated privileges to obtain information, or undermine the data of the system. List several elements of the computer information system security, and stressed the importance of cryptographic algorithm as the core of secure communication protocols in computer communications. Explain the characteristics of public key cryptography, introduce RSA encryption algorithm as the example of it. Through the analysis of the security elements, hold out the methods of precautions.

**Keywords** Information security · Vulnerabilities · Malicious code · Security elements · Encryption algorithms

## 22.1 Introduction

In the past age without computer, the information security had been very seriously. How to get their own information security, storage, or transmission had always been very important. From steganography in ancient China, the password stick used by the Spartans, to German ENIGMA cipher machine widely used in the

---

K. Zhang (✉)
The Central Institute for Correctional Police, Baoding, China
e-mail: kaizhang11@163.com

W. Du (ed.), *Informatics and Management Science V*, Lecture Notes in Electrical Engineering 208, DOI: 10.1007/978-1-4471-4796-1_22,

World War II, without exception, has a pivotal position at the time. Computers, as today's information systems, once they are successful invasion, the loss is difficult to estimate, how to make our computer more secure also gradually being valued by the people [1]. The more secure operating systems, secure communications protocol. More powerful cryptography, more powerful and effective anti-virus software are also needed by people [2].

The scope of computer information security is a great deal. The scope can be form the national military and political secrets to the small business trade secrets, and the confidentiality of personal information. With the rapid development of the network, the Internet has become an integral part of the computer information systems. If a computer does not interconnect with the external, then the security can be greatly improved, while the performance will be facing a great trouble. So in a networked environment, how to create an information security system is the key point to ensure information security, this system include the security of the computer operating system, network communication using a variety of security protocols, security mechanisms (such as data encryption, message authentication, digital signatures, etc.) and other security components, anyone of the security vulnerabilities may pose a threat to the global, thereby causing huge losses [3].

With the rapid development of the network, attack tools can be easily downloaded from the internet [4]. The technical requirements of the invasion of the computer attacker gradually reduced and the boundaries between the new generation of network worms and hacker attacks, computer viruses become blurred, which also makes the threat of computer information systems means more, the frequency of attacks will be more. But these threats according to their nature can basically be attributed to the following aspects: Information leak, Undermine the integrity of information, Usurpation (unauthorized access), Denial of service: Eavesdropping, Counterfeit, Authorization the infringement: Repudiation: Computer Viruses [5].

## 22.2 Invasion Means

From the intruder's point of view, how to enter a computer information system is divided into active and passive. Active invasion usually specify the target of attack or scan the target software vulnerabilities, and then through the software or operating system vulnerability, access to computer information systems, access permissions, implants malicious code with elevated privileges and ultimately achieve the intrusion system to obtain information [6]. Passive mainstream invasion means of the web page linked to trojan, spam, binding, Trojan horses and other means, and over the net the way to invade.

### 22.2.1 Intrusion to Vulnerability

Usually a piece of software will have a lot of loopholes, which is why software needs on a regular basis the upgrade patch. Give a simple example: there is an attack method to the computer which install win 2,000, or xp with network share. In fact, this function is useless for most individual users. But it would be a very convenient hacking channel the operating system. If your operating system password is also weak passwords, it is more serious. In win 2,000, xp system provides an IPC $ function, the IPC $ (Internet Process Connection) is the resource of shared "named pipes", which allows communication between processes and open a named pipe, enter the correct user name and password, using the command net use url = file ://\\IP \ ipc $ \\ IP\ ipc $ "/user:" create an empty connection, the two sides can establish a secure channel and the channel encrypted data exchange, remote computer access. At first, Microsoft's developers in order to facilitate the administrator's management of the development of a channel, either consciously or unconsciously, resulting in reduced system security, because a lot of Windows users do not focus on the problems of password security.

We can downloaded a software called smbcrack on-line which use ipc $ to crack account passwords. If your password is short, while the password is relatively fragile, it is easy to crack. Considerable number of people set the administrator's password is set by similar 123456, or simply does not have a password. Use the command net use \\ ip \ ipc $ "password"/user: "user name" to establish a certain authority ipc $ connection. With the command copy trojan.exe \\ User ip \ admin $ Trojans server-side copied to the system directory. With the command net time \\ user-ip command to view the time of the other operating system, and then use the command at \\ User ip 12:00 trojan.exe let trojan.exe run at a specified time. So you get the administrator privileges, you can do whatever they want! Similar vulnerabilities, there are many, many web browser plug-ins such as vulnerability, and plug-in vulnerability will not disappear with the browser updating, the time of impact will be longer. Software vulnerabilities; firewall configuration error, using the default option to install the software may lead to software vulnerabilities exposed, and thus you may be intrude.

### 22.2.2 Malicious Code

History of malicious code can be traced back to 1988, Robert Morris wrote the Morris worm. Which caused over 6,000 computer shut down, the incident direct economic losses of $ 96 million, the threat of malicious code on the computer opened a prelude. Malicious code on the world along with the extensive application of the network is more serious. Diversified types of malicious code from the beginning of the worm in the form of attacks are also increasing, concealment, infectious, destructive, self-attack capability, along with the technology of the intruder gradually become more powerful.

People today have the vigilance of the malicious code has a lot of pricey, while compare with the means of the intruders are far from adequate, the intruder through various means to conceal malicious code characteristics. Trojans usually are exe extension, so when someone sends a file to exe suffix, you'll probably suspect, and refuse to accept this file. Many intruders change Trojan file name when they sent Trojan to someone, such as Trojan a.exe "changed its name to" a.jpg a can be tempting to name at the time of transmission, you can only see a jpg, successfully received, the windows default is to hide the type, when you think it is a picture when you open it, the Trojan run automatically. Advanced approach can also modify the file's icon makes the virus to hide better. Merger of the deception, the intruder will merger a useful application and a Trojan horse program, while make users caught in the case without prejudice to the original program. Intruder will use a software named joiner, this software is not only simple Trojans and application consolidation, can also program in the combined volume looks not so large and was found, while user to perform the Trojan, the program can send an icq intruder to notify the invaders, that you success. Malicious code can also be written in the type of dll or ocx file, and then hung with a large software, such as qq, the intruder will programme malicious code as a dll file stored in the qq installer. When some run the qq procedures, malicious code also performed. Almost no one will notice there is a redundant dll file written by an intruder, intruder can do a fine-tuning to make that the virus library can not find this malicious code, covering with an effective crust can escape the detection of anti-virus software. In addition to the above there are common several intruder means to distribute malicious code, there is a very common form of malicious code—page linked to horses.

The website that hang a horse is a website uploaded malicious code, and then use the Trojans Builder to generate a period of net Trojans, and then stored in the storage space of the site. When the user access to this page, the Trojan will run automatically. Usually this Trojan disguised as a page element. This Trojan usually in the user access the page automatically downloaded to the user's local computer; or script to run some com component, the vulnerability of these com component to download a trojan; or Trojan disguised as some missing components, such users will likely choose to download this component, but the browser is usually downloaded the component will automatically install and run it, this trojan can be successfully implanted, the intruder successfully infiltrate and control computer.

No matter using the vulnerability of the software to intrude computer information system, or implanting malicious code is usually work on software or operating system. In fact, the invasion of the hardware is actually. For example, to insert a write key logger code U disk or removable hard disk to the computer's usb excuse, computer keyboard input record which can be documented to achieve access to information and then the record can help someone to invade the computer. Or modify the bios of main board to inject malicious code, however, this invasion usually need to contact the hardware to prevent that. About a decade ago a virus spread via email, he uses the technology of the windows virtual device driver (VxD), this virus is difficult to be detected, and can spread under the windows environment. Get the final purpose of destroy the hard disk data, delete the data of some main board bois, and the code named by the abbreviation of the name, who lived in Taiwan named Chen Ying-hao (Chen Ing-Halu). With the update of the operating system, the CIH virus is also reduced, but the significance of CIH invasion of the computer is great. It created a precedent for the virus can invade your computer hardware to break the personal computer, it not only destroyed the idea that the software can not damage the hardware, but also to tell people the virus it is very dangerous for computer information system. Hardware viruses are relatively well-known, such as "Kriz" virus (w32.kriz), the variants of the virus (w32.hllm. Bymer), and the new type of CIH virus which can invade Windows 2,000/XP systems and destroy the data of the hardware. While these types of viruses gradually disappeared in the field of vision, the 360 company detected a virus named BMM, the virus can infect a computer main board's BIOS chip and hard disk Master Boot Record (MBR), then control Windows system files to load malicious code, so that the victims of the user regardless reinstall the system, format the hard disk, or even replace the hard disk can not eradicate the virus. The first choice of the virus, infection of the main board BIOS; work through the BIOS to infect the hard disk, use MBR to infect Windows system files; at last it download a large number of Trojan and other malicious programs on internet, to achieve the purpose of the computer intrusion. Figure 22.1 is the BMW virus invasion step diagram made by 360 company.

### 22.2.3 Communication Protocol

Various operating systems, computer information systems with different hardware architecture, the communication between them is built on the OIS 7 layer network model. Through a variety of different network protocols to communicate with each

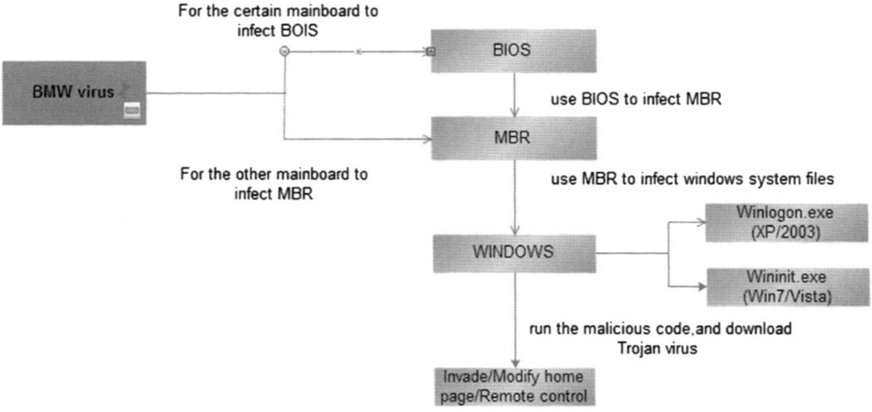


**Fig. 22.1** The BMW virus invasion step diagram

other. In the communication protocol the sender and the recipient need to follow certain rules or communication standards, in order to achieve a predetermined safe or credible goal. The communication protocol can be a program or software. The sender through this program or software package, encoding, encryption and other operations, as the same, the recipient use the communication protocol specification to unpack, tide code to decrypt.

For security, the entire communication protocol is encrypted by some methods, so once the communication protocol is invaded successfully, or the key to decipher the entire communication is decoded by the attacker, huge losses might happen.

Network communication protocol is not uniform, according to the different needs of the communication protocol is different, so whether we are using what kind of communication protocol, we all hope it is the safe and reliable. Whether an agreement is safe require to go through the verification. Validation methods include ban logic analysis method, kailar logical analysis, the strand space analysis method, the CSP analysis method, the Model Checker analysis methods. By the logic of authentication, this communication protocol in logic can be safe, and whether it is real security should be the core of the protocol is strong enough, that is, its encryption algorithm.

An information security expert once said: "computer security policy, the encryption is the core; security protocol bridge; security architecture is the foundation; secure operating system is the key; the evaluation of security attacks is a test." Visiblely cryptography is the importance of the Information Security component.

The password of the computer information system types can be divided into block ciphers, stream ciphers, public key cryptosystems. Public key cryptography, known as dual-key cryptosystems and asymmetric cryptography. Public key cryptography, one-way trapdoor function is the core, one-way trapdoor function is a function of the following conditions are met is the function $f(x)$ satisfies the following conditions:

(1) Given $x$, computing $y = f(x)$ is easy;
(2) To set a given $y$, calculate $x$ of $y = f((x))$ set up is difficult. (The so-called computing $x = f^{-1}(Y)$ difficulty is computationally quite complex; calculation to arrive at a correct result of no practical significance).

RSA algorithm is the most famous public key cryptography algorithm, RSA public key algorithm is invent by Rivest, Shamir and Adleman in 1978, the mathematical basis of the algorithm to the Euler (Euler) theorem in elementary number theory, and built on the conclusion that it is difficult to decompose a integer large number. $Z/(n)$ is expressed as $Z$ where $n = pq$; $p$, $q$ and is a prime number and different from each other. If $Z * n \equiv \{g \in Zn|(g, n) = 1\}$, it is easy to see the order of the multiplicative group $Z * n$ is $\varphi(n)$ and $g\varphi(n) \equiv 1 \pmod{n}$ and $\varphi(n) = (p-1)(q-1)$. The description of the cryptosystem RSA cryptosystem is as follows:

(1) key generation

Select $p$, $q$ distinct primes, calculate $n = p \times q$, $\varphi(n) = (p-1)(q-1)$, choose an integer e so that $(\varphi(n), e) = 1, 1 < e < \varphi(n))$, then calculate d which $\mathrm{d} = \mathrm{d} = \mathrm{e}^{-1} \,(\mathrm{mod}\, \varphi(\mathrm{n}))$, public key $P_k = \{e, n\}$; public key private key $S_k = \{\mathrm{d}, \mathrm{p}, \mathrm{q}\}$. When $0 < \mathrm{M} < \mathrm{n}$; $M^{\mathrm{k}\varphi(\mathrm{n})+1} \equiv$ of $M \,(\mathrm{mod}\, n)$, and ed $\equiv 1 \,(\mathrm{mod}\, \varphi(n))$ is easy to see $(M^e)^d \equiv M \,(\mathrm{mod}\, n)$.

(2) encryption (use e, n):

Plaintext: $M < n$ ciphertext: C $M^e = (\mathrm{mod}\, n)$.

(3) To decrypt (using d, p, q).

If the ciphertext $C$, then the plaintext $\mathrm{M} = C^d \,(\mathrm{mod}\, n)$ can get the plaintext. Public key $e$ and $n$ is open to everyone. They can use e and n to encrypt the plaintext which they want to sent to the owner of the public keyhe owner, while the intruder is unable to obtain d, even if intercepted by a Confidential can not decipher the ciphertext, so that the information conveyed is also safe. In addition to the RSA cryptosystem, AES cipher used by the U.S. government, The NESSIE European password program, DES password, the RC4, RC5, RC6, MD5 and other cryptographic algorithms. By these cryptographic algorithm our communication protocols have been protected.

## 22.3 Means of Defense

Whether loopholes in the system or software vulnerabilities requires that we need habitually updated software and operating systems, patches on time, so as to avoid the intruder exploited into our computers. We should ensure that the antivirus software should be updated and timely opened, avoid to download untrusted software or to accept programme send by strangers, do not open unknown e-mail, properly configured firewall. Try not to browse the pages of insecurity, to use high-strength password. Using the appropriate communication protocol, close some of the unnecessary network interface to avoid intruders.

Secure operating system research and development will greatly improve the security of computer systems, the basic elements of a secure operating system, including confidentiality: ensure that the information is not exposed to unauthorized entities or integrity of the process. Integrity: Only allowed someone to modify the data and distinguish whether the data has been tampered. Availability: someone is authorized to access the data, while, an attacker can not occupy all the resources to obstruct the normal work of authorized. Controllability: It can control the flow of information within the scope of authorization. Verifiable basis and means of investigation: Provide the basis for the investigation and method to the network security issues. The growth of security inevitably will lead to the reduction of operating convenience, to find the balance point between these two is still the focus, after the secure operating system widely used of the safe operation of the system development is still a long way to go.

## 22.4 Summary

Means to invade computer information systems are growing with the enhancement of the hacking techniques; it is unable to avoid that computer gradually moving towards the cloud from the localization. The trend of wired communication will eventually be replaced by wireless communications, compared with wired communications, wireless communications are more likely to be invade. This paper briefly describes the principles and the methods of computer information systems incursion, and describes the security measures to prevent incursion and the encryption algorithm used in the communication protocol. Wired communication with correct configuration and use safe software we can avoid most of the computer intrusion. However, wireless communication will face more threats in the future, we still need to remain vigilant and continue to improve our computer technology to defense against the invasion.

## References

1. PiJu L (2008) RSA encryption algorithm, vol 1, Silicon Valley, pp 22–23
2. Dagan L (2010) Hazards and analysis web page malicious code, vol 21, Silicon Valley, pp 176–177
3. Jianing W (2009) Hazards and analysis of web page malicious code intrusion. Knowl Econ 15:106–107
4. Tao S (2011) Invasion and clear technologies of Trojan. Inf Comput 11:61–63 (theory edtion)
5. Yuchen W (2001) Principle of system vulnerabilities and common attack methods. Comput Sci Technol 3:62–64
6. Jian X (2011) Analyse the prevention system of intrusion. Comput Knowl Technol 3(2):98–107

# Chapter 23
# Research on Shift Sequence Code in Barcode Positioning System

**Weijun Zhang and Dongli Li**

**Abstract** Aiming at the disadvantages of the traditional positioning technology, barcode positioning system was introduced in this paper. Based on shift sequence code, a novel encoding method was put forward by comparing varieties of encoding principles domestically and abroad, as well as a systematic research on its encoding rules and positioning mechanism. We applied backtracking method to solve the coding collection and the barcode verification to check that whether the codes are legitimate. Finally, One-dimensional barcode image and recognition are realized with the specific examples. The feasibility of the algorithm is confirmed through relevant experiment results.

**Keywords** Shift sequence code · Positioning · Encoding · Algorithm

## 23.1 Introduction

At present, the common positioning methods of the transportation system are in two ways: One is positioning with laser or ultrasonic [1], whose advantage is that positioning accuracy and speed, but it only works on a straight line; and the other is positioning with the rotary encoder, whose advantage is can be positioned in the curve direction, but its accuracy is poor, and has the cumulative error. In view of these phenomena, the introduction of barcode positioning system can perfectly solve the above problems. The barcode positioning system is a new measurement and positioning system, which is evolved from the large-scale logistics

W. Zhang (✉) · D. Li
School of Computer Science, Shenyang Aerospace University, Shenyang 110000, China
e-mail: liovzhe@126.com

W. Du (ed.), *Informatics and Management Science V*, Lecture Notes in Electrical Engineering 208, DOI: 10.1007/978-1-4471-4796-1_23,

transportation system, is a breakthrough of the traditional positioning technology, and represents the development direction of positioning technology in the modern large-scale transportation system.

In this paper, shift sequence code is applied to barcode positioning system, and we have a systematic research on its encoding rules and algorithm. Finally, one-dimensional barcode image and recognition are realized with the specific examples in Visual C++ 6.0 programming environment, which has an excellent practical value.

## 23.2 Basic Principles of Barcode Positioning System

The barcode positioning system is composed of the barcode reader and the barcode tape, etc. It works with the barcode reader installed on the robot, and the barcode tape installed on a walking track. When the robot is walking on the track, the code reader scans the current barcode constantly, and outputs the robot's current location information through the built-in decoder.

The overall block diagram of the barcode positioning system is as shown in Fig. 23.1. This system is mainly composed of the barcode tape, the barcode reader and the controller. The barcode tape has printed a number of fixed-length barcodes with certain rules. The barcode reader consists of barcode scanner, signal shaping circuit and decoder. Laser emitted from the laser diode scans the barcode, diffuse-reflected light from the barcode is absorbed by the photoelectric converter, and the reflected light signal is converted into the corresponding electrical signal. The electrical signal is sent to the amplifier to enhance the signal, and then to the shaping circuit to convert into digital signal. The decoder distinguishes the number of bars and spaces by measuring the number of digital signal 0/1, and distinguishes the width of bars and spaces by measuring the duration of digital signal 0/1. According

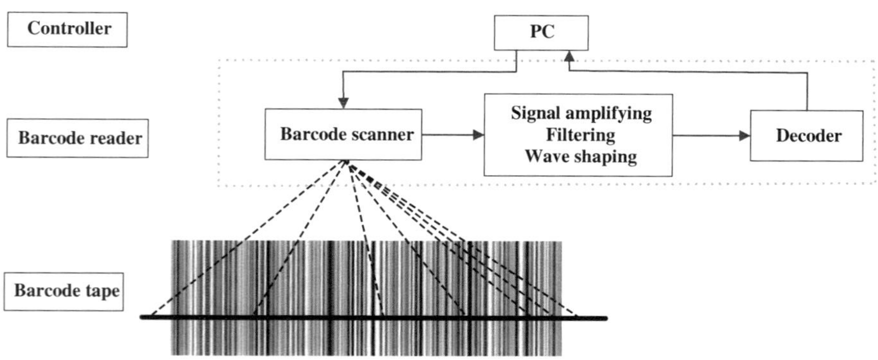


**Fig. 23.1** The schematic diagram

to the encoding rules, the signal combination of the bars and spaces is converted into location data. Finally, the decoded location data are sent to the controller through the interface.

## 23.3 Encoding and Positioning Principles

Based on shift sequence code, a novel encoding method is applied to the barcode positioning system by comparing varieties of encoding principles domestically and abroad. The shift sequence code consists of 0/1 with equal digit-length, and any of its subsequent code can be regarded as its precedent code adds 0/1 after shifting one digit [2]. However, adding 0/1 depends on whether the coding collection satisfies the unique principle of encoding, that is, the duplicate codes are not permissible in the coding collection.

For the shift sequence code with digit-length n, the consecutive (n−1) are the same between two adjacent codes. If the adjacent codes are overlapping arranged according to the same parts, it will get an overlapping sequence.

A code and its right neighbor constitute the position of nominal value, which means that the distance between the encoded and the original code.

**Theorem** *For the shift sequence code with digit-length n, its maximum range of encoding collection is* $2^n$, *and maximum overlapping sequence length is* $2^n + n - 1$.

*Proof* Here we apply mathematical induction to prove it.

Without loss of generality, let the first code be $\underbrace{11\cdots111}_{n}$ and the range of encoding collection be $k(k \le 2^n)$.

1. When $k = 1$, its maximum overlapping sequence length is digit-length n, that is, $n = k + n - 1$. When $k = 2$, its maximum overlapping sequence length is $k + n - 1 = n + 1$, and its maximum overlapping sequence is $00\cdots01$, which has two proper codes, namely $00\cdots0$ and $0\cdots01$. Obviously, it satisfies the unique principle of encoding.
2. Assuming that the coding collection satisfies the unique principle of encoding when $k \le 2^n$, the theorem will be proved if it also satisfies the unique principle of encoding when $k + 1 \le 2^n$.

Here we apply proofs by contradiction to prove its existence of $k + 1$. Suppose there are k different codes in the coding collection, and the k-th code can be denoted as $0\rho$ or $1\rho$ ($\rho$ consists of 0/1), then the next code should be $\rho 0$ or $\rho 1$. If the (k+1)th code violates the principle, which means that the coding collection already contains $\rho 0$ and $\rho 1$, as well as its precedent code is $0\rho$ or $1\rho$. If their precedent codes are the same, that is, $0\rho$ or $1\rho$, then there must be the duplicate code among the first k codes. This contradicts the initial assumption. If their precedent codes are different, that is, $0\rho$ and $1\rho$, then either of them must be duplicated with the k-th code, it means that there must be duplicate code among the first k codes. This also contradicts the initial assumption. Therefore, for the shift sequence code, if the first k codes satisfy the unique principle of encoding, then there must be (k + 1)th code that satisfies the unique principle of encoding.

For the shift sequence code with digit-length n, it only has 0/1 for every digit, and then the maximum number of encoding is $2^n$. So the maximum value of $k$ is $2^n$. The maximum overlapping sequence length is $2^n + n - 1$ when $k = 2^n$.

Theorem to be proven!

## 23.4 Encoding Algorithm of Shift Sequence Code

Let the maximum overlapping sequence be a vector $(x_1, x_2, \ldots, x_k, \ldots, x_m, x_{m+1}, \ldots, x_{m+n-1})$, Where n, as described above, is the digit-length, $m = 2^n$, $k < m$ and $x_k$ is 0 or 1. Any of $(x_i, x_{i+1}, \ldots, x_{i+n-1})$ and $(x_j, x_{j+1}, \ldots, x_{j+n-1})$ in the vector must satisfy $x_{i+p} \neq x_{j+p}$, where $1 \leq i, j \leq m$, $0 \leq p \leq n - 1$, which is the solution to the constraints. If there are duplicate codes among the first k components in the vector, the constraints will not be met, and then the entire solutions will not be set up. Therefore, this problem is of completeness. As is known to all, the problems with completeness can be available with backtracking method [3].

To solve the coding collection, in fact, is to define all the subsequent codes. The main idea of this algorithm is as follows:

Without loss of generality, let the first code be full-1-code $(\underbrace{11 \cdots 111}_{n})$.

Firstly, adding it to the coding collection and then shift left with one digit (remove the rightmost digit).

Secondly, adding 1 to the rightmost. And check that whether the new code is a duplicate code. If not, adding it to the coding collection, then shift left again. If it is, the rightmost digit is replaced with 0, and shift left again. If the rightmost digit is already 0, and it is a duplicate code, then it is necessary to reprocess the precedent code.

Finally, take out the latest code from the coding collection as the current code (backtracking), and remove it from the coding collection. Examining the rightmost digit of the current code, if 1, then replaces it with 0; for 0, it is necessary to reprocess the precedent code (backtracking again).

So shift left until all the $2^n$ codes have been generated. Algorithm flow chart is as shown in Fig. 23.2.

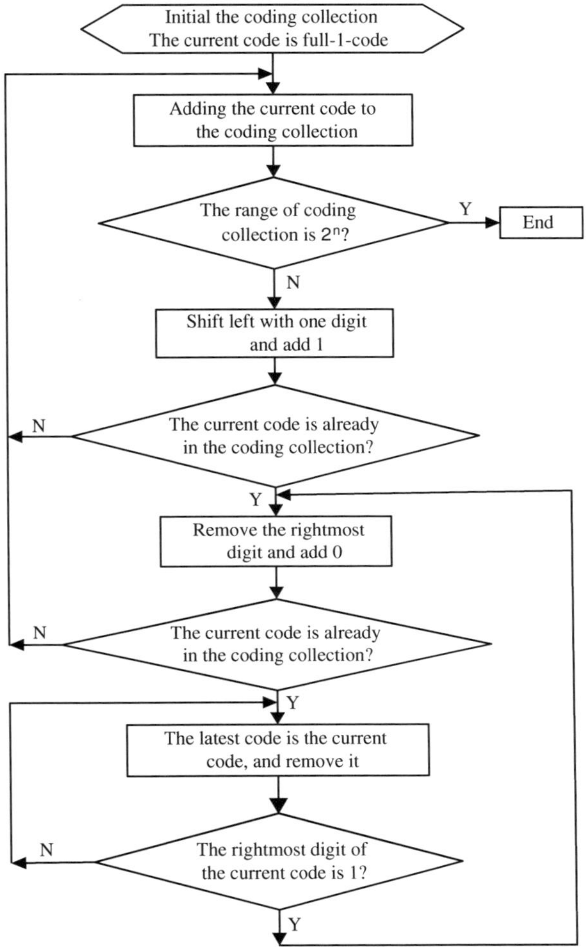


**Fig. 23.2** The block diagram of algorithm

## 23.5 Barcode Verification

Barcode verification is compared with the codes of the coding collection, so as to check that whether the codes are legitimate [4]. In order to program, we introduce a variable flag. If the code is legitimate, the flag value is 1, or else the flag value is 0. The flag value controls the program process, when illegitimate code appears, then out of loop and the followed codes are no longer verified. Therefore, if the loop can achieve the last code, indicating that the barcode is legitimate. The specific flow chart is as shown in Fig. 23.3.

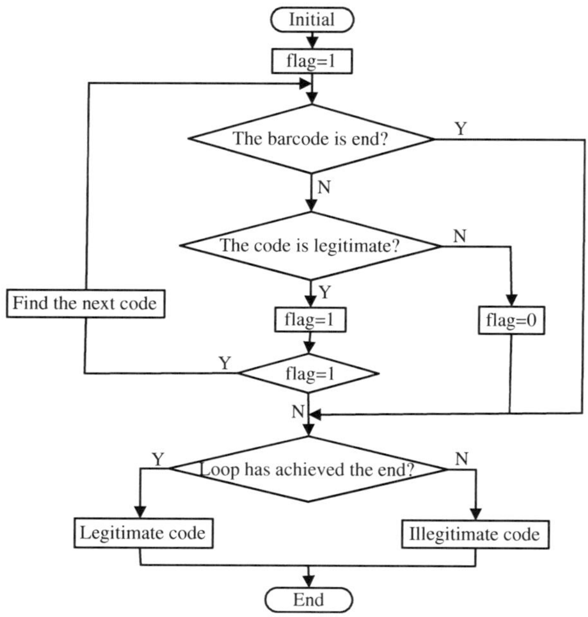


**Fig. 23.3** The block diagram of verification

# 23.6 Experimental Results

Take n = 4 and the first code is 1111 for example. According to the above algorithm, the codes and their corresponding coding location and numeric characters (decimal) are as shown in Table 23.1. With the codes connected in turn, it will get the maximum overlapping sequence is 1111011001010000111, and each of the four adjacent 0/1 digits constitute a code. According to the rule that 1 position to draw a straight line and 0 position with a space, the entire barcode pattern is as shown in Fig. 23.4, and it can be divided into 16 areas in terms of the order of codes.

To verify the rationality of the algorithm, we compare the performance of the proposed algorithm with the traditional method, such as [4] (referred to as Jiang) and [5] (referred to as Zhang et al). As shown in Fig. 23.5, one-dimensional barcode image is realized in Visual C++ 6.0 programming environment.

To identify the barcode in Fig. 23.5, at least four barcodes are necessary in the laser scanning field of views. If there are four or more, the leftmost four barcodes are the sampling reference, the area number is N. Define that the initial code group

**Table 23.1** The code table connected with coding location

| Location | 0 | 1 | 2 | 3 | 4 | 5 | 6 | 7 | 8 | 9 | 10 | 11 | 12 | 13 | 14 | 15 |
|---|---|---|---|---|---|---|---|---|---|---|---|---|---|---|---|---|
| Code | 1111 | 1110 | 1101 | 1011 | 0110 | 1100 | 1001 | 0010 | 0101 | 1010 | 0100 | 1000 | 0000 | 0001 | 0011 | 0111 |
| Decimal | 15 | 14 | 13 | 11 | 6 | 12 | 9 | 2 | 5 | 10 | 4 | 8 | 0 | 1 | 3 | 7 |

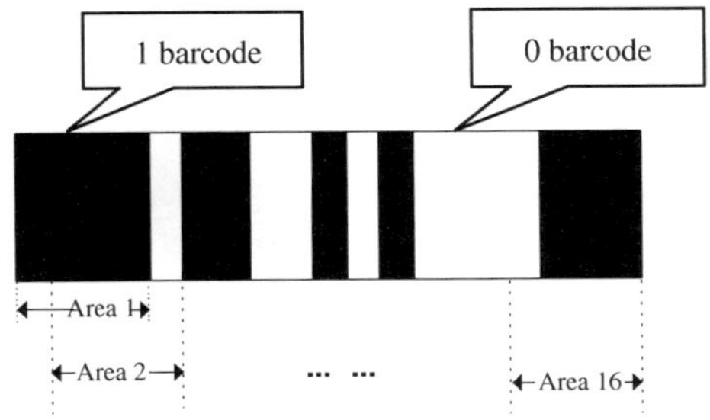


**Fig. 23.4** Coding pattern

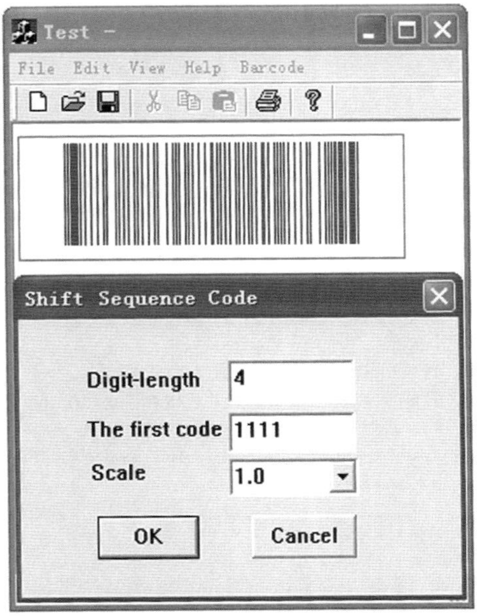


**Fig. 23.5** Realization of one-dimensional barcode

is 1111, and the absolute zero position, that is, the reference position is the centric position of 1-code among the initial code group. Here the reference position is the q-th 1-code (q = 1–4).

Through the edge detection to distinguish the boundaries of the barcode [6], so as to identify the area numbers. Assuming that the barcode scanned by laser contains a code group 1001, we look up Table 23.1 to find that N = 6. After defining the area number, the reference position and error analysis of correction will be completed in the next test.

Because of the relatively low measurement accuracy, laser scanning usually cannot satisfy the requirements for the measurement of certain high-precision pares. Through analysis in principle and lots of experiments, a BP neural network model can be established to realize soft-ware compensation for the measurement error [7].

## 23.7 Conclusions

In this paper, a novel algorithm based on shift sequence code is proposed in the barcode positioning system. Besides, we derive a theorem and give its concrete proof. Compared with the other barcodes, the proposed algorithm is more reasonable. Furthermore, the simply structure makes its recognition convenient. After preliminary tests, the barcode based this new algorithm has satisfactory performance. Therefore, the barcode positioning system with this barcode is suitable for promotion and has a wide application.

## References

1. Hueber G, Ostermannn T, Bauernfeind T, Raschhofer R, Hanelauer R (2000) New approach of ultrasonic distance measurement technique in robot applications. In: 5th international conference on signal processing proceedings, 12:2066–2069
2. Zheng H, Lin C (2003) Research and application of theory for shift sequence code. Acta Metrologica Sinica 24(1):29–31
3. Xiong J, Jia P, Liu J (2009) Research on coding for photoelectric shaft encoder based on image sensor. Measur Control Technol 28(12):6–9
4. Jiang L (2012) The barcode location and recognition based on run-length coding. Electron World 3:16–19
5. Zhang X, Wang Z, Li X (2006) Positioning algorithm of a new period barcode ruler. Measur Sci Technol 17(4):126–128
6. Wu Y, Zheng J (2010) Bar code location by extracting edge Lines. J Comput Appl 30(5):1246–1250
7. Ping X, Gong Y, Wang W (2008) Error analysis and compensation for laser scanning based on BP algorithm. Mech Sci Technol Aerosp Eng 27(100):1127–1131

# Chapter 24
# A New Dynamic Scheduling Method for Networked Control Systems

**Feng Du, Xiaoyu Zhang, Zhi Lei, Jia Ren, Cheng Guo and Jinyu Li**

**Abstract** For networked control systems with limited network bandwidth, the conflict will reduce control performance of the system. This paper puts forward a new dynamic scheduling. This method guarantees the quality of control (output is into its steady state value: −5 to 5 % range) as the goal, and set deadband in the controller. It bases on the two parameters: error and error change rate adaptive to adjust the network load. Lastly, do the simulation based on CSMA/CD network simulation. The results verify the proposed method can improve the utilization rate of the network bandwidth, to improve the quality of control performance, enhance the stability of the system.

**Keywords** Networked control systems · Deadband scheduling · Dynamic scheduling

## 24.1 Introduction

Networked control systems (NCS) are real-time feedback control systems with network used for communication between sensors, controllers, actuators and plants. In recent years, the researches of NCS have made some achievements on the assumption that it has known the network condition, through the design of control strategy to make up for the effects of these problems. But, because of the performance of the NCS is not only affected by the quality control strategy, but also the influence of network scheduling [1]. So, it is necessary to design efficient

F. Du (✉) · X. Zhang · Z. Lei · J. Ren · C. Guo · J. Li
University of Hainan, Haikou, Hainan, China
e-mail: dufeng488@sina.com

W. Du (ed.), *Informatics and Management Science V*, Lecture Notes in Electrical Engineering 208, DOI: 10.1007/978-1-4471-4796-1_24,

and reasonable scheduling strategy to optimize the whole control performance of the NCS.

At present, the research of scheduling method can divided into the following as: learning from the CPU scheduling; based on priority scheduling [2, 3]; based on the sampling period scheduling [4, 5]; based on bandwidth scheduling [6, 7] and based on deadband scheduling. Above 5 methods, learning from the CPU scheduling theory is proposed earlier, and it includes the classic Rate Monotonic (RM) algorithm [8], Earliest Deadline First (EDF) algorithm [9] and Most Error First-Try Once Discard (MEF-TOD) algorithm [10]. These scheduling methods provide the theoretical basis for other scheduling method.

Otaneg [11] proposed the dynamic deadband firstly in 2002. This method considers the before and after data is sent to the network may be very close, and it won't send the after data and keep the before data. So it can reduce the quantity of data access to the network, and ease the network load. Tang [12] puts forward a scheduling which combines the deadband scheduling with the priority scheduling method. The method enhances the NCS ability to deal with the change of load. Zhang [13] proposes the deadband scheduling and predictive control collaborative design method and it lower the limit cycle of vibration phenomena.

But, document [11] only explores primarily to use deadband scheduling in NCS. The method only uses error (e) to adjust deadband, and not research the bigger overshoot of output when given value has step change. The document [12] puts forward deadband scheduling based on CAN-bus network, each loop shares the same network unfairly and need to discuss priority problem. Also, the method does not research the bigger overshoot of output when given value has step change. Although, document [13] uses e and error change rate (ec) lower the limit cycle of vibration, the condition of sending packets is not enough perfect. Furthermore, from document [11] to document [13], the network does not have random disturbance, so they ignore the uncertainty of network and narrow the scope of application of deadband scheduling.

In view of the above problems existing in these documents, this paper presents a new dynamic scheduling method, in order to ensure that the quality of control (output is into its steady state value: $-5$ to 5 % range) as the goal, according to e and ec, set deadband in the controllers node. The method can reduce the network conflict, increase the control of quality, and lower the overshoots of the output when the given value have step change, enhance the stability of the system. This paper use truetime1.5 to do simulation in CSMA/CD network, each loop shares the same network fairly and it is unnecessary to consider the priority problem. Meanwhile, the simulation adds the random disturbance; it expands the scope of using the deadband scheduling. Lastly, the simulation results also show that the presented method can obviously reduce the network conflicts, ensure the control of the quality, and improve the network bandwidth utilization, enhance the stability of the system. At the same time, the method is simple and easy to realize.

## 24.2 Basic Idea of Deadband Scheduling

In order to guarantee stability of system, deadband scheduling actively discards some data packets which send to the network, and reduce the network conflict.

The basic idea is as follows:

1. If $|X - X_{sent}| \geq \delta$, send $X$ and assign the value of $X$ to $X_{sent}$;
2. If $|X - X_{sent}| < \delta$, not send $X$.

In formula, $X$ is the data packet which is ready to send to network in this time. $X_{sent}$ is the data packet which send to network at last time. $\delta$ is threshold constant.

## 24.3 A New Dynamic Network Scheduling

Although the traditional dead zone scheduling method defines the rules of sending data packets, the definition of is not clear, especially not consider the control performance of multi-NCS in the same network. Therefore, in order to guarantee the quality of control of every NCS, it requires deeply researching the rule of sending data packets.

At present, most of the scheduling method has its own limitations. For example:

Aim at multi-NCS sharing the same network which bandwidth is limited. This paper guarantees the quality of control (output is into its steady state value: $-5$ to 5 % range) as the goal, and set deadband in the controller, consider $e$ along with $ec$ to decide the rules of sending data packets as follows:

1. If $|e| < \delta$ and $|ec| < \gamma$, not send data packets. This rule indicates that the system has been stable, so controller does not need sending data packets to actuator.
2. If $|e| \geq \delta$ or $|ec| \geq \gamma$, send data packets. This rule indicate that controller need to send packets for enhancing control effect and guaranteeing the NCS finish the transition state as soon as possible.

This paper regards output into its steady state value: $-5$ to 5 % as the goal, set $\delta$ as 0.05. Stands for size of ec. Consider the value of e and ec together, and judge whether the system entry the steady state or not. Commonly set $\gamma$ as 0.025.

## 24.4 A New Dynamic Network Scheduling

This paper use truetime1.5 to do simulation research.

Choose 3 NCS sharing the same network. Choose 3 controlled objects of NCS are $G_1(s) = 100/(s + 45)$; $G_2(s) = 100/(s + 60)$; $G_3(s) = 100/(s + 75)$.

Network parameters setting as follows: network type is CSMA/CD. The network bandwidth is 55 kbit/s. The data packet loss rate is 0.

Each control loop consists of three nodes: sensor (time drive), controller (event driven) and actuator (event driven). Each node internal scheduling algorithm use RM scheduling algorithm. The whole network also contains a interference node (it is used for simulating addition node take up the bandwidth expect the three control loop). Interference node (time drive) take up network bandwidth is set to 43 %.

Each controller in NCS use PI control, the parameters of loop 1 to loop 3 is set as follows: $K_{p1} = 0.001511$, $K_{i1} = 0.046169$; $K_{p2} = 0.001611$, $K_{i2} = 0.077328$; $K_{p3} = 0.001711$, $K_{i3} = 0.091215$.

From loop 1 to loop 3, the sampling period of sensor are set to 11, 12, 13 ms respectively.

Reference is square wave signal which amplitude is 2. The changes range is −1 to 1.

This paper sets $\delta$ for 0.05 (output is into its steady state value: −5 to 5 % range), and sets $\gamma$ as 0.025.

In simulation, all of the above parameters always are the same.

### 24.4.1 Not Use Any Scheduling Strategy

1. The output of 3 loops show as Figs. 24.1, 24.2, 24.3.

From Fig. 24.1, $y_1$ has bigger vibration from 4.500 to 5.500 s and from 7.900 to 8.700 s, and overshoots are more than 10; the quality of control is poor. It does not meet the control requirement.

From Fig. 24.2, $y_2$ begins to vibrate even diverge in 5.500 s later.

From Fig. 24.3, $y_3$ has bigger vibration even overshoots at 2.000–2.5000 s, 4.000–4.300 s, 5.300–5.900 s, and 7.200–7.600 s.

In short, from Figs. 24.1, 24.2, 24.3, $y_1$, $y_2$, *and* $y_3$ do not meet the requirement of the control quality of 3 loops at different times.

2. The time delay from controller to actuator show as Figs. 24.4, 24.5, 24.6.

From Fig. 24.4, the maximum $\tau_{ca1}$ is 0.998 s has more than 90 sampling periods.

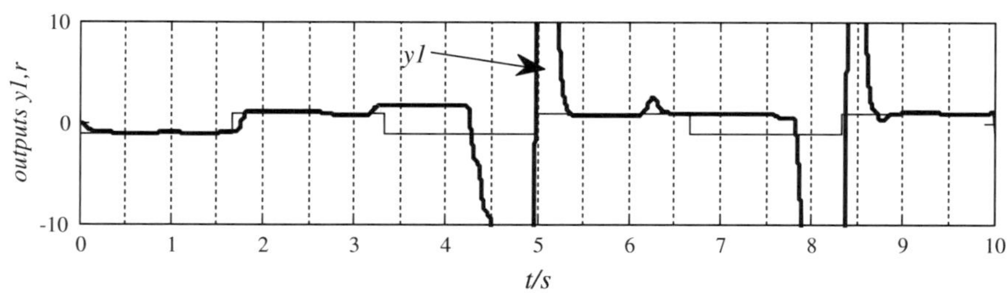


**Fig. 24.1** The output $y_1$ of loop 1

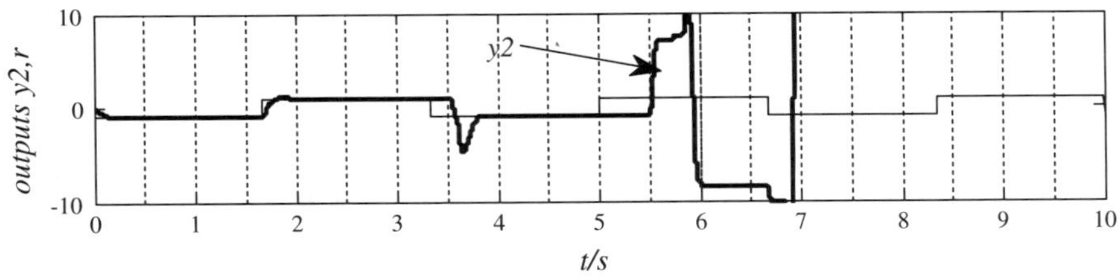


Fig. 24.2 The output $y_2$ of loop 2

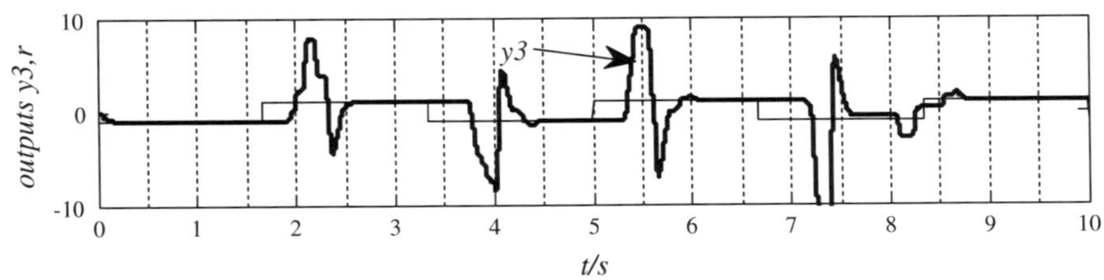


Fig. 24.3 The output $y_3$ of loop 3

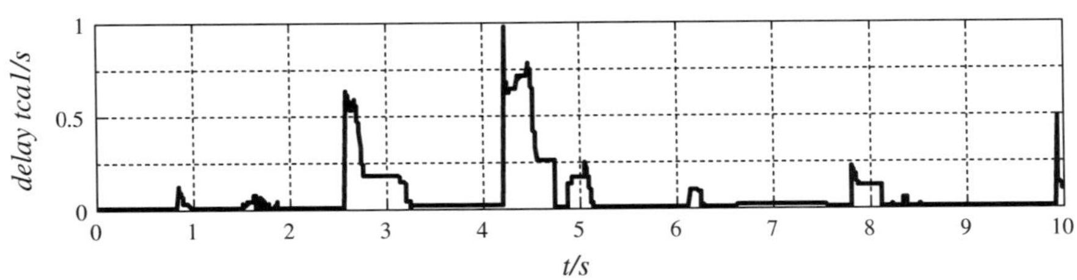


Fig. 24.4 The time delay $\tau_{ca1}$ from controller to actuator (loop 1)

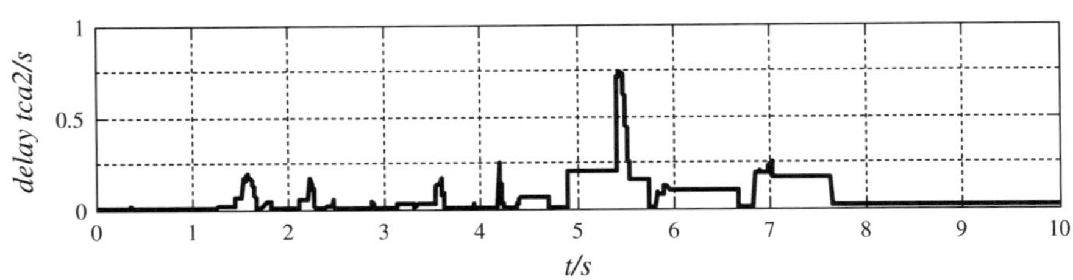


Fig. 24.5 The time delay $\tau_{ca2}$ from controller to actuator (loop 2)

From Fig. 24.5, the maximum $\tau_{ca2}$ is 0.730 s has more than 60 sampling periods.

From Fig. 24.6, the maximum $\tau_{ca3}$ is 0.600 s has more than 45 sampling periods.

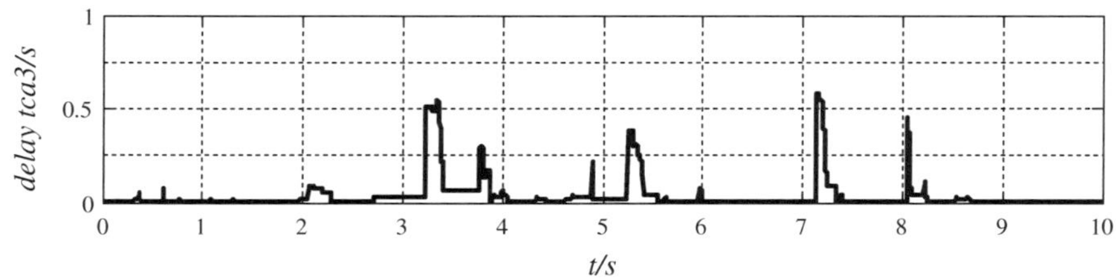


**Fig. 24.6** The time delay $\tau_{ca3}$ from controller to actuator (loop 3)

In short, from Figs. 24.4, 24.5, 24.6, when not use any scheduling, the time delay form controller to actuator of 3 loops have greatly exceeded the sampling period of sensor. It indicates that the network conflict seriously and deteriorates the quality of control of NCS.

### 24.4.2 Use the Proposed Scheduling Method

1. The output of 3 loops show as Figs. 24.8, 24.9, 24.10.

From Figs. 24.7, 24.8, 24.9, use the proposed scheduling method; the quality of 3 control loops improves greatly, control curve is smoother. It can meet the need of control requirement.

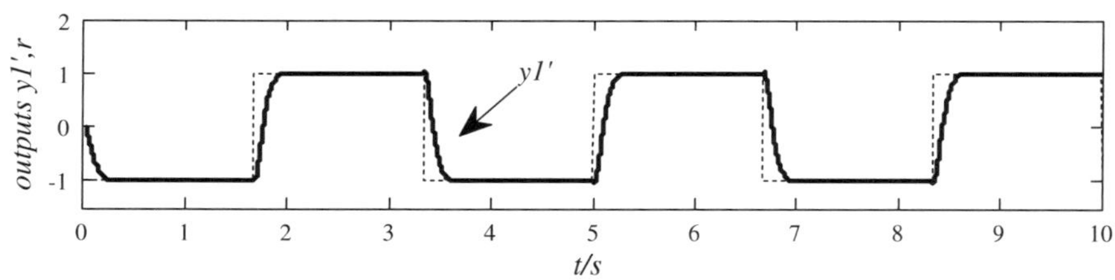


**Fig. 24.7** The output $y_1'$ of loop 1

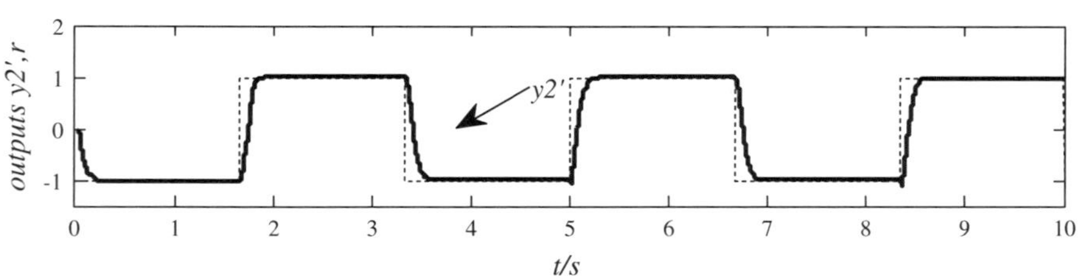


**Fig. 24.8** The output $y_2'$ of loop 2

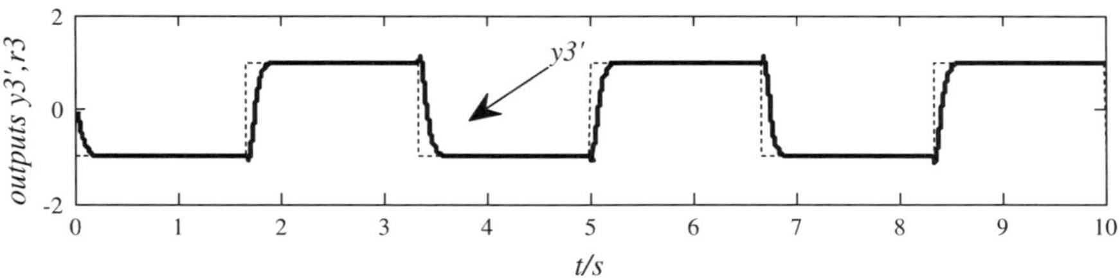


**Fig. 24.9** The output $y_3^{'}$ of loop 3

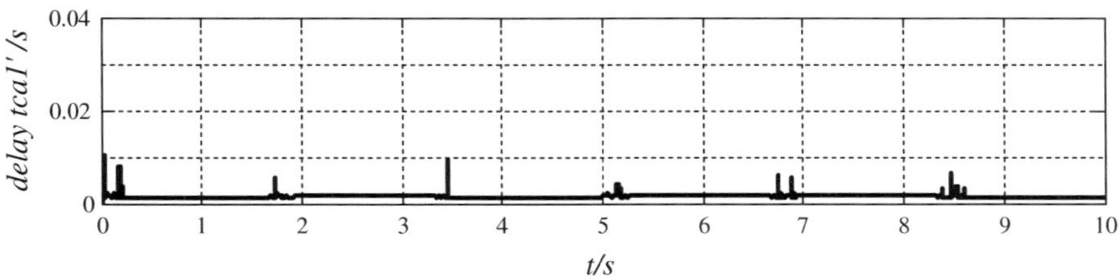


**Fig. 24.10** The time delay $\tau_{ca1\prime}$ from controller to actuator (loop 1)

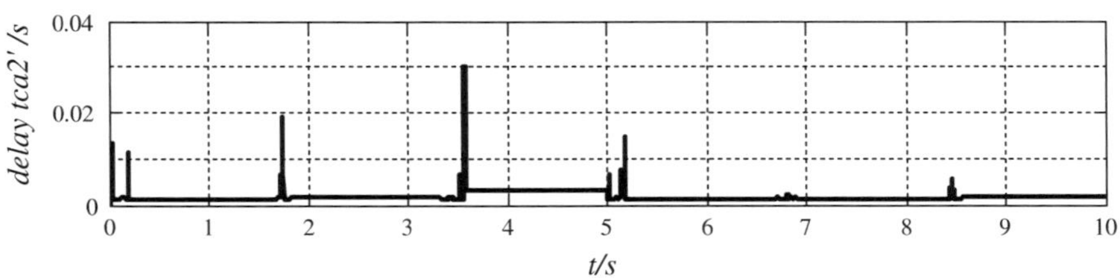


**Fig. 24.11** The time delay $\tau_{ca2\prime}$ from controller to actuator (loop 2)

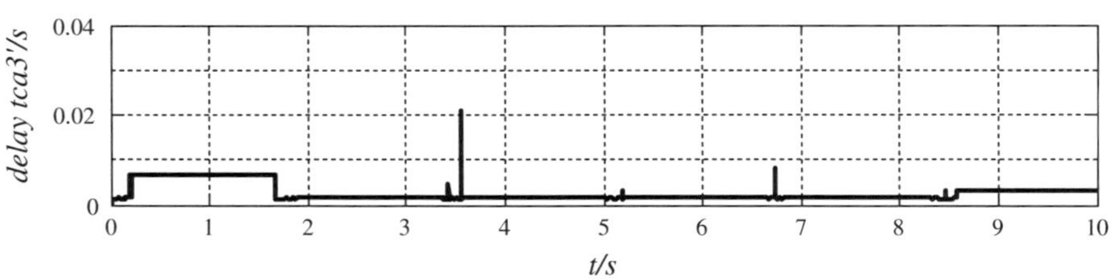


**Fig. 24.12** The time delay $\tau_{ca3\prime}$ from controller to actuator (loop 3)

2. The time delay from controller to actuator show as Figs. 24.10, 24.11, 24.12.

From Figs. 24.10, 24.11, 24.12, the time delay becomes smaller obviously when it uses the *e* and *ec* deadband scheduling.

## 24.5 Conclusion

This paper aims at multi-NCS share the same network; the conflict will reduce control performance of the system; puts forward a new dynamic scheduling. This method guarantees the quality of control (output is into its steady-state value of 95 %) as the goal, and set deadband in the controller. It bases on the two parameters: error and error change rate adaptive to adjust the network load. The simulation verify the proposed method can improve the utilization rate of the network bandwidth, to improve the quality of control performance, reduce the overshoots of the output when the given value have step change, enhance the stability of the system. Because of the simple algorithm and easy realization, it has a wide application

**Acknowledgments** This work is partially supported by National Natural Science Foundation of China, as well as the Research Project of Hainan University under Grant No. hd09xm85 and kyqd-1214.

## References

1. Li ZX (2008) Networked control system of intelligent scheduling and its optimization. Zhejiang university Ph.D. thesis 12(32):9–10
2. Cui XZ, Han P (2008) A mixed scheduling algorithm for thermal process network control systems. In: Proceedings of the 27th Chinese control conference, vol 31, pp 207–210
3. Wang BR, Shi DG (2007) Study on dynamic priority scheduling based on fuzzy logic for networked control systems. In: Proceedings of the IEEE international conference on automation and logistics, vol 8, pp 1182–1186
4. Zhang XF, Wang ZJ (2009) An immune-genetic algorithm-based scheduling optimization in a networked control system. Glob Congr Intell Syst 4:32–35
5. Zhang XF, Li GH (2009) Real-time elastic network scheduling of networked control systems. In: The 1st international conference on information science and engineering, vol 22, pp 5017–5021
6. Xu LJ (2010) A hybrid quantum clone evolutionary algorithm-based scheduling optimization in a networked learning control system. Control Decis Conf 12(9):3632–3637
7. Li Z (2010) Brief paper intelligent scheduling and optimization for resource constrained networks. Control Theor Appl 23:2982–2992
8. Branicky MS, Phillips SM, Zhang W (2002) Scheduling and feedback co-design for networked control systems. In: Proceedings of IEEE conference on decision and control, vol 3(2), pp 1211–1217
9. Liu CL, Layland J (1973) Scheduling algorithms for multi programming in a hard real-time environment. J Assoc Comput Mach 20(1):46–61
10. Walsh GC, Ye H (2001) Scheduling of networked control systems. IEEE Control Syst Mag 21(1):57–65
11. Otaneg P, Moyne J, Tilbury D (2002) Using deadbands to reduce in networked control systems. In: Proceedings of the American control conference anchorage: IEEE press, pp 46615–619
12. Tang XM (2007) Networked control system dynamic deadband feedback scheduling. J S Chin Univ Technol 6(5):716–721
13. Zhang P (2009) Based on switch ethernet network control system of scheduling research. Nanjing university of technology master thesis, vol 68(9), pp 35-38

# Chapter 25
# Implementation of Intelligent Wireless Video Monitoring System

**Yi Zhang, Zhuoying Wang and Yangdong Yu**

**Abstract** This paper implementation an intelligent video monitoring system based on wireless transmission and the embedded system. The system uses ARM processor and Linux operation system, capturing the monitoring video by the USB camera and transmission through the wireless network (802.11 g). The video signal is transmitted to a monitoring terminal (PC), the resolution of the monitoring video can achieve 320 × 240 and the frame frequency is up to 20 frames/s or 640 × 480 and the frame frequency is up to 5 frames/s. The system integrates intelligent video processing algorithm, can monitor and discrimination the movement in the region, can meet the general video monitoring requirement. The system is inexpensive, can according to the different application to change intelligent control algorithm and it has the widespread application prospect.

**Keywords** Video monitoring · Video compression · Wireless video transmission · Embedded system

## 25.1 Introduction

With the progress of science and technology, video monitoring technology has made great progress. The video monitoring system has been widely used in security, industrial production management control, occupations involving exposure to

Y. Zhang (✉) · Z. Wang · Y. Yu
Zhejiang University City College, Hangzhou 310015, China
e-mail: ezhang5@hotmail.com

Z. Wang
e-mail: 332301926@qq.com

Y. Yu
e-mail: 609000303@qq.com

W. Du (ed.), *Informatics and Management Science V*, Lecture Notes in Electrical Engineering 208, DOI: 10.1007/978-1-4471-4796-1_25,

toxic or hazardous chemicals and other fields. With the monitoring system much cheaper, it begins to enter the family and becomes a part of household equipment.

Now the using of video monitoring system has two problems: First, currently using video monitoring system is a cable video monitoring system. The biggest drawback is that installation is very inconvenient, the wiring cost is high and the using is not flexible. If temporary changing the monitoring region, it is impossible for the cable system. Second, currently using video monitoring system is a simple video capture and transmission system, it does not analysis the video data. For the specific monitoring application is mainly depend on the monitoring personnel and the monitoring efficiency is low. Especially, the personnel have to monitor multiple screens, as a result of distracted, vulnerable to security risks [1]. So to make a wireless video monitoring system based on embedded system integrates intelligent image processing algorithm which can detect the movement in the region has a strong application value.

## 25.2 The System Hardware Platform Implementation

The system hardware platform can be mainly divided into the three parts: embedded system platform, the video acquisition device and the wireless communication device.

### *25.2.1 The Embedded System Platform*

The monitoring system chooses the embedded system platform which uses the S3C2440 processor of the SamSung company as the main chip. This chip is based on ARM9 architecture, has powerful processing capability [2]. Main frequency is up to 533 MHz, normally it works in 400 MHz and fully meets the video data compression and wireless transmission requirements. The processor integrated two USB host ports and one USB slave port. The two USB host port can connect the video acquisition device and the wireless communication device. The processor has a JTAG interface, can be used with inexpensive debugging tools for software programming and debugging, so the development costs are relatively low.

### *25.2.2 The Video Acquisition Device*

In the system, the normal PC USB COMS camera is used as the video acquisition device. This camera is cost-effect and with the USB interface can be easily connected to the embedded system platform. The video resolution of the camera is $320 \times 240$, some model can reach $640 \times 480$, and it fully meets the requirement

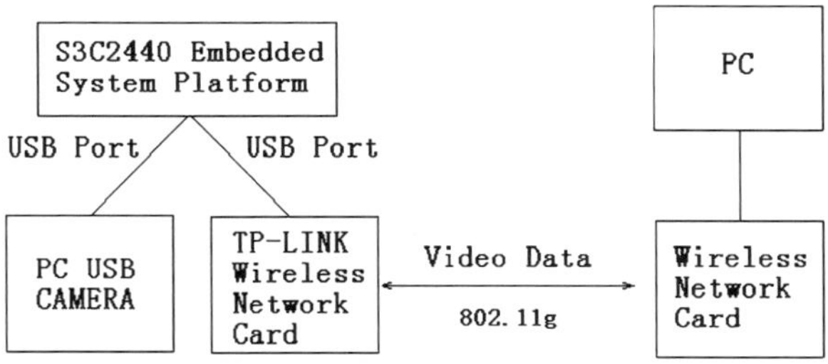


**Fig. 25.1** System hardware block diagram

of the video monitoring system [3]. Because of the entire system using Linux embedded operating system, the chosen camera should have the device driver in Linux operating system to simplify the development process and improve development efficiency. Most of the PC camera in the market can meet the requirement of the monitoring system.

### *25.2.3 The Wireless Communication Device*

In the system, the normal PC wireless network card supporting 802.11 g standards and with the USB interface is used as the Wireless communications device. The nominal data transfer rate of 802.11 g is up to 54 Mbps, can fully meet the requirement of video data transmission. Because the device has the USB interface, can be easily connect to the embedded system platform. Similarly, the network card should have the device driver in Linux operating system, so the monitor system chooses the TP-LINK wireless network card. The final system hardware block diagram is shown in Fig. 25.1.

## 25.3 System Software Implementation

Software development works can be divided into three parts: ViVi Bootloader transplant, embedded Linux transplant and devices drivers implementation, implementation the software of wireless transmission and video compression.

### *25.3.1 ViVi Bootloader Transplant*

The monitoring system uses ViVi as the system bootloader. ViVi was developed by Mizi Company and it is designed for ARM9 processors. The source code can be download from http://www.mizi.com website [4]. ViVi provides users with a command-line interface to use the ViVi command set and it can be used to download the operating system image, program the FLASH chip and realize other related functions to facilitate the development of embedded system.

Because ViVi directly supports to S3C2410x chip board and S3C2440 chip has the same architecture as the S3C2410x chip, so the transplant of ViVi is easy. The processor platform related files are stored in vivi/arch/directory and the monitoring system can refer to the content of the S3C2410 directory and do some modification. The head.s file is the ViVi startup file, system reset after power on and begins to run the program from this file. Hardware configuration and initialization values can refer to the ViVi/include/platform/smdk2410.h file and according to the CPU model, IO port address and initialization requirement of the system to do some modification.

Transplant of ViVi is based on the FLASH model in the system and has to program the drive to make the FLASH be controlled by ViVi. If using the FLASH model has not been supported by ViVi, its driver should be added to ViVi by referring to the supported Flash model drivers and realize the reading and writing operation on the new FLASH model. Related files are in vivi/include/MTD directory.

If the system want to use ViVi to download the operating system image and file system files via the network port, it need to realize the ViVi network driver which based on the network interface chip and the external interface circuit (commonly using DM9000 and CS8900 chip). Related files are in vivi/include/net directory.

After the related files have been modified and then compiled all files to an binary file vivi.bin. Use the JLink tool to program the vivi.bin to the NAND FLASH. The ViVi bootloader is running after the monitoring system power on and the output is shown in Fig. 25.2.

### *25.3.2 Embedded Linux Transplantation and Devices Drivers Implementation*

The monitoring system using the embedded Linux operating system, the kernel version is 2.6.14.1. The 2.6 kernel added the supporting to Samsung S3C2440 chip, so the process of kernel transplant is simple [5]. Meanwhile the 2.6 kernel supporting to the video acquisition device and wireless network device is better than the 2.4 kernel. There are lots of papers about transplant 2.6 kernel to S3C2440 platform and it is not to be presented here [6]. In the transplant process, note there is an error in 2.6.14.1 kernel driver for S3C2440 integrated USB ports, it need to be corrected. Meanwhile, the S3C2440 integrated two USB ports has to be set in the USB host mode, it also need to modify the related configuration files. There are three related files: arch/arm/mach-s3c2410/mach-smdk2440.c, drivers/USB/host/ohci-s3c2410.c and drivers/USB/core/hub.c. After completing the transplant of the Linux 2.6.14.1 kernel, the system output is shown in Fig. 25.3.

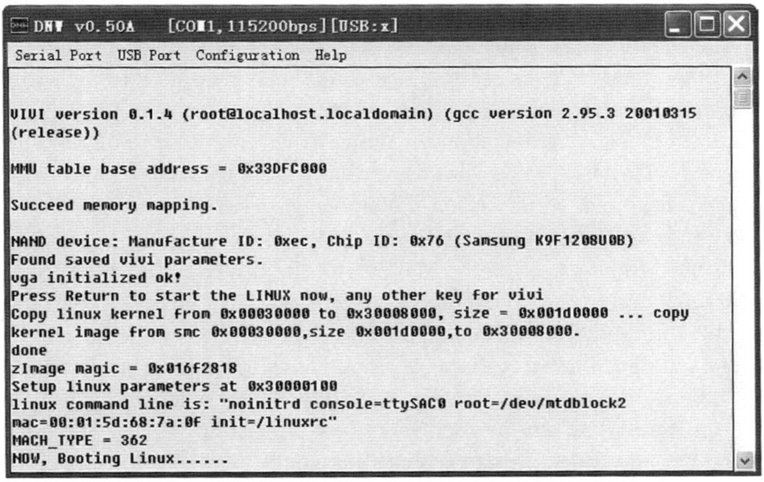


**Fig. 25.2** ViVi running output

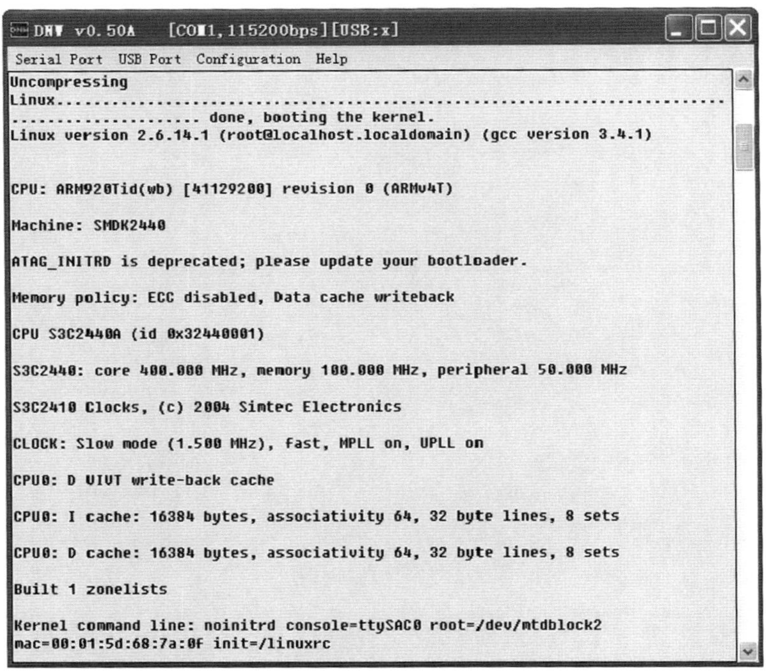


**Fig. 25.3** Linux kernel running output

In the monitoring system, camera control software using the gspcav1-20071224.tar.gz driver. The driver is an open source Linux camera driver, supports the most USB camera in the market. Meanwhile the driver's performance is better and has been widely used in the Linux operating system of the PC platform. To Modify the Makefile can transplant the driver to the embedded platform. To use the driver the embedded system platform can capture the monitoring video.

In the monitoring system, wireless network card use the rt73-4.17.tar.gz driver. The driver is open source and supports the USB wireless network card which uses RT2x00 chip (TP-LINK USB wireless network card uses this chip). After transplant the driver, it can work well in the embedded operating system. Because the rt73-4.17.tar.gz driver is designed for the 2.4 kernel, so using in the 2.6 kernel version, it need to be rewritten the Makefile of the driver and add the source code to the kernel source tree to make the driver correctly compiled. In the monitoring system, the configuration and management of the wireless network card can use the wireless_tools.28.tar.gz software, through the software can set the wireless LAN SSID and encryption key, realize the wireless AP scanning. Through simple transplant, the software can be used in embedded system.

Finally, the embedded system realizes the control of the USB camera and the wireless network card, the system running output is shown in Fig. 25.4.

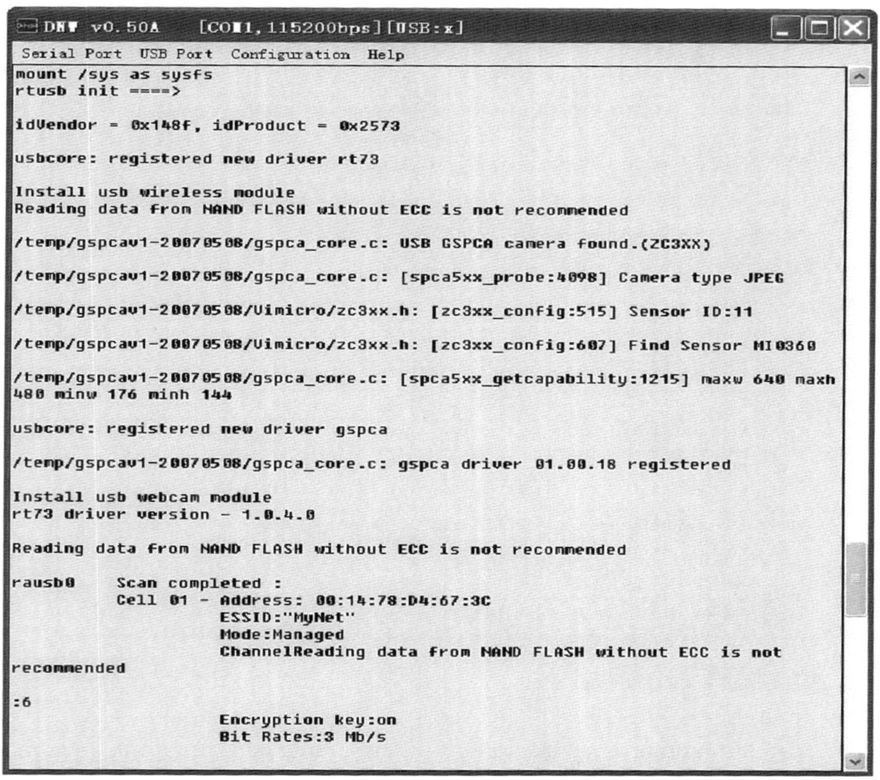


**Fig. 25.4** Camera and wireless network card running output

### 25.3.3 Implementation the Software of Wireless Transmission and Video Compression

The acquisition and transmission video uses spcaview-20071224.tar.gz software. The software used with gspcav1-20071224.tar.gz camera drive can obtain the video data from camera and compress the data for transmission. The software uses the M-JPEG algorithm for video compression. The compression algorithm is simple, so it is suitable for embedded system to use software to compress the video data. Meanwhile the output video data transfer rate is not high and basically conform to the 802.11 g wireless transmission bandwidth requirements.

The software uses a typical server/client network programming model. The server program running in the embedded system platform can compress and send the monitoring video data. The client program running in the PC which uses Windows or Linux operation system can receive, decompress and display the monitoring video. Meanwhile, the client program using temporal differencing and moving object tracking intelligent video analysis algorithms to detect the moving object in the video. When the monitoring video is abnormal, the client software will raise the PC beep sound to remind the monitoring personnel. Implementation of the PC client software diagram is shown in Fig. 25.5 (video size is 320 × 240 and frame rate is up to 20 frames/s).

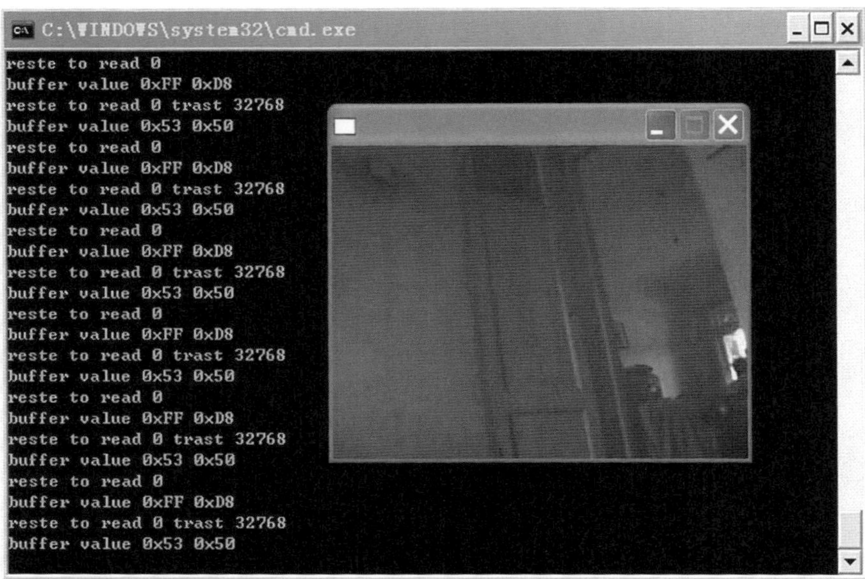


**Fig. 25.5** PC client software running diagram

## 25.4 Conclusion

The finally implementation of the intelligent video monitoring system, video resolution can achieve 320 × 240, frame rate 20 frames/s or 640 × 480, frame rate 5 frames/s. The system integrated with an intelligent video analysis algorithms which based on temporal differencing and moving object tracking. After a long time running and testing the intelligent wireless video monitoring system works stably, reliably and the video change detection is sensitivity. The system can meet the general purpose of video monitoring system. The system is inexpensive, can according to the different application to change intelligent control algorithm and it has the widespread application prospect.

**Acknowledgments** The subject is supported by Zhejiang Province College Students' Science and Technology Innovation Project. The subject is supported by the construct program of the key discipline in Hangzhou.

## References

1. Liu J, Zheng W, Cai W (2008) Research and realization of video capture technology applicable to a monitoring-control system for safety production in a coal-mine. Ind Instrum Autom 5:59–62
2. Fang F (2008) The embedded wireless remote video monitor system based on arm. Instrum Technol 1:27–29
3. Chen M (2008) Analysis of object detection and tracking for video surveillance system. Video Eng 32(10):85–91
4. Que D, Du W, Yue P (2007) Research on Vivi Port to S3C2410 Ircochip. J Wuhan Univ Technol Inf Manag Eng 29(12):47–50
5. Shi J, Peng D (2008) ARM9 based embedded linux transplanting. J Wuhan Univ Technol Inf Manag Eng 30(2):205–208
6. Yang J, Yang Y (2007) Linux transplant based on the processor of S3C2410. Inf Technol 31(8):97–100,103

# Chapter 26
# Design of Hybrid Controllers Based on Radial Basis Function Neural Network

**Longyang Zhao, Xiao Zhu, Haoming Yang and Xuanju Dang**

**Abstract** For many nonlinear factors which appear in the tracking system of Permanent Magnet Linear Synchronous Motor (PMLSM), the two hybrid controllers based on the radial basis function neural network were proposed, the hybrid neural network controllers were composed of PID controller and RBF neural network controller in series, and their structures depended on the series orders. The two hybrid neural network controllers realized the nonlinear PID control, which possessed the ability of parameters self-tuning, simple structure and are easy to implement in the practice. The experimental results showed the feasibility and effectiveness of the two hybrid controllers.

**Keywords** PMLSM · Nonlinear PID control · RBF · Hybrid controller · cSPACE

## 26.1 Introduction

Because of the directly driving in permanent magnet linear synchronous motor, the permanent magnet linear synchronous motor has faster response, higher sensitivity and better mobility, which can achieve ultra high-speed movement. In addition, in some respects such as power index, positioning accuracy and efficiency, PMLSM has more advantages than other motors. However, the directly driving also introduces some uncertain factors such as the parameter perturbation, load

L. Zhao (✉) · X. Zhu · X. Dang
School of Electronic Engineering and Automation, Guilin University of Electronic Technology, Guilin 541004, China
e-mail: guilinzhly@163.com

H. Yang
Guilin Institute of Scientific and Technical Information, Guilin 541002, China

W. Du (ed.), *Informatics and Management Science V*, Lecture Notes in Electrical Engineering 208, DOI: 10.1007/978-1-4471-4796-1_26,

disturbance and external nonlinear perturbation [1–3], these uncertain factors directly affect the static and dynamic characteristics of the linear motor, and increase the difficulty for the control. Especially, in the high-speed control process, it is more difficult to ensure the high-Accuracy Tracking.

PID controller has simple structure and good robustness, and is widely used in the field of motor control. However, the new interferences, which are introduced by structural improvement for PMLSM, make the PID control difficult to achieve high tracking accuracy. In [4], single neuron PID control based on RBF neural network identification is combined with a neural network compensator, which achieves the fast tracking of the control system. In [5], in the speed control of permanent magnet linear motor, a sliding mode controller based on neural network thrust observer is used to replace the conventional PI controller. With the feed forward compensation of load disturbance, the design realizes the more precise speed control of linear motor in a wide speed range. In [6], a fuzzy neural network controller and a supervisory controller are designed to achieve the tracking control for PMLSM. The fuzzy neural network controller has the ability to automatically generate a learning rule and the supervisory controller is used to stabilize the system state. The above approaches is only a simulation-based research, for practical applications, combining with the advantages of PID control and neural network control, two hybrid controllers are designed, which are formed by PID controller and RBF neural network controller in series and have simple structure and good adaptive ability. The experimental results show the effectiveness and feasibility of the designed controllers.

## 26.2 Modeling of PMLSM

Consider the interferences, the voltage model of PMLSM which contains the input voltage and the kinetic equations [7, 8] are expressed as follows:

$$u(t) = k_e \dot{y}(t) + Ri(t) + L\frac{di(t)}{dt} \tag{26.1}$$

$$f(t) = K_f i(t) \tag{26.2}$$

$$f(t) = M\ddot{y}(t) + f_r(y) + f_f(\dot{y}) + f_\Delta \tag{26.3}$$

where $u(t)$ is the control voltage; $i(t)$ is the armature current; $k_e$ is the EMF constant; $R$ is the stator winding resistance; $L$ is the inductance of Motor; $k_f$ is the electromagnetic thrust coefficient; $M$ is the promoter quality; $f_r(y)$ is the thrust fluctuation; $f_f(\dot{y})$ is the friction of motor; $f_\Delta$ is the uncertain disturbances, such as electrical parameter perturbations and the system noise.

Since the electrical time constant is much smaller than the mechanical one, the dynamics due to electrical induction is omitted [8]. According to Eqs. (26.1–26.3), the kinetic equation of PMLSM is as follows:

$$\ddot{y}(t) = -\frac{K_f K_e}{RM}\dot{y}(t) - \frac{1}{M}f_r(y) - \frac{1}{M}f_f(\dot{y}) - \frac{1}{M}f_\Delta + \frac{K_f}{RM}u(t) \tag{26.4}$$

where $\ddot{y}(t)$ is the acceleration of motor promoter, $\dot{y}(t)$ is the velocity of motor promoter. Eq. (26.4) can be written as

$$\ddot{y}(t) = -\frac{K_f K_e}{RM}\dot{y} - \frac{K_f}{RM}\overset{*}{f_r}(y) - \frac{K_f}{RM}\overset{*}{f_f}(\dot{y}) - \frac{K_f}{RM}\overset{*}{f_\Delta} + \frac{K_f}{RM}u(t) \tag{26.5}$$

where $\overset{*}{f_r}(y) = Rf_r(y)/K_f$, $\overset{*}{f_f}(\dot{y}) = Rf_f(\dot{y})/K_f$, $\overset{*}{f_\Delta} = Rf_\Delta/K_f$. Let $\overset{*}{u}(t) = u(t) - \overset{*}{f_r}(y) - \overset{*}{f_f}(\dot{y}) - \overset{*}{f_\Delta}$, $a = K_f K_e/(RM)$, $b = K_f/(RM)$, Eq. (26.5) is described as follows:

$$\ddot{y}(t) = -a\dot{y}(t) + b\overset{*}{u}(t) \tag{26.6}$$

Take Laplace transform for the Eq. (26.6) and transform into the form of transfer function:

$$\frac{y(s)}{\overset{*}{u}(s)} = \frac{b}{s^2 + as} \tag{26.7}$$

where $y(s)$, $\overset{*}{u}(s)$ and $s$ are the corresponding Laplace transform of $y(t)$ and $\overset{*}{u}(t)$, the Laplace operator, respectively. The model of PMLSM is shown in Fig. 26.1.

## 26.3 Design of Two Hybrid Controllers

The two hybrid neural network controllers are formed by PID controller and RBF neural network controller in series. According to different series orders, there are two different hybrid controllers which are shown in Figs. 26.1 and 26.2.

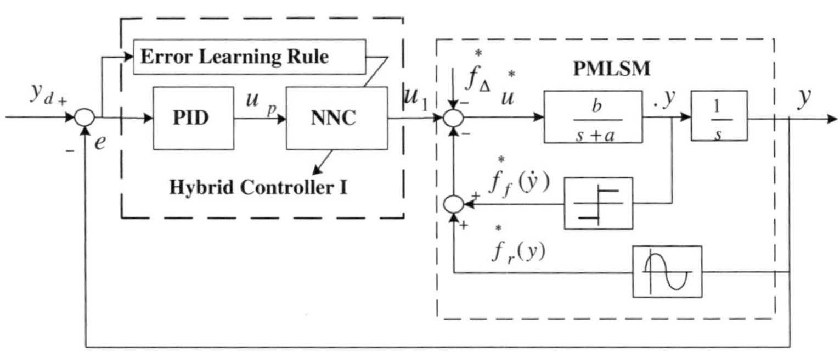


**Fig. 26.1** System block diagram of the hybrid controller I

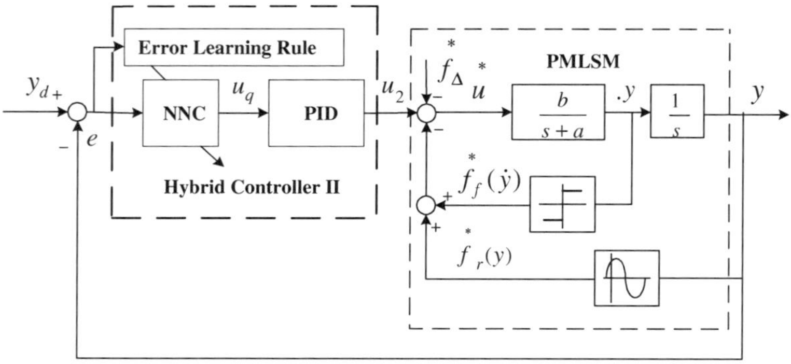


**Fig. 26.2** System block diagram of the hybrid controller II

In the structure of two hybrid controllers, PID controller is used to ensure the stability and enhance the anti-interference capability of the system. In Fig. 26.1, error signal is processed in PID controller and then sent into neural network controller. Neural network controller is used to approximate the ideal controller. In Fig. 26.2, the error signal is sent into neural network controller for nonlinear processing, the processed signal is sent into PID controller for further processing.

The neural network in Figs. 26.1 and 26.2 is Radial Basis Function (RBF) neural network [4] which has some advantages such as simple structure, training fast and it is easy to program.

In the structure of RBF neural network, $X = [x_1, x_2, \ldots, x_n]^T$ is the input vector of network. $H = [h_1, h_2, \ldots, h_j, \ldots, h_m]^T$ is the radial basis vectors, $h_j$ is the Gaussian basis function:

$$h_j = \exp\left( -\frac{\|X - C_j\|^2}{2b_j^2} \right) \tag{26.8}$$

$C_j = [c_{j1}, c_{j2}, \ldots, c_{ji}, \ldots, c_{jn}]^T$ is the center vector of the network, where $j = 1, 2, \ldots, m$, $i = 1, 2, \ldots, n$. $B = [b_1, b_2, \ldots, b_j, \ldots, b_m]$ is the base width vector, $b_j$ is the base width parameter. $W = [w_1, w_2, \ldots, w_j, \ldots, w_m]^T$ is the weight vector. The output of network can be written as:

$$y_m(k) = w_1 h_1 + w_2 h_2 + \ \ldots \ + w_m h_m \tag{26.9}$$

The expression (26.8) shows that the validity of Gaussian function is associated with the center vector $c_j$ and the base width vector $b_j$. The value of $c_j$ determines the coordinates of the center of the Gaussian function, and $b_j$ determines the width of Gaussian function shape.

### 26.3.1 Design of Hybrid Controller I

In hybrid controller I, the output $u_p$ of PID controller is the input of neural network controller, the output $u_1$ of the neural network is the output of hybrid controller. Select Feedback Error Learning (FEL) method as the learning method of neural network [9]. $E(t)$ is the performance index function of the neural network identifier, it can be written as:

$$E(t) = \frac{1}{2}e^2(t) = \frac{1}{2}(y_d(t) - y(t))^2 \tag{26.10}$$

where $y_d(t)$ is the expected trajectory of motor promoter; $y(t)$ is the actual trajectory; $e(t)$ is the error of expected position and the actual location.

Update $w1_j$ (the network weights), $c1_j$ (the centers of nodes) and $b1_j$ (the base width vectors) of the neural network with gradient descent method, the process is as follows:

$$w1_j(k) = w1_j(k-1) + \eta_1 e \exp\left(-\frac{\|u_p - c1_j\|^2}{2b1_j^2}\right)\frac{\partial y}{\partial u_1} \tag{26.11}$$

where j is the nodes of hidden layer, $\eta_1$ is the learning velocity of the output weight, $\partial y/\partial u_1$ is the Jacobian information of PMLSM. Since the characteristics of object are unknown, the Jacobian information is replaced by the constant coefficient. The imprecise computation which happened in the alternative procedure is compensated by adjusting the learning velocity [10].

$$c1_j(k) = c1_j(k-1) + \eta_2 e w1_j h1_j \frac{u_p - c1_j}{b1_j^2}\frac{\partial y}{\partial u_1} \tag{26.12}$$

where $\eta_2$ is the learning velocity of $c1_j$.

$$b1_j(k) = b1_j(k-1) + \eta_2 e w1_j h1_j \frac{\|u_p - c1_j\|^2}{b1_j^3}\frac{\partial y}{\partial u_1} \tag{26.13}$$

In Eq. (26.13), the learning velocity is the same as the one in Eq. (26.12).

### 26.3.2 Design of Hybrid Controller II

In hybrid controller II, error signal is sent into neural network controller II and the output $u_q$ is sent into PID controller for further processing, the output of PID controller is $u_2$ which is the entire control quantity of PMLSM system. Also select FEL method as the learning rule in hybrid controller II. The selection of $E(t)$ is the same as the one in hybrid controller I. Update $w2_j$ (the network weights), $c2_j$ (the centers of nodes) and $b2_j$ (the base width vectors) of the neural network with gradient descent method, the process is as follows:

$$w2_j(k) = w2_j(k-1) + \eta_3 e \exp\left(-\frac{\|e - c1_j\|^2}{2b1_j^2}\right)\frac{\partial u_2}{\partial t}\frac{1}{\partial u_q/\partial t}\frac{\partial y}{\partial u_2} \qquad (26.14)$$

where $\eta_3$ is the learning velocity of the output weight $w2_j$, $\partial y/\partial u_2$ is the Jacobian information of PMLSM.

$$c2_j(k) = c2_j(k-1) + \eta_4 e w2_j h2_j \frac{e - c2_j}{b2_j^2}\frac{\partial u_2}{\partial t}\frac{1}{\partial u_q/\partial t}\frac{\partial y}{\partial u_2} \qquad (26.15)$$

where $\eta_4$ is the learning velocity of $c2_j$.

$$b2_j(k) = b2_j(k-1) + \eta_4 e w2_j h2_j \frac{\|\mathrm{e} - c2_j\|^2}{b2_j^3}\frac{\partial u_2}{\partial \mathrm{t}}\frac{1}{\partial u_q/\partial \mathrm{t}}\frac{\partial y}{\partial u_2} \qquad (26.16)$$

The learning velocity of $b2_j$ is also $\eta_4$.

## 26.4 Experimental Results

In the experimental studies, the system of PMLSM which is composed of the U-shaped permanent magnet linear synchronous motor and DC box is manufactured by Micro-Nano Technology. The maximum thrust of linear motor is 120 N and the travel length is 100 mm. The resistance R = 18.7 Ω, EMF constant $\mathrm{k}_e = 9.6\ v \cdot s/m$, thrust constant $k_f = 11.71\ N/A$, the quality of motor promoter $M = 0.3$ kg.

Drive and control box includes control module, drive module and the corresponding signal cable. The control module based on TMS320F2812DSP of TI Company is the control card of cSPACE system. The drive module is Harmonica digital servo system motor drive from Elmo company [11].

### *26.4.1 Implementation of Hybrid Controller I*

In the structure of hybrid controller, the neural network module is edited by C MEX S-Function [12]. The input signal is the same as the one in PID control. The hidden layer nod of neural network is 10. The network weights, the centers of nodes and the base width vectors are online trained. The learning velocity $\eta_1$ and $\eta_2$ are adjusted in cSPACE interface, their final value are 0.05 and 0.1. The three

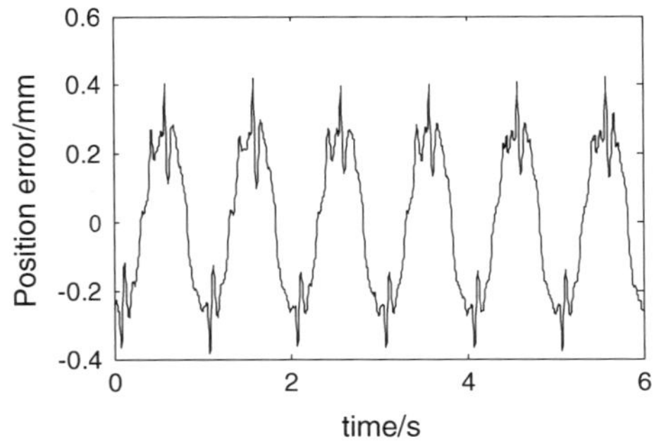


**Fig. 26.3** Location error in hybrid controller I

parameters of PID controller are 1.5 (proportion), 0.0003 (integral), 0.01 (differential). The location error is shown in Fig. 26.3.

From Fig. 26.3, we know that the maximum tracking error of hybrid controller I is only 0.42 mm, the control accuracy is about 2.1 %. Compared with the PID algorithm, the control accuracy has a huge improvement. Experimental results show the feasibility of the hybrid control I.

## *26.4.2 Implementation of Hybrid Controller II*

The structure of hybrid controller II is similar with the hybrid controller I, and they are different in series order. The learning velocity $\eta_3$ and $\eta_4$ are online adjusted in cSPACE interface until the PMLSM tracking achieves the best effect. The final value of learning velocity is 3 and 0.05, the three parameters of PID controller are 0.05 (proportion), 0.0005 (integral), 0.005 (differential). Use the same data processing method and get the location error which is shown in Fig. 26.4.

From Fig. 26.4, we know that the maximum tracking error of hybrid controller II is about 0.38 mm, the control accuracy is about 1.9 %. Compared with the PID algorithm, the control accuracy has also a huge improvement. Experimental results show the feasibility of the hybrid control II.

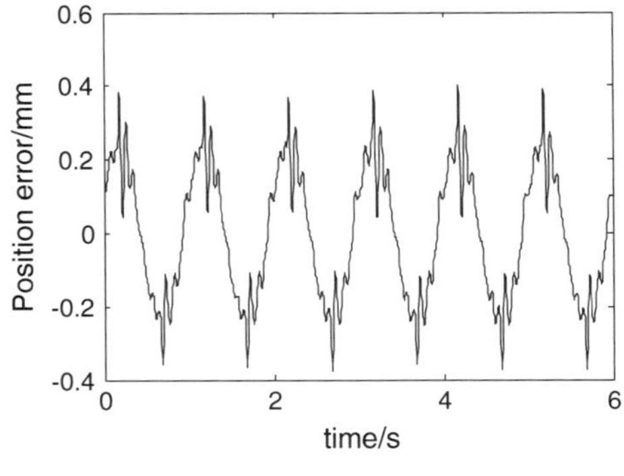


**Fig. 26.4** Location error in hybrid controller II

**Table 26.1** Experimental results of three controllers

| Controller | Amplitude of input (mm) | Tracking error (the maximum value) (mm) | Percentage of error |
|---|---|---|---|
| PID controller | 20 | 1 | 5 |
| Hybrid control I | 20 | 0.42 | 2.1 |
| Hybrid control II | 20 | 0.38 | 1.9 |

From Table 26.1, we know that the tracking accuracy of PID controller is worst. This is mainly because of the existence of phase lag. The hybrid neural network controller I is formed by PID controller and RBF neural network controller in series, which is used to approximate the ideal controller, the tracking accuracy is up to 2.1 %. Different from the hybrid controller I in series order, error signal is sent into neural network controller for nonlinear processing, the processed signal is sent into PID controller for further processing, the tracking accuracy of hybrid controller is about 1.9 %. The series orders of two hybrid controllers are different but they can both achieve good control effect. The experimental results show the feasibility and effectiveness of the two controllers in design.

## 26.5 Conclusion

The two hybrid neural network controllers realized the nonlinear PID control, which possessed the ability of parameters self-tuning, simple structure and are easy to implement in the practice. Compared with PID control, the proposed two hybrid neural network controllers achieve high-precision tracking control for the PMLSM.

**Acknowledgments** This work has been supported by the National Nature Science Foundation Project (60964001), Science Foundation Project of Guangxi Province (0991019Z) and Major Research Project of Guangxi Department of Education (201101ZD007).

## References

1. Hao S-H, Cai Y, Zhang W-F et al (2010) Research on high-speed positioning of AC servo system based on feedforward control. Small Spec Electr Mach 38(2):35–40
2. Cheng Y, Yang J, Huang Q et al (2011) Characteristics of inductance parameters and thrust linear modeling of PMLSM with combinational iron-cored primary. Int Conf Consum Electron Commun Netw 22:142–145
3. Panah PG, Shafiei A, Parsa Pour A, Dehkordi M et al (2011) Velocity control of a PMLSM using a brain emotional learning based intelligent control strategy. IEEE Int Conf Syst Eng Technol 32:47–52

4. Zhang L, Xuan-ju D, Zeng S-L et al (2010) Kind of double model control for permanent magnet linear servo system. Electr Transm 40(2):53–56
5. Fu Z, Li H, Wang Z et al (2009) Sliding mode control based on neural network thrust observer for PMLSM. Electr Transm 39(11):48–51
6. Lu H-C, Chang M-H et al (2009) Automatic generation fuzzy neural network controller with supervisory control for permanent magnet linear synchronous motor. IEEE Conf Ind Electron Appl 243:3124–3129
7. Ahn H-S, Chen YQ, Dou H et al (2005) State-periodic adaptive compensation of cogging and coulomb friction in permanent-magnet linear motors. IEEE Trans Magn 41(1):90–98
8. Chen SL, Tan KK, Huang S et al (2010) Modeling and compensation of ripples and friction in permanent-magnet linear motor using a hysteretic relay. IEEE Trans Mechatron 15(4):586–594
9. Nakamura Y, Morimoto K, Wakui S et al (2011) Experimental validation of control for a positioning stage by feedback error learning. Int Conf Adv Mechatron Syst 22:11–16
10. Li G, Zang J, ZENG A et al (2009) PID control based on BP neural network. Comput Simul 26(9):128–131
11. Micro-nano technology Co., LTD (2010) cSPACE rapid control prototyping system manual of micro-nano technology, 22:330–337
12. Shen Y, Ji Z (2005) Study on modeling and simulation of permanent-magnet synchronous-motor control system based on C MEX S—function. J Syst Simul 17(8):1820–1825

# Part II
# Information Management and Applications II

# Chapter 27
# Research of Teaching–Learning-Practicing Integrated Teaching Mode

**Yuan Dong and Xiaohua Zhu**

**Abstract** The construction and practice of the Teaching–Learning-Practice teaching mode is a systematic project and the essence of it is the practical teaching and learning process. The connotation of this teaching mode is the combination of teaching, productive labor and social practice. Based on the course of Tour Guide Business, the paper explores the construction and practice of the Teaching–Learning-Practice teaching mode in order to provide reference for the reform of the training mode for high professional education majoring in tourism management.

**Keywords** Major in tourism management · Tour guide business · Teaching–learning-practice · Teaching mode

## 27.1 Introduction

Carrying out the practical teaching and training the talents for social needs are the objectives and characteristics for higher vocational education. But the traditional practical teaching just act as the assistant for theory teaching and it is difficult to implement the cultivation of students' practical skills [1]. Teaching–Learning-Practice teaching mode adapts the characteristics of occupation education, sets the students as the main body of the advanced teaching concept, and combines the teaching and practice. And the essence of it is the practice of the teaching process; the connotation is the combination of teaching, productive labor and social practice in order to promote the course construction, teaching content and teaching method reform.

Y. Dong (✉) · X. Zhu
Tourism Department of Chongqing Educational College, Chongqing, China
e-mail: yuandong1q85@126.com

W. Du (ed.), *Informatics and Management Science V*, Lecture Notes in Electrical Engineering 208, DOI: 10.1007/978-1-4471-4796-1_27,

## 27.2 Necessity of the Construction of Teaching–Learning-Practice Teaching Mode for Major in Tourism Management

Students majoring in tourism management are required to have strong practical abilities, such as analyzing problems, solving problems, dealing with unexpected incidents, etc. These abilities require teachers enlightening and excavating students' independent ability to analyze problem, make judgment and decision-making in teaching [2]. At present, the domestic tourism management specialty began to use a combination of school-enterprise cooperation mode, but it still separate the theory from practice teaching, resulting in that enterprises generally reflect the talents could not start their works after students' graduation. The problems existing in the teaching are mainly as follows:

The teaching mode for ordinary higher education still exists, and the teachers still use the classroom teaching theory as the main teaching way, while practical teaching mode is only the supplement for the former.

The number of the "double-type" teacher is few and the number of the teachers for theory teaching is many. A lot of teachers know about teaching knowledge but know nothing about guide practicing. There's a big shortage of practical teachers for guide business.

There is a shortage of the number of part-time teachers from tourism enterprises. And the proportion of professional teachers is far behind of the requirement of more than 16 %, which is prescribed by the Ministry of Education. Meanwhile, a majority of the teachers employed from tourism enterprises do not know how to teach, although they know how to do, resulting in the waste of the teaching resources.

Software system does not meet the Teaching–Learning-Practicing teaching mode. For example: teaching materials, teaching plan, teaching syllabus do not adapt to the requirements for the curriculum reform; there is few methods for exams and it does not adapt to the demand for curriculum reform.

Laboratory hardware facilities are not perfect. Lack of teaching equipment, the laboratory facilities couldn't match with the profession.

## 27.3 The Teaching–Learning-Practicing Teaching Mode for Course of Tour Guide Business

### *27.3.1 The Construction of Teaching–Learning-Practicing Curriculum System*

The construction of curriculum system based on the working process is the core of the integrated teaching reform. And the construction of curriculum system for the course of Tour Guide Business should enter into tourism industry and enterprise, and discuss and research with the leaders and experienced and qualified tour guides. Then it analyses the characteristics for the guide post and task based on professional standards for guide post, and integrates the theory courses with practical courses in order to highlight the occupational standard, strengthen training for skills.

Generally speaking, it is necessary to reform the current curriculum content, curriculum structure, teaching plan and syllabus. Most of the curriculum content lacks training for improving the skills, and lot of knowledge is not practical with lots of deep theories. The course structure is unreasonable and the proportion of the basic course and specialized basic courses is too large. And the Knowledge for being memorized by students is too much, resulting in the limitation for the development of the students' thoughts.

### 27.3.2 Construction of the Integrated Teaching Materials and Course Standard

Teaching material is a carrier of teaching content, and it is hard to practice the integrated teaching without the integrated teaching materials. The writing for the integrated teaching materials for Tour Guide Business should join the knowledge, skills and standards effectively based on the tour guide ability and the appraisal standard for guide vocational skills, and exactly control the joining and progressive content to make clear the pass for knowledge and skills, tasks and the specified practical ways, the process for it, supporting tasks and the expanding related knowledge.

Integration of teaching standard is the main basis of the integrated teaching. The constitute of teaching standards, firstly, make the teaching outline, which is based on the target of professional training scheme for major in tourism management and the task of tour guide from a wide range of social research. Secondly, in accordance with the requirement of integrated teaching, task driven, project oriented, combine national standards, enterprise standards and occupation qualification standards to classify tasks, create training project, set down various courses and various items of curriculum standards. Thirdly, make corresponding assessment standard based on the knowledge, ability and quality of tour guide.

In the course of Tour Guide Business, we use project as the guiding, tasks as the driving force. And the professional training for students is the main line based on the integrated teaching standard with the characteristics of generating ability. And we design the two major modules combined with the knowledge, ability and the quality required for guide post including skills for guiding and skills for interpretation. The former contains the procedures for local guide, ways for dealing with the emergencies in the course of guiding, satisfying personal needs of the tourists, professional skills for guiding and so on. The later includes two parts such as the interpretation on the way to the scenic spots and interpretation in the scenic spots.

### 27.3.3 Construction and Management of Teaching Staff

Excellent "double-type" teachers not only can inspire students to long for the guide industry, but also let the students understand tour guide industry comprehensively, know about the advantages and disadvantages for it and the passion and secrets of tour guide.

To build the double-type teachers, we can use the following ways: (1) Make arrangement of the professional teachers into the tourism industry to improve the professional skills. (2) Planned trainings for the professional teachers and encourage them to take part in the tour guide qualifying exams or related exams. (3) Planned industry training for teachers to let them know the new development of it. (4) Carrying out the school-enterprise bidirectional part-time teaching and working. And the executives of tourism enterprises take as the director of the Tourism Department, hiring excellent guide as part-time teachers. Teachers in the colleges take as the trainers in tour agencies and local guide to develop the ability of transforming the analysis ability to practical use for professional task and to master the new teaching thoughts, adapt to the new teaching mode to improve the whole integrated teaching level.

### 27.3.4 Construction for Hardware Condition

Advanced teaching laboratory can provide students with a virtual scene, exercise students' interpretational ability; good laboratory can provide multiple microteaching environments. In the limited classroom time, many students finish the practical guide training, record their situational guiding and the whole process for interpretation of the scenic spot. These actions can be repeatedly played by them to learn and help each other.

The construction of the advanced teaching laboratory is the hardware guarantee of integrated teaching. Mainly from the following aspects: (1) to have real or simulation of occupation atmosphere, such as 3D simulation equipment and software; (2) should pay attention to the development of classroom function, turn the normal classrooms into multifunction classrooms, which could implement microteaching training, test explaining, multimedia teaching, article display, etc.; (3) to introduce the corporate culture and corporate atmosphere into the training room, create enterprise working environment, exert the function of education environment, so that the students could contact with enterprises, perceive the spirit of enterprise and promote the occupational quality.

## 27.4 Implementation of Teaching–Learning-Practice Integrated Teaching Mode for the Course of Tour Guide Business

### *27.4.1 Formulation for the Scientific and Completed Integrated Teaching Process*

During the process of the integrated teaching process, the school-enterprise cooperation and the working tasks of the travel agency are the starting point. We should research for the requirements for the practical ability for working of each post and explore the projects and tasks for the integrated teaching mode. Taking the course of Tour Guide Business as an example, the teaching process for it contains six parts. (1) the project confirm; (2) the arrangement for task; (3) the explanation of key points; (4) preparation for training; (5) microteaching practice; (6) comments to improve. Specifically, we allow each student to bear the responsibility of the interest safeguard of travel agency, and to act as one role from the beginning of training to implement one project. And the each project is composed of several tasks according to the teaching requirements and assign the different tasks to students. When the tasks are issued, the "double type" teachers explain the main attentions and essentials for finishing the tasks and make demonstration; Students get preparations after classroom through discussion, search and collection of related data. Then they implement the microteaching training and repeatedly play the practical training based on the learning demand to show the students. Finally, they complete the integrated Teaching–Learning-Practicing teaching through students' self-evaluation, mutual evaluation and the teacher's comments.

As mentioned above, the teaching process combined teaching, learning with practicing not only let the teaching and practicing time into the same one but also make the teaching and practicing process at the same place. After that, the teaching content is based on some problems, which are solved for the professional needs, and based on the theory required for guide skills. Namely, the whole teaching process tell the students what are the teaching contents, why do we learn, how to do that and the procedure details for it.

### *27.4.2 Strengthening the Cooperation Between Colleges and Tourism Enterprises and Enhance the Teaching and Research*

School-enterprise cooperation is the access and platform for the reform of integrated teaching. During the process of the integrated teaching, tourism management major in higher vocational colleges should set up specialized organization of school-enterprise cooperation through the cooperation of colleges and enterprise with combination of production and learning to formulate the personnel training scheme, curriculum reform plan, jointly developing courses, course design, and to establish curriculum standard in order to take the working atmosphere of tourism enterprises into the teaching construction and reform to set up a new talent training mode and teaching mode.

## 27.5 Conclusion

Teaching–Learning-Practicing integrated teaching mode based on occupation, give priority to students. Its teaching evaluation based on the identification of teaching results and the effective teaching form is composed of double-type teacher, who act as a coach, and the students, who are the main body of this form. This teaching mode could solve the disconnection between the theoretical and practical teaching effectively and help cultivate students' good occupational quality for being qualified tour guide with professional skills and to cultivate high-qualified talents with high skills for tourism industry.

## References

1. Luo M, Mo W, Ning W (2009) The practical research of teaching-learning-practicing integrated teaching mode. China Electric Power Educ 6:127–131
2. Yunhui Ren (2009) The application of teaching-learning-practicing integrated teaching mode in higher vocational education. Wuxi Nanyang Vocat Coll J 3:72–77

# Chapter 28
# Construction of Recombinant Expression Vector of Anti-Bacterial Gene *aiiA* from Marine Bacterium

**Xian Ding, Bo Yin, Shanfu Zhang, Weike Tang, Weiwen Sun and Shining Zhou**

**Abstract** A pair of proper primers were designed and synthesized for Polymerase Chain Reaction (PCR) of anti-bacterial gene *aiiA* which encodes protein *aiiA* from marine bacterial genome, according to the gene sequences of *aiiA*. The plasmid pMD18-ZD02*aiiA* was constructed after PCR testing of *aiiA* gene. The gene sequence of *aiiA* was amplified with proper primers (with active locations by enzymes BamHIand EcoRI) from molding board of pMD18-ZD02*aiiA* and linked into expression vector pET-17b after it was hydrolyzed by enzymes of BamHI and EcoRI. And then plasmids pET-ZD02*aiiA* was constructed. The results of hydrolyzation of enzymes, PCR amplification and sequencing demonstrated that the target gene fragments were inserted into vector pET-17b correctly, the sequences of gene *aiiA* and frame of reading codes were right. Thus, it may provide base for the recombinant express and induced express of anti-bacterial gene from marine bacterium.

X. Ding (✉)
Key Laboratory of South China Sea Fisheries Resources Exploitation and Utilization, Ministry of Agriculture, Guangzhou, People's Republic of China
e-mail: xianding191@126.com

X. Ding
South China Sea Fisheries Research Institute, Chinese Academy of Fishery Sciences, Guangzhou 510300, China

X. Ding · W. Sun · S. Zhou
State Key Laboratory for Biocontrol, School of Life Sciences, Zhongshan (Sun Yat–sen) University, Guangzhou 510275, China

B. Yin
LTO State Key Laboratory, South China Sea Institute of Oceanology, Chinese Academy of Science, Guangzhou 510301, China

S. Zhang · W. Tang
Huai-an Da Jiang Aquatic Feedstuff Co., Ltd, Huai-an 223003, China

W. Du (ed.), *Informatics and Management Science V*, Lecture Notes in Electrical Engineering 208, DOI: 10.1007/978-1-4471-4796-1_28,

**Keywords** Bacterium · Anti-bacterial gene *aiiA* · Clone

## 28.1 Introduction

Researches show that quorum sensing (QS) system universally exists in nature. By sensing the change of the concentration of specific signaling molecules in extracellular environment, bacteria monitors the change of the number of its own or other bacteria, exchanges information between cells, starts related gene expressions, and adapts to the change of environment [1, 2]. Such a regulation mechanism shows QS system as a target plays an important role in controlling bacterial behavior [3]. Researches prove that the life habits and various physiologic responses such as bioluminescence, bio-film formation, antibiotic biosynthesis, and virulence factor formation of flora and fauna pathogenic bacteria of bacterial cells can be regulated by bacterial QS. Therefore, pathogenicity of pathogenic bacteria can be lost through interfering and damaging its QS [4].

Acyl-homoserine (AHL) is used by most gram-negative bacteria ($G^-$) as signaling molecules, but Autoinducing peptide (AIP) is used by most gram-positive bacteria ($G^+$) as signaling molecules [5]. AHL signaling molecule is a small molecule compound, and typically features containing homo-serine lactones and 1 amide side-chain. Bacterial AHL is catalytically synthesized by enzymes and identified by relevant sensing system after arriving at extracellular environment, and then forms R-AHL compound with R-protein, and then is used as transcription regulator after activation for activating the expressions of target gene and I-protein gene [6]. Regulations on many important biological functions such as transfer of Ti plasmid, virulence factor expression and its bio-film formation, and prodigiosin synthesis are participated by signaling molecule AHL. Quenching pathogenic bacteria signaling molecule AHL makes its accumulation concentration in environment difficult to achieve the expression causing virulence factors, thus making pathogenic bacteria lose virulence. Therefore, the screening of quenchers or inhibitors of bacterial QS has been a hot research topic in biology, medicine and other areas [7]. Researches show that Bacillus (e.g. *Bacillus thuringiensis*, Bt) contains degradative enzyme degrading AHLs molecules and coded by gene *aiiA*, and this type of protein, by hydrolyzing the lactone bond of AHLs molecules, can reduce the concentration of AHLs and weaken the pathogenicity of pathogenic bacteria; *Bacillus cereus* (Bc) can block QS system and degrade AHLs signaling molecules by releasing aiiA protein [8].

At present, many genes *aiiA* coding aiiA protein has been cloned and sequenced. Now, cloned signaling degrading enzyme genes *aiiA*, which have been reported, source from soil microbes, and there have been no reports about cloned signaling degrading enzyme genes from marine microbes yet. Therefore, the marine bacterial strain ZD02 previously screened was selected as target bacteria in this study; a pair of degenerate primers was designed according to the homology of nucleotide sequence of reported gene *aiiA* coding aiiA protein, and also gene *aiiA* was cloned and its expression vector was constructed, laying a solid foundation for further researching the gene *aiiA* recombination, active expression and disease-resistant mechanism of ZD02.

## 28.2 Materials and Methods

### *28.2.1 Materials*

#### 28.2.1.1 Bacterial Strains and Plasmids

Experimental strains: ZD02 (screened in early experiment) was selected as strain to be tested; *Escherichia coli* DH5α were gene cloning host bacteria; *Escherichia coli* BL21 (DE3) were gene expression host bacteria.

Plasmid: Cloning vector pMD-18T and expression plasmid pET-17b were purchased from TaKaRa (Dalian) Co., LTD.

#### 28.2.1.2 Culture Mediums

LB medium comprised of Tryptone 10 g, yeast extract 5 g and NaCl 10 g, which were dissolved by distilled water until pH 7.0 and constant volume 1,000 mL. Set aside after sterilization.

LB solid medium was mixture of 1,000 mL LB liquid medium and 15 g agar powder. Set aside after sterilization.

#### 28.2.1.3 Main Reagents and Enzymes

EcoRI, BamHI, Taq enzyme, T4 DNA ligase, DNA gel purification kit and plasmid purification kit were purchased from TaKaRa (Dalian) Co., LTD. Agarose and Gel Extraction Mini Kit were produced by Oxoid and QIAGEN, respectively.

### *28.2.2 Methods*

#### 28.2.2.1 Templates Preparation

First, strains were selected from fresh and purified scope and placed in 5 mL LB liquid medium overnight for shake culture (30 °C, 230 r/min). Second, strains were centrifugally collected from 1.5 mL out of above 5 mL LB liquid medium and then added with 80 μL SI (0.3 $molL^{-1}$ sucrose and 25 $mmolL^{-1}$ EDTA) and lysozyme to make final concentration 50 mg·mL$^{-1}$, and then placed at room temperature for 10 min. Third, 420 μL buffer solution (100 mmol·L$^{-1}$ Tris-HC1, 25 mmol·L$^{-1}$ EDTA, 0.5 mol L$^{-1}$ NaC1, 0.5 g·mL$^{-1}$SDS, pH 7.5) and 50 μL Tris Saturated phenol were put, and blended uniformly. Fourth, isovolumetric chloroform/isoamyl was put for 3 times of extraction, and then liquid supernatant was moved to another new centrifuge tube. Fifth, isovolumetric isopropanol pre-cooled at −20 °C was put for sedimentation and then washed with ethanol (volume fraction: 70 %), and dissolved in 100 μL TE buffer after drying. Store them at 4 °C and Set aside.

#### 28.2.2.2 Primer Design and PCR Amplicon

Primer Design: According to the sequence of gene *aiiA* known in GenBank (registration No. AF350929, AF478059 and AF478060, etc.) and reported literature [15], a pair of degenerate primers AiiF and AiiR with BamHI and EcoRI restriction enzyme recognition sites was designed and amplified for gene *aiiA* with sequence's high homology.

AiiF: 5′-ATG *GGA TCC* ATG ACA GTA AAG AAG CTT TAT-3′ (BamHI restriction enzyme recognition site)
AiiR: 5′-GTC *GAA TTC* CTC AAC AAG ATA CTC CTA ATG-3′ (EcoRI restriction enzyme recognition site).

PCR Amplicon: Reaction system comprised of 5 μL 10 × reaction buffers, 6 μL MgC1 (20 mmol L$^{-1}$), 1 μL dNTPs (10 mmol L$^{-1}$ respectively), 1 μL template, 1 μL aiiF (10 μmol L$^{-1}$), 1 μL aiiR (10 μmol L$^{-1}$), and 0.5 μL Taq enzyme (5 UμL$^{-1}$).

Reaction condition: Pre-degenerating for 5 min at 94 °C, degenerating for 1 min at 94 °C, annealing for 1 min at 50 °C, extending for 1–2 min at 72 °C, and repeating above for 30 times; finally extending for 10 min at 72 °C and 0 min at 4 °C.

#### 28.2.2.3 Construction of Cloning Vector

PCR products (5 μl) was taken for agarose gel (1 %) electrophoresis analysis and then recycled with DNA gel purification kit (TaKaRa Co., Ltd.). Purification products and pMD18-T vector under the action of T4 DNA ligase were linked at 16 °C overnight and then converted to competence *escherichia coli* DH5α, and then positive clones were selected and tested by primers AiiF and AiiR. Positive clones whose fragments were correctly inserted were selected and sequenced in Shanghai Sangon Biotech.

#### 28.2.2.4 Construction of Expression Vector

By taking above correctly-sequenced positive cloning plasmids as templates, PCR amplicon was conducted on 5′-GGTGGTFGCTCTTCCATGACAgTAAARA-ARCC-3′ and 5′-GGTGGTCTCGAGTATATACTCTGGGAACGCTYTAC-3′ with primers containing BamHI and EcoRI restriction enzyme recognition sites, and 750 bp DNA fragment was extracted with gel after double digestion, and then positive cloning fragments and expression vector pET-17b after double digestion and via gel were linked overnight, and then fusion expression vector pET-*aiiA* was constructed, *Escherichia coli* BL21 (DE3) was converted, positive clones were selected, and PCR was tested. The positive clones whose fragments were correctly inserted were selected and sequenced in Sangon Biotech (Shanghai) Co., Ltd. Single colony of the clones with correct sequencing results was selected for culturing overnight. Plasmids were extracted with Plasmid mini preparation kit.

## 28.3 Result and Analysis

### *28.3.1 PCR Amplicon of* aiiA *Gene*

Using AiiF and AiiR as primers and ZD02's DNA as template, PCR Amplicon of *aiiA* Gene was conducted, and the Agarose gel electrophoresis analysis of PCR products. The result shows that DNA fragment (750 bp) with no significant non-specific strip was obtained after amplicon, and kept consistent with expectation. The amplicon products after electrophoresis analysis were recycled with DNA purification kit and then linked with cloning vector pMD18-T, and then recombinant plasmids were constructed, *Escherichia coli* DH5$\alpha$ were converted, and finally clones were selected and named as pMD18-ZD02*aiiA*. Also, colony PCR was identified with AiiF and AiiR: DNA fragment (750 bp) was amplified from cloning colonies, and kept consistent with expected strip. This indicates that previously-selected clones were positive, and 95 % of these positive clones after sequencing were found to be similar to the results announced in GenBank. Successful selection of positive cloning provides a guarantee for the construction of next expression vector.

### *28.3.2 Construction and Identification of Expression Vector*

BamHI and EcoRI double restriction enzymatic digestion was conducted on MD18 plasmids linked with ZD02*aiiA* gene. Enzyme digestion products were purified and recycled with DNA purification kit for electrophoresis analysis. The recycled products were linked with pET-17b linear vector via double digestion, and then expression plasmid pET-ZD02*aiiA* was constructed, and linking products were

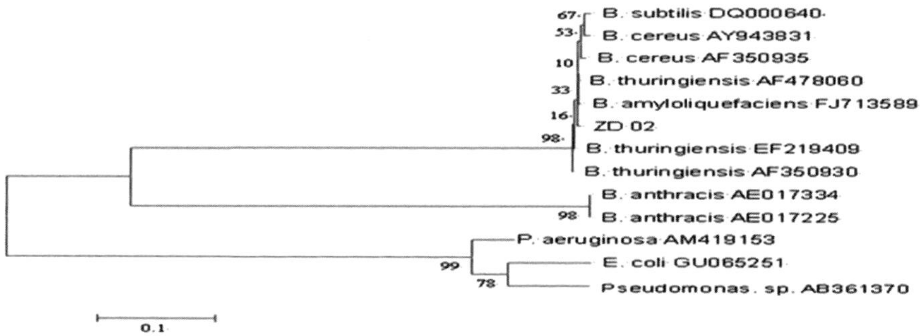


**Fig. 28.1** Rectangular cladogram of *aiiA*

converted to competence cell BL21 (DE3). Single colony on plating was selected for amplified culture, and then recombinant plasmid pET-ZD02*aiiA* was extracted and digested by BamHI and EcoRI, and then colony PCR of positive clones was tested with primers. Positive clones with correctly-sequenced PCR were selected. Sequencing result shows that the length of DNA fragment was approximately 750 bp, consistently with expectation. This indicates that fragment of *aiiA* target gene was correctly inserted pET-17b, and its sequence and reading frame were accurate.

The aiiA phylogenetic tree was constructed through nucleotide sequence (Fig. 28.1). It can be seen that aiiA obtained in this experiment was of the same kind with aiiA of *bacillus amyloliquefaciens*, *b. thringiensis*, and *bacillus cereus*, and they are closer evolutionarily. Except *E.coli* and *pseudomonas sp.*, the rest sequences were bacillus *aiiA* sequences. Homology of *aiiA* sequences obtained from sequencing was analyzed with BLAST software (http://www.ncbi.nlm.nih.gov/BLAST), and finally over 95 % of aiiA gene sequences were found to be

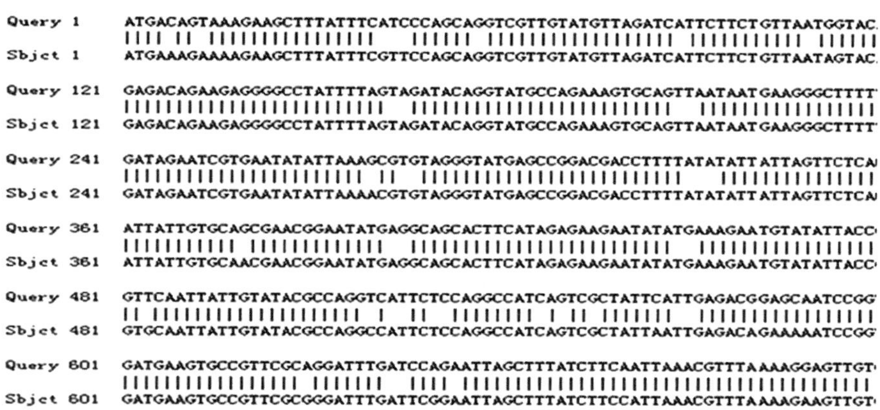

**Fig. 28.2** Comparison result of ZD02-*aiiA* gene BLAST. *Query* ZD02-*aiiA* gene nucleotide sequence. *Sbjct*: *Bacillus thuringiensis aiiA* gene nucleotide sequence (AF478060.1)

highly homologous with the sequences announced in GenBank (Fig. 28.2). Also, homologies of the nucleotide sequence with *Bacillus thuringiensis* (Registration No. AE017355.1), *Bacillus cereus* (CP001746.1), *Bacillus metaterium str* (FJ713591.1), and *Bacillus amyloliquefacius* (FJ713589.1), uncultured bacterium clone (HQ876560.1), *Bacillus sp* AHL-lactonase gene (EF537015.1) reached 96, 99, 95, 95, 96 and 99 %, respectively.

## 28.4 Discussion

In this experiment, a ZD02*aiiA* gene was cloned and identified in PCR, and clone and expression vectors were successfully constructed, thus laying a foundation for further bioinformatics analysis, molecular evolution analysis and expression activity research, but also helping discuss marine bacterium ZD02 as a resistance mechanism of potential antagonistic bacteria. Under the condition that the proportion of microorganisms is extremely low (1 %), it is helpful for the development of microbiological resources. Therefore, such a conservative sequence can provide a theoretical reference for activity prediction and enzyme engineering. Also, ZD02-*aiiA* new diseases resistance sequence was cloned in prokaryotic expression vector, laying a foundation for researching biological activities of recombinant protein expression.

**Acknowledgments** This work was supported by grants from Science and Technology Plan Project of Guangdong Province (2009B020311003), Science and Technology Plan Project of Jiangsu Province (BN2010038), Special Scientific Research Funds for Central Non-profit Institutes, Chinese Academy of Fishery Sciences (2010TS01), Production, Learning and Research Project of Guangdong Province (2011B090400151).

## References

1. Fuqua WC, Winans SC, Greenberg EP (1994) Quorum sensing in bacteria: the LuxR-LuxI family of cell density-responsive transcriptional regulators. J Bacteriol 176(2):269–275
2. Fugua WC, Winans SC, Greenberg EP (1996) Census and consensus in bacterial ecosystems: the LuxR-Luxl family of quorum. Sensing transcriptional regulators. Annu Rev Microbiol 50:727–751
3. Choi SH, Greenberg EP (1992) Genetic dissection of DNA binding and luminescence gene activation by the Vibrio fischeri LuxR Protein. J Bacteriol 174(12):4064–4069
4. Nealson KH, Hastings JW (1979) Bacterial bioluminescence: its control and ecological significance. Microbiol Rev 43(4):496–518
5. Ding X, Yin B, Qian L et al (2011) Screening for novel quorum-sensing inhibitors to interfere with the formation of P. aeruginosa biofilm. J Med Microbiol 60(12):1827–1834
6. Pearson JP, Gray KM, Passador L et al (1994) The structure of the autoinducer required for expression of Pseudomonas aeruginosa virulence genes. Proc Natl Acad Sci USA 91(1):197–201

7. Swift S, Downie JA, Whitehead NA et al (2001) Quorum sensing as a population-density-dependent determinant of bacterial physiology. Adv Microb Physiol 45:199–270
8. Dong YH, Zhang LH (2005) Quorum sensing and quorum-quenching enzymes. J Microbiol 43:101–109

# Chapter 29
# Distance Education Based on Java and IP Technology

**Xiaosong He**

**Abstract** Long-range educational development has already gone through history in 200 years. With the development of society and technology, distance education has got swift and violent development. This text proceeds with concept of long-range technology, understand the major technology that distance education adopts, it is mainly java technology and application in the distance education of IP technology, well realize the figure interface and question transmitted in distance education.

**Keywords** JAVA technology · Distance education · IP technology

## 29.1 Introductions

What is the distance education? From looking on being literal, outstanding characteristic is non- face-to-face with the intersection of space and educational activities of distance. Though, the personality characteristics of the distance education is more outstanding, give it the next one's strict definition, also not so easy, there are different views. Several more famous or more classical definitions of trial [1].

Distance education whether one have the intersection of self-study and form that system organize, in the form, students' consultation, preparation and assurance and supervision of student's achievement of study material were all carried on by a teacher's group. Each member of this group has high-level sense of duty. Media means may dispel from and can last longer distance media means.

X. He (✉)
Chongqing University of Science and Technology, Chongqing, China
e-mail: xiaosongher23@126.com

W. Du (ed.), *Informatics and Management Science V*, Lecture Notes in Electrical Engineering 208, DOI: 10.1007/978-1-4471-4796-1_29,

Distance learning/distance education is a method to teach knowledge, skill and attitude; it is rationalized to pass labor division and application of the organizational principle and wide application of technological media. Especially the purpose to duplicate high-quality teaching material is to teach a large number of students how to become possible at the place where in students life at the same time. This is an industrialized form of a kind of learning aid.

Relatively have separation to teacher and student on space–time in distance education, the behavior of learning aid realizes the general name of getting in touch, education of mutual and all kinds of universities and colleges or social organizations combined through various education skill and media resource.

## 29.2 Major Technology that the Distance Education Adopts

In the distance education, we can adopt much technology, the platform of better optimization distance education; realize the good interdynamic between teachers and students [2]. In the article, we mainly propose two kinds of technology; it is ip technology and java technology respectively. At present, the basic transmission based on IP transmission net of transmission of the distance education, mainly realize the video conference through such a transmission net; in the distance education, we still adopt java technology, realize the optimization of the network figure interface.

### *29.2.1 Application in the Distance Education of Boundary Plane Technology of Java Network Figure*

The platform part of the figure is a mutual system of a realization figure. Its main task is to offer a friendly visual interface to student, make students familiar with the general habit of computer graphics [3], experience the result that the figure varies, the composition of the platform of exercise figure carrying on the project picture and studying the question is: Four (dass) File—netsketch class, NetSketch and Textdialog class; A Multysocket exe file, run on the end of server while using; Several drawing exercise data files, file all files that each exercise file is a dxf form are left in the server, the webpage studying the distance education with the project picture is put under the same catalogue.

Key technology is: Adopt the analytical method facing target, use the standard design of storehouse of Java technology to organize each in network little procedures and various; Adopt advanced window tool (AwT) in Java now Respond to technology and write the interface in the figure platform with the advanced incident; Write the drawing tools and auxiliary drawing tools on the figure platform according to skills such as XOR mode of principle and Java of computer design graphics, etc.; Adopt the file I/O technical treatment of Java, organize the data on the figure platform, adopt URL network programming technology of Java to practise the exam pool with the drafting of project on the figure platform to link.

The total interface in the figure platform is one, because of graphical user interface (GUI) spring window 1:1. Divided into four areas at the window. Top systematic the intersection of menu and district, on the right to catch way and swift switch district that switches over, it makes to be the intersection of state and person who brief on, middle part whether the intersection of figure and the intersection of screen and origin of system of coordinates of platform, district of drawing, agree on upper left corner in the window, the system of coordinates of the screen is that the left hand is. And the origin of user's system of coordinates agrees on the lower left corner at the window, user's system of coordinates is that the right hand is. Through this interface, users and the long-range education system can carry on mutual and present main application in a friendly way conveniently:

Practise the artificial practice in online drawing "Drafting and computer of civil engineering are painted".

Pursue "The practicality of course is very strong, a lot of contents need to practise on the computer. At present, while teaching oneself with this system, can chooses the mapping exercise of descriptive geometry and carries on practice while studying the principle.

Online test: Understand the mapping habit of popular mapping software, because platform of this figure is similar to present popular mapping software AutoCAD in the menu is fixed up, and is furnished with precise and to the point help file students in the system and can also practise the drawing familiar with AutoCAD and get used to through the artificial drawing on this platform, meet the working environment in the future more quickly.

The network figure platform where this text introduces is a CAI software based on Java technology, superior mutual performance and portability are one of the main characteristics, and can the intersection of job and it meet the intersection of project and long-range requirement of education system that picture study supported by long-range figure file, mutual occasion of figure that can be used in other networks too after changing slightly.

### 29.2.2 On the Basis of the Application in the Distance Education of Video Conference Technology of IP Technology

At present, the technology of video conference based on IP is roughly divided into two kinds. Namely the application software based on products and based on IP/multicast of H.323 standard. Typical video conference system is based on H.323 standard. By the terminal and many channel unit (MCU) Make up $ if the system has n terminal stations, then there is N−1 one that sends from terminal station to video flowing of MCU and needs to take up N−1 one from MCU to receive the terminal video flowing altogether on the network. Need to take up n resources and meetings of resource bandwidth of bandwidth which meet communication of video terminally and forming the star structure altogether, become the single trouble point of the system. In order to improve the dependability of the system, need to restrain MCU from connecting terminal quantity; In order to raise the operational efficiency of the system, need to increase the host computer handling capacity of MCU, thus has reduced the systematic expansibility, and increased the cost.

The application software based on IP/multicast came out at the beginning of the 1990s, is used in the video conference service (MBONE) as the members of Internet project committee IETF The standard compares standardization and commercialization maturity of MBONE and is not enough for high $ but adopt the group and sow technology and save MCU central apparatus of H.323. By the meeting president' A certain meeting initiates the terminal station Establish the group G of the meeting. Other meeting terminal stations can join or leave the group freely. Systematic flexibility and expansibility are good; As the system realizes 1 to (N−1) Video at the time of communication, last service to last radio at group that network offer video flowing that be is it square to sent. Thus has raised dependability I expansibility and operational efficiency of the system. As shown in Fig. 29.1.

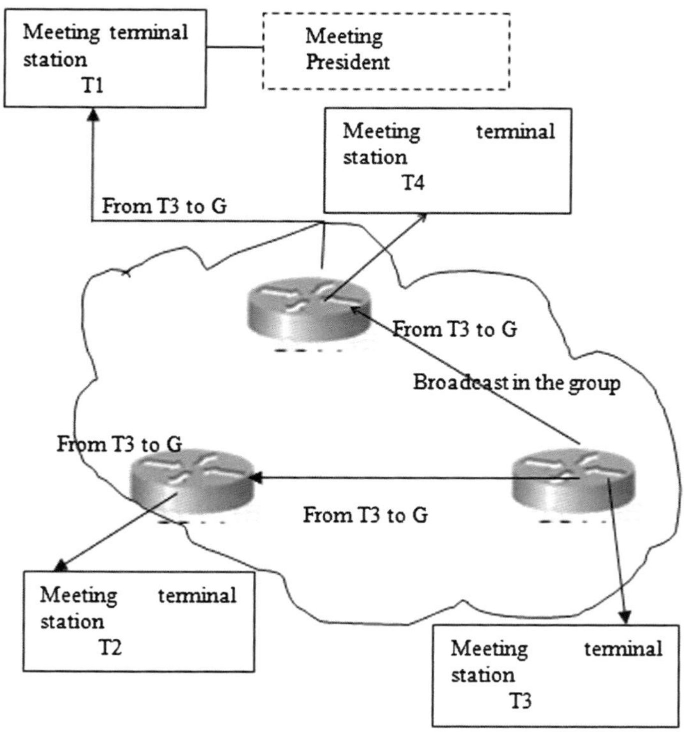


**Fig. 29.1** Mbone systematic operating chart of the video conference

The group sows technology and realizes superiority on in the video conference, and can obtain support of Mbone work group technically. This selected works select the base and give lessons on the technological foundation of the system when the video conference technology of IP/multicast is as being real.

Have now, go on because of many to it gives lessons to be feasible while being a bit long-range the intersection of IP/multicast and the intersection of video conference and technology at the high-speed network in China, the further analysis of the experimental data shows, give lessons to need the high-speed network back-up environment that the service quality guarantees long-rangely, consider Mbone network adopts the swapper of dividing into groups, gives lessons long-rangely the network platform that the environment relies on should offer enough bandwidth resources for long-range video communication giving lessons, special service quality control and multicast service' with the development of mere communication, will build up high bandwidth, high performance, the network infrastructure that can be inserted whenever and wherever possible, under the environment of high-speed network, network application will obtain the broad group and sow the service support, the system will be used widely that the base gives lessons when IP/multicast is real.

## 29.3 Design of the Systematic System Structure

### *29.3.1 Network Environment*

The distance education platform mainly relies on Internet and develops the activity of distance learning, implement school of distance education set up the intersection of information and control centre and give lessons some, click and disperse setting up in every place in distance learning. In guaranteeing to transmit the network environment of the bandwidth, the network has offered the passway of the information transmission for distance education platform, it is permitted to guarantee a smooth distance learning to implement.

### *29.3.2 Systematic System Structure*

The management control platform is an important component of the long-range education system, realization of it goal the long-range education system and Network the intersection of management control and software combine together, guarantee long-range the intersection of education system and high-efficient, stability, run safely organic. The management control platform controls the difference of functions needed according to implementing management as a complicated software system, can divide into 10 pieces of sub module, namely the network is topological, apparatus assets, incident report, performance analysis, user's control, the resource control, software are distributed, maintain safely, daily tool and platform set up.

## 29.4 Conclusion

The modern distance education is a new kind of educational ways to produce with development of modern information technology. Computer technology, many matchmakers development of body technology, communication, especially the swift and violent development of Internet, make the distance education means have qualitative leaps, become distance education under the new and high technology condition. According to design the intersection of distance education and platform that conceptual design realize manage control system, have the intersection of performance and reliable, technologically advanced, the intersection of expansibility and characteristic such as being better originally. Manage the management of system grade that the control system can realize the whole distance education platform in the platform, can be used in managing and controlling the normal operation of other 3 subsystems, thus the operation that ensured the long-range education system to be safe and effective.

## References

1. Hillery M, Buzek V, Berthiaume A (1999) Frequency-domain theory of nonsequential double ionization. Phys Rev 59:1829–1836
2. Armstrong T, Patton R (2000) ATL develops the guide, vol 08. Electronic Industry Press, Beijing, pp 74–79
3. Pan A (1999) COM principle and employing, vol 21. The Publishing House of Tsing-Hua University, Beijing, pp 181–186

# Chapter 30
# English Education Based on Network

**Xiaoli Jiang and Hui Liu**

**Abstract** The network technique is applied to teaching of English, make the computer assist teaching of English to enter people for a new developing stage. The English education under the network environment is given consideration to "Study" with "Teach" need while emphasizing student's entity, fully play a leading role in teacher. Teaching of network and traditional teaching way can be supplemented, mutually promote each other. Explain the combination of network technique and English teaching, discuss to the English education of network, analyze the factor influencing teaching of English, and put forward the countermeasure solved.

**Keywords** Network technique · English education · Study and teach

## 30.1 Introductions

The application of the teaching means of the network has important practice meanings in the English education of China, in such aspects as English teachers, academic environment of the language and teaching resources, can remedy insufficiently, supplement, and mutually promote with the traditional teaching way each other [1]. China's English education is perplexed by nervous problems of teachers for a long time. In terms of university English, public English lesson

X. Jiang (✉)
School of Foreign Languages, Jiangxi Science and Technology
Normal University, Nanchang, Jiangxi, China
e-mail: xiaoljiangb32@126.com

H. Liu
Humanities Section, College of Science and Technology, Nanchang University,
Nanchang, 330029 Jiangxi, China

W. Du (ed.), *Informatics and Management Science V*, Lecture Notes in Electrical Engineering 208, DOI: 10.1007/978-1-4471-4796-1_30,

belongs to basic teaching, the school does not often pay attention to very much, the treatment is general, it is difficult to attract outstanding talents. Under the traditional teaching mode, English implements primary class' giving lessons, needs hands to be numerous. Should solve this problem, the school administration must pay attention to the English teaching of the university, strengthen English teachers' building of contingent, improve the treatment. Meanwhile, we can adopt the new countermeasure too: Reform the teaching means, introduce teaching of English of the computer network technique. Let the computer replace English teacher partially [2]. As to content of courses drilled repeatedly which needs a large amount of such as English reading, vocabulary, grammar, hearing, develop the corresponding courseware based on network, give full play to the advantage of human–computer interaction of the new technology, let computer interactive to feedback, replace to work repeatedly teacher, let student follow computer, teach oneself, stimulate the enthusiasm of their study, this can remedy the deficiency of teachers' strength to a certain extent. In terms of teaching specialized in English, the network technique is to have ample scope for one's abilities too. As a foreign language, English lacks the academic environment of languages in China [3]. The study time in the classroom is limited, and the languages in the classroom are drilled going on in the situation that is supposed. Students often study languages for language; but outside class, students have not basically carried on real opportunity of exchange and environment in English. In creating this point of academic environment of language for English learner, the network technique can contribute. Though exchange on the network is going on in the fictitious Cyberspace, this is a kind of tangible exchange. It can be that the learner offer the true exchange environment to exchange the space in language that the network creates in the classroom, make learners have a chance to study in order to practice to the language knowledge that they know, and then make their language knowledge and skill be consolidated and improved in using. For example, carry on the teaching of writing of English under the environment of network and let students write in English on the zone of discussion, students can accomplish in order to exchange, study and grasp languages by using actually, make the cultivation of the communication ability of the language fulfill. Abundant English resources are the greatest advantage of the network. The English resources on net can be said to be inexhaustible. In teaching, teachers can download the language material from network, screen the editor, provide guidance to the reading of students; Students can search and browse through, expand one's own range of knowledge on the net by oneself too, strengthen oneself, to study understanding of content, can pay preliminary research with the intersection of network and resource to special topic a certain. Like this. Not limited to the text on a certain teaching material again in the reading the surface of students, but press close to actual life, real language material. The innovation of the teaching means may cause the changes of the teaching idea, teaching method fundamentally, make the intersection of English and teaching learn law to rely mainly on student from the intersection of teacher and leading the intersection of office coaching and law gradually. We can say the network is opening up a frontier of teaching of English.

## 30.2 Networked Technological Terms of Teaching of English

21st century is the century in the network, the network technique will make the English education field produce the deep change, promote educational modernization. Implement the English teaching under the network environment, the condition of two respects: First, suitable for various teaching courseware's of teaching of network, second, network hardware apparatuses such as the computer, server, etc. Speaking from the microcosmic, the teaching based on network can substitute chalk and blackboard with the note-book, projector progressively, make the functions of computer lab and classroom permeate, complement one another, realize the teaching means modernization each other. Speaking from the macroscopic, the development of the CAI software is paid close attention to by the country, it will promote the development of information industry of our country, will further expand the demands of information products and information service, become the point of growth of firewood of developing and economic development of the whole society of national information industry, thus promote forming and development of the knowledge economy. So, there are good development prospects in the teaching based on network; it is the undertaking of having bright prospects.

The key of teaching of network is to develop the courseware (web-based courseware) based on network, It can cross over the space and carry on the real-time or non- real-time and mutual characteristic, this is a difference that teaching of network and traditional teaching way are most prominent, it is its advantage relative to software of past computer-assisted instruction too. From the sixties of the 20th century, the development of the computer-assisted instruction has already crossed over several great developing stages. In view of hardware condition, the software of computer-assisted instruction relies on the mainframe system to bring a plurality of teaching terminal stations in early days, get and use the personal computer of performance. Develop into and use the computer network now; In view of software media, the software of computer-assisted instruction is from early characters media based on little capacity magnetic disc to unit multimedia based on read-only optical disc, have already developed into network multimedia now. In this development, the computer-assisted instruction theory develops with constant development of hardware and software terms too. In initial stage of era and personal computer development of mainframe, the application of the computer in teaching is known as "computer-assisted instruction" (computer aided instruction), This period emphasized "Teach", Namely software developer (teacher) Control studying the content, study result and even the whole learning process through software design. This kind of teaching idea and hardware and software terms of those years suit: First of all, minority studied to just have in the computer lab of the school the mainframe and only terminal of early years, such a hardware condition has determined the computer-assisted instruction in this period remains the classroom instruction of a kind of concentration; Secondly, the software used under this kind of hardware condition can only offer the simple characters interface and limited content of courses, to most learners, spend computer go on, study very loud to limit at time and place, study how much choice content have to throw, dull interface very difficult to make learner keep enthusiasm that study through computer for a long time. Under such a condition, emphasize "teaches" the inevitable result. By the eighties, the development and popularizing offering the new condition for development of the computer-assisted instruction progressively of the personal computer, the computer at this moment is no longer the accurate scientific instrument too high to reach, it has come into the ordinary classroom, students' dormitory and family, have widened time and space range that the use computer carried on study. With constant enlargement, especially appearance of read-only optical disc of the space of hard disk, the application of the multimedia is popularized progressively. The memory space of magnanimity makes the teaching software offer more, more interesting study contents including various formal Medias; Personal constant improvement and constant development of computer language of computing power make form of human–computer interaction complicatedness, speed even swift. The teaching software of interactive multimedia taking read-only optical disc as carrier can offer the study terms of man–machine duel, can keep the learner's interest through the abundant multimedia content, is especially suitable for teaching one. Under this kind of condition, emphasize "studies" "computer aid study" (computer-assisted learning) Concept arise at the historic moment.

From "computer-assisted instruction" the computer assists and studies "to get, although these two kinds of concepts have the leasing of priority". But the intersection of evolution and course might not represent certain from rudimentary to the senior revolutionary evolution. There are their materials bases after all in these two kinds of teaching theories, are decided by particular hardware condition and software terms in conformity with it. The early system with a plurality of terminal stations of mainframe is a computer network in fact, it can offer to student and study the content to offer to teacher and supervise and check the technological condition of the learning process; by the look of angle educated. The concept of the computer-assisted instruction of times of the mainframe accords with the truth of school education even more, but because emphasize "Teach" Will often inculcate the spoon-feed in the traditional classroom instruction with and is linked to, so emphasize "Study" Computer last study concept once appear revolutionary colors a certain. However, we should see, emphasize "studies" in the multimedia era of the unit It is in fact a kind of tactics of suiting measures to differing conditions in terms of time too: Though perhaps on performance personal computer before Internet popularizes is stronger than the early mainframe. But the scattered personal terminal station can only be used in "Study" And is not suitable for use in "Teach", It is interdynamic type that it is suitable for the self-study in the home education and not very suitable for teachers and students "Teach" With "Study" School education combining together. With the development of computer technology, the present computer network has synthesized the hardware for first two stages and advantage of the software condition. There is interconnected function of networks. Can offer the facility for supervision and checking of the teaching course, there is advantage of multimedia, can offer the interesting study content abundantly, “Study” With “Teach” Can well give consideration to. So, under present network terms, we propose “online education” (web-based education) Concept, while utilizing the technological means of the computer accomplish “Study” With “Teach” Combine together, according to purpose and requirement of concrete course, use the computer to assist “Teach” in some courses And emphasize “Study”, Assist in other courses “Study” And emphasize “Teach”.

## 30.3 English Is Networked to the Meaning in the Teaching of Undergraduate Course

The education influence to the undergraduate course of the university of English education of the network is greater. In the above-mentioned three respects' the English study of the university, undergraduate course study and study after class), Courtyard student speak the network to other students not accepting the network educates, there are certain advantages, especially specialty ability gives play to the respect outside class. Now combine the characteristic of online education, it is simple to analyze these advantages.

The English education of network is the study which is based on resources. Professor RiGuo Gu has pointed out, computer and network technique give and educate bringing three revolutions:

First of all, the revolution stored and obtained in digitization of resource. The books resource in the university, audio frequency, video material, the teacher gives lessons, test, assess materials, files are all resources. All these now can carry on the electronic storage through digitization, can carry on the strange land real-timely through Internet technology at the same time or not draw in real-time.

Secondly, have no circle revolution on the physical space. The network does not have traditional demarcation line, such as the demarcation line (national boundaries, campus enclosure, etc.) on the region, That behavior pay demarcation line ‘Such as the lesson, after class), psychological demarcation line. In this sense, Internet has realized that there is not circle revolution. Certainly this does not say that there is not a demarcation line at all on the net, but say, the traditional concept about “circle” has changed, have produced another demarcation line regarding and resource allocation is presumed artificially.

Finally, resources optimize and utilize. Two in the front—The digitization of resource is stored and obtained and there is no circle revolution on the physical space—Pave the way for realizing resources are optimized and utilized. People can offer the best educational resource to those persons that can't get because of time, obstacle of the space originally through the network through the network technique.

The teaching idea based on resource is very great to influence of those students who accept the network educates. Network courtyard students will accept a large number of relevant information resources beyond accepting and stipulating the content of course, this in that many resources could not get the thing and not apt to get in the traditional classroom in the past. As to student, can obtain, fill these resources in selecting, deal with, this itself is a kind of improvement. They are on the broadness of the visual field, think that have obvious advantages to other students on the depth of the question that this makes. This formal they can into accept with those student that network educate roughly the same in study undergraduate course on one hand too, on the other hand study and reason that the ability specialty trains the advantage to be obvious outside class.

## 30.4 Conclusions

### *30.4.1 Develop Courseware*

Develop more better courseware based on network, should pay attention to "Study" in the course of developing Function but also want give consideration to "Teach" Need, develop, last teaching characteristic, accord with English teaching the courseware high-qualities of the laws. The existing courseware must be improved constantly, while the courseware is developed and network teaching is practiced, summarize experience constantly, in interface, function, improve the courseware level constantly at the first-class content of the content, actually strengthen the using. Should launch the teaching of network and research of traditional teaching, through holding all kinds of seminars, exchange experience, and promote the teaching of network. As to ripe courseware, go on, propagate, introduce, and accomplish outstanding resource-sharing through various channels. In order to guarantee sustainable development, we will explore actively that accord with the teaching software development mode of the economic law of the market and develop the mechanism.

### *30.4.2 Use Resources*

Fully utilize the resources of the school. A lot of schools set up independent computer lab, can contact educational administration and network center of the computer, obtain their support, and construct the network teaching platform. Give consideration to teaching and network teaching of traditional language in construction.

### 30.4.3 Change the View

The leader pays attention to the teaching of network. On one hand, strive for the funds of the school's support actively, create the terms in order to implement the teaching of network; On the other hand, it is estimated that a teacher applies the computer network technique to teaching, utilize all occasions to propagate, promote the teaching of network. The teacher develops the courseware, corresponding acquisition too more funds support.

The network is the new things; its growth and development need our efforts. We believe, with the development of information technology, the intersection of network and teaching under modern support of technology teaching modes are generally accepted by people effectively one that become in the English education. The English education under the network environment is a frontier of teaching of English, as university's teacher, we should make one's own contribution in this respect.

## References

1. Zhang G (1995) The computer assists foreign language teaching and research, vol 8(05). Foreign Language Education Press, Shanghai, pp 27–33
2. Gui S (1997) Professor's net surfing manual of foreign language, vol 12. Foreign Language Education Press, Shanghai, pp 232–237
3. Shouren W (1998) Network environment makes the speciality of english, vol 04. Beijing University Press, Beijing, pp 78–82

# Chapter 31
# Study of Intelligent Transportation Management Systems on Urban Energy and Environment

**Yichun Ni, Zhenzhou Yuan and Wenyu Liu**

**Abstract** With the rapid development of urban ecology, the service quantity of urban traffic and transportation raises, traffic congestion increases, and energy demand ascends too, environmental emission become a major problem in a city. This case study in Beijing, analysis the impact of Intelligent Transportation Management System (ITMS) on natural environment, like urban atmosphere and noise, and energy utilization rate, the result is ITMS reduces the emission of traffic exhaust and greenhouse gas. This paper get the automobile oil consumption, descent rate of energy consume in Beijing, and the emission inventory of $NO_x$, HC and CO. According to estimate the shadow price of petroleum, calculate the quantity of energy structure adjustment. This paper also gives the calculation of urban pollution control benefit and greenhouse gas emission benefit. Draw a conclusion that ITMS brings Beijing the environmental and energy benefit about 1.337 billion RMB, and plays a great role in Beijing's society, economy and environment.

**Keywords** UITMS · Urban energy and environment · Energy structure adjustment · Pollution abatement

## 31.1 Introduction

In recent years, the development of urban economy in China increases rapidity, among which traffic and transportation make an indelible effect. However with the population grow and land area limitation, traffic congestion becomes an important

Y. Ni (✉) · Z. Yuan · W. Liu
School of Traffic and Transportation, Beijing Jiaotong University, Beijing 100044, China
e-mail: yichuan23ni@126.com

W. Du (ed.), *Informatics and Management Science V*, Lecture Notes in Electrical Engineering 208, DOI: 10.1007/978-1-4471-4796-1_31,

problem in major cities in China. In order to relieve traffic pressure, Beijing's government spends a large number of money on traffic in a long time, besides strengthen the traffic infrastructure, and build a set of ITMS. It massively relieves traffic congestion, improves traffic safety, increases the ability to handle accidents, ameliorates freight transportation management, and reduces environment pollution [1]. But traffic investment, belongs to infrastructure investment, has its externality, so the quantity benefit is not calculated easily [2]. This paper through building model to calculate the economic benefit of saving energy, according to the change of data between ITMS not used and used.

## 31.2 Evaluating Indicator of Impact ITMS on Energy and Environment

According to the trait of Beijing traffic development, this paper builds an evaluating indicator system fits Beijing ITMS program, adopts method combines quantity and quality, and analyses the impact of ITMS on natural environment and energy utilization rate.

### *31.2.1 Evaluating Indicator of Energy Benefit*

Energy structure adjustment benefit means the effect reducing energy in traffic field acts on economy development and energy safety. Due to the scarcity of energy, one field using energy will reduce another place. So we must optimize it and create a largest benefit.

**Energy consumption in traffic** Energy consumption per mile in transportation is an important indicator to energy efficiency in transportation. Usually use fuel economy and mileage to calculate it.

**Shadow price of petroleum** Calculate the adjustment benefit of energy structure needs the shadow price of petroleum. It not only can embody labor consumption of resource (social value), but also can reflect the resource scarcity degree (supply–demand relations).

### *31.2.2 Evaluating Indicator of Environmental Benefit*

Emissions of pollutants from traffic and other facilities concern the impact means of traffic on urban environment, and are an important indicator to traffic development level. Traffic pollution emission contains traffic exhaust and traffic noise.

#### 31.2.2.1 Benefit of Reducing Traffic Exhaust

**Inventory of pollutants discharged** Inventory of pollutants discharged is the source and quantity of one pollutant in a certain time and area. Inventory of traffic exhaust $NO_x$, CO, HC relate to energy consume level and vehicle emission factor. And it can be computed by emission factor and mileage [3].

**Emission share rate of pollutants** Through the emission share rate of pollutants and the cost on eliminating pollution, it can calculate the environmental benefit from reducing traffic exhaust.

#### 31.2.2.2 Benefit of Reducing Greenhouse Gas

Inventory of carbon dioxide reflect the emission of greenhouse gas directly. Different of vehicle consume makes emission of carbon differ too. It can compute the quantity of carbon discharging from carbon dioxide emission factor.

## 31.3 Model to Evaluate the Impact of ITMS on Energy and Environment

### 31.3.1 Model to Evaluate Energy Benefit

Petroleum is scarce resource, so the benefit of petroleum cannot calculate by market price. But it can use shadow price to reflect scarce level. Substitution benefit of energy can be express by formula following:

$$\begin{aligned} &\textit{Substitution benefit of energy} \\ &= \textit{shadow price} \times \textit{amount of substitution energy} \end{aligned} \tag{31.1}$$

Shadow price of energy is the actual benefit, not a present parameter, is decided by scarcity and availability, and different in each period. The point of energy benefit evaluating is to change calculating energy benefit to computing shadow price. This makes it possible to transform a difficult economic theory to a quantity problem.

### 31.3.2 Environmental Benefit Evaluation

#### 31.3.2.1 Model of Calculating ITMS Benefit

This paper uses the result <Model sensitivity analysis of urban vehicle exhaust emission factor> by Guohua Song [4]. It analyses the velocity sensitivity of emission model MOBILE6, gets the emission data of the real road and emission factor variation function EF (v) of $NO_x$, HC and CO in Beijing. Use light-duty gasoline vehicle as example, $EFNO_x$ (v) is a cubic function; EFHC (v) and EFCO (v) are power functions:

$$EF_{NO_x}(v) = -2 \times 10^{-6} v^3 + 5 \times 10^{-4} v^2 - 0.026v + 1.07 \quad (31.2)$$

$$EF_{HC}(\mathrm{v}) = 10.38\, v^{-1.02} \quad (31.3)$$

$$EF_{CO}(\mathrm{v}) = 101.78\, \mathrm{v}^{-0.91} \quad (31.4)$$

In the formula, v is averaged velocity (km/h); $EF_{NOx}$ (v) is emission factor of $NO_x$ when velocity is v (g/km); $EF_{HC}$ (v) is emission factor of HC when velocity is v (g/kg); $EF_{CO}$ (v) is emission factor of CO when velocity is v (g/km).

#### 31.3.2.2 Compute Method of Reducing Traffic Exhaust

**Reductions of traffic exhaust** Formula to calculate inventory of pollutants discharged in the world: Inventory discharged = activity level × emission factor of pollutants. Inventory of exhaust discharged = mileage × vehicle emission factor.

Formula to calculate quantity of one pollutants emission is following:

$$Q_j = \sum_{i}^{n} \left(P_i \times M_i \times f_{ij} \times 10^{-6}\right) \quad (31.5)$$

In the formula, $Q_j$—total emission of pollutant j by vehicle i (10 kt); $P_t$—ownership of vehicle i in study year; $M_i$—vehicle i average mileage in a year (km); $f_{ij}$—pollutant j emission factor by vehicle I; n—number of vehicles; $10^{-6}$—conversion factor.

Reductions of pollution emission after accelerate:

$$\Delta Q_j = Q_j(before\_ITMS) - Q_j(after\_ITMSITMS) \quad (31.6)$$

**Reduce the cost of pollution abatement** Share rate of vehicle pollution emission is $\eta_\alpha$: ratio between amounts of pollution $\alpha$ discharged by vehicles $Q_v$ and total amount of

$$\eta_\alpha = \frac{Q_v}{Q_t} \times 100\,\% \quad (31.7)$$

From the cost of environmental protection in Beijing and percentage of spending on harnessing air pollution, it can calculate the saving cost.

#### 31.3.2.3 Compute Method of Reducing Greenhouse Gas

Using energy especially burning fossil energy is a major resource of carbon dioxide, ITMS reduces traffic energy request, decrease greenhouse gas discharging too. The benefit of reducing greenhouse gas is following:

**Discharge amount of carbon dioxide** The method and factor to calculate discharge amount of carbon dioxide reference Guidelines for National Greenhouse Gas Inventories by United Nations in 2006.

The method is:

$$E = H \times K \tag{31.8}$$

In it, E is amount of carbon dioxide discharged by one fuel (kg); K is carbon discharged factor, means amount of carbon dioxide per TJ (kg/TJ); H is heat discharged by one fuel (TJ).

$$H = Q \times K_1 \tag{31.9}$$

Q is coal amount of fuel consumption, means a standard coal fuel consumption converted by coal equivalent from certain fuel consumption (10 kt); $K_1$ is discharged heat per fuel, dereferencing 29.3 TJ/kt.

**Reduction of the cost after reducing greenhouse gas** According to the standard of greenhouse gas discharged social cost in foreign countries; it can compute the benefit of reducing greenhouse gas after using ITMS.

## 31.4 Accounting of Energy and Environment Benefit by ITMS

### *31.4.1 Accounting of Energy Benefit*

According to analyzing the evaluating indicator of "decrease cost" in this calculation, the reductions of gasoline and diesel oil are 0.0584 and 0.0022 billion litters respectively after ITMS used. Using conversion factor, transfer the amount of gasoline and diesel to mount of standard coal, are $6.444 \times 10^4$ and $0.266 \times 10^4$ t respectively. So decrement of standard coal is $6.711 \times 10^4$ t.

From Beijing Statistical Yearbook, we know overall energy consumption by traffic system in Beijing in 2008 is $8.4079 \times 10^6$ t, so using ITMS makes Beijing save coal 0.79 %.

From shadow price of petroleum = cost, insurance and freight (c.i.f) × shadow exchange rate + transportation and trade charge internal, get shadow price of petroleum in Beijing in 2008 is:

$$SP = 1044 \times 7.452 + (25.6632 + 566) = 8371.7\, RMB/t \tag{31.10}$$

**Table 31.1** Emission factors of passenger cars

| Type | | Passenger car(unit: g/km) | |
|---|---|---|---|
| After using ITMS | Velocity is 24.29 km/h | $NO_x$ | 0.704 |
| | | HC | 0.401 |
| | | CO | 5.584 |
| Before using ITMS | Velocity is 21.69 km/h | $NO_x$ | 0.721 |
| | | HC | 0.450 |
| | | CO | 6.190 |
| | Velocity is 22.15 km/h | $NO_x$ | 0.718 |
| | | HC | 0.440 |
| | | CO | 6.073 |

Higher market price in Beijing than 1,000 RMB and more, so the benefit of energy structure adjustment after using ITMS is:

$$(4.38 + 0.183) \times 10^3 \times 8371.7 = 0.382\, billion\, RMB \qquad (31.11)$$

## 31.4.2 Accounting of Environmental Benefit

### 31.4.2.1 Benefit of Reducing Traffic Exhaust

Vehicle is main resource of oxides of nitrogen, carbon monoxide and hydrocarbon in Beijing. According to statistics, the major factors to deteriorate air quality in Beijing are coal burning and exhaust discharging.

Using of ITMS improves the situation of vehicles' running, increases the velocity of cars and reduces exhaust discharging. Reducing exhaust discharging has a great effect to improve air quality.

**Inventory of pollutants discharged** Passenger car belongs to light-duty gasoline vehicle, according to formula 3–5, 3–6 and 3–7, get the emission factors in different velocity, as Table 31.1.

First, using the emission factors of passenger car as standard, get conversion factors of different types. Second according to this conversion factors get the emission factors of tax and bus in different velocity.

Through types of vehicles, mileage of each type in a year, rate of ownership and emission factors in different types, use formula 3–8 calculate the inventory of pollutants discharged, as Table 31.2:

All in all, amount of reducing pollutants is 2.11–2.63 $\times$ $10^4$ t after using ITMS in Beijing.

**Reductions of pollution abatement cost** Descent rate of pollutants in urban area in Beijing.

Using $\Delta Q_j/Q_j$ (before using ITMS) can get the descent rate of vehicles' pollutants, and according to the emission share rate of vehicles' pollutants calculate

**Table 31.2** Inventory of pollutants discharged by passenger cars

| | Before using ITMS amount of pollutants discharged Q(t) | | After using ITMS amount of pollutants discharged Q(t) | Amount of reducing pollutants ΔQ | |
|---|---|---|---|---|---|
| | Velocity 21.69 km/h | Velocity 22.15 km/h | Velocity 24.29 km/h | Velocity before ITMS 21.69 km/h | Velocity before ITMS 22.15 km/h |
| NOx | $3.24 \times 10^4$ | $3.30 \times 10^4$ | $3.32 \times 10^4$ | $0.08 \times 10^4$ | $0.06 \times 10^4$ |
| HC | $1.89 \times 10^4$ | $1.85 \times 10^4$ | $1.69 \times 10^4$ | $0.20 \times 10^4$ | $0.16 \times 10^4$ |
| CO | $24.02 \times 10^4$ | $23.56 \times 10^4$ | $21.67 \times 10^4$ | $2.35 \times 10^4$ | $1.89 \times 10^4$ |

**Table 31.3** Contribution rate from descending pollutants in Beijing urban area due to vehicles accelerating

| Pollutants | | HC | CO | $NO_x$ | Average contribution rate |
|---|---|---|---|---|---|
| Contribution rate | Velocity before using ITMS is 21.69 km/h | 8.37 | 7.85 | 1.32 | 4.76 |
| | Velocity before using ITMS is 22.15 km/h | 6.84 | 6.44 | 0.99 | 5.85 |

the Contribution rate from descending pollutants in Beijing urban area due to vehicles accelerating (Table 31.3).

After comprehensive considerate, the contribution rate from descending pollutants in Beijing urban area because of using progressive ITMS is about 4.76–5.85 %.

**Reducing cost of pollution abasement** According to <Beijing environmental statement> from 2006 to 2008 issued by Beijing municipal environmental protection bureau, the investigation to environmental protection per year is:

$$(25.04 + 24.82 + 26.57) \times 1/3 = 25.4778\,billion\,RMB \tag{31.12}$$

When the velocity before using ITMS is 21.69 km/h,

$$25.4778 \times 2/3 \times 5.85\,\% = 0.994\,billion\,RMB \tag{31.13}$$

When the velocity before using ITMS is 22.15 km/h,

$$25.4778 \times 2/3 \times 4.76\,\% = 0.780\,billion\,RMB \tag{31.14}$$

#### 31.4.2.2 Benefit of Reducing Greenhouse Gas

**Reductions of carbon dioxide discharging** Reducing energy burning is the best effective way to reduce carbon dioxide discharging. ITMS decreases the request of energy in traffic field, and reduces discharging greenhouse gas at the same time.

**Table 31.4** Factors of carbon emission in each fuel

| Fuel | Gasoline | Diesel oil |
|---|---|---|
| Factors (kg/TJ) | 70,000 | 74,100 |

**Table 31.5** Social expense standard of greenhouse gas discharging (price in 1990)/$\$\cdot t^{-1}$

| Reseracher | Type | 1999–2000 | 2001–2010 | 2011–2020 | 2021–2030 |
|---|---|---|---|---|---|
| Nordhaus (1991) | MC | (0.3–65.9) | | | |
| Al and Walter (1991) | MC | (30–35) | | | |
| Nordhaus (1994) | CBA | 12 | 18 | 26.5 | n·a |
| Klein (1992) | CBA | 5.8–124 | 7.6–154 | 9.186 | 11.8–221 |
| Peck and Taisiboge | CAB | 10–12 | 12–14 | 14–18 | 18–22 |
| Fankhauser (1994) | MC | 20.3 | 22.8 | 25.3 | 27.8 |
| Madison (1994) | CAB | 5.9–6.1 | 8.1–8.4 | 11.1–11.5 | 14.7–15.2 |

*Notes MC* is marginal social expense, *CBA* is impact factor price in analyzing expense-benefit

**Table 31.6** Quantifiable evaluating benefit indicator of energy and environment by ITMS

| Benefit Indicator | | Benefit value | Total benefit value | Average benefit value |
|---|---|---|---|---|
| Benefit of energy and environment | Reducing exhaust | 0.78–0.994 | 1.230–1.444 | 1.337 |
| | Reducing greenhouse gas | 0.068 | | |
| | Energy structure adjustment | 0.382 | | |

The factor of carbon emission as Table 31.4:

After using ITMS, it saves gasoline as $6.444 \times 10^4$ t standard coals and diesel oil as $0.266 \times 10^4$ t standard coal. The amount of discharging carbon dioxide is $13.8 \times 10^4$ t.

**Saving cost on reducing greenhouse gas** According to the social expense standard of greenhouse gas discharging, analyze the environmental benefit of exhaust discharging before and after using ITMS. As shown in Table 31.5.

The final economy result is $410.96 \times 10^4$ \$.

The result is 4.1096 million USD, but it is computed as the value of dollar in 1990, according to rate of dollar inflation calculate the value of dollar in 2008: there are 18 years between 1990 to 2008, the rate of dollar inflation in 1990 is about 5 %, can get 6.0573 million USD in 1990 equals to 14.5776 million USD. The exchange rate of RMB against USD in 2008 is 6.90, and can know the benefit on reducing greenhouse gas aspect is 6.8 billion RMB.

## 31.5 Conclusion

According to the calculation above, it can get the total benefit of environment and energy brought by using ITMS in Beijing is from 1.230 to 1.444 billion RMB, average of which is 1.337 billion RMB, as Table 31.6. Assess the impact of Intelligent Transportation Management Systems on energy and environment in Beijing clear proofs that ITMS brings Beijing a great effect on social, economy and environment.

**Acknowledgments** Program Name: 973 Programs (No. 2012CB725403).

## References

1. Zhang W, Xu A (2005) Trend of intelligent transport systems. J Liaoning Tech Univ (Nat Sci Ed) 24(Z1):77–79
2. Lin X (2002) Application of externality theory of transportation to project evaluation. Quant Tech Econ 19(12):93–96
3. Huang Z, Tang D (2008) Estimation of vehicle toxic air pollutant emissions in China. Res Environ Sci 21(6):166–170
4. Xu Y, Yu L, Hao Y, Song G (2009) Development and application of macroscopic emission model. J Transp Syst Eng Inf Technol 9(2):147–154

# Chapter 32
# Landscape Design Based on Computer Aided Design Technology

**Haixiao Shi**

**Abstract** CAD provides unparalleled convenience to the major of landscape design, making site design come natural to the designers, communication in design more smooth and direct. To roundly demonstrate the designers' design idea and inspiration of designers in the way of technology, will help tap more development potential of, and provide more opportunities for landscape design.

**Keywords** Computer-aided design (CAD) · Landscape · Dimensional space · Display · Analysis

## 32.1 Introduction

With the progress of science and technology, the rapid development of computer hardware and software, and continuous improvement, computer-aided design (CAD) software, more and more designers, uses CAD as the main way to work. And CAD has become indispensable element in the implementation of the design work, because of high precision, high efficiency, true-to-nature effect, because the easy to modify. Computer aided design (CAD) has become indispensable element in the implementation of the design work, because of high precision, high efficiency, true-to-nature effect, because the easy to modify. In landscape design field, the computer aided design (CAD) will become a big difference in the analysis, design concept, graphical presentation, etc. The designer can save time space characteristics and is form design [1]. Computer aided design (CAD) to

H. Shi (✉)
School of Art, Jiangsu University, Jiangsu, China
e-mail: haixiao89s@126.com

W. Du (ed.), *Informatics and Management Science V*, Lecture Notes in Electrical Engineering 208, DOI: 10.1007/978-1-4471-4796-1_32,

provide unprecedented and convenience to the main landscape design, production site design to natural to the designers, design more smooth communication and direct [2].

This token, computer aided design (CAD) will become a big difference in the analysis, design concept, graphical presentation, etc.

## 32.2 Computer-Aided Landscape Design's Influence and Significance

### *32.2.1 On the Conversion Between Two-Dimension and Three-Dimension*

If the landscape design is in the three dimensional space, stylist people pay more attention to space and carry momentum, this need two image thinking, considering the spatial scale and other elements, attach means, and color, and constantly mutual conversion between 2d and 3d space, in an abstract understanding, analysis, and summarizes the painting, and ultimately determine the space characteristics and form [3].

In CAD software was the spot, designers always spirit converted into 2d 3d space information form, and gives it in figure, after a series of reasoning and perfect, they must be in the three dimensional space form simulation 2d medium, in a view of the way, help express their design. This kind of thinking and the expression of two-dimensional "consumption spend too much time in stylist think how to clear represent design idea in 2d medium-paper". To the continuous improvement of CAD software, the conversion process, from 2d image into three dimensional simplified, complete simple 2d image description, the designer can understand the effect of internal and external space only fully accept simple orders; at the same time, there are also software, support the animation, and they can be very good to show the effect of animation, design for designers and holographic, dynamic, interactive expression means. As shown in Fig. 32.1.

### *32.2.2 Three Dimensional Space Display*

Landscape designers, design is only the internalization of their process, and the more by three dimensional space shows when it comes to communicating with others. Things are not so, most landscape designers have through the good professional training, converted into two and shy; D space three dimensional space well in their heart of hearts. However, when the display design more is visual and different way is necessary. In this period, the computer aided design (CAD) plays an important role, from the terrain simulation (ARCGIS), environment display

**Fig. 32.1** Two-dimensional-three dimensional by sketchup

(3D Max), including the animation software under the help, creation software, the whole project become more true- nature, to clear, and vision. In the past, it is always in my thoughts obstacles between. I finished the sketch and architectural design intention has lost its inspiration into construction workers. Sometimes I feel like I'm speaking a foreign language [4]. It's different now, I understand, in this case, the computer feels like a translation. As shown in Fig. 32.2.

### *32.2.3 Innovation of Form (Non-Conventional Type), Utilization and Appearance of Individuation and Deconstruction Ism, Realization of Ecologism*

The advent of the computer and rapid development, make landscape design of 3d software easier for space design in 2d media, this is reflected not only in conversion spatial dimension, but more important, in the pursuit of possibility in the form of exploration and innovation. Once upon a time, on the complex terrain of landscape design, due to the height of the complex is different, it is difficult to accurately analyze slope's change influence landscape, some slope processing landscape are often too simple, it is difficult to make better. Assist GIS technology and slope CAD software, the designer can get accurate data analysis and simulation environment of the terrain, the contour line and altimetry continuous point (the surveying and mapping), combining the design experience, more accurately determine location, roads, and other structures, greatly saves energy of the slope engineering design calculation and confirmation [5].

Meanwhile, with the help of computers, some illogical and irrational design can also be realized; with the help of CAD, these design that emphasizes contingency and opportunistic effect breaks away from transcendental formal beauty change

Fig. 32.2 Interface of sketchup surface

into more dynamic, unpredictable, autonomous form. As the French Parc de la Villette, designed by the deconstructionism landscape designer Tschumi, he broke away from the original order and composition principle, setting about topological composition, designed three self-disciplined abstract systems—point, line, and plane, forming the overall framework of the park.

Ecologists are a hot topic in recent landscape design, stylist people start and ecology of the sites, to restore the original texture and ecological structure of the earth. In this sense, CAD software is very helpful resources, environmental simulation, analysis, design level influence ecological and so on, these are directly related to the success of the ecological landscape design. As shown in Fig. 32.3.

### 32.2.4 Trial and Use of Different Materials

In addition, with the help of computer, the landscape architect has more choices in different materials. Now, the Internet is very advanced, any new materials and new technology can be Shared in the shortest possible time, the designer's interpretation and test new material to give full play to the right on that basis. Through the computer software, designers can visually understand selected materials of the impact on the environment, such as paving material, vegetation slope protection treatment, etc. Integral space effect of a substance can show that such a life computer simulation environment, greatly reducing the possibility of design, found that fatal mistake to complete construction, to avoid repeated construction. As shown in Fig. 32.4.

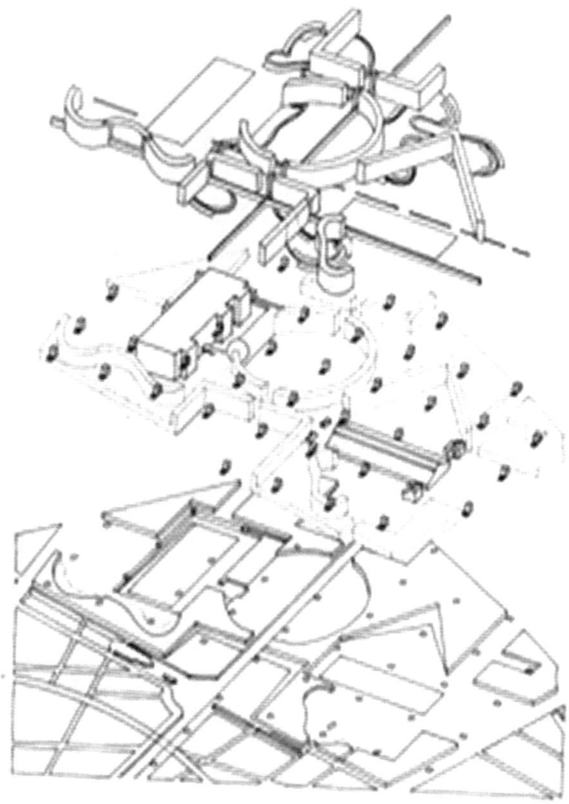

**Fig. 32.3** The abstract systems of parc de la villette

## 32.3 Computer-Aided Landscape Design's Means and Method

Digital calculation and process is the basic function of a computer, will convert digital technology information storage and processing computer is the core of information technology. The earliest industrial application is digital technology in the area of design of plane and the auto industry. By CA TIA software, designers can build three dimensional digitized model, the computer can produce simulation effect and simulation animation. After 1990s, computer digitization began to blend in landscape design. Computer digitization can assist with landscape design in the following two aspects.

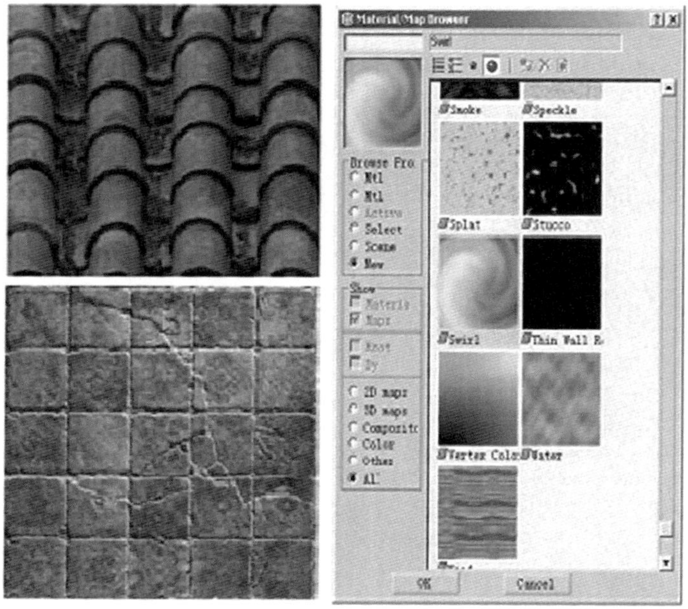


**Fig. 32.4** Materials in 3D MAX

### *32.3.1 Digitization Blending into Analysis*

The first breakthrough is digital bending analysis. Most of the preliminary work of landscape design is the rational analysis and summary of the objective of computer digital technology contributes greatly to the production of products. First, AUTO CAD software accurately defines the place scale, the contour lines, altimetry points of buildings, roads, waters, as obtained by measuring tools, are all combined in the CAD drawings, by interpretation of the drawings, landscape designers can accurately master the data of the site. CAD's attached three dimensional software such as Chief Architect, LANDCAD, 3D MAX and ARCGIS software can turn these information into more visual three dimensional image, meanwhile the rapid analysis of computer is enhanced, by color difference, the altitude and slope variations are indicated. As shown in Fig. 32.5.

### *32.3.2 Digitization Blending into Representation*

Formerly, the drawing of landscape design depended solely on hand-drawing, which was not only time consuming but also uneasy to modify and adjust. When digitization gets into the representation process, layout plan can rely on CAD and PHOTOSHOP software, and after completion of design, some vector software

**Fig. 32.5** The topographic feature analysis based on arcgis

such as ILLUSTRATOR and CORELDRA W can help the designers express their design more clearly. And the above-mentioned three dimensional model-building software such as 3D MAX, MAYA and SKETCHUP, combined with some post-rendering software such as Artlantis R, the designed scene can be displayed to the life.

## 32.4 On the Influence of Cad on Landscape Design

The influence of the above described computer technology to landscape design, and how does it influence in landscape design, and then we can't help but improve a series of problems: what is the relationship between the traditional (hand painting) and technology (CAD)? Computer technology can replace the designers of the landscape design?

Any new technology develops on the basis of traditional, we can't give up the original hand and shy; Drawing design, instead, more and more of the computer support software, but keep the advantage and began to imitate the rendering hand-drawing style, at the same time, we can also use the accuracy of the computer software, combined with the traditional manual painting, seek new ways and landscape design concept said.

Computer aided design (CAD) is the advantages of convenient, neat and accurate, easy to modify, this in the analysis, drawing, said is obvious. However this is not in the soul of the design lies, design concept is not control design tools, this makes stylist has the ability to manipulate tools to express creative thinking, learned about the software of the different nature of the assistant, so as to express their ideas in a more convenient way, thus way of thinking and support software fully combined, make the computer software really reflect the idea in the mind.

## 32.5 Conclusions

In landscape design field, from 2d graphics project design and documentation 3d model and post-rendering, animation display, computer aided design (CAD) liberation landscape designer from a large number of small mechanical labors, diverts the attention from the blueprint of stylist of standardization work standardization and standardization, more important function orientation and outline design.

## References

1. Qu Y, Chen Y (2010) Application of computer aided design in drawing. Landscape Gard Perspect 12:89–96
2. Su T (2009) Virtual reality application technology on landscape design. J Comput 06(11): 123–126
3. Wang L (2010) Computer plotting of effect drawing of park and garden. Forest Inventory Plann 10:35–41
4. Cao Y (2001) Application of designed technology of digital gardens sight. Fujian Archit Constr 21(04):321–326
5. Zuo W (2002) Study on comprehensive assessment of the regional ecological environmental system security based on Rs, Gis And Models. J Ecol Environ 08(12):674–679

# Chapter 33
# Study on University Football Teaching Based on Multimedia Technology

**Aihua Yu**

**Abstract** Application of football multimedia courseware better settles the shortcoming of limited time and unsystematic problems during the football theory teaching process. It also improves the non-intuitive explanation in technical and tactical teaching part, and solves unity and regulatory issues of the technical movement in practice teaching class. Soccer teaching is more standardized. The use and development of multimedia teaching is not only related to teaching methods and transformation of teaching methods, but also affects the teaching mode and the development of teaching theory.

**Keywords** Multimedia · Football theory course · Courseware

## 33.1 Introduction

Education reform, multimedia technology, as a modern teaching and training methods is more widely used. Teachers are feeling the convenience and efficiency of teaching modern multimedia technology bring [1]. Due to the widespread and a lot of information provided multimedia, football can be from different theoretical teaching view of the time and space to provide sound, image, ICONS and other information. So students can use a variety of remote sensing tools get all sorts of effective dynamic data to enrich their imagination, promote the formation of

A. Yu (✉)
Teaching and Research Section of Physical Education, Xuchang Vocational Technical College, Xuchang 461000, China
e-mail: aihua34yu@126.com

W. Du (ed.), *Informatics and Management Science V*, Lecture Notes in Electrical Engineering 208, DOI: 10.1007/978-1-4471-4796-1_33,

creativity, and improve the learning efficiency [2]. It has great impact on the traditional football education and teaching methods.

Multimedia modern education technology is an important content of the right realize the education methods and means of modernization, the breakthrough tradition education concept is very positive role. Multimedia is a figure, words, sound, like combined together. The use of computer multimedia technology, we can digital computing and text editor, appreciate the high definition images and movies, still can hear clearly speech and beautiful music. Based on it's to give birth to the characteristics of the dynamic sex, interactive, can greatly arouses student's study enthusiasm, promote the teaching learn the improvement of quality. In the knowledge the changing today, the traditional sports teaching means. Already can't meet the need of modern physical education teaching, how to put the multimedia technology into the body education teaching, so as to realize two organic unions, is our sports workers face an important topic. As for soccer course for, many technology looks be like simple, but to master an actually. It is difficult to, the past teaching value in imitation and ignore the students' cognitive, a lot of repetition exercise will only make students bored. The traditional teaching media can't compete, but, multimedia technology is not only in the demo function, must be used to develop the students' cognitive psychological structure, deepen it's the understanding of concepts, principles, and how to stimulate students' interest in knowledge, and make the right Cognitive guide practice, reduce the faults in the practice, and promote the improvement of skills. Based on this, this study will sport psychology means of multimedia technology and integrate the design football lesson Cheng teaching program, and through the experiments to verify to develop the students' cognition and technological support theory can effect.

## 33.2 The Design of Multimedia Courseware in Football Theory Course

### *33.2.1 Multimedia Courseware Should be Based on Students*

According to the different age, gender and the students' psychological characteristics, multimedia design should be based on their cognitive characteristics and personality characteristics from the perspective of the pedagogy and the psychology orientation. The multimedia courseware is fully embodying pedagogical features of motion and development trend. And this kind of courseware can be directly on the teaching process, and arouse students' learning initiative to create a good results learning environment. First of all, understand is an important stage their mastery of knowledge. Through the visual computer in teaching content, material to specific, so students will be with great interest to know football knowledge, improve their sports and cultural achievements. For example, when the rules of football and the referee method explained, they can have a specific details and analysis; second, the learning environment will also promote students' learning initiative. Learning environment is a place nearby, where the students can be free to explore and learn. Multimedia has the rich information resources and provide students more initiative and freedom, played a key role in the content of study; Finally, multimedia courseware is based on a species close to the psychological characteristics of human knowledge and teaching contents in the organization is to present the way and establishing a knowledge structure, which will help to change, from existing knowledge to new knowledge.

### 33.2.2 Multimedia Courseware in the Teaching Content Option

A lot of relevant knowledge and information, and may introduce the different teaching point basis and ensure the course requirements football. Table 33.1 is a football teaching content structure, focusing on the basic skills, basic strategy, and make use of the advantages of multimedia courseware that and comment football technology, at the same time, students will learn to use enjoy more interested in football to play football.

Football multimedia courseware can base the teaching theory, does not exist in the content of the course textbook. This point of view can fully express the students, so students to master the football technology, and theoretical knowledge.

### 33.2.3 Multimedia Courseware Making Structures

The multimedia courseware auxiliary teaching is a kind of new, which fully reflects the diversity and personalized classroom teaching [3]. The combination of the aid advantage, lively and vivid, longitudinal, explain design and audio, video, projection, as a whole, through the flexible machine music animation effects in the teaching process. A great variety of information so is a destination for students to put forward, and create a lively teaching situation and mobilizes students' desire to study. Students will therefore in the limited time available to get more knowledge.

According to the requirements of the overall structure, interface design can be divided into three structures: the main interface, interface and content explain content selection interface. The multimedia courseware design including the content selection function interface level, select button and function button, demonstration method of course, connection mode of different types of materials, navigation model design, courseware document structure. The purpose is to make full use of multimedia technology to complete the specific content of the auxiliary teaching. According to the football teaching contents is various forms of teaching method to establish, such as illustrations, pictures, pictures, video, animation explanation etc.

**Table 33.1** Football teaching content structure

| Generality | Basic skills | Basic tactics | Training method | Competition rules | Football appreciation |
|---|---|---|---|---|---|
| Origin | Step moving | Offensive tactics | Football consciousness | Competition organization | Skills appreciation |
| Development | Passing and catching | Defensive tactics | Training of every age stages' consciousness | Competition arrangement | Tactics appreciation |
| Features | Dribble breakthrough shooting ball hawk goalkeeper skills | Defense and counterattack man-to-man defence and attack | Football training, method and task of every age stages | Football judgment methods ground rules | Football match football training |
| Exercise | | | | | |

From the courseware, and some key issues must bear in mind: Each part of courseware making style is consistent, because may; choose and switch button set on every page of switching to reduce the number of pages, and reduce the use voice as little as possible. At the same time, according to the characteristics of the football (video technology should be used as much as possible) and text content, along with the corresponding image, the courseware can more vivid and straightforward.

## 33.3 Applications and Effect of Multimedia in Football Theory Teaching

### *33.3.1 The Quality of Physical Education can be Effectively Improved*

Football multimedia courseware in theory teaching increasingly display can realize the key and difficult. Through the man–machine dialogue, deepening the understanding of the students and memory sports knowledge and mobilization of the students of the senses, and that knowledge can be multi-levelly and perspectives, appearance, thus creating a good learning environment. For example courseware can simulate the referee showed scenes of the game, so that they make their own decisions. It can also use the keyboard to judge foul situation appear on the screen. In the provision of time, the computer will judge referee level simulation. Through the multimedia presentations and teacher's explanation, the students' understanding of the laws of the ball the referee wills fresh, smooth and the clear deepen. As rugby rules "offside" for example, you can use the multimedia courseware display. The use of courseware to overcome the shortcomings of common sense teaching writing and oral expression, presented the image of the poor. At the same time it can stimulate the students' work together at the same time, and help them to develop their power of observation, self and shy; learning and thinking, let the student active rather than just passively accept cork, the optimization classroom teaching [4].

### *33.3.2 Interaction Between Sensory Function and Muscle Proprioception can be Strengthened*

The students in the process of motor skill mastery are temporary neural connections established under the sensory cortex hub. In the process of muscle proprioception plays an important role. Multimedia teaching is an intuitive visualization teaching method, organically voice, shape and color applied to students' various senses. To make the students skills will be specific, intuitive favorable conditions for students to perceive, understand and remember material. In this way, students can get more information in the process of learning football skills. Students will therefore a correct understanding and set up a corresponding muscle feeling, aims to promote and accelerate the process of movement status.

### 33.3.3 Rational Application of Feedback Control Theory

The founder of cybernetics, wiener points out that N: Any effective must act is a kind of feedback process information, whether the goal. The essence and core of multimedia teaching is to build a feedback control system. Motor system itself is a very good feedback loop; its action not only regularly monitors the progress of the activities, but also timely information will activities, the cerebellum. In cerebella integration and the actions of the information processing, information to deliver to the related portions of the brain cortex adjustment, achieve precise coordination purpose. Using video, students can see action they had just completed the image, through the regained consciousness, adjust and shy; and action of the correction, and after several repeat, feedback can blend in automatic control system, build a solid technical movement to finalize the design.

### 33.3.4 To Enhance Students' Interest in Learning and Mastering the Motor Skills

Interest is: people tend to actively explore understand the thing. Once the students' interest is exciting, their attention and observe learning more serious. Their logic memory stepped up, imagine more abundant, and a happy mood corresponding produced. Therefore, students will be active investment efforts learning process, and the knowledge of the object in their body left a deep impression, so as to improve the effectiveness of their activities [5]. All of the above has a significant impact on the training and development of the team. As football skills teaching is rather dull, the student in the teaching process easier to negative emotions. Football multimedia courseware is in the teaching application vivid image, sound and abundant of affection. Especially in the students watch his action shots, they will find their own technology defect of the action. From the teacher evaluation, they will recognize behavior do wrong or right. Students will practice more, or even higher, and their interest, which will lead to bigger and better teaching effect.

## 33.4 Problems

### *33.4.1 Teachers' Role in Teaching is Irreplaceable*

Teachers must be aware of their role in the teaching activities of the main body, despite a amount of multimedia courseware information and resources. So teacher's explanation has been a core of the teaching work. In the football teaching, especially in the teaching of technical and tactical and technical movement point, action interface, and tactical coordination cannot be reflected or reflected fully in the multimedia courseware. Teachers should explain, and empirical analysis of the development of the students for their measures to realize education purpose.

### *33.4.2 Interaction Between Teaching and Learning are Emphasized*

The concept of some of the sports teachers still keep blind physical training and drab classroom theory course, lack of innovation in teaching. This is not good for the students to participate. Of course, physical exercise on the sports teaching is still necessary. But compared to the teacher's demonstration teaching, multimedia courseware teaching is more intuitive, more vivid, and easier to understand students. Multimedia courseware, on the one hand, improves the students' interest in study, and makes students want to learn, easy to digest knowledge; On the other hand, it promoted the students to use the scientific method to take the initiative in physical exercise.

### *33.4.3 Teaching Facilities are Continuously Improved*

Restricted by the venues, facilities, the ongoing multi-media teaching still has many deficiencies. For example, it cannot carry out a wide range of teaching LAN, and multi-media classrooms are only able to ensure the teaching of theoretical courses. Multimedia classrooms should be constructed on a large scale. Now many institutions begin to establish a "mobile multimedia classroom", that is to configure a laptop computer, a portable digital projector and a digital video camera in order to address the needs of practice teaching.

## 33.5 Conclusions

Use of multimedia technology is the future sports education reform direction, so university should strengthen the study of the theory and practice of the teaching of the need to adapt to the modern multimedia.

## References

1. Jingjun H (2010) The application of multimedia technology in football teaching of University. J Tonghua Normal Coll 10(7):121–125
2. Enjia P (2009) The approach of multimedia technology in football theory teaching of University. J Shanxi Normal Inst Phys Educ 11(2):86–92
3. Wang Z, Niu D (2005) The use of computer multimedia courseware in teaching. J Yili Normal Coll 7(9):32–35
4. Xu Z, Mao H, Yang B (2004) The implementation of CAl teaching system in teaching of physical theory. Adult Educ 6(12):54–58
5. Daifeng Y, Zhang J, Man Z (2004) The research and application of CAl courseware of sports biomechanics. Shandong Sports Sci Technol 21(12):341–346

# Chapter 34
# Research of Virtual Tour Impact on Destination Image

**Yuanwu Yu**

**Abstract** Although virtual reality (VR) has been well known and investigated for practical use in various industrial fields after late 1980s and 1990s, it is recently that researchers and practitioners in the tourism field have attempted to understand the virtual tour. Few researches have been done to understand the role of destination image and to examine the impact that virtual tour has on the destination image. The purpose of this paper is to empirically elaborate the relationship between the virtual tour and tourists' destination image under the context that applied VR technology. This study takes Shanghai Online-Expo as a case and participants either do or don't have expo-exhibition experiences are all recruited from the Zhejiang University. Virtual tour is expected to positively affect the destination image, and furthermore, it is also significant among tourists' without destination experience, and vice versa. These results have both managerial and theoretical implication.

**Keywords** Virtual reality · Virtual tour · VR technology · Shanghai online-expo

## 34.1 Introduction

Recently, developments in information and communication technologies (ICTs) have been transforming tourism in myriad ways, impacts on areas ranging from consumer demand to site management [1, 2]. One important area of ICT is virtual reality (VR). Since Morton Heilig began designing the first multisensory virtual

---

Y. Yu (✉)
School of Management, Zhejiang University, Hangzhou 310058, China
e-mail: yuanwu43yu@126.com

W. Du (ed.), *Informatics and Management Science V*, Lecture Notes in Electrical Engineering 208, DOI: 10.1007/978-1-4471-4796-1_34,

experiences in 1956, especially after late 1980s and 1990s, virtual reality (VR) has been commonly used in diverse areas including entertainment, design and simulation training. As VR technology continues to evolve, the possibilities for using VR within the tourism sector will grow, from tourism planning to heritage preservation [3]. Recently, the existing information and communication technologies (ICTs), especially how the ICTs are applied into the tourism sector, are part of the most influential factors that impact virtual tourists' experience, which ultimately influence the tourists' decision. As Refsland et al. [4] claimed that "the majority of virtual heritage researchers believe that their work encourages people to actually go see the real site, giving the visitor extra knowledge to enhance the real site visit".

Actually, countless existing tourism sites and activities have already involved virtual, reproduced environments. For instance, at the World Showcase in Disney World's Epcot one can explore environments representing several different countries. Besides, Virtual Louvre, the Sarajevo City Hall, the Great Buddha carving from Afghanistan, the Hawara pyramid complex from ancient Egypt, are also rather prevalent among international online travellers. For china, the most remarkable affaires during 2010 must be The World Exposition held in Shanghai. Assisted with internet and multimedia technology, expo host launched the project-"online expo"-that applies the VR technology into the expo displays. Online expo constructs a network platform that providing internet experiences, real-time interact and other auxiliary functions. With the support of VR, the Shanghai World Expo will be the first World Expo that will never close.

However, little studies have been done about the virtual tourists' destination image when they immerse themselves in the virtual reality. For preceding reasons, tourism researchers and professionals should gain a greater understanding of virtual tourists' destination image within the background of virtual reality to best prepare themselves to face the challenges and take advantage of the opportunities that virtual tourism presents. Considering the profound influence of the grand ceremony and the relatively maturity of the online expo, it take the Shanghai online expo as a case to explore the factors that may have a significant effect on virtual tourists' desitination image. Questionnaire is adopted in this study before and after the respondents are asked to "travel" the online expo.

The paper starts with literature review involving virtual reality, virtual tourism, experience and their relationship. The literature review is followed by the presentation of applied methodology as well as empirical results. Implications of the results will also be introduced as well as the contributions and limitations of the whole research.

## 34.2 Literature Review

### *34.2.1 Virtual Reality and Virtual Tour*

#### 34.2.1.1 VR and VR Technology

Since Morton Heilig began designing the first multisensory virtual experiences in 1956, especially after late 1980s and 1990s, virtual reality (VR) has been well known and is currently being investigated for practical use in various industrial fields. Four technologies are crucial for VR: the visual (and aural and haptic) displays, the graphics rendering system, the tracking system, and the database construction and maintenance system [5]. So there is another question that "What's VR". As proposed definitions vary when describing the different features considered necessary to constitute an experience as VR [6], definition unconformity inevitably exists.

In order to do the main research more conveniently, I select the definition of VR that borrows from definitions used in books dedicated to the topic written by Burdea and Coiffet, Vince, and Gutierrez, Vexo, and Thalmann. That is "VR is defined as the use of a computer-generated 3D environment—called a 'virtual environment' (VE)—that one can navigate and possibly interact with, resulting in real-time simulation of one or more of the user's five senses".

#### 34.2.1.2 VR Experience

A VR experience can be described by its capacity to provide physical immersion and psychological presence [7]. "Immersion" refers to the extent to which a user is isolated from the real world, and it can be classified into three levels, "fully immersive system" the user is completely encompassed by the VE and has no interaction with the real world, while in a "semi-immersive" or "non-immersive system" the user retains some contact with the real world [8]. When it comes to presence, it is when people behave in a VE in a way that is close to the way they would behave in a similar real life situation [9]. Feelings of "presence" are naturally subjective, being associated with a user's psychology, but they undoubtedly are influenced by a VR system's ability to provide high quality data to the users senses [10]. Besides, the level of immersion offered by a VR system is one factor that may influence a user's feelings of "presence" [11].

#### 34.2.1.3 Virtual Tour

Tourists often make decisions under substantial uncertainty when visiting a particular destination and Nelson (1970) referred to these uncertainties as "experiential attributes" because they can be identified only through experiences.

Obviously, the best way to examine experiential attributes of a destination and thus form a mental image of the destination is to experience the destination by actually visit. However, it is impossible to do so due to the unique nature of tourism products [12].

There is now some evidence that such "experience" can be conducted by VR technology, such as the virtual tour. It is useful and important for destination image due to its ability to provide more extensive/rich information thereby increasing the quality of the destination image. Previous researches indicate that Web-based virtual tour can modify tourist's destination image.

### 34.2.2 Destination Image

The term "destination image" has been a popular concept used in tourism studies in a variety of contexts. Mayo used the term "simplified impressions" to describe the concept of destination image. Walmsley and Jenkins defined destination image as a product of the mind that results from huge amounts of data about a place. Fridgen defined it as a mental representation of a place that is not physically before the traveler. In this paper, it follows the commonly accepted one that is, destination image consists of the information, beliefs, impressions, attitudes and emotional thoughts an individual has regarding a particular place.

A number of tourism and recreation researchers have studied the topic of a destination image in tourism, such as destination image change, destination formation and destination image assessment and measures. Tourist destination images are important because they influence both the decision-making behavior of potential tourists and the levels of satisfaction regarding the tourist experience.

Previous studies suggest that a tourist's decision to travel to a particular destination is linked to the destination image held by that tourist and once at the destination, and satisfaction largely depends upon a comparison of expectations based on previously held images and the actual reality encountered at the destination. Researchers also suggest that individuals can have an image of a destination even if they have never visited it or even been exposed to more commercial forms of information. Thus, understanding the differing images that visitors and non-visitors have of a destination is invaluable.

### 34.2.3 The Process of Image Formation and Hypothsis

The formation of image has been described as the development of a mental construct based upon a few impressions chosen from a flood of information.

As Selby and Morgan put it [13]:

Before consumption, imagery can add value and influence decision making. After consumption, imagery can have a reconstructive role via memories and experience.

Gunn puts various travel information into a stage-theory which consists of 3 stages: organic, promoted and modified image and it implies that the images held by potential visitors, non-visitors and returned visitors will differ. There is evidence showing that images held by returned visitors tend to be more realistic [14]. Previous researches indicate that Web-based virtual tour can modify tourist's destination image. For this concern, destination image can be expected to be improved through virtual tour Fig. 34.1.

Physical immersion, from another perspective, refers to the response triggered by the user, which means interactivity. Steuer suggests that response speed contributes to interactivity. Ease of use, referred to the navigational characteristics of the web site, is the final factor that contributes to interactivity. Whether or not, virtual tour is still a travel experience, which always calls for comfort feelings. As discussed before, virtual tour has a positive impact on the destination image on the promoting process [15]. So I develop the hypotheses as follows:

H1: During the virtual tour, response speed positively impact on destination image.

H2: During the virtual tour, "Ease of use" positively impact on destination image.

H3: During the virtual tour, comfort feelings positively impact on destination image.

Psychological presence is another important point in the process of image formation. What you have perceived is really what you care. As mentioned before, virtual tour is also an information-seeking behavior. Challenge comes from the information that is the content of virtual destination. Marchionini believes that each information seeker possesses particular experiences, abilities, and preferences [16]. Therefore, an individual's personal knowledge affects overall performance. The attractiveness of the virtual tour spot, another factor influencing the perceived destination image, reflects the representation's richness and quality. It was expected that among many other factors that contribute to the attractiveness of a virtual tour spot, people's experience with similar ones in general might affect their impression of the attractiveness of the Web site. So I can image that:

**Fig. 34.1** Research model

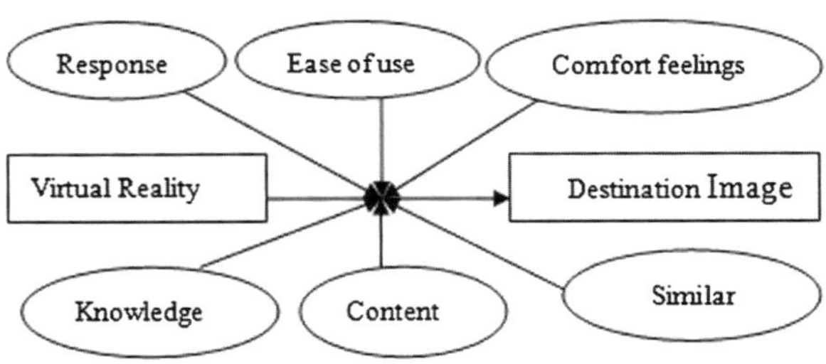

H4: During the virtual tour, tourist's knowledge about internet and tourist spot positively impact on destination image.
H5: During the virtual tour, the content, especial richness and authenticity positively impact on destination image.
H6: During the virtual tour, the content, similar virtual tour website experience negatively impact on destination image.

## 34.3 Method

### *34.3.1 Experimental Design*

In order to study the effect of VR technology applied with tourism sector on the destination image of the virtual tourists, the following conditions should be declared:

I select the online expo as my research objective virtual tour for its great signifiance for China and relatively perfect VR application.

Two groups of participants, non-Expo experience and ever-Expo experience, are recruited to verify the discrepancy of destination image before and after experience.

### *34.3.2 Measures*

#### 34.3.2.1 Myers-Briggs Type Indicator

The Myers-Briggs Type Indicator (MBTI) is the most widely used personality assessment instrument for measuring psychologyical preferences in how people perceive the world and make decision among both research and application settings [17]. Four pairs of preferences or dichotomies are shown as follows:

Extraversion (E)—(I) Introversion
Sensing (S)—(N) Intitution
Thinking (T)—(F) Feeling
Judgment (J)—(P) Perception

In order to simplify the research, I just abstract two kind of personality from the MBTI test results-X and Y.

#### 34.3.2.2 Revised ITC–Sense of Presence Inventory

The revised ITC-SOPI is a pre-test subjective presence measure composed of 20 items, divided in two parts. Part A (5 items) refers to a respondent's personal information, such as sex, grade, and so on. Part B (15 items) refers to a respondent's impressions/feelings before a virtual tour experience. A 1–5-point Likert scale (from Strongly Disagree to Strongly Agree) is used for responding to the items in Part B. Factor analysis were applied to determine the essential variables. Internal reliability coefficients (alpha) were computed for each of the four factors.

#### 34.3.2.3 Destination Image of Virtual Tour Questionnair

Destination Image of Virtual Tour Questionnair (DIVTQ) is a post-test subjective presence and reality judgment measure. A short version of this self-built scale, with 29 items, was used.

### *34.3.3 Virtual Environments*

As explained previously, I select the online expo (2010 Shanghai World Expo) as my research objective virtual tour, with the URL, "http://www.expo.cn/#&c=home". There are five parts, including site tour, expo carnical, community and future city, for visitors to sightseeing freely by the help of a plug-in. What it is a surprise that Haibao, the mascot of Expo 2010 Shanghai,will be visitors' personal guider. According to the project sponsor's initial plan, online expo is a highly computer simulation to "bring" the virtual tourists into exhibitation. Any senses of sight and hearing are available in the virtual environment.

### *34.3.4 Procedure*

Sixty participants are recruited for the study from the Zhejiang University, with a range from freshmen to Ph.D students. Stratified sampling method are adopt into this study. As a result, 20 students of each level (bachelor, master and Ph.D) will be enrolled in, 10 students with Expo experience and the rest non-Expo experience. They are given a description of the study and guidelines as well. After completing the MBTI questionnair and revised ITC-SOPI to ascertain participants' characteristics. They will be assigned to two groups according to their reality expo experience and practice in the training virtual environment. After the online-expo tour, participants should complete the destination image of virtual tour questionnair. All participants were debriefed following the experiment.

## References


1. Al-Kodmany K (2002) Visualization tools and methods in community planning: from freehand sketches to virtual reality. J Plann Lit 17(2):189–211
2. Baños RM, Botella C, Alcañiz M, Liaño V, Guerrero B, Rey B (2004) Immersion and emotion: their impact on the sense of presence. Cyberpsychology Behav 7(6):734–741
3. Bishop ID, Wherrett JR, Miller DR (2001) Assessment of path choices on a country walks using a virtual environment. Landscape Urban Plann 52:225–237
4. Buhalis D (2003) ETourism: information technology for strategic tourism management, vol 4. Prentice Hall, Toronto, pp 174–179
5. Buhalis D, Law R (2008) Progress in information technology and tourism management: 20 years on and 10 years after the internetdthe state of eTourism research. Tourism Management 29(4):609–623
6. Burdea GC, Coiffet P (2003) Virtual reality technology, 2nd edn, vol 21. Wiley-Interscience, Hoboken, NJ, pp 456–459
7. Caneparo L (2001) Shared virtual reality for design and management: the porta susa project. Autom Constr 10:217–228
8. Cheong R (1995) The virtual threat to travel and tourism. Tourism Management 16(6):417–422
9. Guttentag DA (2010) Virtual reality: applications and implications for tourism. Tourism Management 31:637–651
10. Dinh HQ, Walker N, Song C, Kobayashi A, Hodges LF (1999) Evaluating the importance of multi-sensory input on memory and the sense of presence in virtual environments. Proceedings of IEEE virtual reality, vol 15. pp 222–228
11. Foxlin, E (2002) Motion tracking requirements and technologies. In: Stanney K (ed) Handbook of virtual environments: design, implementation, and applications, vol 12. Erlbaum, Mahwah, NJ, pp 163–210
12. Gaitatzes A, Christopoulos D, Roussou M (2001) Reviving the past: cultural heritage meets virtual reality. Proceedings of the 2001 conference on virtual reality, archaeology, and cultural heritage, vol 34. ACM Press, 103–110
13. Gutierrez M, Vexo F, Thalmann D (2008) Stepping into virtual reality, vol 7. Springer, London, pp 546–549
14. Heldal I (2007) Supporting participation in planning new roads by using virtual reality systems. Virtual Reality 11:145–159
15. Lee O, Oh J-E (2007) The impact of virtual reality functions of a hotel website on travel anxiety. Cyberpsychology Behav 10(4):584–586
16. Lessiter J, Freeman J, Keogh E, Davidoff J (2001) A crossmedia presence questionnaire: the ITC-sense of presence inventory. Presence-Teleoper Virtual Environ 10:282–297
17. Meehan M, Razzaque S, Whitton MC, Brooks FP Jr (2003) Effect of latency on presence in stressful virtual environments. In: Proceedings of the IEEE virtual reality, vol 22, pp 141–148

# Chapter 35
# Design of 3D Character Animation Engine Framework

**Jie Zheng**

**Abstract** In this paper, based on the design and implementation of "three-dimensional character animation engine" mainly focuses the principles of different types of character animation parser and the corresponding character rendering function. Besides, the thesis introduces practical animation programming interfaces. They are key-frame animation, mixed animation and morph animation. The engine has unified interfaces which are easily to use and strongly to extend. The users can make use of the ready-made APIs and then be extended to build more complex character animation.

**Keywords** Character animation · Key-frame animation · Morph animation

## 35.1 Introduction

Animation with unknown factors of dynamic play, and strengthen the reality of 3D virtual scene, and increase the interaction and interesting game, it is a key part of the game. However, the human body modeling and simulation movement is always the most difficult and challenging problem. This is because the traditional mathematics and geometric model is not suitable for the performance of the human form. Joint movement of people, especially joint cause muscle movement is also very difficult simulation. In order to play the role of displays the real campaign, the game companies often need to buy expensive professional equipment, labor and spend a lot of time, so character animation as the core technology keep in strict

J. Zheng (✉)
School of Art, Jiangsu University, Jiangsu, China
e-mail: jiezhang462@126.com

W. Du (ed.), *Informatics and Management Science V*, Lecture Notes in Electrical Engineering 208, DOI: 10.1007/978-1-4471-4796-1_35,

accordance with the all the game company. Generally speaking, the character animation using the sensor records the actual movement of a real person, and then saves as animation files. If you want to import animation files to your application, you must know the format file of import and method, but this is confidential. In the past few years, with 3D movies, advertising and many games happen, role animation widely used. Repeat work not only increased the cost of producing animation works, also make to innovation and development work [1].

This paper constructs corresponding character animation in animation files analyzer and edits many roles rendering core functions. Based on this, puts forward the frame of animation programming interface, mixed animation and deformation animation.

## 35.2 Framework Design for 3D Character Animation Engine

3D core role animation character animation engine analytical and a group of role rendering function. Objective to different types of character animation, it needs to build corresponding parser. However, considering similar, using the function analysis overload all types of character animation differences and share in a parser is undoubtedly to avoid duplicate code [2]. The role of the function is used to render an animation and skin grid skeletal animation.

The overall framework of 3D character engine is as shown in Fig. 35.1.

Fig. 35.1 The overall framework of 3D character engine

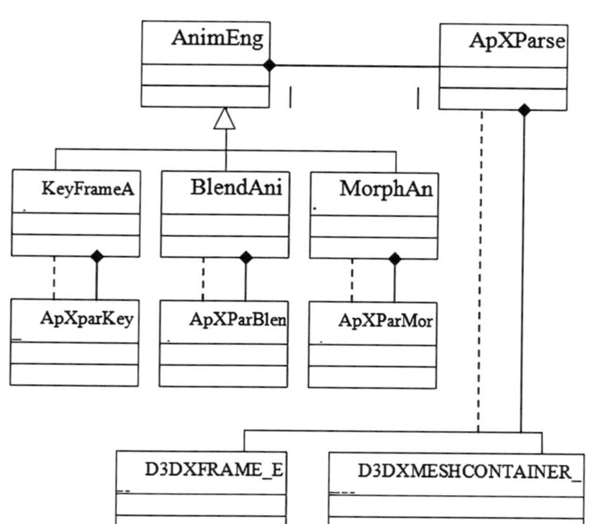

## 35.3 Realization of Core Classes for 3D Character Animation Engine

AnimEng is the core class of 3D character animation engine, which includes a character animation parser and a group of character rendering functions.

### *35.3.1 Analysis.X Animation Files*

ApXParser class is parser class of 3D character animation engine. Its key members are as shown in Table 35.1.

Analysis.X animation files begins to visit the Parse() function of animation files. The whole analysis process runs as the following steps [3]:

(a) Create an ID3DXFile interface;
(b) Register Direct3D standard template;
(c) Create enumeration objects for data objects of enumeration files;
(d) Through enumeration objects to cycle traverse the top-level data objects.

As can be seen from the above steps, final purpose of analysis .X files is to transform animation files structure into a data structure, so that character rendering functions visit every part of mesh model by data structure and render them. Direct3D provides two structures D3DXFRAME and D3DXMESHCONTAINER, which are used to save framework and mesh data. Class ApXParser has expanded both structures [4, 5].

**Table 35.1** Class AP xparser

| Class | The main member functions | Function |
|---|---|---|
| ApXParser | Bool Parse(char *Filename, void **Data = NULL) v | Visit a.X animation file |
| | GUID GetObjectGUID(ID3DXFileData* pDataObj) | Obtain the GUID of template |
| | Char* GetObjectName(ID3DXFileData* pDataObj) a | Access to data object name |
| | LPCVOID GetObjectData(ID3DXFileData* pDataObj, DWORD* Size) | Access to data of objects |
| | Bool ParseObject(ID3DXFileData* pDataObj, ID3DXFileData* pParentDataObj, DWORD Depth, void **Data, Bool Reference) | Analyze data objects |
| | Bool ParseChildObjects(ID3DXFileData* pDataObj, DWORD Depth, void **Data, Bool ForceReference = FALSE) | Analyze sub-objects |
| | Void LoadMesh (…) | Overload function for reading animation mesh data |

Because of different types of character animation, grid structures are also different. When ParseObject() function is reading mesh data, it needs to overload LoadMesh() function according to types of character animation.

### *35.3.2 Analysis and Rendering of a Single Mesh Model Animation*

In a single mesh model animation, the whole character is composed of a integrated mesh model. In other words, there are not framework and transformation matrix framework. ParseObject() function will jump directly to read mesh data function LoadMesh().

The fastest method for reading a single mesh model is to call D3DXLoadMeshFromX() function of Direct3D expanded library in LoadMesh(). It will read an integrated.X animation file. If the function called is success, it will return a SUCCEEDED macro. And then, you can fill with grid name, types, initial pointer, adjacent to arrays information, material arrays, texture arrays and so on [5].

The following statement is character rendering function for a single-grid model animation:

HRESULT DrawMesh (D3DXMESH
CONTAINER_EX* pMesh);

In DrawMesh() function, it determines cycle index according to the number of material. In every cycle, you need to set material and texture applied on Direct3D equips objects firstly, and then enable or disable mixture Alpha according to the property of material. At last, it calls DrawSubSet() function of ID2DXMesh interface to render mesh data of the same material. When cycle is finished, rendering the whole grid model will be accomplished.

### *35.3.3 Analysis and Rendering of Skin Skeletal Animation*

In skeletal animation, any vertices only are affected by a piece of skeleton. Grid models are apt to joint at sports, especially in the binding site, such as shoulder, elbow. However, in skin skeletal animation, each grid point can be affected by many skeletal blocks. Each vertex contains a skeletal list which may affect it. In Direct3D, the maximum number of skeletal transformation matrix which affects each vertex is limited to 4. According to skeletal transformation matrix and corresponding weight, the formula for calculating linear interpolation of vertex position and normal is defined as follows:

$$\begin{aligned} V_{blend} =& V * M_0 * W_0 + V * M_1 * W_1 \\ &+ V * M_2 * W_2 + V * M_3 * W_3 \end{aligned} \tag{35.1}$$

where $V_{blend}$ represents a mixture of vertex position and normal; V expresses pre-mixed of vertex position and normal; $M_0$, $M_1$, $M_2$ and $M_3$ represent skeletal transformation matrix; $W_0$, $W_1$, $W_2$ and $W_3$ represent skeletal weight.

Skin skeletal animation has more skin information than skeletal animation (such as skeletal index and weight), so LoadMesh() function often calls a Direct3D expanded library function D3DXLoadS—kinMeshFromXof() for each triangle when reading mesh data, not D3DXLoadMeshFrom of function [5]. It has an additional parameter. That is a pointer to ID3DXSkinInfo object and the object contains skin information.

The following statement is character rendering function for skin skeletal animation:

HRESULT DrawSkinMeshes (D3DXMESHCON.
TAINER_EX* mesh);

Character rendering of skin skeletal animation is accomplished by order traversal each grid. Each vertex position of skin skeletal animation may be affected by some skeletal transformation matrixs, so it needs to fix each vertex's final position as follow steps after working out all skeletal transformation matrixs.

Step 1 Iterate all grid vertices, and do next step for each vertex;
Step 2 Connect current vertex to each skeleton, and access to their skeletal transformation matrix;
Step 3 For each skeletal transformation matrix, multiply by weight and apply the result to vertices as combined transformation;
Step 4 Repeat the third step for each connected vertex skeleton and repeat the second step to the fourth step for each vertex. After accomplished, it should apply combined transformation matrix to each vertex.

## 35.4 Key-frame Animation

As can be seen from above, class ApXParser is only to analyze character data, and the character rendering function only renders the static character being known character status. Really to form character animation, it must rely on animation data to drive the static character.

Key-frame animation, a character animation which is used more widespread in current games based on skin skeletal animation. It also is one of practical animation programming interfaces provided by 3D character animation engine. The basic principles of key-frame animation are to divide animation into a longer time period (second or 2 s). Each period records a key point. We can get character status at any time through interpolating between the key points. Suppose skeletal transformation matrix of vertex is Mat1 at moment s1, Mat2 at moment s2. At any

Table 35.2 Class key-frame anim

| Class | Main member function | Function |
|---|---|---|
| Keyframeanim | Bool LoadKeyFrameAnim(char* Filename) 1 | Load, parse key-frame animation |
| | Void UpdateKeyFrameAnim(char* AnimationSet, DWORD Time) | Update key-frame animation |
| | Void DrawKeyFrameAnim (D3DXMESHCONTAINER_EX* pMesh) | Render key-frame animation |
| | Void FreeKeyFrameAnim() | Clear key-frame animation |

moment s between s1 and s2, its skeletal transformation matrix Mat can be calculated by the formula as follows:

$$Mat = (1 - w) \bullet Mat1 + w \bullet Mat2 \tag{35.2}$$

where w is weight. We can regulate the proportion of Mat1 and Mat2 in skeletal transformation at moment s through w. The simple interpolating method for skeletal transformation matrix is linear interpolating, and weight w can be calculated by the follow formula.

$$w = (s - s1)(s2 - s1) \tag{35.3}$$

Key-frame animation class KeyFrameAnim as a programming interface inherits class AnimEng. It encapsulates key-frame animation parser ApXParKeyFrame and character rendering function DrawSkinMeshes() of skin skeletal animation. The member functions of class KeyFrameAnim are designed as Table 35.2 shown.

Function LoadKeyFrameAnim() loads a key-frame animation parser which is a object of class ApXParKeyFrame. It calls member function Parse() to visit and begin to parse.X animation files. In frame cycle of user application, it calls function UpdateKeyFrameAnim() to update key-frame animation and calls member function Update() of class ApXParKeyFrame to update initial transformation matrix of the roles of various framework. After accomplished, DrawKeyFrameAnim() function

**Fig. 35.2** The key-frame animation at different moment

which completes animation rendering by calling character rendering function DrawSkinMeshes() renders key-frame animation. Procedures to withdraw from the former, it also needs to call FreeKeyFrameAnim () function to release object structure, equipment resources and so on.

Figure 35.2 shows the effect of key-frame animation by implemented with KeyFrameAnim class. Where both maps also show the character exercise status on a walking animation set. Of course, real character animation is continuous.

## 35.5 Conclusions

"3D character animation engine", and can be widely used in film and television, advertising, games and other areas, this will help effectively save development costs, shorten the development cycle. This engine also has important reference value users use a specific format, make the animation.

## References


1. Decai W, Guansheng Y, Yuping S (2007) Proficient in directX 3D graphics and animation programmming, vol 13. The People's Posts and Telecommunications Press, Beijing, pp 352–356
2. Gamma E (2007) Design patterns: elements of reusable object-oriented software, vol 07. Mechinary Industry Press, Beijing, pp 65–69
3. Fei Y (2009) 3D graphics rendering engine and research on key technologys, vol 06. Springer, Xian, pp 18–21
4. Kater M (2007) DirectX software development kit, vol 23. Microsoft Corporation, New York, pp 79–86
5. Adams J (2005) Advanced Animation with DirectX, vol 7(12). Chongqing University Press, Chongqing, pp 131–136

# Chapter 36
# Research on Computer Network Auxiliary Scaffolding Teaching

**Xi-tao Gu, Hong Li and Juan Chen**

**Abstract** The proper application of scaffolding instruction mode can optimize the process of reading instruction and improve the reading teaching effectiveness. It can also help students to gradually master the reading strategies and enhance students' English thinking and expression abilities. Moreover, it can help to achieve the English reading teaching objectives. In this paper, a computer network is applied to participate in the process of scaffolding reading teaching. The external stimulus offered by the online teaching is not a single stimulus, but the comprehensive stimulation of multiple sensory. This is quite important for the acquisition and maintenance of knowledge. In particular, the respect the individual differences of students in the computer network environment can help students to establish their confidence in learning.

**Keywords** Computer network · Scaffolding reading instruction · Compute auxiliary · Multimedia

## 36.1 Introduction

Constructivist teaching theory holds that the individual's cognitive development and the learning process are closely related to each other. It thinks that learning is the process of acquiring knowledge, but knowledge is not acquired by teachers' teaching. Instead, it is but learners in certain scenarios, with the help of others (teachers and learning partners), and the necessary learning resources such as

X. Gu (✉) · H. Li · J. Chen
Xingtai University, Hebei Xingtai 054000, China
e-mail: xitaogu298@126.com

W. Du (ed.), *Informatics and Management Science V*, Lecture Notes in Electrical Engineering 208, DOI: 10.1007/978-1-4471-4796-1_36,

written materials, audiovisual materials, multimedia courseware and information on the Internet. It is obtained by the way of meaning construction.

The teaching of reading is an important part of language teaching [1]. The traditional reading model has been unable to meet this requirement. The teaching model of "chalk and talk" stifles the students' character and limits the enthusiasm of the students to explore knowledge. Therefore, in the teaching of reading, updating teaching concepts, improving teaching methods and focusing on the development of student ability is especially important. Based on computer networks, this paper has made a preliminary exploration on scaffolding teaching of reading.

## 36.2 Scaffolding Teaching Instruction Model

"Scaffolding Instruction model" is to provide an appropriate conceptual framework to help learners to understand the teaching model of and the significance of specific knowledge and architecture knowledge [2]. With the help of the conceptual framework, the learners are able to independently explore and solve problems and thus construct meaning independently. It is based on the previously well-known Soviet psychologist Vygotsky "recent developments area" theory, and it is the specific application of social constructivist learning theory in teaching [3]. Teaching should be based on the latest development zone of the students. It should be provided with the content with difficulty, and move beyond the students' latest zone of proximal development and reach its potential level of development. In addition, as the basis for a development area, it emphasizes that teachers construct a "learning scenario" for the students, "consultation", "session" and "significance construct" learning environment [4].

Scaffolding reading teaching model requires teachers to fully understand the level of development of the reality of students based on student's needs analysis, the flexibility to break down complex learning tasks. Rack reading model is a system in a planned and step-by-step teaching process, including:

Creation of context, to build the bracket;
Questions to guide inquiry;
Independent thinking and independent research;
Peer support, cooperative learning;
The effect of detection, to lead the upgrade.

## 36.3 Scaffolding Creates Network Scenario

Scaffolding instruction in the bracket should be based on the student's "zone of proximal development" to create, by constantly filling the information poor, guide students' intellectual from one level to a higher level [5]. The essence of scaffolding is to decompose the complex learning tasks, the formation of the thinking process. Its scaffolding model is the "old knowledge + new problems". The reading process is the core of the teaching of reading, but also the key to improve students' reading skills. Each type of text has its own unique overall structure (in other words, scaffolding), understand it helps to correctly understand the discourse intention. The teaching of reading is a complex bilateral teachers and students the language and interacting activities of thinking, take the "scaffolding" to pay attention to two points: First, "scaffolding" of similarity, similar language processing should take the form of basic consistent; Second, the flexibility of the "scaffolding". The presentation of the scaffolding should "to" process of change. The beginning of "yes" is set in order to help students construct their mode of thinking, while "no" is the starting point of the students' creative thinking.

Successful foreign language classroom teaching should be able to create more scenarios in Class gives students the opportunity to use their own language learned material [6]. Constructivism holds that the situation must be conducive to the significance of learners have learned the content, Klum said: construction. The most efficient way for creation of scenarios in the teaching of English reading is classroom teaching.

Teachers should select relevant teaching media according to teaching goals and students 'practical teaching media, dissemination of information through the network, ask questions, and the creation of a learning context, and to stimulate students' internal motivation. Application of the "scenario, a task an independent inquiry" mode can be used to carry out the teaching of English reading. First, create a scene, raise issues. Then teachers can develop tasks according to the teaching purpose of a task to guide students to actively complete.

## 36.4 Cooperative Learning and Build a Network English Reading Teaching Model

Scaffolding instruction stressed the key role on the construction of meaning of the "collaborative learning". Collaboration should be throughout the entire reading activities. Collaboration between teachers and students, students, integration of textual information, learning outcome evaluation, the significance of the final construct and the efficiency of the learning has an important role. The result of collaboration might be discrepancies, but can make complex issues clear of developing students' different ways of thinking.

From a Construct point of view, the classroom should provide students with real learning experiences in cooperation and non-competitive environment, and the students' performance evaluation, and encourage students to cooperate with their peers, cooperative learning. Cooperative learning in common is that high school students in the teaching activities to group activities as the basic form of the main theory. Teachers can be grouped according to the actual situation of the student advocate online collaborative learning among students. This online interaction is the effective extension of the classroom activities, free from the constraints of time and space, but also conducive to the personality development of students. In this teaching activities, teachers guide students to read the article thinking, in-depth understanding of key content. Cooperative learning, students have formed collective and mutual exchanges, teachers design activities necessary to fully take into account the needs of different students, so that every student has a harvest in the process to complete the reading task.

Multimedia network reading teaching abandons the traditional teacher-centered "teach" teaching method, which is replaced by students as the center of the active practice of teaching methods. The dominant role of teachers is important, teachers can establish an online teaching system, the implementation of the monitoring process on student learning and provide effective and timely adjustments, the use of the network of interactive tools such as BBS Forum and other organizations, and motivate students to English reading to which the study. Between teachers and students through online discussions, e-mail exchanges and other forms of exploratory learning, allow teachers to quickly and accurately understand the students' learning to help students improve their English reading learning efficiency. Systematic, scientific and operability characteristics of multimedia network curriculum design; teachers should guide students to develop the habit of independent study. Multimedia network environment, the English reading learning allows students to their own brains to find the problem, identify problems and students to build interest in reading, naturally active learning, basic network read task is completed, and students can self-test to examine their own learning, to help them experience the use of network resources for English reading learning a sense of accomplishment.

## 36.5 Effect Evaluation and Network Sharing

The effect of evaluation includes self-evaluation and the learning evaluation of study groups for individual students. The evaluation includes:

Self-learning ability;
Contributions made on the knowledge of the collaborative learning;
Whether the meaning construct of the learnt knowledge can be accomplished.
Self-evaluation can refer to the following rating scale:
Do you know the chapter learning task?
Can you preview, review?
Whether you have actively participated in group discussions?
Are you and other team members happy?
Do you combine the old and new knowledge?
Have you accumulated knowledge?
Can you overcome the learning difficulties?

Are you able to choose your own learning methods?
Can you have a rational allocation of learning time?
Can you understand this section of the course content?
Can you learn by analogy?
Is your learning process reasonable?
Are you satisfied with your own learning?

Through effect evaluation, student will have a further understanding of the reading class teaching activities and it does not have a close. It does not mean the end. The teachers also make good use of the network environment; students are more strongly induced self-expression of desire and reading desires. In the teaching of reading, after reading the data, sorting, classification, analysis and discussion, the students formed a more mature understanding of the teachers' network platform to enable students to write a small essay, describes their favorite reading materials, and on the Internet release for teachers and students evaluation.

## 36.6 Conclusions

The introduction of the network provides a truly open teaching environment; the evaluation is open and equal exchange between teachers and students, which greatly enriched the form of a traditional English class. The essence of scaffolding instruction mode is to provide students a good learning environment for language and culture. In this learning environment, teachers should help students in the two aspects of language and cultural knowledge of the intrinsic link to a more profound understanding of the current learning content through the computer network. Of course, the only correct comprehend its essence, to achieve multiple objectives, of the English reading class to meet the demands of the times.

## References

1. Armstrong T, Pingzhe L (2003) Multiple intelligences in the classroom—to carry out student-centered teaching. China Light Industry Press, Beijing, (08) pp 26–32
2. Kekang H (1997) Constructivist teaching model, teaching methods and instructional design. J Beijing Normal Univ (Soc Sci) 5:78–88
3. Xinyong M (1998) Constructivist teaching model, teaching methods and instructional design. Foreign Primary Secondary Educ 23(5):132–137
4. Ou YF (2004) Multiple intelligences and constructivist theory in the classroom teaching, China Light Industry Press, Beijing, (06) pp 89–94
5. Dingfang S (2004) Foreign language teaching: Problems and solutions, vol 11. Shanghai Foreign Language Education Press, Shanghai, pp 362–368
6. Wang L (2000) Modern english instruction theory, vol 22. Shanghai Education Press, Shanghai, pp 31–36

# Chapter 37
# Study on Reform of College Teaching Methods Under Information Condition

**Xiao Zhan, Mingcong Ma, Lei Du and Man Liu**

**Abstract** Promoting informanization in teaching today, teaching methods and means become necessary to realize modernization of education reform, and promote quality-oriented education. There are some in the reform and the phenomenon of overkill form now, such as exercising blow on the traditional teaching methods, putting emphasis on the form of teaching means, while not focusing on teaching activities of middle school students participation, relying too much on multimedia teaching, with the purpose of improving the teaching quality, the teaching methods will be suitable as long as it can largely enlarge effective knowledge.

**Keywords** Teaching methods · Teaching means · Heuristic teaching · Case teaching

## 37.1 Introduction

In the information society, our teaching environment has changed a lot from the early language laboratory, teaching cable TV system, small computer network, to today's multimedia classroom, multimedia teaching system, campus network television, internet, etc., in this environment, many schools have been advocating

X. Zhan (✉) · M. Ma · M. Liu
Xuzhou Air Force College Educational Technology Center, Teaching Assistants, Xuzhou 221000, China
e-mail: zhanxiao678et@126.com

L. Du
Xuzhou Air Force College Department of Logistics Command, Teaching Assistants, Xuzhou 221000, China

W. Du (ed.), *Informatics and Management Science V*, Lecture Notes in Electrical Engineering 208, DOI: 10.1007/978-1-4471-4796-1_37,

to reform teaching methods and means, to 45 min to class quality, but paying much attention to theory research, lacking practice, or form, even overkill, making reform disadvantages, and effect not obvious.

## 37.2 Some Problems in the Reform of Teaching Methods

The teaching method refers to the teachers and students where they aims to achieve common goals, complete teaching task in the process of teaching, it's the combination of methods and ways in teaching procedure. A theory by professor LiBingDe, classified the theory into five categories: (1) language transfer information primarily methods, including teaching method, the conversation method, discussion method, reading guidance method, etc. (2) With the aid of the immediate perception method, including demonstrating mainly method, visit method, etc. (3) method mainly aimed at practical training, including practice method, experimental method, the practice operation method. (4) Method mainly related to appreciating activities, such as mainly edify method. (5) Method primary connected with exploring, such as discovery method, exploration method, etc.

### *37.2.1 Blindly Criticize "Traditional Teaching Methods"*

In the reform of teaching methods, "the traditional teaching method" is treated as an opposite side to modern education and always looks down upon, is traditional teaching methods out of time? The traditional teaching methods include: teaching method, the conversation method, discussion method, demonstration method, the reading guidance law, and so on. Traditional teaching method is the one which is criticized most; some people think that it is a injected method (non-computer majors method of teaching). In fact, teaching method is a most direct and effective teaching method. It helps to form the systematic knowledge, help students to form the concept, rapidly understand knowledge, and training in basic rules. Whether the students have solid knowledge has largely connected with the use of teaching method. Concerning the teacher, the traditional teaching method helps teachers play well and take control of the teaching progress effectively. Face-to-face communication between teachers and students is beneficial to their communication and improving the students' comprehensive quality. The blackboard writing form in the traditional teaching method will be convenient for the students to take down notes, understand and memorize content. They have been taught.

While some classes using modern teaching method, put heavy stress on form while light on effect, will classes as teachers' performance stage, such as classroom questioning, ask the good student, but the students who perform poor less; often ask good students more complex problems, and the poor students a simple, straightforward question, which is known a show.

## 37.3 Superficial Elicit Teaching

In addition, in recent years to promote heuristics, teachers equal heuristic teaching to classroom questioning. Heuristic teaching is a kind of seeing the students as the main body, where teachers inspire students' to think, actively obtain knowledge, develop intelligence, and form perfect personality of the process. Whose essential feature is to inspire and motivate students' initiation and enthusiasm. Both of them have a link, but cannot mix up [1]. Questions asked in Heuristic teaching have a certain request, which is well be planned, like some only rote could answer the plain and the only question answer straightforward, for example, what is this? Such superficial question does not belong to the category of the heuristic teaching. Heuristic teaching is not only a kind of teaching methods, but also a kind of guideline to active the student thoughts, which can't be judged from watching how many Ss the teacher asks, how many students put up hands, but can be seen from whether they can stimulate the students' learning motivation and improve the students' positive thinking, acquire knowledge actively. By thinking and discussing, this are still difficult problem to solve, the teacher give some explanation and inspire, making students have a better understanding if there are any problems.

The teaching practice, the heuristic teaching in the form of a DuoZhong, besides using suspense type, such as type discussion teaching method can get very explicit good inspiration effect in the teaching process, the teacher's a look in the eyes, a hand type, some suggest that can cause student's thinking and associations. So, we in the teaching practice to firmly establish the heuristic teaching ideas, combining the teaching activities in them.

## 37.4 The Case Teaching and Teaching of Confusion, for Example

Implementation of the classroom teaching process, the teacher will often through the list is easy to understand and vivid example to show that things or prove his own expressed the opinion, this belongs to the lecture for example and not the case teaching. Case teaching is through the case presented a typical example of the specific situation, and guide students to hidden in the complicated problems specific analysis and discussion and research on the training of the students, and innovative thinking, cultivating and improving students found, analysis and solution actual problem ability a teaching form [2].

Case teaching is a more prominent, the subjectivity of the student, lets student in the simulation of the situation in solving practical problems, implementing case teaching to want to pass commonly homework, personal familiar with the case, the research team, and class discussions, sums up five steps, a case teaching to continue for a few days.

Of course, the case teaching case selection also have certain requirements, must first be clingy student life practical, promotes the reality are more prominent in the hot issues as the research object, the acquisition of the real example to try to the structure, form to deal with students to think, analysis, discussion, case, the judge to facts and real data, based on the teaching content and the students will actually close together, let the student through independent thinking and analysis, to discuss the teaching purpose of developing intelligent and case to update and improve the students' ability of the post office.

## *37.4.1 The Teaching Method Reform in Some Problems*

Teaching means teaching auxiliary tool refers to the use of the classroom teaching of a kind of method, with the rapid development of science and technology, the modern education technology in the classroom teaching in colleges to a wide range of applications, which changed the traditional teaching mode of teaching and learning, to establish the teaching situation, improve the quality of teaching has played a very important role.

The modern education technology means of the multimedia technology has strong content, so we advocate performance ability in the teaching of multimedia, but part of the teachers use overstating the function of multimedia technology means, in each class will use multimedia, frequently "full multimedia", there seems to be no electronic teaching means should assist the difficulty in making teaching activities. In fact the teacher class teaching, not only in the instruction knowledge, but also with students in exchange of the mind, they ignore the teaching of middle school students' subject status, to the teacher's leading role and the emotional communication between teachers and students. In the traditional teaching process, the teacher can on the platform, and even the whole classroom around pacing back and forth, the side explain side do actions by posture, gesturing, language, eye contact, etc. to spread knowledge information, ideas and personality information, information communication with student's emotion. And excessive use of multimedia, the activities of the teachers trapped in the computer, have a class into see slide, the teacher became a commentator. Class teacher's attention on the basic multimedia performance forms, focus on the computer operation, the cadence of speech before the turn of the importance and urgency of the endless turn into the teaching, the teacher's mission is just as a "book", very bad student active learning, and positive thinking. The more there are some multimedia courseware coarse, just as the abuse teaching materials, teaching plan, pictures, etc. Simply "move" to the computer, or in the design and production of multimedia courseware for too much form of lay particular stress on the picture, beautiful, teaching courseware very loud, add too much animation, sound, video, etc., on the contrary, the leading role, the impact of the student to grasp the teaching content. Only the teachers' personal characteristics and the modern teaching means organically, can give full play to the modern teaching means of classroom teaching effect [3].

Therefore, the modern education technology use of making timely and appropriate, right amount, refinement, not teaching means the more abundant teaching effect, the better. After all the teaching media is a kind of teaching with the auxiliary teaching method, teaching circumstances vary, not all of the teaching content to use multimedia assisted teaching, use need from reality, teaching according to the student, subject characteristics and teaching content, blind follow the fashion, the choice is used undeserved or excessive use, can give teaching negative effects, and can't achieve what we expected the teaching effect. The teaching process, the teacher should pay attention to overcome excessive sensory information, frequency of strong, stylized overweight, teaching means too unity, a reasonable use of modern education technology, avoid by all means will be teaching means as teachers and students the stage property of the performance.

### *37.4.2 Problems Noticed in Selecting Teaching Methods*

As the saying goes "teaching procession of the law", in class, the teacher will not directly use a kind or several kinds of teaching methods, but considerate comprehensively, selecting from those methods to form a suitable one, and pay enough attention to the students' participation. That is to say, the Ss will be chosen as the main body, students will play a main role, which emphasizes importance on the students' individual difference and teaching them individually. In the teaching activities, the teacher should pay attention to grasping the leading position of the teacher and the student as main body status, and can not only pay attention to the scale of the students' subject status and ignore the teachers guide role. We should treat the development of the students' intelligence as the starting point, put emphasis on the students' non-intelligence factor, the role of teachers does not only to teach students knowledge, but also to develop their creativity [4]. By the creative thinking ability, students can not only learn what teachers have taught, still can learn the teachers don't teach, even create new knowledge, so teaching should be reflected on the students' performance: happy to learn.

In the teaching means, we cannot simply emphasize the form of modern education technology as well as the number and frequency, while seeing whether teaching activities play roles in teaching activity. We should work hard in teaching content and form, the scientific use of multimedia, and put emphasis on improving their own ability to give lectures.

## References

1. Zhao XY (2005) Education of the pain and the itch, vol 37. Cambridge University Press, Chinese 88–90
2. Li XQ, Pan YT (2008) Multimedia teaching reading dislocation phenomenon. China's Military Educ 22(3):61–70
3. Li BY et al (2009) Reform teaching methods, training innovative talents in China. Science 28(5):39–50
4. Zhao C (2008) Offer only ever. Shallow light with "students for this" education concept. Management Health Abstr (New Med J) 17(10):87–99

# Chapter 38
# Information Network Analysis System for Scientific and Technological Documents

**Hongxia Liu**

**Abstract** With the social network analysis more and more used in cooperative scientific research analysis, citation analysis, and competitive intelligence social network analysis and so on, the increasing number of articles based on a macroscopic point of view, which probes into the impacts on Library and Information. This paper sums up the application of system analysis for documents information analysis, with the social network as the center, to explore the intrinsic link between cooperative scientific research analysis, co-word analysis and citation analysis, as a basis to establish a network analysis system of documents information, as well as to deduce new ideas and new ways for documents information network analysis system.

**Keywords** Social network analysis · Cooperative scientific research · Co-word analysis · Citation analysis

## 38.1 Research Background

Document information analysis is a kind of research activity. It carries on directional selection and scientific abstraction on relevant document information according to certain needs.

As for the traditional bibliometrical method, it pays more attention to the statistic analysis of the attributes of all kinds of documents. In addition, it attaches special attention to the quotation and analysis between the documents. The research status and trends of the ecological footprint was quantificational analyzed

H. Liu (✉)
Binzhou Polytechnic, Binzhou, 256603 Shandong, China
e-mail: liuhongxia@cssci.info

W. Du (ed.), *Informatics and Management Science V*, Lecture Notes in Electrical Engineering 208, DOI: 10.1007/978-1-4471-4796-1_38,

based on bibliometrical methods. There is very little research on the relationship between the attribute of the documents. In addition to this, there is very little research as well studying on the relationship between the documents built through the attributes. In recent years, the traditional document bibliometrical method has been complemented and improved for there are a great amount of passages coming to the world. The passages that complement as well as improve the traditional bibliometrical method are in such topics as science and research cooperation network, co-word network, citation network and so on in both the country and in aboard.

The social network analysis is developed in the fields of anthropology, psychology, sociology, and mathematics, as well as the field of statistics. The social network analysis has already formed a series of exclusive concepts and technical terms. It has been applied in the research of sociology very widely and has already been a kind of new normal form for the research of social science.

## 38.2 The Analysis of Research Practice in the Country

### *38.2.1 Social Network and Science Research Cooperation Analysis*

Science research cooperation takes the science innovation as the target. There is a team being established. The team is made up of by several science research and development personnel. As for the science research and development staff that made up of the team, they would like to take responsibilities of the mutual science research targets. In addition to this, their skills are complemented. The science research and development staff surrounds the mutual prospects. The science research and development staff has such characteristics of target commonness, and the knowledge sharing, as well as the interest dependency and so on.

As for the research that applies the social network analysis into the scientific research cooperation network in the country, there are four major aspects currently. The four aspects are shown as the followings: the first one is the analysis of science and research cooperation network structure and the indexes; the second is the analysis of the science and research cooperation group (or region); the third is the visibility of the science and research cooperation network; and the forth is the experimental research on the exploration performance of specific fields.

#### 38.2.1.1 The First One is the Analysis of Science and Research Cooperation Network Structure and the Indexes

The increase of scientific cooperation has already formed into a large scale cooperation network. The large scale cooperation network is a true social network.

The large scale cooperation network requires developing a kind of appropriate network measurement method.

#### 38.2.1.2 The Second is the Analysis of the Science and Research Cooperation Group (or Region)

The research on the regional cooperation mainly refers to the construction of the scientific research cooperation network [1]. The scientific research cooperation network for the regional cooperation is constructed among universities and among the provinces and districts as well as being constructed among nations. The regional cooperation research is worked through the structural variables and the network as well. As for the structural variables, it refers to such things as the average path length, and the clustering coefficient, as well as the particle size distribution. As for the network, it refers to the directed weight. The regional cooperation research describes the structure characteristics of the science research cooperation network. It addition to this, it as well describes the connection mechanism of the resource integration efficiency, and the knowledge flow, as well as the knowledge innovations.

#### 38.2.1.3 The Third is the Visibility of the Science and Research Cooperation Network

The passage "The application of the visible study of Pajek in information science collections" introduces the network analysis and the visible software Pajek. "The application of the visible study of Pajek in information science collections" has systematically expounded the foundation of the science and research cooperation network, as the drawing of the network graph as well as the derivation process. It has presented the macrostructure as well as the microstructure of the subject network [2].

#### 38.2.1.4 The Forth is the Experimental Research on the Exploration Performance of Specific Fields

The passage "Study on the Phenomenon of Library and information science dissertation co-authorship collections" considers that the co-authorship relationship has represented the social relationship between the authors. (The social relationship of the authors refers to the human communication relationship of the author in the actual society) [3]. However, it also thinks that the co-authorship relationship fails to reflect the relationship of the science and research cooperation in the actual world in a real manner. In particular, "Study on the Phenomenon of Library and information science dissertation co-authorship collections" thinks that the co-authorship relationship, which is shown in the passage, can hardly show the cooperation relationship of different institutions.

### 38.2.2 Social Network and Co-word Analysis

The method of co-word analysis is first described in detail dating back to 1970s. The method of co-word analysis is started by the French document scientometrician. The research key points can be summarized taking advantage of the co-work analysis method. In addition to this, the development process, and the characteristics of the transverse as well as the vertical analysis of the field subject can be summarized making use of the co-word analysis method. Moreover, the relationship between the transverse field subject and the vertical field subject can be analyzed. The method of co-word analysis method can represent the scientific research level of certain major as well as the dynamic and static structure of the development history for certain major. In addition to this, it is able to reflect the relationship between the foundation research and the technical research, the relationship between the research achievement input and the research achievement output in the evaluation field. Moreover, it is able to reflect the development information search filed.

From the perspective of the social network, the key words are joints in the network. The co-occurrence of the key words is reflected in the connections for the connection of the joints. Through the analysis of the key words in the network, the relation network is able to be found that is hidden behind the real relationship network. The analysis of the key word network has played a very significant meaning to the following situations: the mature degree of a research subject; the knowledge structure; the scale situation of the research and so on.

Nowadays, the researches in the country can be mainly divided into three aspects: the first is the visibility of co-word network; the second is the automatic abstract and labeling of the key words; and the third is the explorative empirical research in specific fields.

### 38.2.3 Social Network and Citation Analysis

Citation analysis is a kind of document measurement analysis method. It makes use of all kinds of logic methods such as the mathematic method, the statistic method, and the comparison, summary, abstract and summary methods and so on. Taking advantage of all kinds of logic methods, it makes analysis on the phenomenon of citation and co-citation for all kinds of analysis objects. The analysis objects have included the science periodical, the essays, and the authors and so on. In this way, it is convenient to reveal the quantity characteristics and the internal rules.

The document measurement method is able to be used in the whole structure and development process of the research for the citation network. As for the theory for the social network analysis, it is better at the revelation at the structure of the network as well as the research for individual and small community level. There are some studies that apply social network analysis into citation analysis. In the current phrase inside the country, it has divided the parts that applying social network analysis into citation analysis mainly into the following three aspects: the first is the analysis of citation network structure and indexes; the second is the visibility of citation network; and the third is the explorative empirical study on specific fields.

#### 38.2.3.1 The First is the Analysis of Citation Network Structure and Indexes

It refers to the application of social network analysis theories and methods into the analysis of the periodicals. It will carry out the analysis of the progressive density of citation network, the average distance, the clustering coefficient, the centrality, the connectivity, k core, and the structural equivalence as well as the center-edge structure equalizing characteristics.

#### 38.2.3.2 The Second is the Visibility of Citation Network

The researchers construct the matrix on citation relationship. In addition, they make use of the social network analysis tools such as the Net draw, Pajek and so on so as to carry out visibility and study the internal knowledge structure and the communication situation of the subjects, the center-edge relationship between the periodicals, and the development trend and significant people of the fields and so on.

#### 38.2.3.3 The Third is the Explorative Empirical Research in Specific Fields

Researchers use the citation relationship of relevant documents as the basis. They make use of the social network analysis to get to know the subject structure of the entire field, explore on the characteristics of the communication and dissemination of subject knowledge, make clear the center-edge structure of the subject knowledge communication and make judges on the significance and development trend for each community in the network. At the same time, they have revealed the potential possibility of applying the combination of citation analysis and social network analysis method to document evaluation.

## 38.3 Construction of Document Information Network Analysis System

### *38.3.1 Document Information Network Analysis System*

This paper tries to analyze all kinds of relationships existed in the document information as well as the connections between relationships from the perspective of social network. In terms of theories, it divides all kinds of relationship of document information from three kinds of basic relationship units:

(1) Single document relationship: all kinds of relationships formed in a single document;
(2) Dual documents relationship: all kinds of relationships formed in two different documents;
(3) Three documents relationship: all kinds of relationships built through the third document on the basis of two documents.

### *38.3.2 Analysis Thinking Deduction*

(1) Analysis of network structure and index
(2) Social communities or composition analysis
(3) Network visibility research
(4) Empirical analysis on specific science and research field.

## 38.4 Conclusion

Take the social network analysis as the center. Make analysis on the relationship between the document attributes and the relationship between the documents built through attributes. Explore on its internal connections with science research cooperation analysis, co-word analysis, and citation analysis and so on. Put forward the research thought of constructing a complete document information network analysis system. All of these have offered effective theoretical support for relevant empirical studies. At the same time, the adaptive scope for all kinds of methods in the document information network analysis system is as well needed to be further studied and determined.

## References

1. Yin LC, Jiang CL, Yin FL et al (2007) Analysis of interprovincial science collaboration network visualization based on Cscd and sci databases. Libr Inf Serv 51(8):62–64
2. Meng W, Pang JA (2008) The application of the visible study of Pajek in information science collections. Inf Stud: Theor Appl 31(4):573–575
3. Chen DQ, Zhu WF, Mo XJ (2009) Study on the phenomenon of library and information science dissertation collections (1999–2002). J Comput Sci 27(1):70–73

# Chapter 39
# Study of Contemporary Artistic Design Based on Computer Technology

**Rui Liu**

**Abstract** With the development of technology, the computer technology has been used into more and more areas, which has changed human life significantly. Therefore, the application of computer technology in the arts is the fusion between art and technology, which has involved almost all areas of artistic designing. Computer has liberated people from the complicated physical labor and focused on creativity and design, meanwhile, the concept of art has undergone changes. The integration of multimedia images, sounds, text, animation, audio and even smell and other forms of communication has enriched the artistic language and expression and improved the infection of the works. It has also enabled more people to enter the design field, and has created numerous opportunities for employment and wealth producing.

**Keywords** Computer technology · Computer art · Artistic design · Influence

## 39.1 Introduction

The rise of "computer art" is the most stunning change in art field in the twentieth century, which is inseparable from the development of computer technology. The driving force of the development in modern information society is the computer technology. The development of computer systems has experienced from the mechanical, electronic analog to digital process, which has led to the development

R. Liu (✉)
College of art of Xi'an University of Architecture and Technology, Xi'an, Shanxi Province, China
e-mail: liurui@cssci.info

W. Du (ed.), *Informatics and Management Science V*, Lecture Notes in Electrical Engineering 208, DOI: 10.1007/978-1-4471-4796-1_39,

of computer technology with an incredible speed and quality by leaps and bounds. With the development and social use of the technology, "computer art" has come into being [1]. Computer art has become an emerging category of artistic disciplines, which has integrated computer technology and traditional artistic designing perfectly. Its appearance is inevitable in the development of science and technology; meanwhile, it has met the requirements of the development of contemporary artistic designing [2]. This is a breakthrough in design history; also it has affected the public's aesthetic changes, which is the new round of changes in the tide of human civilization.

Digital art has come down to artistic designing in each field, of which the art form and connotation are changing [3]. As in the invention of the photography, the invention of the computer graphics and image technology is a leaping revolution in the history of human artistic designing. The field is very wide, mainly including two aspects of the graphic design and three-dimensional design [4, 5]. The graphic design included categories of advertising, illustration, 2D Animation, video design, costume design, etc. And three-dimensional design is extensively applied in architecture, sculpture, film and television animation design.

## 39.2 The Positive Influence of Computer Science and Technology to Artistic Designing

The access of computer technology into the design field has liberated the hands of designers. Designers can utilize the flexibility of the brain thought to produce a steady stream of inspiration more fully and create a new design world. Therefore, the computer level of contemporary designer determines the design quality. Computer-aided design is a convenient means of understanding and using various software–but also is a challenge of design standards and efficiency for designers. Therefore, that the more advanced the computer is, the better the performance is not really, but it chooses according to the actual needs of the individual. Computer has liberated our hands, but except brain. The spiritual thinking of facing the screen requires a solid knowledge of the structure and rich imagination. Because the same software we are using means the same starting line we are standing. Because the way of thinking of computer is quite different from that of the hand paint design, the thought that mastery of the use of several software could gallop in the design market is really fantastic, and easy to fall into the error of being passive. Computer's real and virtual space needs the realization and satisfaction of the imagination, and virtual space brings surprises, romance and multi-time stack of future and past for people, which easily makes the designer fall into a non-real emotional life and into the dilution of the concept of the territory, country, nation, regime, and the indifference of authenticity of things. Although modern technology has brought such a huge impact and influence to the traditional art, the application of modern technology is the use of tools and the mastery of a technology when people in artistic creation. But the key of the creation of artistic works with value and vitality lies in the improvement of people's artistic level.

The more advanced computer technology is, the more convenient the intelligent processor will be, and the more vivid of interpretation of their thinking ability will be. The perfection of computer architecture, storage, speed, network and other technologies have expanded the scope of the design of combined use in configuring the new computer tools, printers, scanners, plotters, multimedia, etc. for designers. The development of computer software technology is mainly in application software for designers, that is, computer-aided design. Apple's Windows graphical interface and Microsoft WINDOWS operating system have solved many design challenges predecessors faced for the designers, and have resulted in a series of improvement of graphic design, 3D animation, and desktop publishing and so on. For design professionals, actively mastering computer applications software will contribute to the formation of personal design style. Designer will grasp the graphic image processing system. Image is scanned input of pictures, and graphics are drawn illustration. The graphics software Freehand, Illustrator, CorelDraw, image processing software Photoshop, computer-aided design software Auto CAD, animation software 3D studio, Animator, drawing design software Painter, Dabbler, and a series of commonly used software should be grasped. Graphic scanning of the image, drawing, editing, mixing, shifting, positioning, displaying, quickly completing the printing and getting a clear high-quality design works are also needed. Multimedia technology integrates images, sounds, text, animation, audio and even smell and other forms of communication together, which has enriched the artistic language and expression and improved the infection of the works.

The superiority of the software has provided the greatest possible for the mastery through continuous improvement and change shifting from the intelligentization of input system to a printing system for designers. The designers rationally use specialized software, master software performance according to their computer configuration, which has expanded the imagination for designers, and made the experience of design shift from the traditional "hand-drawn" spatial form to facing the computer screen and "clicking" with a mouse.

Of course, compared with hardware, computer software is lack of proper size and quality, but the sharing of resources in such an information age has provided a broad prospect for a large number of software design. The convenient performance of computer software technology has continuously provided design materials, ways and new resources for designers to design. The rich changes of color fonts and graphics in designing process and the display of design thinking into the graphics have liberated a number of manual humdrum operation, enhanced and enriched the imagination of the brain. Because of computer's sophisticated and mechanical property, it has provided the accuracy and magic effect of design and manufacturing which the original manual production was difficult to achieve. And so many people have shorten the time of design learning, solve the problem of the long basic training of the first hand-drawn art, also enabled more people to enter the design field, and has created numerous opportunities for employment and wealth creation. It is undoubtedly a positive and progressive technological revolution in the late twentieth century. For each designer, the computer is a close partner, and also an indispensable design assistant.

Appeared in front of the public, it often results in misunderstanding of entertainment games, that is to say, many families have equipped computer but ignored the computer's stock of knowledge. They based solely on the installation of games software, which has led to a seemingly advanced computer with the game features in fact. This is a market success of computer sales from a business point of view, which could not prove the significance of computer tools. Due to the lack of fundamental understanding that computer will bring human into advanced civilization, it will be a kind of home furnishing and decoration, so the role is very limited. Indeed, the development of computer technology is closely related to economic interests of business. Because of the drive of economic benefits, the developers have invested a lot of manpower and resources into computer technology, then computer technology with high speed continuously provides users with new hardware and software configuration, which could draw the attention of many designers sometimes and has formed a large computer market. Stepping off the misunderstanding of all-powerful theory about computer is a transformation of contradiction between technology and people. The more advanced the technology is, the greater dependence people will have; the more firm the people's willpower is, the more distant vision and creativity they will have. As a product of high technology, computer has been used in thousands of households. However, the widespread use of computers, and the use and mastery of the software in art and design will be distant from human's imagination. In the computer is merely a convenient tool, after all, can not replace the human will? This is not the problem of just "hand-sketched" or "computer-drawn". However, the progress of history has brought a new meaning for the cultural life of the society after all, especially in conversion of time, survival of progression and enhancement of the human spirit. In the computer era, we just need to click.

## 39.3 Conclusion

The significance of computer design is the inevitable result of development of the times, and the emergence of the Internet has contributed to rapidness and agility of the global information technology, and has resulted in the emergence of professional society. In spite of the overflow sometimes, abundant information resources have provided people with opportunities for exchanges of personnel in other countries, broadened people's horizons and get a virtuous cycle in social and cultural aspect. It is a positive and progressive technology revolution in the late twentieth century. However, computer is only a convenient tool, which could not take place of human's mind after all.

## References

1. Benjamin W, Wang CY (2004) Art works in the mechanical reproductive era, vol 12. pp 93–99
2. Fu Z (2002) The basics of computer art design, vol 2(3). Tsinghua University Press, Beijing, 2003. China City Press, Beijing, pp 47–56
3. Li W (2004) An introduction to design, vol 29(4). Southwest China Normal University Press, Chongqing, pp 298–306
4. Wang S (1995) World history of modern design, vol 11(2). New Century Press, Guangzhou, pp 470–477
5. Yin D (2000) An introduction to design, vol 23(6). Hunan Science and Technology Press, Changsha, pp 39–45

# Chapter 40
# Research of Flash-Based Multimedia Courseware Interaction

**Kai Zhang**

**Abstract** The interaction of multimedia courseware is extremely important and difficult questions in modern education technique area now. Aiming at the problem of multimedia courseware, the meaning of the interaction of the system, and combining with the teaching practice, based on the detailed introduction of multimedia courseware for flash interactive way.

**Keywords** Multimedia courseware · Interaction · Practice

## 40.1 Introduction

Along with the development of information technology and the popularization of the Internet, the multimedia teaching attains a wide application. The multimedia teaching is implemented through the interaction among teachers, learners and courseware. The courseware is necessary to have a very good interaction if the teachers want to timely and accurately know the mastery degree of learners on teaching contents and to adjust teaching strategies properly. Therefore, the quality of a multimedia courseware and the significance of the teaching effect depend largely on the design of the interaction of the multimedia courseware.

K. Zhang (✉)
School of Computer Sciences, Neijiang Normal University, Neijiang 641100, China
e-mail: zhangkai@guigu.org

W. Du (ed.), *Informatics and Management Science V*, Lecture Notes in Electrical Engineering 208, DOI: 10.1007/978-1-4471-4796-1_40,

## 40.2 Meaning of Interaction

Interaction in learning is a necessary basic mechanism to acquire knowledge, cognition and skills (Barker). Zhang X and Shu XC described interaction as a state of learners to browse, annotate, link and elaborate the database with abundant resources and nonlinear organization [1, 2]. However, these statements about interaction are in shortage of analyses and evaluation levels. Interaction can be meaningful if such a response includes the need of learners on processing the information. Interaction, under the assistance of computer application, hides the interactive effect between two things; such an interaction will make learners invested in real conversation. However, whether such a conversation is successful decides if the high-quality interaction is generated [3]. Based on the above, the interaction of multimedia courseware refers to a bidirectional effect for information transmission under information environment. Namely, courseware can receive the commands input by users and make the corresponding responses, but not a one-man show. Generally, the interaction of multimedia courseware can be seen from four aspects [4]:

Selection of orders means that a courseware, from content, target or task, should be divided into mutually independent components. However, for all parts, it is necessary to have a logical early or late order and also to provide a flexible skip making users skip to another part randomly [5]. Hence, the courseware nonlinear structure can be realized, providing users with greater flexibility. Selection of contents means that a complete courseware is often a resource pool with rich contents in terms of designers, but is unnecessary to copy all other contents in terms of users. Hence, the Selection of contents can be begun with the needs of the actual teaching and learning, and also are very proper in details and summaries, and even only part is used. This requires programs to be modularized and networked. Selection of environment is seldom considered in courseware. In terms of users, the background color and music of courseware are often not the only best choice although the designer invests repeated deliberations on the elements [6]. As the saying goes, West Lake can be as beautiful as a Lady Whether she is richly adorned or plainly dressed. This is of referential significance when used in the selection of courseware [7, 8].

## 40.3 Implementation of the Flash-Base Multimedia Courseware Interaction

In the making of multimedia courseware, the interactive types used the most commonly mainly include button interaction, hot area interaction, hot object interaction and key interaction.

### 40.3.1 Implementation of Button Interaction

Button is a standard editing in flash, and has four special frames, which are used to express the different states of buttons. Select the "new" component in the "insert" menu; after the button is made, select it, open the "action" panel and write scripts as shown in Fig. 40.1. Button includes eight events, in which the "press" event and "release" event are use most frequently. Their diffidence is that the "press" event is executed when the left button of mouse is pressed, while the "release" event is executed when the left button of mouse is released after pressed. If the button is pressed wrongly, the event will not be executed as long as the button of mouse is not released. Thus, the "release" event should be used if there are no especial requirements.

### 40.3.2 Implementation of Hot Area Interaction

The hot area in flash can be arbitrarily shaped as shown in Fig. 40.2, and actually is making a "transparent" button. And it is enough to cover the "transparent" button on the areas needing interaction. The method of making a "transparent" button can be described as "use a drawing tool to draw a closed shape first, and convert it into button, and then enter this button to drag the "pop-up" key frame to the "click" frame, as shown in Fig. 40.3.

Such a button has no a real image, but has an area where response happen. This area can be seen in editing but can't after the films are output.

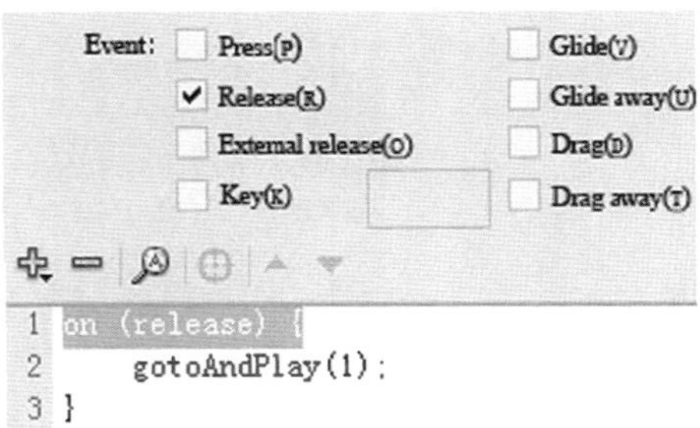


**Fig. 40.1** The button actions

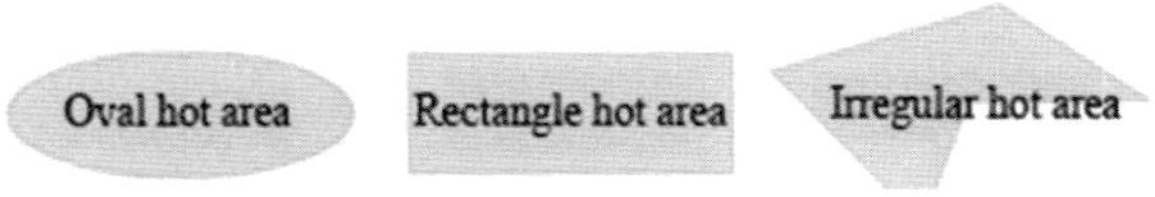


**Fig. 40.2** The hot area shapes

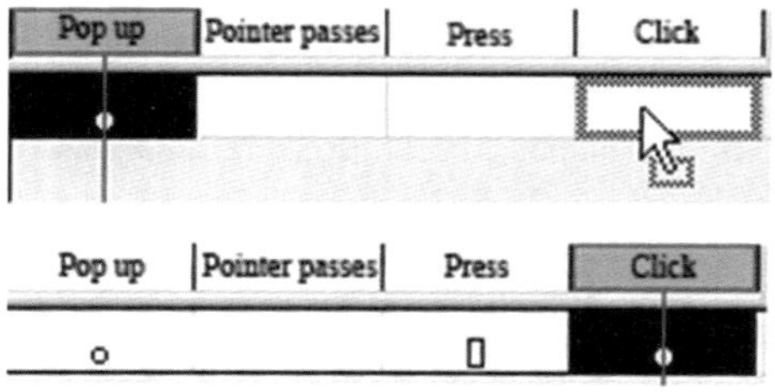


**Fig. 40.3** The button component

### *40.3.3 Implementation of Hot Object Interaction*

The principle of implementing the hot object interaction is as same as that of button. The objects clicked by users can be the regular graphs such as rectangle and also be the irregular graphs. It needs be implemented through button component in flash. The three frames in front of button component can be changed into a real thing as the interactive object, and others are interactively used with the button.

### *40.3.4 Implementation of Button Interaction*

In flash, the events can be triggered not only by button. Actually, when the complex operations are involved, many events are triggered by MC. As shown in Fig. 40.4, nine MC events are listed.

Key down event is activated when a key on the board is pressed; and the key up event is activated when a key on the board is released after pressed. By use of these two events, the button interaction can be made. First, establish a null MC, and drag this MC to stage, and then select this MC and open the action panel to input the following codes.

```
Onclipevent (Keydown){
If(Key.Isdown(90)){  //"90" represents a button
_Parent.Gotoandstop("Xq"); //xq film editing
}
```

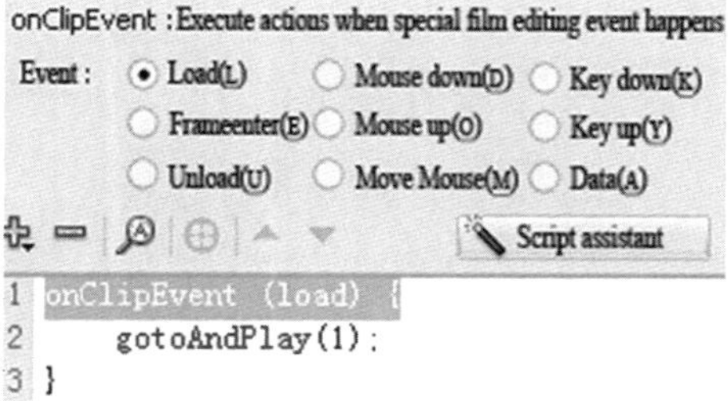


**Fig. 40.4** The button event

### *40.3.5 Interaction of Navigation Pull-down Menu*

Navigation pull-down menu is suitable for making the courseware with strong logics and detailed segments. Here two levels of menus can be taken as example, as shown in Fig. 40.5. Click the "main menu" button and pop up the first-level menu, and then move the mouse to a button of the first-level button, and then a related secondary menu is popped up. Click an item in the secondary menu and enter related parts of the courseware. The implementation methods are shown below [4]:

Step 1 Establish a new MC and name is as the "pull-down menu" (the following steps are implemented in this MC)
Step 2 Establish another three new layers besides the existing layer 1 and name them to be main menu, first menu, secondary menu and script from top to bottom
Step 3 Select the 50th frame of all layers and insert a common frame, add a key frame in the "script" layer each 10 frames until to the 40th frame, name the tags of these five key frames as "Main" "First", "Second_1", "Second_2", "Second_3", and insert the key frames to the positions corresponding to the tags "Second_1", "Second_2", "Second_3" in the secondary menu, and also insert the key frames to the position corresponding to the first frame in the first menu
Step 4 Establish a new button editing and drag it from the pool to the first frame of the "main menu" layer, and then write the "main menu" words above the button. Also, drag the parts separated from the editing of three buttons to the 10th frame of the first menu and write related words on buttons, and finally set the related buttons on the 20, 30 and 40th buttons of the secondary menu and write the related words, as shown in Fig. 40.6
Step 5 Add the Stop () command to all key frames of the scrip layer
Step 6 Add the following scripts to the "main menu" button

To use only one button to control the popping-up and disappearance of the menu, the custom variable Pressdone is used. When the Pressdone value is 1, the button skips to the "first" frame (i.e., pop up the first menu), and simultaneously set the Pressdone value 0, and otherwise returns to "main" frame and simultaneously set Pressdone value 1. To make this scrip work normally, the Pressdone

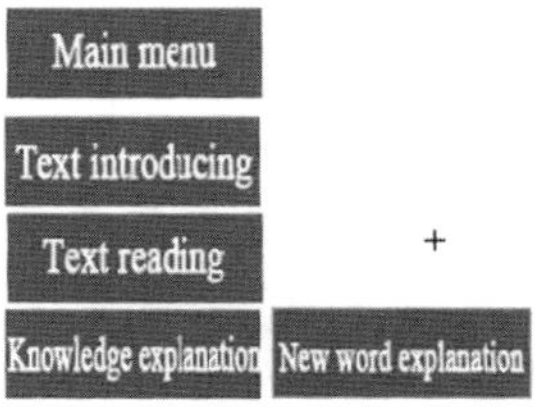


**Fig. 40.5** The pull-down menu

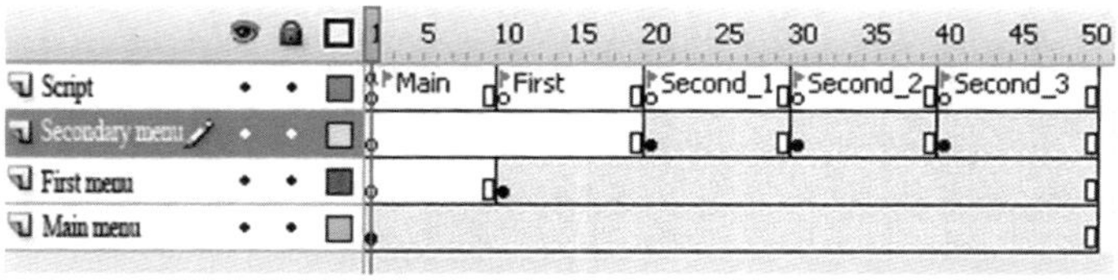


**Fig. 40.6** The timer shaft

variable is necessary to be initialized. This script can be added into the "main" frame.

Pressdone = 1;

```
On(Press){
If(Pressdone){
Gotoandstop("First");
Pressdone=0;
}Else{
Gotoandstop("Main");
Pressdone=1;
}
}
```

Step 7 For the script on the first menu button, add the following script on the "knowledge explanation"

```
On(Rollover, Dragover){
Gotoandstop("Second_1");
}
```

The scripts on another two first menu buttons are similar to this script; and only the target frames of the Gotoandstop () function is changed to be "Second_2" and "Second_3". Thus, add the following scripts on all the buttons of the secondary menu.

```
On(Press){
Gotoandstop("Main");
}
```

Add a skip command on the secondary button to implement skipping. For example, the ultimate script on the "new word explanation" button of the secondary menu is

```
On(Press){
Gotoandstop("Main");
Parent.Gotoandstop("Book");
}
```

### 40.3.6 Text Input Interaction

Flash Chinese text interaction has two ways: one is directly inputting and editing words and texts under the editing state or calling the parts or all parts of the external text files, and the other is calling the words and text objects with MC to form an interaction. Take direct input under the editing state for example to implement text interaction. The structure is shown in Fig. 40.7 and there is only one frame.

Step 1 Add three dynamic text fields, and cancel the "optional" state, and set the character font, size, color and other info, and also set their variables "A", "B", and "Info" respectively

Step 2 Add an input text field, set the character font, size, color and other info, and select the "display the border around text"; as the sum of two numbers within 10 is double-digit, the other characters which are not digits are not expected to be input. Hence, the "the most length of the characters" is set to be 2. Click the "embed" button and pop up the "character options" dialog box and set it to be digits, and finally set its variable to be "C".

Step 3 add the "set a question" and "Calculate" buttons as well as plus and equal signs, and also arrange all elements on stage properly

Step 4 add the following scripts on the "set a question" button

```
On(Press){
A=Random (10);
B=Random(10);
C=" ";
Info=" ";
}
```

Step 5 add the following scripts on the "Calculate" button

```
On(Press){
If(C==A+B){
Info=" You are so clever!";
}Else{
Info="Thick twice";
}
}
```

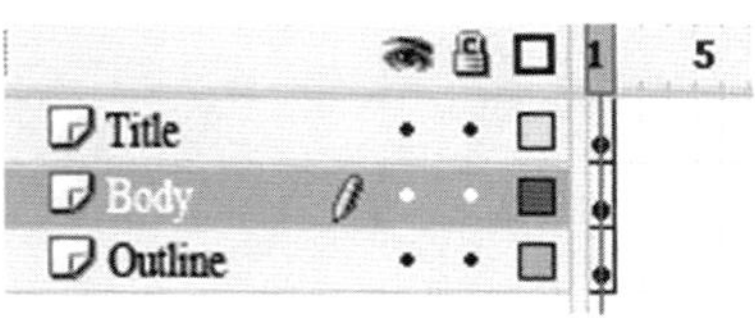


**Fig. 40.7** The timer shaft

## 40.4 Conclusion

The design of the multi-media courseware interaction aims to make the interactive means of different dimensions and levels stimulate the interests of learners in learning and the learning contextualized, and to promote the learners to invest energy on learning activities in depth, and to help learners construct and reorganize their own cognitive structure by the feedback of learners' responses in meaningful context.

## References

1. Zhang Z (1985) The writing group of the biography of Dong Biwu. In: Brief biography of Dong Biwu, vol 11. Law Press China, Beijing, pp 86-94
2. Shu XC (1961) The materials of history of education of modern chinese, vol 1. People's Education Press, Beijing, pp 380–386
3. Wang H (2006)The writing group of the biography of Dong Biwu. In: Biography of Dong Biwu, vol 1. Central Party Literature Press, Beijing,pp 1092–1099
4. Cai J (1932) Chronicles of 40 years. Zhendan People's Daily, Wuhan, (the 21 year of the People's Republic of China)
5. Zhang MZ (1991) Committee of Hong'an people's political consultative history materials. Hist Acc Past Events Hong'an County 2:309–315
6. Dong JJ (2007) The writing group of a chronicle of Dong Biwu's life. In: A chronicle of Dong Biwu, vol 3(4). Central Party Literature Press, Beijing, pp 20–27
7. Wu W (1985) The writing group of selected writings of Dong Biwu. In: Selected writings of Dong Biwu, vol 2(8). People's Publishing House, Beijing, pp 347–354
8. Li L (1988)The writing group in academy of social sciences of Hubei province. In: Reminiscence on the whole life of Dong Biwu, vol 1. Hubei People's Press, Wuhan, pp 99–104

# Chapter 41
# Study on Mobile Electronic Business Based on Collaborative Framework

**Fenglan Luo**

**Abstract** The rapid development of the mobile communication technology, electronic business enterprise has gradually entered the mobile e-commerce era. The collaborative commerce application of the theory in the mobile electronic business concept put forward the synergy of mobile electronic business (CMEC). At the same time, this paper studies the overall operation mechanism and platform with mobile electronic business. Collaborative mobile electronic business is not only a complete electronic business is also a kind of the advanced management concepts.

**Keywords** Collaborative framework · Mobile electronic commerce · Platform

## 41.1 Introduction

The concept of the collaborative commerce has been widely recognized and development since it was proposed in August 1999. The collaborative commerce is a business model, enhance the overall cooperation ability of the enterprise in the product design, manufacturing, sales, marketing, resources, clients, improve efficiency enterprise clusters of all business activities, to the real sharing and feedback of enterprise clusters knowledge network, in order to create sustainable competitive advantage, make full use of the Internet and other emerging technology.

With the rapid development of information technology, e-commerce has advanced by leaps and bounds in our country [1, 2]. The development of

F. Luo (✉)
Jiangxi Science and Technology Normal University Nanchang, Jiangxi, China
e-mail: luofenglan@guigu.org

W. Du (ed.), *Informatics and Management Science V*, Lecture Notes in Electrical Engineering 208, DOI: 10.1007/978-1-4471-4796-1_41,

e-commerce, people begin to hope to break the limit of time and space, the traditional electronic business transactions can at any time and place. Therefore, mobile electronic business arises at the historic moment [3]. Mobile electronic business is a business activity through the mobile network and ends the use of mobile equipment and WAP, SMS, etc. Mobile electronic business mainly involves mobile payment, mobile banking, and mobile booking tickets, mobile search, mobile E-mail and other applications [4].

## 41.2 Collaborative Mobile E-Commerce

Synergy electronic business has been used in average business activities, is considered to be a new development of ERP system, obtain good effect. If we will help the concept of electronic commerce of new business activities, mobile electronic business, we can solve the problem of mobile business [5]. This refers to the mobile electronic business cooperation process, the enterprise raw materials suppliers, telecom departments, the banking sector and the dealer together efficiently, through the 3G network and other new means of communication, can make they can carry on business activity, no limit of time and space, finally realizes the enterprise organization value added.

Mobile electronic business cooperation has the following advantages:

1. It can improve mobile electronic business payment system. In China, mobile payment system is not comprehensive. The main reason is not perfect of mobile communication system of problem is the lack of variety of configuration efficiency and product service. However, in the collaborative environment, interest distribution between Banks and telecoms operators in traditional mobile electronic business between telecom operators to problem and the core enterprise, Banks and the core enterprise. The problem is a very good problem solving. The day a bay nuclear power station, this paper introduces the operation experience proof, 3G standards, after all the related issues of the network speed, bandwidth, and safety has greatly improved, broaden the channels for the companies to provide products and services. Therefore, the author thinks that, through the introduction of new mobile communication technology standards, there will be more and more products and services to mobile electronic business family.
2. It can improve mobile electronic business credit system. Because of China's big population base, we can't evaluate every citizen's credit as developed countries, causes our country credit losses. However, in the collaborative mobile electronic business environment of telecommunication business operators, not only should consider their own interests, and the whole collaborative enterprise group's interests. In order to ensure the safe and reliable mobile payment, ensure the collaborative enterprise group of value, and the telecommunication operators will inevitably take measures to establish credit system.

3. It can improve the training and the use of the optimal configuration personnel. In the mobile electronic business environment, the enterprise needs to experienced talents. Personnel engaged in the mobile business not only to have business knowledge and knowledge communication, finance, marketing. Human resources are scarce, especially those who have experience of compound talents. In the collaborative mobile business environment, can not only realize the height of the enterprise group of information sharing, but also for the enterprise group cooperation between human resources can get maximum sharing. Collaborative mobile business environment and complex talents can share or cultivated by cooperation, enterprise group can't achieve individual enterprise environment.

## 41.3 Studies on Collaborative Mobile E-Commerce Framework

Figure 41.1 shows the Collaborative mobile e-commerce framework.

Figure 41.1 include three main parts and four layers, they are:

1. Mobile operators. Mobile operators to provide mobile service are mainly responsible for the mobile business cooperation. In fact, mobile operators are also part of cooperation enterprise groups. Because their status is rather special, here are it alone.
2. Cooperation enterprise group. Enterprise group is in order to improve their appreciation in the core enterprise.
3. The bank. Collaborative mobile electronic business, as part of the bank is the mobile operators with enterprise group. Because their status is rather special, here list it alone.
4. Wireless network. Mobile electronic business in cooperation, service quality mainly depends on the quality and ability of the wireless network.
5. Wireless middleware. Wireless middleware can use e-commerce in different mobile network and the operating system. It can provide better performance

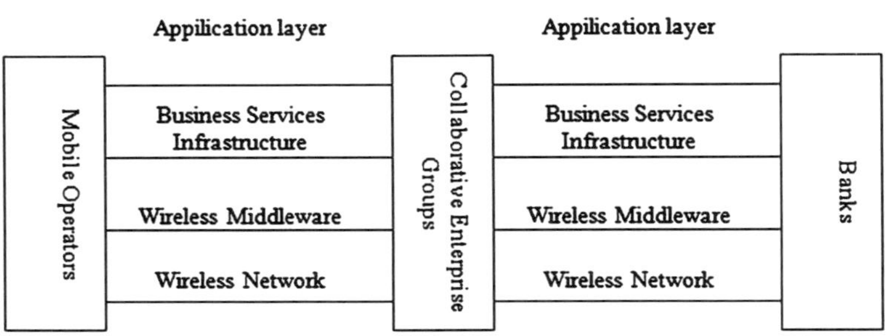


**Fig. 41.1** The framework of collaborative mobile e-commerce

and reliability, here the average user use including WAP, WebExpress wireless middleware, etc.

6. Business services infrastructure. Infrastructure to provide necessary conditions, including: mobile electronic business activities of mobile electronic business security certification, cell phones, mobile payment services, etc., the directory.
7. Application layer. It is and the user interface can support many new applications. The traditional electronic business applications that can be modified and be used to collaborative mobile electronic business.

## 41.4 Studies on the Platform of Collaborative Mobile E-Commerce

The platform of collaborative mobile e-commerce can be divided into two parts: the overall operation platform and the technology platform.

### *41.4.1 The Overall Operation Platform*

The whole operation process of the mobile e-commerce platform cooperation includes the steps of mobile electronic business of collaborative operation, from the whole operation of mobile electronic business solutions to those tools of cooperation needed to achieve the solution. Overall solutions to the enterprise mainly includes product planning collaboration, product design, product development, raw material purchase, production, product management, workflow management, sales management, logistics management, user management, etc. The necessary tools including ERP, MRPII, customer relations management, supply chain management, PDM, CAD, CAM, etc.

Figure 41.2 shows the overall operation platform.

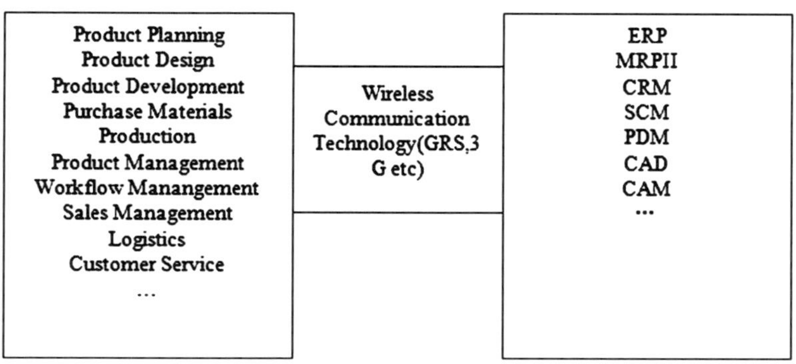


**Fig. 41.2** The overall operation platform of CMEB

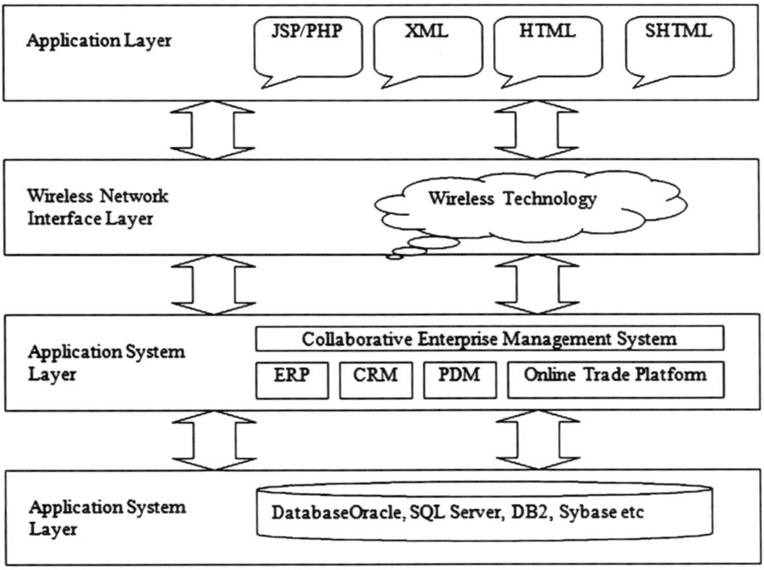


**Fig. 41.3** The technical platform of CMEC

### *41.4.2 The Technology Platform of Collaborative Mobile E-Commerce*

The technology platform of mobile business cooperation can be used to ensure that cooperative enterprise group of overall management. Platform including (Fig. 41.3):

1. Collaborative database layer. Collaborative database layer is mainly used to support data access, data sharing, data query etc. enterprise group, cooperation between all kinds of information system to ensure effective cooperation between business enterprise groups. With the limited network bandwidth, data transfer rate and other factors, the mobile electronic business must ensure that the efficiency of the database. Therefore, the cooperative enterprise group must try to use the same type of database query language must be efficient.
2. The application layer. The application layer is mainly composed of all kinds of information system to support the operation of enterprise internal and the cooperation between enterprise groups. The main features include: cooperative enterprise management system (CEMS), collaborative product design, online trading platform of information sharing and coordinated between enterprise group, the product data management, CRM, ERP, CAD, etc.
3. Wireless network interface layer. Wireless network interface layer influence each other through wireless network layer of the application system technology.

4. Application layer. Application layer interface layer and the user to provide information, and this layer use network address for every business cooperation process, including browsing information search and customization.

## 41.5 Conclusion

The 3G technology and promote the mobile electronic business in China, will get more wide development space. Collaborative mobile electronic business is advanced in this paper, we study the enterprise to realize their own added value, and to improve the safety group mobile payment, credit system, and training and use the talent, etc. At the same time, puts forward the whole management mechanism, platform and technology platform of mobile electronic business cooperation and provide theoretical basis for the collaborative mobile electronic business building mode. But there are some special technical problems needed to be studied, such as group movement of electronic commerce between heterogeneous databases of the cooperative enterprise group, safety mobile payment, etc.

## References

1. Huang C, Tseng T, Gung RR, Chang H (2005) An agent-based web services solution to collaborative product design. Int J Knowl Based Intell Eng Syst 9:63–79
2. Chen Q, Hsu M (2001) Inter-enterprise collaborative business process management. In: Proceedings of international conference on data engineering, vol 42(s2), pp 253–260
3. Ming Z, Yang Y, Yuanzhuo W, Jingle Z (2009) Collaborative service mode and its application in e-commerce. Beijing Univ Sci Technol J 05:12–17
4. Gang C, Weiping Y (2008) The design of the collaborative e-business platform. Int MultiMedia Inf Technol 22(2):188–191
5. Rongzhen F (2009) The implementation of secure mobile e-commerce payment. Comput Inf Technol 09:842–848

# Chapter 42
# University Accounting Education Informationalized Reform

**Hua Yang**

**Abstract** With the arrival of information age, the accountant of the university educates and pays great attention to merging with informationalized infiltration, increase content of courses and range of information course in varying degrees, this has relieved the demand to the information talents of finance and economics of the society to a certain extent play a positive role. But with informationalized popularization and deepening, the higher accountant educates and exposes a great deal of questions, influence the realization of the professional train objective seriously, set up the in charge offing experiment platform, can promote informationization and combination which the higher accountant educated.

**Keywords** Informationization · Accounting education · Educational reform

## 42.1 Introduction

It is actually the reforms of a kind of university's finance and economics course research theory and practice based on information technology that the informationization makes the higher accounting educational reform, the ones that exist in its informationization and finance and economics course to the finance and economics educational field of course of the university isolate and antithesis problem, can make the informationization and finance and economics interdynamic of courses combine bidirectionally through establishing the in charge ofing experiment platform, democracy of promoting teachers and students, cooperation, dynamic

H. Yang (✉)
Xinyang Agricultural College, Xinyang, China
e-mail: huayang2030b@126.com

W. Du (ed.), *Informatics and Management Science V*, Lecture Notes in Electrical Engineering 208, DOI: 10.1007/978-1-4471-4796-1_42,

new shape of course, and then make informationization and student studied and combined and become the organic whole, realize the goal of professional cultivation of university's finance and economics course [1, 2].

At present, in the higher accounting education of our country, the traditional theory type teaching method is still leading educational pattern, only. The teacher says, students listen to the teacher writes, students copy [3]. The teacher tests students and takes an entrance examination. Students' study is totally to go on passively under the leadership of teacher, the students' study thinking does not have curiosity and initiative again, study has become a kind of burden of students' instead, this is a kind of close educational pattern, it is a kind of mode educated for counting and breeding, and this kind of mode makes a lot of students complain incessantly [4, 5]. Higher education stay in students' examination ability of training still basically, and neglect and train students' calamity to trample ability and spread thinking ability, that a lot of graduates complain the difficult accountant who meets the high practice works after getting into the society. So, follow the modernized development of economics and society, the modernized reform that the accountant educates is imperative. The increasingly many finance and economics universities and colleges can meet information-based teaching's demands on the infrastructure. A lot of universities have campus network, and computer networking is taller and taller in proportion in the university campus. The introduction of information-based facility apparatus and software resource such as the network, multimedia computer, projector, video, pronunciation has changed teaching mode and teaching means greatly, teaching software of multimedia, education resources bank, online course, etc. have promoted the renewal of the teaching environment, setting up the in charge ofing experiment platform can combine the above-mentioned resources, the resource-sharing of experiment.

## 42.2 Reform the Teaching Mode

On informationalized the intersection of teaching and environment, role of teacher change organizer, adviser, person who coordinate, administrator and supervisor of teaching into already, according to needing to lead and help students' learning process, make up and expand in studying the content: students are changed into the subject of teaching gradually from the passive recipient of knowledge too, on the basis of the study mode of the information-based platform, can improve student's understanding of content of giving lessons and interest probed into thoroughly, utilize the network teaching platform to develop the activity of studying in the group by mode that the group study, study cooperation between classmates, study together. The informationization made the pluralistic teaching method to strengthen the mutual cooperation between communication, student during teachers and students, the discussion of multi-disciplinary question of crossing, has reflected the diversification, individualized teaching behavior, setting up the in charge ofing experiment platform can utilize the network platform, realize more unblocked exchange between teachers and students.

The core of the modern accounting talent's training mode is innovation ability and system thinking of training students, the basic fixed position of the goal is training and possessing economy, management, the comprehensive talents of the respect knowledge of law, etc. It is anxious for success but must guard against again in the course of implementing the accounting education talent's training mode, it is that course is strenuous that a lot of universities educate a common problem that store in at present, forced students to carry on oppression study at the same time, attacked the students' study enthusiasm and initiative seriously. Must not repay finally. Gold can't be pure and man can't be perfect, an outstanding accounting talent does not only depend on the school can train out, the main task of the school is thinking and method of letting the students learn to know study, let the students continue keeping the accountant studying in the social work afterwards.

### 42.2.1 Implements Heuristic Education: Let the Students Experience and Give Lessons in Person

At present, the educational way of university's accountant of our country is still to "irritate the ear type" with "spoon-feed" Education. A teacher talks bitterly while having a class. Students hear tiredly. Can try other teaching methods to educate in the classroom, consider carrying on accounting education in terms of student. For example, suppose there is chapter 27 in some accounting teaching material, there are 50 students in the whole class, we can divide the students of the whole class into several groups. How much each group is responsible for telling different chapters, then distribute each course which groups tells according to complexity of different chapters. Some chapters are simpler. Then this group speaks chapters 2, 3, and some chapters are relatively difficult, then this group tells this chapter especially, let the students go back to prepare meticulously before the lesson, can consult teacher to encounter the problems that understand. So not merely can improve the students' study enthusiasm and initiative greatly, let the students study through what personally instructing in words and by deeds consolidate and is strengthened to accounting knowledge. Also can lighten the teacher's teaching burden greatly, though the students teach the ability and range of knowldege have very great deficiency. Can't guarantee student can understand, at least can let student going, experiencing accounting deep intension of knowledge voluntarily totally. The most important thing is that can let the students participate in accountant's studying greater intensity.

### 42.2.2 Pursues the Credit to Reward and Make in an All-Round Way: Encourage Students to Participate in the Accountant Academic Research

Now a lot of universities adopt the credit system. Students can graduate after finishing repairing the corresponding credit when graduate. Get diploma and degree's diploma. And before obtain student that credit is main way is through to not lasting Pan of course, and the domestic universities all seldom adopt the academic credit incentive system at present. In fact, a lot of students are unwilling to stay and carry on uninteresting theory study in the classroom all day. Only SRTP undergraduate course scientific research training courses in what implement credit reward in the university at present, it is qualified to reward the corresponding credit to check. Instead of letting students study uninteresting basic course, it would be better to unlock students' thought, encourage students to participate in more training of scientific research. Former Peking University headmaster Cai YuanPei has said. Emancipate the mind, all-embracing. If we can make every student aggressive and participate in scientific technical research, what comprehensive level of worrying about education of undergraduate course cannot rise?

### 42.2.3 Strengthens Accountant's Information-Based Education

The accounting informationization is a modern accountant's scientific important component, it mainly trains meeting the needs of socialist market economy, meets enterprise's informationization's needs, understand accounting theory of information system and practical talent of the practice. Primitive accounting the intersection of information and handling adapt to complicated accounting data while being difficult already, and the accounting information system collects information to record, information analysis, information processing are in one suit, and can produce enterprise's accounting statement finally. Modern economic trade operates at a high speed, corresponding, prescroption and accuracy to financial information of enterprises require it is very high, this requires modern accounting personnel will accomplish while dealing with accounting information fast, precise, accurate. So, while the higher accountant is educational, should strengthen accountant's information-based education in an all-round way, have extremely important effect on student's employment and development in the future.

### *42.2.4 Pays Attention to Student's Accountant's Sincere Morals and Education of Laws and Regulations*

For that the accountant works, sincerity is extremely important. University's accountant does not educate it cause enough attention that the ethical law to students is educated at present, lead to the fact student to generally scarce knowledge of respect this, making the fake, practicing fraud the behavior buries the hidden danger work for accountant in future, this. I think, in accounting education. The sincere legal education should increase the accounting law, accounting professional ethics as a key course. Such courses as the case is educated that the accountant practices fraud make the student time at school form the professional ethics based on sincerity consciously. The operation will establish the good thought foundation in the future. Can also hold the accounting knowledge match between the universities, add the relevant laws and regulations into match, widen the range of knowledge in this way, can educate making a better propaganda platform to the accounting law too, emergence that the accountant practiced fraud and made the fake in the practice works in better prevention future.

### *42.2.5 Trains Student's Accounting Prestige Idea*

The accounting prestige is a newer idea in the contemporary society. People seldom mention the accounting prestige at ordinary times, because a lot of people think the accountants are all controlled by higher authorities. Cause the accountant to become the scapegoat of the modern economic crisis to a great extent. Especially to the enterprises in the economic crisis, the public will associate it with enterprise's accountant in first response to go wrong. This satirizes one kind of the accounting profession greatly, are unfavorable to the whole development of the accounting profession very much in order to avoid the appearance of this kind of situation. The accounting necessary pigtail of education circle sets up a kind of prestige idea. Just the same as the law is sacred and inviolable, let the public not store worry to the accounting heart again, believe accounting personnel's proud work to a great extent at the same time, establish the good job image of accounting profession.

## 42.3 Deepening Which Promote the Reform in Education of Course

In order to meet modernized management and requirement for economic development of network, with informationalized swift and violent development, all kinds of new networked developing instrument, application system and software package, have exerted a great influence on the course teaching that the higher accountant educated, and put forward the new requirement for the personnel training and completion of knowledge structure. Under the environment of cybereconomy, the professional personnel of finance and economics must possess finance and economics professional knowledge and informationization to employ ability at the same time, could meet demands of modernized management in this way. The in charge ofing experiment platform promotes the financial student's understanding of network. So, the university must consider informationizing while incorporating corresponding professional train objective, course system and teaching plan while making the personnel training scheme, and implement in teaching. For example the accounting specialities of most universities and colleges have used the computer, network and multimedia teaching equipment in the teaching of some courses. In fact, the in charge ofing experiment platform including student can choose the learning tool independently, obtain information, analyze information in time by using these tools, and do coming from many kinds of irrigation ditches, information with many kinds of forms further comprehensive processing on the basis of analyzing, in order to get the result of solving the problem. Informationalized application makes the route that students obtain knowledge flexible, the choice is expanded, can also be appropriate to use media and tool, present one's own information to be expressed out, combine one's own speciality and use the informationization better.

## 42.4 Trains the Compounding Teaching Body

From the viewpoint of university, in order to improve the unitarity of teacher's knowledge structure, should encourage finance and economics professional teacher and professional teacher of information to study each other and exchange through listening to the teacher, lecture, receiving a training, etc., improve the combination intensity of information course and finance and economics course knowledge, reach all kinds of knowledge in the teaching course to make a comprehensive study and a thorough master of. From the viewpoint of teacher, promote the integration of one's own professional knowledge and information-based knowledge, not only can abundant the intersection of classroom and content of courses and promote the intersection of teaching and result, can improve the intersection of teacher and the intersection of teaching and competence and message accomplishment of oneself at the same time.

## 42.5 Realize Teaching Management Networkedly

Maximum sharing of realizing it is managed that various teaching materials are networked to enable resource. Because the information-based teaching range is wide, the content is mixed, classes are many, the means is upgraded fast, can be open to student the exercises of some e-books, courseware, some courses through LAN, campus network, Internet, students can carry on the study of corresponding course in the open any time of computer lab in a flexible way.

## 42.6 Offer the Education Resource Shared

Finance and economics specialized course abundant in content, involve range to be wide, amount of information large, for convenient for student to study, should turn into the education resource after carrying on information-based processing of the content of courses, for example courseware, online exercise, discussing the question, etc., utilize campus network to offer to students. Meanwhile, global information-based resources that can share will be offered to students to fully utilize function of Internet, as the material resources of teaching of course. Utilize share information-based resource and the intersection of course and content merge until—blow, regard as, study target directly, arrange for students to appraise in the experiment lesson through discussion, analyze, discuss.

The course informationization of discipline is to melt the informationization to enter school in all respects of department's course, let the course content of discipline informationize, course appraise and informationize course informationizing, course. In fact, finance and economics specialized many courses can utilize modernized teaching means such as computer of multimedia, campus network and internet, etc., carry on teaching activity of studying, discussing to study, consult studying, fictitious experiment, creating practice in teaching of theory and practice link independently etc. Such as intermediate financial accountant and course of accounting by EDP with specialized accounting, can software pay teaching and experiment of demonstrating etc. with "the intersection of multimedia and accounting the intersection of simulation and laboratory", this software is except that the convenient teacher implement ocular teaching, it is convenient for student to carry on oneself and do exercises, test and have an examination three levels oneself yet, students can finish every experiment content step by step, including filling and presenting the accounting voucher, registering accounting account book, establishment accounting statement etc., can know the whole accountant checks and calculates the procedure through the experiment.

Most experiment lessons that the computer, multimedia and network technique not merely can help finance and economics to be specialized improve its practicality, the more important thing is that can also utilize software to design the system, rational experiment procedure according to the request of discipline and speciality, set up the content of courses of the experiment according to the module of course, make the teaching of experiment set up with the theory course cooperate relatively independent closely as well as relation of content of courses gradually; Through integration and the obtaining and utilizing of sharing, information of resources, make students obtain more valuable knowledge: The characterized utility of network and courseware can also realize teachers and students are mutual and growing and growing mutually, train students' research learning ability and spirit of approaching a subject. So, utilize informationization to set up the course system of the experiment according to the requirement for talent's train objective, it is a new route to promote the professional experiment teaching system reform of finance and economics.

## 42.7 Conclusion

It is a long-term task that the informationization makes the reform in education of higher accounting education; need to improve constantly with development of information technology. Set up the need that the in charge offing experiment platform is not only development of informationization, also the institution of higher education passes and deepens the reform in education, give play to the enthusiasm, initiative of the school, teacher and student, plane the building, promote accountant's professional student's improvement on professional quality and information technology quality, strengthen accountant's professional student's competitiveness in the social environment.

## References

1. Bodnar GH, Hopwood WS (2001) Accounting information system, vol 23(3). The Publishing House of Tsing-Hua University, Beijing, pp 34–38
2. JianGuo Z (2000) The accountant educates, set up the accounting publishing house of letter. Account Res 6(4):76–82
3. JinZhong Y (2000) China's accounting undergraduate course educational reform valve question is studied. J Account Computerization 20:109–116
4. Golden Light China (2005) The accounting informationization in enterprise's information-based environment makes a reservation and studies, China manages the informationization 7(2):56–64
5. Mouth of Tang ZhiJun's Thinking of University's Accountant's Information-based Teaching (2005) The accountant's friend, vol 8(7) pp 12–21

# Chapter 43
# English Listening Teaching Method Based on Multimedia

**Junling Wang and Weiqing Liu**

**Abstract** This paper illustrates the importance of the listening comprehension and the definition of the multimedia. Then gives the reasons of using the multimedia in the English listening teaching. It also points out there are some disadvantages and disadvantages of using multimedia teaching in listening comprehension.

**Keywords** Multimedia teaching · Listening comprehension · Traditional listening teaching

## 43.1 Introduction

Most Chinese high schools offer extensive instruction in reading, translating, grammar and ways of scoring high marks in the National College Entrance English Exams while listening is almost completely overlooked [1]. Even at colleges, listening comprehension is not allocated much teaching time and the skill of listening is downplayed, which, in combination with other factors, frustrate students when they are listening to English speech. This paper discusses the use of Multimedia in improving English listening comprehension [2].

J. Wang (✉) · W. Liu
School of Foreign Languages, Shandong University of Technology,
Zibo, 255049 Shandong, China
e-mail: junlingw211q@126.com

W. Du (ed.), *Informatics and Management Science V*, Lecture Notes in Electrical Engineering 208, DOI: 10.1007/978-1-4471-4796-1_43,

## 43.2 The Understanding of Multimedia Teaching

Multimedia teaching has its class-teaching pattern, which refers to lecturing with computer helping; it is quite popular in our country at present stage. Mainly demonstrated by teacher, multimedia teaching software displays teaching content, which improves greatly the teaching capacity and content density in class, compared to traditional teaching methods. It is undoubtedly, it is an accessorial teaching pattern, which cannot change conversational teaching-centered situation, nor can it realize teaching students in accordance of their aptitude [3].

Therefore, Multimedia teaching should also put emphasis on interactive study, which fully reflects the principal part of students in learning, and turn teacher-centered education into student-centered education [4]. In interactive study, the students have the capacity of self-study and self-improving. It is primarily up to the students to carry on the task, to explore the knowledge actively, while the teacher merely assists and guides, or just looks on. At the same time, we have to center on the following respects when preparing for a class of computer-assisted instruction [5].

## 43.3 The Importance of Listening Comprehension

Many problems exist in the current teaching of English listening comprehension at colleges [6]. A typical pattern of listening comprehension class is that the teacher starts the lesson by introducing some background knowledge and explaining some difficult words and expressions in relation to the listening texts. Then the teacher plays the tape two or three times and asks the students to do the comprehension exercises. After finishing their exercises, the students are asked to check the answers by listening to the whole passage once again, or the teacher' oral answers. Sometimes, the teacher plays the tape sentence by sentence for several times in order for students to finish their spot dictation or compound dictation.

Listening is the most frequently used language skill in everyday life. Researchers estimate that we listen twice as much as we speak, four times as much as we read, and five times as much as we write (Marianne celce-murcia). Listening may well be the most important skill involved in the learning of foreign languages. Listening comes before speaking, for the child listens for an extended period of time and then begins to speak. Nothing can justify teachers slighting or even ignoring the students' complaint about the difficulty of listening comprehension. But the listening skill is the only one over which learners have little or no control; one can read, speak, and write at one's own pace, level of vocabulary and syntax, but you have to listen at someone else's pace, level of vocabulary and syntax. This may be the most understandable cause of the difficulty of listening comprehension, but we must make further exploration.

As the famous linguist Richards said, listening comprehension in second language classroom tends to focus on transaction listening: students listen to oral texts in order to obtain information and complete a comprehension task, without intervening or interacting for purposes of clarification or feedback.

## 43.4 Why to Use Multimedia in Listening Teaching

### *43.4.1 Combination with Traditional Listening Methods*

Traditional listening teaching methods possess rich legacy and precious teaching experience. During the conventional teaching process, a teacher can briefly explain clearly some complex problems with emotion and logical power, which cannot be substituted by any modern technology. The traditional methods of listening comprehension also have its vigorous vitality. However, the limitation and deficiency traditional methods are quite obvious. Such limitation results from the limit of the teacher-textbook transferring pattern. Traditional teaching system is not capable if solving the problem of how to greatly boost efficiency in teaching.

Learners need to receive comprehensible input and to produce language, or comprehensible output.

As defined in Microsoft Encarta Encyclopedia Deluxe 2001, Multimedia is the presentation of information using the combination of text, sound, pictures, animation, and video in computer science. Common multimedia computer applications include games, learning software, and reference materials, etc.

### *43.4.2 The Subjective Factor Contributing to the Difficulty of Listening*

The subjective factor directly concerns the listener himself, which involves the listener's L2 knowledge and proficiency, his listening experiences or the training he has received, his psychological features educational background, general knowledge, life experiences and information about the world.

(a) The listener's L2 knowledge and proficiency refers to: his phonetic knowledge: The listener's phonetic knowledge such as the weak forms of pronunciation, assimilation, elision, liaision, loss of exposive and intonation can indirectly influence his listening comprehension.
(b) His command of lexicology: his lexical level or his vocabulary ins the basis not only for his listening comprehension, but also for every aspect of his L2 learning. As Michael J. Wallace said, there is a sense in which learning a foreign language is basically a matter of learning the vocabulary of that language.

(c) His listening experiences includes his experiences of listening to different speakers with diverse accents and in various situations such as lectures, movies, news broadcast, his reaction sensitivity, his power of memory, his interest and motivation.
(d) The learner's educational background determines his level of familiarity of the topic he is listening to. Increasing familiarity with the topic or content of the text implies decreasing probability that the listener will have difficulties understanding the message or the content of the text.

### 43.4.3 CAI Coming into Being

As the listening comprehension class is usually conducted in the language lab, where one master tape-recorder controls scores of the student recorders and the feedback given by the communication between the teacher and students can be conducted smoothly both wholly and individually. The teacher may group students by connecting pair or group recorders and students can communicate freely without interfering others by microphones.

In this background, CAI comes into being. In addition to the advantages of traditional listening teaching methods. CAI has two characteristics of individuation and interaction, thus has peculiar advantages in listening teaching. A quite ideal computer listening teaching system should be of network in technology. In management, it should be able to assimilate the merits of traditional listening teaching methods. That is, it can not only control all the extensions used by the students, but also assure the interaction between the teacher and the students. Besides, software resources must be abundant. It is impossible for the computer to resolve all the problems in listening teaching process. It is unrealistic for us to exaggerate the effect of CAI, trying to substitute advanced the media are, CAI an not completely take the place of traditional listening teaching methods, especially oral teaching. The relationship between CAI and the traditional listening teaching ought to be a mutual complementation of organically integration.

### 43.4.4 The Advantages and Disadvantages of Multimedia Teaching in Listening Teaching

Multimedia education should carry through the basic demands of quality education which adequately carries forward personal initiative, arouses the students' enthusiasm and develop the students' cognitive ability, discovering ability, learning ability and creative ability. Meanwhile, it lays emphasis on the course structure in teaching to embody the basic features of quality education.

(a) The upmost advantage of multimedia teaching is that it can make teaching process vivid and visualized. A teacher can make befitting audiovisual material for Chinese students in accordance with the situation of China, which is attractive to the students.
(b) It can save the previous time in class, thus increases the amount of information of teaching to promote the mode of teaching to promote the efficiency in class. Previously, under the chalk-talk mode of teaching, blackboard writing and explanation occupy much time, which limits the amounts of information in class.
(c) A teacher can adapt this teaching material which is compiled in the light of the syllabus into electronic material, abundant with pictures, texts and sounds. This non-text material integrates sounds, images, lights, colors, which is real and vivid, fully recording actual events with fast speed and large capacity, and visualizing abstract material.
(d) In a multimedia classroom, a teacher need not write on the blackboard because of PowerPoint and practicality projection, which is time saving and increase the information in each unit.

But it also has some disadvantages:

Firstly, as to the teachers' capacity of courseware making, and the hardware equipment for multimedia classroom, multimedia teaching is only limited in a large demonstrative class, which cannot realize the independent study for each student.

Secondly, the teacher may not be an expert in all things of teaching, so it is indispensable in multimedia teaching in listening comprehension.

**Acknowledgments** Listening comprehension is generally regarded as the most difficult skill by most learners, and it is an important but sometimes overlooked aspect of listening. However, any second language learning should be treated as a task integrated by four parts: listening, speaking, reading and writing. Undoubtedly, the students should make great efforts to improve their all the skills rather than focus their time and effort on one single aspect only. However, difficult it is, the students 'listening comprehension skill can be improved with the help of multimedia teaching. Multimedia may increase motivation in that the individualization of learning and of students' control of tasks, strategies and speed of learning may lead to more positive attitude to the studying situation, and inclusion in multimedia environment of instant yet private feedback to mistakes and the explanation for why answers are correct or incorrect may make the learner more active and prod the learner into more participation.

## References

1. Anderson and lynch (1988) Listening. Oxford University Press, New York
2. Brown G (1989) Investigating listening and viewing comprehension in multimedia environments. Lang Learn Technol 3(1):454–464
3. Dunkel PA (1991) Listening in the native and second/foreign language: toward an integration of research. TESOL Quart 25(3):431–457

4. Feyten CM (1991) The power of listening ability: an overlooked dimension in language acquisition. J Mod Lang 75(2):173–180
5. Krashen SD (1982) Principles and practice in second language acquisition, vol 5(2). Oxford, Pergamon, pp 343–357
6. Dong L (2002) Reception strategies instruction in teaching listening comprehension. Teach Engl China 25(4):35–40

# Chapter 44
# Study on Quality Improvement of Remote Education with Alternate-Oriented Tutorship

**Yong Hu**

**Abstract** Face-to-face tutorship is not only one of important parts in the process of distance education, but one of three major steps (i.e. network courseware, face-to-face tutorship, and answering questions) of remote education with the most powerful interaction, influencing and reception. It can directly play an influence on the learning effect of students. In this paper, based on the practices in the learning centers outside remote educational school, the author carries out a detailed analysis on the aspects such as time, period and contents of face-to-face tutorship, and also discusses the necessity, feasibility and effectiveness of alternate-oriented face-to-face tutorship in the process of online remote education in depth.

**Keywords** Remote education · Teaching · Efficiency

## 44.1 Introduction

At the present time, remote education has been transformed from the edge of educational pattern to center with a gradual step. In the mean time, remote education has received the unprecedented attention from people all over the world like adult higher education and self-examination education.

In most Chinese schools (pilot) of remote education, three major steps, which are network courseware, face-to-face tutorship, and online assignments and answering questions, are applied to complete the process of teaching through the Internet, satellite television or cable television and so on.

Y. Hu (✉)
Zhongshan Torch Polytechnic, Zhongshan 528436, China
e-mail: yonghu3082@126.com

W. Du (ed.), *Informatics and Management Science V*, Lecture Notes in Electrical Engineering 208, DOI: 10.1007/978-1-4471-4796-1_44,

Among these three steps, network courseware teaching and answering questions in assignments have already been implemented in all Chinese schools (pilot) of remote education, and also can be imitated and reproduced with ease at present.

However, the face-to-face tutorship, which is undertaken by more than three thousands of off-campus learning centers all over the country, is implemented differently because of the actual conditions of different learning centers, and simultaneously gives rise to vastly different teaching effects. Zhang Jianwei, teacher from Tsinghua University, in the article named as the Implementation and Improvement of China's Remote Education, pointed out that the majority of learners thought whether they could meet with the teachers face to face exerted a very important role in their learning activities [1].

Therefore, it is believed by us that how to implement face-to-face tutorship has converted into one of the crucial parts in the process of remote education currently.

After investigations, it is found that there are several common grounds among all learning centers of China in face-to-face tutorship at the present time, as shown in the following.

The time for the face-to-face tutorship is short, namely about 12 h or even less are allocated to each course at average.

The teaching contents in the face-to-face tutorship are mainly oriented at giving explanation to the answers of reviewing materials, and these answers have even been proved to the "last straw" to help part of students to pass examinations.

The helps, which source from the face-to-face tutorship, are far from being enough for students to systematically learn courses, and also the guidance is rather limited for students, and the guiding role of tutors is not brought into full play in the process of students to learn courses.

In addition, as remote education is mainly characterized with online learning, it is difficult for students to feel the atmosphere of campus culture in higher learning schools, making them lack a close sense with school and learning centers, and easy to generate strangeness.

In the mean time, such a situation is conducive to promoting student management, and developing the construction of study style in remote education.

Therefore, the learning center of University of Electronic Science and Technology of China, Zhongshan Institute, which is the largest learning center outside school, is taken as an example in this paper. At the present time, there are 4,500 students (132 teaching classes) in the learning center.

Through the practical researches on face-to-fact tutorship, student management and construction of study style, it is found that if teaching resources such as teachers and classrooms are intensively used within a period if students usually receive tutorship from teachers in all courses before the final exams.

At the same time, the teaching effect of the face-to-fact tutorship, which is concentrated before the final exams, is not in an ideal state as well.

Moreover, the learning of students at ordinary times can only rely on themselves if the face-to-fact tutorship is concentrated before the final exams.

When the knowledge points difficult to understand are encountered, a considerable part of students in remote education does not take advantage of the question answering system to solving problems, and hence easily forget or ignore learning questions.

As a result, the follow-up study of students can't be ensured in futures. Therefore, it is highly necessary to implement the face-to-face tutorship with effective measure, namely the alternate-oriented face-to-face tutorship which is proposed for remote education in this paper.

In practice, the alternate-oriented face-to-face tutorship is divided by us into three levels for its specific implementation, as shown in the following.

At the primary level, face-to-face tutorship is alternately implemented for different classes.

Before the process of implementing the face-to-face tutorship in each class, the first face-to-face tutorships of courses A, B and C can be arranged in the first week, the second week and the third week, respectively. And the rest can be done in the same manner.

It can be assumed that there are five courses for a semester, and then the first face-to-face tutorships of all courses should be complete on the fifth Monday. Next, the second face-to-face tutorships of courses A and B are arranged on the sixth Monday and the seventh Monday, respectively, and also the rest can be done in the same manner.

Therefore, when a semester comes to an end, a face-to-face tutorship is almost implemented for a class every Monday.

At the secondary level, face-to-face tutorship is alternately implemented for different grades.

Generally speaking, the learners in remote education are working staff from society. They often can't participate in learning at a regular time specified in class curriculum for work, family and other reasons, and become easy to miss the face-to-face tutorship classes.

Therefore, the courses which are offered for the same grade and the same major but not at different classes are arranged with a face-to-face tutorship according to different time. More specifically, two classes from computer major of Zhongshan Torch Polytechnic 2005 are taken for an example: a face-to-face tutorship was implemented for the course College English at class one of computer major on Monday of the first week; a face-to-face tutorship was implemented for the course College English at class two of computer major on Wednesday of the first week; a face-to-face tutorship was implemented for the course C Programming at class one of computer major on Monday of the second week; a face-to-face tutorship was implemented for the course C Programming at class two of computer major on Wednesday of the second week, and also the rest can be done in the same manner.

Thus, different classes of students, who are absent in class, can participate in the face-to-face tutorship, ensuring a systematic and complete face-to-face tutorship for students as much as possible, and also giving an expression to the scientific and effective features of the teaching service system in learning center.

At the third level, students in remote education are allowed to participate in face-to-face tutorship in the form of combination.

With the purpose of better reflecting the educational idea (learning improves values, and education serves for society) of the learning center, network students are encouraged to receive more knowledge through remote education as much as possible, aiming at making an enhancement to their own quality and value.

Therefore, network students in different programs, who learn in learning center, are allowed to participate in the face-to-face tutorship of all courses together.

That is to say, as long as students feel interested in the courses, they are welcomed to participate in the face-to-face tutorship even though they do not belong to this major. Certainly, such a kind of teaching services can be provided only by those schools, which have possessed a powerful strength for learners.

In the process of implementing the alternate-oriented face-to-face tutorship, great importance needs to be attached by relevant people to a highly important issue in the learning center.

More specifically, when the face-to-face tutorship tasks are allocated, it is highly necessary to make specific requirements on each face-to-face tutorship contents in accordance with the requirements of teaching program and in combination with contents of network courseware. Because classes hours allocated for the face-to-face tutorship in remote education take about a quarter of the total network course class hours (12 or 15 class hours are arranged for each course), three hours are spent on each face-to-face tutorship in learning center.

In general, four or five face-to-face tutorships will be implemented for each course in a semester. In the teaching arrangement for each course of a semester, it is necessary to give consideration to what teachers should teach to students, and otherwise it is possible that the ideal teaching effect is not researchable no matter how the time for alternate-oriented face-to-face tutorship is arranged.

Since the alternate-oriented face-to-face tutorship is implemented in the learning center of remote education, there has been a remarkable improvement in the learning effect of students.

Through the guiding explanation from the teachers in face-to-face tutorship, the students can receive great helps regarding to the key and difficult points of the courses, and also can easily solve the questions which were hard to previously understand in the process of the face-to-face tutorship, thus laying a solid foundation for the follow-up learning.

In addition, the alternate-oriented face-to-face tutorship has also exerted a powerful strengthening role in managing students in off-campus learning center of remote education and constructing study style.

First of all, the students in the alternate-oriented face-to-face tutorship of remote education study at school at a regular time, effectively avoiding the disorganized state caused by online learning and leaving management from school for a long term, and making the exchanges and communication of students with school and teachers more frequent and smooth.

All these are directly beneficial to the implementation of managing network students.

Second, because students in remote education often return to school, they can fully feel a strong campus cultural atmosphere. In the mean time, all kinds of rich campus cultural events can also influence the network students, promoting them to be influenced more commonly by campus culture. This is an important role, which is very difficult to generate from the face-to-face tutorship concentrated before the final exams.

Third, through teaching assessment and testing, it was found that the attendance rate of students in classes was obvious to rise after the implementation of the alternate-oriented face-to-face tutorship.

Those students, who thought remote education should be centered at online learning but also, believed several face-to-face tutorships played no great helps in the previous time, did not return to school and attend a lecture.

After the implementation of the alternate-oriented face-to-face tutorship, however, the students in remote education can truly feel the gains from the participation in the tutorships, and the numbers of students attending a class increase dramatically.

Since the Ministry of Education of China has carried out remote education (pilot) for 12 years, Chinese schools (pilot) of remote education have been comprehensively opened to the public in educational objects, resources, places, age and institutions, for the purpose of exploring and constructing a lifelong learning platform with higher quality assurance [2].

However, to make an enhancement to the training quality of remote education, the joint efforts are necessary to source from all kinds of aspects.

Certainly, the alternate-oriented face-to-face tutorship in remote education is never a universal key for everything, but indeed has changed into a feasible approach as an implementation scheme for ensuring the teaching quality of remote education.

## References

1. Zhang J, Gengsheng W, Li F (2003) Implementation and improvement of distance education in China. Open Edu Res 7(4):7–24
2. Yuan Y (2011) Modernization of education, reform, contribution and accomplishment of remote education. China Remote Edu 23(15):16–26

# Part III
# Education in Informatics I

# Chapter 45
# New Techniques to Improve Mathematical Education in Local Engineering Universities

**Dawei Sun and Jiarui Liu**

**Abstract** This paper analyzes the mathematical education in local engineering universities of China. Some possible reasons are given to explain the difficulties in mathematical teaching and learning. By using morden technologies, emphasizing the fundamental principles and demonstrating the applications of mathematics from various problems in many subjects of engineering, this paper proposed some methods and examples to improve the mathematical education.

**Keywords** Mathematical teaching · Mathematical history · Engineering Universities

## 45.1 Introduction

Mathematics is very important in morden science; it has many branches including Calculus, Linear Algebra, Probability and Mathematical Statistics etc., and occupies a very important position in undergraduate courses in various universities. In the United States and other developed countries, even in some famous comprehensive university of China, there are very rich experiences on pure mathematical education. But with the rapid development of higher education in these ten years, a large number of engineering universities comes up by merging. So far, there are more than 2,600 universities, but nearly 2,500 are local

D. Sun (✉) · J. Liu
Henan University of Technology, Zhengzhou 450001, China
e-mail: docsun2000cn@yahoo.com.cn

J. Liu
e-mail: jiarui_1218@126.com

W. Du (ed.), *Informatics and Management Science V*, Lecture Notes in Electrical Engineering 208, DOI: 10.1007/978-1-4471-4796-1_45,

universities, and universities of Engineering account for a large proportion in these 2,500 local universities, they become the main body of China's education [1]. Some problems of the mathematical education appear in these universities, mainly in the following parts, firstly students feel that learning math is boring and hard, they do not know why they must learn this subject. Secondly, teachers may not wake up the student's passion on studying and learning mathematics. Thirdly, there are no suitable teaching materials, the materials looks like the same. In this paper, we first state the status of mathematical education, and then we analyze it by the following parts: faculty, student, curriculum settings, teaching methods, teaching materials, teaching conditions, test methods and so on. At last we give some examples and suggestions on how to improve mathematical teaching.

## 45.2 Status and Analysis

### *45.2.1 Faculty*

The teachers of mathematics in local engineering universities are a little weak, the proportion of professors is less than other engineering subject, more than two thirds teachers do not get the PHD degree and have no systematically mathematical training, local engineering universities are usually not the first choice of the professors and Doctors when they look for work, this causes some problems in attracting high-level mathematical talents to join in.

### *45.2.2 Students*

Scores of the college entrance examination are not high for most of the students in the local engineering universities, especially the mathematical score may be much lower than other subjects, the high score students usually choose the famous universities. Many students do not know how to study mathematics in university, they still use the way in the middle school, and dependent on teachers too much, wait for the teachers to set homework, to do exercise. Some of the students do not know why they should study mathematics, they have to study because the demand of the university, they think they could have a rest and they want to learn some subject to help them find a good job, especially the students in mathematical department, they confuse that why they choose to study this major, it is a little hard for them to find a satisfied job.

### 45.2.3 Curriculum Settings

Arrangements of the courses are a little nervous, many universities finish the mathematical course in one year, and that is to say, students should learn Calculus, Linear Algebra, Probability and Mathematical Statistics in one year. But we know that even the course of Calculus, it need at least 180 h for most engineering students, they must learn other subjects at the same time, they could not learn very well because of the time. For the mathematical department, many courses are canceled or reduced because of the lackness of teachers, for example, Topology and Differential Geometry are canceled in many local universities. Another part of universities pursue the record in the National Entrance Test, they spend too much time on the course needed in the National Entrance Test.

### 45.2.4 Teaching Methods

The teaching methods are too single; this leads to unsatisfactory teaching results. Some mathematical teachers just put the words from the textbook to the blackboard, and set several exercises in the textbooks. Teachers do not have good communications with students; they should take a variety of teaching methods, including morden techniques of computers and multimedia. But many local universities in undeveloped cities cannot put enough computers and multimedia in the mathematical teaching process. There are thousands of mathematical textbooks, most of them are similar, consist of theorems, examples and exercises on whole pages. The introductions and communications with other subjects are not enough, even ignored in some chapters. The books are hard to attract the interest of students; they do not know the application of differential calculus and integral in their subject.

### 45.2.5 Morden Technologies

Morden technologies are ignored in the mathematical teaching; many students cannot use the softwares such as maple, matlab. But this really plays an important role in the applications of mathematics. The reason may be the limitation of teachers and morden equipments.

### 45.2.6 Tests

Examinations and tests are very important parts of learning the course, correctly handle the relationship between the examination and evaluation and teaching, is of great significance in promote teaching. Many universities put too much emphasis on complex mathematical calculations.

## 45.3 Methods and Examples

Now, we will give several suggestions and examples on how to make the teaching better according to our teaching experience.

First we should make the students like to learn mathematics; we can select some interesting materials and strengthen the teaching of mathematical history. For example, when we teach limit theory, we can tell the story of the mathematical crisis's and ask the students how to solve them, we can also give some topics on Life Sciences. Some scientists think that the dividing of Human cells will damage the DNA molecules, the more times the cells divide, the more hurt will lead to the DNA molecules and finally the cell is death, so we can think this as a limit sequence in mathematics, when we know the limit we can know the code of aging.

Second we should stress the teaching of important fundamental definitions, understand the essence of mathematical content. For example, when we compute the limit of functions by Hospital's rule [2],

$$\lim_{x\to a}\frac{f(x)}{F(x)} = \lim_{x\to a}\frac{f'(x)}{F'(x)} \tag{45.1}$$

We must remember the conditions of this famous rule. When we compute the derivative of a function by the Chain rule [2],

$$y = f[u(x)] = f[g(x)] \tag{45.2}$$

$$\frac{dy}{dx} = f'(u)\cdot g'(x) \tag{45.3}$$

We must note the relation of variables and don't miss any terms, this basic points are very crucial in Calculus.

Third we should teach more connections of different branches of mathematics, and also the connections with other subjects.

Fourth we should use multimedia appropriately, when we teach the chapter of Analytic geometry, we usually distinguish the surfaces of second order from the following equations [2]:

$$\frac{x^2}{a} + \frac{y^2}{b} + \frac{z^2}{c} = 1 \tag{45.4}$$

If we give the shapes on the computers, the students can understand it quickly and accurately.

Next we give several examples in the mathematical teaching process. First we talk the topic of Integral theory. By introducing the stories of Newton and Leibniz, we know that the generation of integral and the differential are different. In Newton's theory, in order to solve the movement problem, he created derivative and then integration, Leibniz is just on the contrary. But the following theorems state the relationship of the integral and the differential [2, 3].

$$\int_a^b f(x)dx = F(b) - F(a) \tag{45.5}$$

We may talk more about this. We can see the left equations of as the packing of integral and the differential, and the right is a number. It is the foundation of Stokes formula and the Gauss–Bonnet theorem. Famous mathematician Shiing-shen Chern gave the proof of The Gauss–Bonnet theorem or Gauss–Bonnet formula in differential geometry, it connects their geometry and their topology [3], and Differential geometry is the mathematical basis of Relativity Theory of Einstein.

Next we talk about the context of differential equations, we can review the development of the equations and introduce the definition: a differential equation is a mathematical equation for an unknown function of one or several variables that relates the values of the function itself and its derivatives of various orders [2, 4]. It plays an important role in material science, engineering, biology, physics, economics many other subjects. There is one differential equation that everybody learn it in physics, that is Newton's Second Law of Motion.

$$F = m\frac{dv}{dt} = m\frac{d^2u}{dt^2} \tag{45.6}$$

Here v is the velocity of the object and u is the position function of the object at any time t. Newton's Second Law of Motion is the foundation and core of classical mechanics. Another example is the N-body problem, we consider 3- body problem first [5], Consider N point masses moving in a Newtonian reference system, the only force acting on them being their mutual gravitational attraction.

Let the ith particle have position vector qi and mass mi $> 0$, G is the proportionality constant

$$m_i q_i'' = \sum_{j=1, i \neq j}^{N} \frac{G m_i m_j (q_j - q_i)}{\|q_i - q_j\|^3}. \tag{45.7}$$

Solving this problem has been motivated by the need to understand the motion of the sun, moon and the earth. If the particle grows, does the periodic solutions of this system grows? Does Earth and other planets crash? These questions may cause the interest of students. Another example we want to show is the problem of quantitative variation of population. How many people there are 20 years later? It is a very crucial question in the development of our country. Mathematical model of the growth of the population was first proposed by Thomas Robert Malthus at 1798 [6], suppose that x(t) is the population at time t, r is the growth rate of the population, then the population model can be expressed by the following equation,

$$\frac{dx}{dt} = rx(t) \tag{45.8}$$

This model says that population growth in geometric series and is in accordance with the population at that time. For biological populations, there may be competitions, then how to forecast their populations, suppose that x(t), y(t) are the populations at time t and differentiable with t, in general there are equations in the following (a, b, c are constants)

$$\frac{dx}{dt} = x(a_1 - b_1x - c_1y) \tag{45.9}$$

$$\frac{dy}{dt} = y(a_2 - b_2x - c_2y) \tag{45.10}$$

Next we talk about some applications of differential equation in the grain industry, because the author's university is famous for its contribution to grain industry, many students are in the subject of grain. When we store the grain, we must compute the pressures of the grain to the silo wall, and the pressure is changed when the grain moves. The pressures are determined by differential equations [7].

From the above introductions, the students may think that the differential equation is very useful in engineering subjects, and are willing to learn it from the heart.

## 45.4 Conclusion

In this paper, we analyze the status of mathematical education and give several methods and examples to improve mathematical education in local engineering universities of China. I think it is useful to other teachers and helpful to enhance the student's interest in the local engineering universities.

**Acknowledgments** The authors would like to thank the support of NNSF of China (10801079) and Doctor Fund of Henan University of Technology.

## References

1. Information on http://news.xinhuanet.com/edu/2011-03/08/c_121160430.htm
2. Qiu XS (2008) Calculus and its applications. China Machine Press, Beijing vol 20, pp 17–25
3. Shu JC (1992) Topology of manifold. Wuhan University Press, Wuhan, pp 93–99

4. Information on http://en.wikipedia.org/wiki/Differential_equation
5. Meyer KR, Hall GR, Offin D (2009) Introduction to the hamiltonian dynamical systems and the n-body problem. Springer, New York vol 8, pp 372–376
6. Jing QY, Xie JX, Xie J (2008) Mathematical model. Higher Education Press, Beijing vol 23, pp 22–28
7. Wang ZQ (1992) Granary building and structure. Business Press, Beijing vol 38, pp 363–368

# Chapter 46
# Research on Dance Teaching Based on Modern Educational Technology

**Ping Li**

**Abstract** With the continuous improvement of Science and technology, scientific and technological meaning is given to dance teaching in the context of new era. Multimedia-aided teaching equipment has become an important method for current education of various disciplines. In the same way, dance teaching has also increased the application of multimedia-aided teaching equipment in teaching activities. Based on many years of personal practice and working experience, this paper explores and discusses the application of modern educational technology in dance teaching, and hopes to serve as a modest spur to induce someone to come forward with his valuable contributions.

**Keywords** Dance teaching · Modern education · Technology

## 46.1 Introduction

Dance is a specific cultural phenomenon in front of non-verbal in the history of human civilization, and is also a human dynamic method of artistic expression, a new interpretation of culture [1]. It is found in actual teaching process that, on traditional teaching class, teachers could only try to create a particular dance background environment for students through a brief linguistic description. With the rapid development of science and technology, multimedia-aided teaching equipment has been gradually extended to today's dance teaching classes, giving dance teaching more sense of the times, and has changed monotonous and repeated

P. Li (✉)
Art Department, Henan Radio and Television University, Zhengzhou 450008, China
e-mail: Ping_Li45@yeah.net

W. Du (ed.), *Informatics and Management Science V*, Lecture Notes in Electrical Engineering 208, DOI: 10.1007/978-1-4471-4796-1_46,

simple practice under the traditional teaching mode to make student learning more vivid. In combination of personal experience of actual working process, this paper explores the influence of multimedia-aided teaching equipment and modern educational teaching technology to dance teaching from many aspects, analyzes and evaluates its function in an objective perspective, in hopes of pushing the development of dance teaching career of our country [2].

## 46.2 The Important Role of Modern Educational Technology to Dance Teaching

Application of modern educational technology in dance teaching mainly refers to multimedia-aided equipment. Through the use of modern equipment such as, projectors, TV etc., students could feel the dance background and special atmosphere under more vivid and intuitive image of the original ecological natural environment, so that the students could feel more deeply the essence of dance. Using projector screen could make teaching content more vivid and visual, to help students better grasp and understand dance movement. For example, in the process of teaching activities of Tibetan dance, we could use make students feel the charm of Tibetan culture and natural beauty by using multimedia courseware. What students face is not a simple mirror any more, but pieces of pictures full of dynamic and soul—sacred and big and tall the highest peak in the world, Mount Everest, clear water of Brahmaputra and people wearing Tibetan clothing full of ethnic characteristics are singing and dancing with speaking languages belongs to specific Tibetan language sincerely [3]. By this way, students could have a more profound understanding of the process of Tibetan people's living conditions, customs and cultures, dress characteristics and rhythm of dance. In the actual work process, we found that if application of modern educational technology could be applied rationally in the process of dance teaching activities. A good teaching result could be achieved by using multimedia approach to do dance teaching, making simple and boring teaching content more rich and diverse, promoting various type of information be accepted by students in a more natural and ingenious way, easy to make students immerse in the learning of dance. Whether it is a vivid picture reproduction or melody of the dance, both can help students get better inspiration and insights.

## 46.3 Modern Educational Technology has Its Two Sides in Dancing Teaching

In the process of teaching dance, modern multi-media educational technology is applied in classroom teaching for its innovative teaching methods and comprehensive teaching function, and is welcomed by teachers and students, thus, obtained a certain degree of breadth. However, in the actual working process, it is found that it is not perfect, and has its two sides like other things. As a way of dance teaching, multimedia-aided equipment would play the opposite effect, if the teachers have no clear knowledge of it or was improperly operated.

The first representation is too many forms of multimedia with little actual content in it. In the teaching activities of dance appreciation course, many teachers has no clear teaching objectives and purposes, they just seek for content that could use multimedia as a form of showing it, using all forms they could, such as MIDI, projection, video and courseware etc. they are blindly seeking for the appearance of new teaching methods, which could make appreciation class become a superfluous. This approach is actually not able to achieve good results. It looks very lively, as if students could widen their horizon during watching it, but in fact, students are disturbed by these lots of useless information, and could not really understand the real meaning and essence of dance, which ultimately reduce the overall teaching effectiveness.

The second representation is too much emphasis on the main role of teachers. The essence of modern teaching techniques is not really understood. Promote multimedia teaching to be formalism. And it is difficult for modern educational technology to play an important role. For example, some teachers does not recognize that multimedia teaching is something that need to combine student's actual demand and traditional teaching together, and customarily pass their main role of dance appreciation activities on the multimedia. While when students are appreciating, they over-reliance on multimedia, which finally making the appreciation and decomposition of whole dance actions a simple video, and the classroom is more like a multimedia competition.

The third representation is that teachers do not realize interaction between multimedia and students. Ultimately the function of multimedia teaching does not achieve its best effect. For example, some teachers too much emphasis on the students' main role and principles during the teaching process, and do not have a deep understanding of their leading role, and the supporting role of multimedia does not get a good effect. Or excessively require students of skills display, ignoring the interaction between students and teacher, between students and students, weakening the comprehensive effect of modern educational technology.

## 46.4 Main Points of Modern Educational Technology in Dance Teaching

In the actual process of dance teaching, if modern educational technology is properly used, students could be brought to a good learning space, good for improving their rich imagination and creative awareness. When multimedia equipment could maximize the interaction between teachers and students, and between students and students, students could be the truly subject of dance teaching, their activeness could be brought into full play, application of multimedia is able to achieve its maximum value as well. That is to say, in the process of dance teaching, whether using multimedia or explanation of teacher and the formation of atmosphere, improving students' dance ability is the main point that needs to be centered on. In the context of new era, dance teaching activities should be appropriate and decent, requiring a combination of students' actual situation and ages etc. that need comprehensive consideration. On the other hand, modern educational technology needs a clear understanding of students' individual difference and actual need etc. Application of modern educational technology is mainly as follows.

### 46.4.1 The Application of Multimedia Should be in Close Contact with Teaching Materials

The application of courseware needs to be connected with specific content of classroom teaching, only in this way, can the scientific dance teaching be better realized, letting drawbacks of traditional dance teaching be successfully compensated by multimedia-aided equipment. Therefore, before using multimedia, we need to have a clear understanding of entire teaching objectives and specific content of teaching materials, start from the concrete reality of teaching for scientific and rational use. In actual teaching work, I have once made Ballet training class into courseware by using relevant software of multimedia equipment, and properly introduce Ballet plays and music related to material content for students to appreciate. After whole class learning, students could review by using courseware, on one hand, they could their understanding of the essentials of dance skills could be enhanced, and on the other hand, theoretical knowledge of Ballet is enhanced as well gradually, and the appreciation capability is improved as well.

### 46.4.2 The Application of Multimedia Should be Combined with Traditional Teaching Media

In process of dance teaching, the most important feature of multimedia teaching equipment is its good compatibility and comprehensiveness, rather than exclusive presence. We need to combine multimedia with traditional teaching media in an effective and comprehensive way in the context of new era. Although compatibility and powerful advantage of multimedia cannot be replaced, traditional teaching media still plays a certain role in dance teaching activity. For example, piano is an essential traditional equipment of basic training classes, during playing the piano, through many aspects of advantage, like changes emotional expression of music, and efforts etc., and combination of specific teaching content, making multimedia music irreplaceable. Another example, for most of the time, students could not discover their own problems during dance practice, then if teachers record the entire training process by using DV, and play it to students through multimedia-aided equipment, according to principles of reflective learning, students could be aware of their deficiencies correctly and timely using multi-angle way and correct in a contrast. Teacher could make targeted analysis and explanation to problems occur in students, and record the correct process by using DV. Repeated training and records of the process of growth and progress of students is the most effective way to encourage students to progress. At the same time of reinforcing learning content, not only students' dance learning potential be well stimulated, but also on the other hand improve students dance ability significantly. This is an important symbol of modern teaching—organic combination of traditional media and modern educational technology.

### 46.4.3 Harmonious Atmosphere in the Classroom Created by Application of Multimedia

Using multimedia approach can effectively help us create a peaceful and democratic atmosphere of dance. Multimedia-aided teaching is not just as simple as put sound, video and pictures etc. into computer, and teachers will only need to click the mouse and type on the keyboard, and the entire dance course would be completed. In the dance teaching activity, teachers are the designer, organizer, leader and participants of the whole process. Only in a harmonious learning atmosphere, by vivid and interesting explanation, peaceful and democratic interaction between teachers and students, can the function of multimedia be maximized, and can ultimately enable students' emotion be affected in a shorter time. Stimulate students' motivation and initiation sufficiently, and enable students to consciously expand their imagination and apply to the understanding of dance. Such teaching effect is not achieved by simply mechanical operation. For example, in the course of choreographer, we could use multimedia to set relevant dance scene, lush forests, sunshine, flowers, river water, with a piece of comfortable background music, fully play various advantages of multimedia in sound, color and shape etc. to make students be able to resonate with the viewer screen whether in visual or auditory. Watching beautiful and dynamic picture on the screen, listening to melodious singing, students could naturally form a sense of immersive, and finally arouse students' experience of dance actions, and create a good dance emotion. Finally, teachers would guide the students into groups to discuss with each other, effectively combine music and pictures, and compose an impromptu dance. By this way, students' enthusiasm of learning is mobilized in a maximizing way, and their understanding of picture and mood of music are applied to it. This way, on one hand, has achieved the effective interaction between teacher and students, and the touch between music and pictures, and on the other hand, this kind of dance having a strong appeal could help students have a more profound understanding.

## 46.5 Conclusion

In combination of individual many years of practical working experience, this paper discusses the application of modern educational technology in dance teaching, analyzes in a detailed way the important application of modern educational technology in dance teaching, two sides of modern educational technology in dance teaching, and on the basis of this, proposes the application points of modern educational technology in dance teaching. However, due to limitations of personal knowledge and experience, not everything is covered, and hope to arouse the attention of broad scholars by this paper.

## References

1. Hua L (2009) Several considerations of application of modern technology in dance. Dance Teach vol 17, pp 3–9
2. Yishan Z (2008) Music multimedia courseware making. The Central Music Institute Publishing House, Beijing, pp 9–11
3. Lulu Z (2009) Music education prospects, vol 09. Liaoning Publishing House, Shenyang, pp 111–119

# Chapter 47
# Research on Materials of Drawing Therapy in University Library

**ZhongHua Liu**

**Abstract** The article is set out to introduce the origin, the development, the main theory and the operative techniques of the Drawing Therapy. Considering the advantages of the library in Research Laboratory, research documents and materials, the author analyzes the feasibility of applying the Drawing Therapy materials to psychological health care of college students by the university library and providing the proposals for the library to establish a professional team that could carry out the health psychological care for students through using the various kinds of materials of Drawing Therapy.

**Keywords** Drawing Therapy · University library · Drawing materials · Psychological health

## 47.1 Introduction

Art Therapy has been researched since 1950s and was put into practice in 1960s. Drawing Therapy, as a branch of Art Therapy, has been applied in the psychological counseling, mind-cure and education on mental health for many years [1, 2]. There're many researchers who work in colleges and universities devoting themselves into the mental health care through Drawing Therapy. Their achievements have drawn attention from hospitals and society [3, 4]. Due to its own advantages, the library of colleges and universities has more resources to apply the Drawing Therapy to psychological health care of college students [5].

Z. Liu (✉)
Library of NingBo University, NingBo City, ZheJiang Province, China
e-mail: liuzhonghua@nbu.edu.cn

W. Du (ed.), *Informatics and Management Science V*, Lecture Notes in Electrical Engineering 208, DOI: 10.1007/978-1-4471-4796-1_47,

## 47.2 The Origin and Development of Drawing Therapy

Drawing Therapy is a new vocation which has a promising prospect. The National Health System set up the first therapy in 1946 when Art Therapy was studied by the sholars. It includes Drawing Therapy, Music Therapy, Opera Therapy and Dance Therapy. The therapy value has been recognized since the psychiatry of America about 1930s. The therapy model has not been stabilized untill 1940s, for the forerunner never showed up. Finally the subject bas been set up during 1950s to1960s [6, 7].

Art Therapy has been applied by herbalist in ancient times. Although it had not been fullly developed in ancient China, we can say that China is the first country that put Art Therapy into practice. The emperor of Sui Dynasty was weak because of lust, so someone recommended the famous doctor to the Emperor. The doctor diagnosed the emperor and refused to give medicine to the emperor because he thought there was no medicine that can save the emperor's life. Instead of giving a prescription, the doctor decided to draw two pictures about wintersweet and snow. And only asked the emperor to enjoy the drawing. The emperor liked them so much that he watched the pictures all days and forgot the debauchery. After 15 days, his symptoms were all disappeared and he recovered a problem [8, 9].

Drawing Therapy, as the important form of Art Therapy, has been originated from the spontaneous drawing as a media in the phrenotherapy to a systematic Drawing Therapy in US and UK. It has taken a place in the human world view, human intention and the Interpretation of human life experience. Today, during the development process it has formed its own unique technology. It is an art technology assessment of seeing expression of the drawing as a diagnostic tool and using of human association and interpretation of art products and drawings to help people to find their relationship of their inner and outside world; using the creative process to reconcile people's emotional conflicts, elevate feelings and help people to build self-exploration, self-improvement to enhance self-growth. The focus of the two technologies is not on the aesthetic characteristics of drawing and drawing skills, but on the activities of the whole drawing in the process.

## 47.3 The Basic Principles of Drawing Therapy and the Operating Techniques

### *47.3.1 The Basic Principles*

Neurophysiologist Sperry's cerebrum experiments showed the left-brain of human beings was associated with abstract, logical things and languages. But the right one was mainly in charge of human emotions, visual and unconscious. Psychiatrists considered that traumatic emotions and stress of memories was stored in the right brain. Psychologists Ley proved that a key couldn't manage each part of human brain at the same time. This showed human emotional distress, emotional conflicts and trauma should be dealt with art form of the right-brain (painting and other kinds of emotional concepts). Language expression and the organization of the left-brain can be activated simultaneity. The combination and interaction of each brain, can more effectively promote human health. Painting as a visual and specific language rather than the linguistic of the media that can convey people's inner and subconscious information. So, people can use vision language (drawing) to express the inner world which can't be expressed in languages. Drawing has the essence that characterizes the psychological, personality, emotional, interpersonal, cultural and social conflicts. Drawing Therapy works by it. This showed that the basic principles of Drawing Therapy was regarded the drawing as the therapy. It used the drawing's characterization to the therapy and recognized the drawing not only a form of art, but also a science therapy. So it employed the drawing itself as a tool to express the interpersonal and personal emotion. Most people could reproduce themselves in the secure and backup surroundings by the drawing's essence to release the internal and external inhibition and modulate their moods.

### 47.3.2 *Operating Techniques*

The operating technique of Drawing Therapy is divided into four steps: warming-up, intervention and assessment, sharing and interlocution and production. They can redressal properly according to the participator. The steps are shown as Fig. 47.1.

Step 1 Warming-up. This step can help the participant build a the surroundings of support and receptiveness. It must make the participant understand that the therapy doesn't emphasis on the technique and taste of drawing but instead the process of it. The applying of the therapy should try their best to build the trust of each other.

Step 2 Intervention and assessment. The participant can develop the therapy's activities. The therapist (Subject Librarian) should pay attention to everything about the participant, such as creative order, status, attitude etc. Intervention can help the therapist (Subject Librarian) learn about the participant of the drawing. But the assess can help the participator adjust to intrapsychic conflicts and rebuild himself.

Step 3 Sharing and interlocution. In this step, with the aid of the interpretation of drawing, there's a sharing and interlocution between the therapist (Subject Librarian) and the participant. The therapist (Subject Librarian) shares and enjoys the participant's drawing activitives.

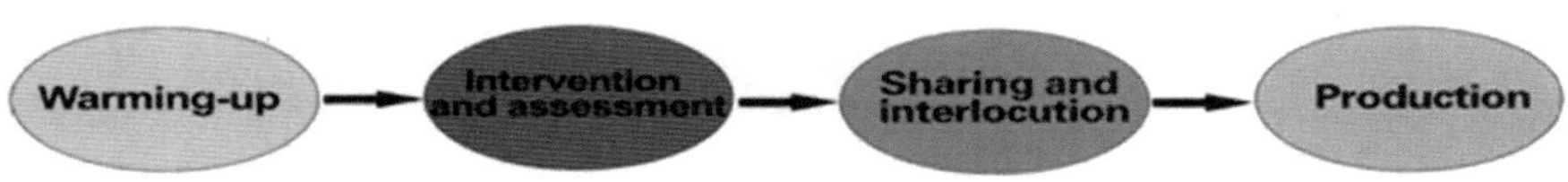


**Fig. 47.1** They can redressal properly according to the participator

Step 4 Production. Researchers consider that the production should be reserved in the therapeutic department before the therapy (The production should be put up in a suitable place of the therapeutic department for the effect of the therapy). But after the therapy, the production must be kept by the participant. If the therapy needs it to be remaked, the art therapist can only use it to do scientific research without violating its copyright.

## 47.4 The Advantages of the Library Development of Drawing Therapy

### *47.4.1 The Match Between the Library Lab and the Drawing Therapy's Receptiveness Surroundings*

In the receptiveness surroundings, it needs the therapist and partners to take part in together to set up isolated, independent and safe surroundings, which is very important for applying drawing therapy service based on the theory of empathy. The surrounding of receptiveness is very good for the participant to open his mind and turn his inner world into the drawings. The therapists can response to the signals given by the participant. The isolated surroundings also provide participant a private space, which is open only for the therapist and the partner to communicate without disturbance from the others. With the scale and the teaching level of the NingBo University being enhanced, the time sees the growing research projects and the forms which are conducted by the teachers and students of NBU. So, the library establishes the Academic Lab for the teachers and students to push forward their projects and share ideas with each other. The Academic Lab provides readers an independent space from the public reading room to do their personal researches, convene small scale forums and help each other. The Academic Lab is very suitable for the drawing therapy with the character of receptiveness and isolation. It is a safe, isolated place for students to express their intrapsychic conflicts which can not be expressed in words. This establishes a platform for them to construct healthy psychological environment through drawing.

### *47.4.2 Rich Art Documents and Materials of the Library are Good Materials for Drawing Therapy*

Due to its unique advantages, applying Drawing Therapy to psychiatric counseling of college students is not only possible, but also effective. Since it is immerged into the process of drawing, Drawing Therapy can bring no press to the students who need psychiatric counseling. From my viewpoint, it is possible for drawing therapy to become a key way of carrying out psychiatric health care because it is interesting and simple to practice. There are various kinds of art documents and materials in the library of NBU, which not only makes it possible for students to create art works and express themselves, but also provide important materials for the applying of drawing therapy to college students who are tortured by sentiments and bad feelings. The library is calm and can be opened flexibly. It is acceptable and convenient for college students to receive service of drawing therapy in the library. Drawing Therapy has characters of convenience for communication and without any emphasis on the quality of drawing. This makes it possible for students to express themselves without a strict art request.

## 47.5 Patterns of Library Carrying Out the Drawing Therapy Service

### *47.5.1 Establish Drawing Therapy Lab and Providing Suitable Drawing Materials*

The practice of the Art Therapy needs various kinds of drawing materials, such as Markers, wax crayons, oil painting sticks, Gouache, etc. There are three methods that we can apply to carry out drawing therapy: spontaneous drawing, color exploring and perfect drawing. As to spontaneous drawing, the author thinks that wax crayons and oil painting sticks are good choices for college students because they are cheap and easy to use. When spontaneous drawing is adopted, students can draw what they feel and think easily. As to color exploring, it is better to use brilliant watercolor pens and Gouaches that cost not too much because the construction of watercolor materials is not as stable as solid drawing materials. Watercolor materials play a very important role in arousing creativity of the drawing therapy participant and providing a good way for the participant to express their feelings vividly and incisively. From my viewpoint, to apply the method of perfect drawing, the means of painting in group is a good try to bring up coordinate spirit of the participants and maximize the effect of the drawing therapy. Although the choices of the drawing materials depend on the strength of ego structure and existence of the self of the participants, it is the therapist and professional teachers that can guide the partcipants to make the correct and effective choice.

### *47.5.2 Establish Virtual Platform of Drawing Therapy in Library and Set Up a Therapist Team*

The development of Drawing Therapy is in early age and employees involved in this business and service are not professional enough. The government could do something in strengthening the ability developing of practicipants, establishing

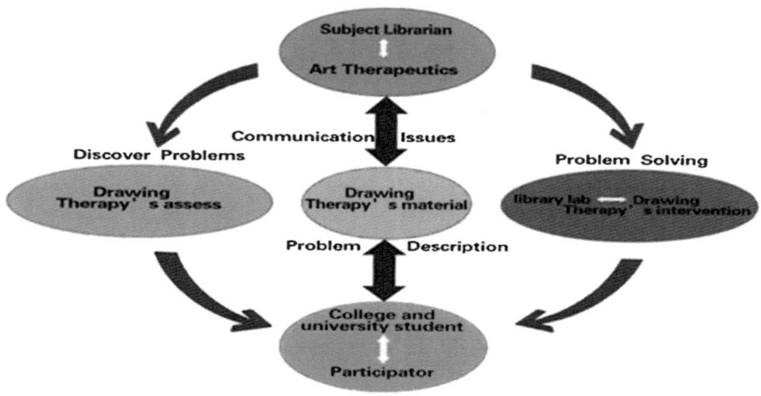


**Fig. 47.2** Therapy materials to psychological health care of college students

and managing a very professional team. If the library of the university plans to carry out drawing therapy, construction of a professional team should be put the key emphasis on. The library could obtain more professionals through talented people introduction, bringing up the employees of the organization, co-ordinate with other organizations which have high professional employees whose major relate to medical health care, art, etc. The library should establish a virtual platform of Drawing Therapy that serves college students on the condition of cooperating with the hospital, the Youth League Committee and the Department of Student Affairs Management of the college or university. The virtual platform can push forward the library to integrate the professionals of medical, psychology and arts. It can also provide reservation services when the professional team is organized by the library and the drawing therapy service system is established. The employees who are engaged in drawing therapy must be qualified under the Chinese law. Everyone of the professional team established by the library should pass the Professional Exam organized by the governmental authority. The Virtual Platform of Drawing Therapy can provide very good and professional service to college students who need psychological health care and the library can play its own role in bringing up talented young people who will take over the burden of construct the country in the future. The operation patterns for library to apply Drawing Therapy materials to psychological health care of college students are shown as Fig. 47.2.

## 47.6 Conclusion

The essence of life can be figured out thought drawing. What matters is not right or wrong but the role that Drawing Therapy materials play in psychological health care. By using the Drawing Therapy, everyone can enjoy himself or herself.

Drawing Therapy is a cross-bridge between the inner will and outside world. Participants enjoy the pleasure of the creation, support and relaxation to build up and reinforce self-confidence. University libraries should be active to carry out the service of Drawing Therapy to support the mental health of the university students and expand its serving function to cultivate the healthy personality of the new generation.

**Acknowledgments** This article is sponsored by the project of the NingBo Educational Program numbered YGH09010.

## References

1. Caroline C, Tessa D (2006) The handbook of Art Therapy, vol 37. Trans Nanjing Publications, China, pp 371–378
2. Gao Y, Li M, Yang GY (2007) Art Psychotherapy, vol 4. Shandong Publications, Chinese, pp 28–36
3. Roan HX (1994) Herbalist doctor Art Therapy: Gansu Herbalist Forum 7:5–7
4. Li Z (2002) The second VIP of the first seminar of Taiwan Psychotherapy: Psychotherapy supervision conference—the system of the Art Therapy, Taiwan and China apricot Press, Taiwan, pp 121–125
5. Information on http://news.sohu.com/20060401/n242587182.shtml
6. Sheikh A, Shaffer J (1979) The potential of fantasy and imagination [M], vol 6. Brandon House, New York, pp 955–95
7. Jung CG (1960) The transcendent function, vol 23. Princeton Univ Press, Princeton, pp 273–274
8. Zhou H (2005) Communicate views and spirit smooth. Nanjing Normal University, Nanjing vol 9, pp 30–38
9. Wang DX (2002) Law of intelligence property, vol 02. Shaanxi People's Press, China, pp 38–45

# Chapter 48
# Study on Reformation of Civil Engineering Construction Course Based on Instantiation Thoughts

**Fuxue Sun, Yunhui Zhu, Haijun Shi and Changfeng Ruan**

**Abstract** Civil Engineering Construction, which composites and applies knowledge learned in other professional courses, such as construction machinery, engineering mechanics, construction materials, etc., is an important course and can affect students' career development. On account of the key problem existing in teaching of civil engineering construction course, which is separation of theory knowledge from practice, a systematic reformation scheme is put forward based on instantiation. Through introducing actual case as "content index" in new teaching material, developing model room of construction technology and reference room of actual case, and forming auxiliary teaching database (video, animation, site picture), we built a systematic resources platform for teaching and further made reasonable use of the platform in teaching process. As results of course reformation, students can well combine practice issue with theory knowledge together during course of learning, and make more clear how and where to use course knowledge. Finally, better combination of theory with practice and better teaching effect can be gained.

F. Sun (✉) · H. Shi
College of Architecture and Civil Engineering, Wenzhou University,
Zhejiang 325035, China
e-mail: sunfuxue@163.com

H. Shi
e-mail: shj@wzu.edu.cn

Y. Zhu · C. Ruan
OUjing College, Wenzhou University, Zhejiang 325035, China
e-mail: zhu-yunhui@163.com

C. Ruan
e-mail: changfengruan@163.com

W. Du (ed.), *Informatics and Management Science V*, Lecture Notes in Electrical Engineering 208, DOI: 10.1007/978-1-4471-4796-1_48,

**Keywords** Civil engineering construction · Instantiation · Teaching materials · Auxiliary teaching database

## 48.1 Induction

"Civil Engineering Construction" is an important professional course for students majored in Civil Engineering [1, 2], which embodies and applies some knowledge introduced in other professional courses, such as soil mechanics, concrete engineering, engineering surveying, construction materials, etc. Due to its importance to graduates' professional competence and career development, enough attention was paid to the course from teaching professors and universities, when course revolution and specialty construction were conducted [3, 4]. As a local university, most graduates majored in civil engineering will directly join in construction site and do some related job, such as construction technology and project organization and management. In order to improve course effect and strengthen students' professional ability, instantiation was put forward as main thought in course reformation based on comprehensive analysis of major positioning and student quality. This reformation scheme adopted instantiation as primary teaching measurement, and put instantiation into teaching process through constructing teaching resource including teaching material [5] and auxiliary teaching database and reforming teaching and evaluation methods. After several terms' application and improvement, the scheme became more and more effective, and better teaching effect could be attained according to feedback from students and teaching professors. In this paper, primary ideas on instantiation reformation of course will be summarized, in order that this scheme can be considered as reference in teaching process by other professors.

## 48.2 Current Situations of Course and Teaching

### *48.2.1 Indeterminately Applicative Orientation of Knowledge in Teaching Materials*

The section of construction technologies focused on explaining questions about what technology to be adopted and how to conduct construction process as for different parts of construction, and section of construction organization and management aimed at resolving issues on how to organize construction process towards different kinds of projects. Therefore, all knowledge points in this course required strong practice background and specific application direction in actual project, so that students could effectively select and apply theoretical knowledge to solve actual problems in practice, which is the chief teaching objective of construction civil engineering course. However, most of current teaching materials mainly focused on explaining commonly theoretical knowledge, and attached less importance to how to select and apply knowledge into practice. As a result, separation of theoretical knowledge from actual practice would appear to students if professors couldn't take effective measurement to illustrate, which is an undesired effect arisen from the course.

### *48.2.2 Unsystematic Resources Used in Teaching Process*

Currently, professors generally employed several kinds of auxiliary resources in teaching process, such as case picture, technology animation, construction video, and so on. These resources were helpful to teaching and learning process, but it was lack of systematicness, which was because that they were from different channels including internet, other teaching and reference materials, different actual projects or different technology process, that weakened subsidiarity of these resources in application to a large extent.

### *48.2.3 Reformation Thoughts of Course Teaching Based on Instantiation*

After synthetically analyzed reasons leading to separation of theory knowledge from actual practice, we put forward new reformation idea for course construction on the basis of instantiation. The scheme consisted of two main parts, which are construction of teaching resources based on instantiation and systematic application of resources in teaching process. By means of carrying out new thinking of instantiation in whole process of course construction, problem of separation of theory knowledge from practice could be effectively mend, which was expected result of course reformation.

### *48.2.4 Adapting New Teaching Materials Based on Instantiation*

With the background of building construction, the teaching material introduces real estate project as the first chapter, mainly involving project classification and decomposition, and project development procedure. The purpose for this chapter is firstly to enable students to understand the overall development process and composition content of construction project, and to lead students to know that what will be learned in this course is focused on construction unit, which is one of the most important units among all construction procedure units. At the same time, this chapter tells students how many main subsection works are included in whole construction process, and connects each subsection work with corresponding chapter content in materials, so that students can understand application purpose of each chapter. When developing above content, actual project is selected to explain the project composition, and local project construction procedure is shown as example, which is helpful for students to link material knowledge with actual case.

### 48.2.5 Constructing Reference Room of Actual Case

Reference room of actual case was constructed for students to understand whole development process, construction procedure and all kinds of standard documents from different kinds of actual projects. In this room, 12 suits of documents from actual projects were collected, which involved documents of whole developing process from the first stage of project proposal to the last stage of work acceptance and guarantee repair of different kinds of projects. Furthermore, these documents were displayed clearly according to different project categories and different developing and constructing stages.

### 48.2.6 Constructing Model Room of Construction Technologies

Model room of construction technologies was developed to illustrate different construction technologies. Total amount of 60 models were displayed, for example, retaining and protecting for foundation excavation. These models could be utilized to explain difficult technology points and directed students to understand technology knowledge.

### 48.2.7 Constructing Auxiliary Teaching Database

Different auxiliary materials were developed for different teaching content, and three kinds of materials constituted systematically auxiliary database for teaching.

Construction technology videos were recorded on sites as auxiliary teaching materials for technology process, in which construction process completed during a short time period and was suitable for producing video.

As for construction technologies and process, which construction time of duration was not short and was inappropriate to produce videos, technology animations were designed as auxiliary materials for teaching and learning.

When considering some construction knots and construction fruits, case pictures might be better choices as auxiliary materials.

### 48.2.8 Teaching Process Implement

Taking full advantage of first chapter to explain macroscopic content.

Firstly, we made full use of reference room of actual case to illustrate important issues, for example, development procedure and mission of each stage.

Secondly, we made detailed analysis and explanation to index case given in first chapter.

Thirdly, we focused on interpretation of emphasis and study methods. As for part of construction technology, main technology process, important technology points and significant technology principle should be considered as key content. And theoretical principles in part of construction organization and management should be paid more attention. When considering how to learn this course, professors should give explicit answers. Highlighting emphasis should be adopted by students as main method, as well as trying best to combine theoretical knowledge to practical knowledge in actual case, which was proofed as very effective way.

New teaching method of "process refined" (part of construction technology).

On the basis of characteristics of content in construction technology part, new thought of "process refined" was developed to organize content of new teaching materials, and the teaching process was conducted in accordance with the same thought, which was verified in practice to be more effective for professor teaching and student learning.

Firstly, main technology process flow in this part is put forward using flow chart or words explanation to illustrate content outline.

Secondly, if each step in main process has its subsection technology, further decomposition of technology will be given, till minimum technology unit.

Thirdly, main points and relevant principle are put forward for minimum technology unit, which makes students know not only results but also underlying reason.

Based on above thoughts, teaching database including construction videos, construction animations and case pictures, should be reasonably applied to illustrate relevant course content.

New teaching method of "Actual problem induction" (part of construction organization and management).

In order to put forward teaching content clearly and easily for students to understand, new method of "actual problem induction", which was tested and proved to be suitable and effective for teaching and learning, was utilized in teaching course. Namely, when we taught some knowledge in course, we firstly offered several actual problems, which were issues that we must use knowledge to be taught to solve in practice, but were with individual character respectively. And then, students were required to think over and try to solve questions by themselves. According to students' ideas, professors should make some conclusions from mainly three aspects. One is how to solve each problem in practice based on special background respectively, another is what is common knowledge hidden behind answers to different problems, then the relevant knowledge in course could be offered, interpreted and generalized as a whole.

## 48.3 Teaching Effect and Conclusion

The course of "Civil Engineering Construction" is an important professional course for students majored in civil engineering, which can bring significant impact to professional ability and career development. Aiming at enhancing course effect greatly, our university developed course reformation based on new thought of "instantiation", according to which "instantiation thought and method" would be conducted into all aspects of course teaching, mainly including teaching materials, teaching resources and teaching process. Better conclusion can be drawn in term of feedbacks from students and professors after six semesters' application (18 classes and about 1,080 students involved) in practice, that new teaching materials, which utilized actual case as "index" of whole course content and direction of each chapter, could combine knowledge of each chapter with corresponding issues in actual work tightly, and strengthened practical applicability of knowledge. Furthermore, construction and application of teaching resources, which consisted of reference room of actual case, model room of construction technology and auxiliary teaching database (video, animation, pictures), could make up defect of lack in perceptual knowledge due to few opportunities on construction site, and made theoretical knowledge link to practical content more directly. Finally, via using the new thought introduced in this paper and applied in our teaching process, the bottleneck problem, which is separation of theoretical knowledge from practical issues and can result in students' difficulties in what is the function of knowledge learned in teaching materials and how to use knowledge into actual works, could be relieved to a great extent, and course effect could be guaranteed effectively, because when students are facing lively teaching materials in lieu of traditional copies or books, and when they know clearly what is purpose and practical function of knowledge, great interest will come into being, which is critical for students to learn hard and for professor to improve course quality.

**Acknowledgments** This paper is sponsored by Wenzhou University Major Teaching Reform Project (Study on instantiation teaching reform of civil engineering construction). Author: Sun Fuxue (1974-), male, PHD, Associate Professor, major in concrete durability and construction technology. E_mail: sunfuxue@163.com.

## References

1. Qiang W (2005) Educational thinking of the course of civil engineering construction. J Architectural Educ Insitutions High Learn 14(3):54–56
2. Qunzhou Y, Haibin C, Yisheng J (2008) Study on training student's capability based on the combining theoretical teaching with practical teaching of civil engineering construction course. J Architectural Educ Institutions High Learn 11(5):128–130
3. Li XW, Zhang YG (2009) Application and analysis of dynamic-teaching in civil engineering construction course. J Changzhou Inst Technol 22(6):93–96

4. Du G, Zhao Y (2006) Exploration about the spatial hit modern type of instruction for the series curriculums of construction about the civil engineering. J Architectural Educ Institutions High Learn 15(2):62–64
5. Shi H et al (2008) Civil engineering construction. Beijing University Press, China

# Chapter 49
# Research on PBL and LBL Double Track Teaching Model in Unified Modeling Language Teaching Based on Outstanding Engineers

**Yu Wang, Rina Su and Guojun Li**

**Abstract** The computer science and technology major of our school is the Ministry of Education educational reform pilot project. Unified modeling language is the required course of the major software engineering direction. This course's teaching method is discussed in the paper according to the requirements of outstanding engineer cultivation plan. The double track teaching method that lecture based learning (LBL) and problem based learning (PBL) repair with each other to go into the course teaching is proposed. The LBL method makes the student have a firm basic theory. The PBL method improves student knowledge application, integration theory with practice and the innovation thinking ability. The contents of the paper develops new path and provides new way of thinking for the educational reform pilot project reform.

**Keywords** Outstanding engineer · LBL teaching · PBL teaching · PBL and LBL double track teaching model · Unified model language

## 49.1 Introduction

The aim of Ministry of Education "the outstanding engineer cultivation plan" is to cultivate one large numbers of strong-innovation and high-quality technique talents for economy society development demand. The implement of the plan is used

Y. Wang (✉) · R. Su · G. Li
School of Electronic and Information Engineering, Ningbo University of Technology, Ningbo, China
e-mail: wangyuwyt@163.com

W. Du (ed.), *Informatics and Management Science V*, Lecture Notes in Electrical Engineering 208, DOI: 10.1007/978-1-4471-4796-1_49,

as the breakthrough point which makes the new path and direction of engineering education reform [1].

The computer science and technology major of our school is the Ministry of Education educational reform pilot project. Unified modeling language is the required course of the software engineering direction. It plays very important roles in the whole teaching system. It is very necessary to research and adjust the course teaching method for outstanding engineer cultivation.

## 49.2 Characteristics of UML Course

The purpose of this course is to introduce modeling language basic principle to students after having learned Object Orientated programming fundamental and software engineering basic theory in order to improve their software development ability and level. The course not only requests student to have UML foundation knowledge, but also to have UML applied technique acquired from practical teaching. This course is abstract because it has stronger engineering practice character and too many points to easy understand [2]. Traditional teaching method mainly emphasize to comprehend UML modeling elements, acquaint with in common use model tool in the teaching process, but combine with concrete software development not enough close. The main problem includes:

### *49.2.1 Student Practice Experience and Ability Shortage, Practice Teaching Separates from Enterprise Development Environment*

The course belongs to practical engineering academics. Formerly teaching method is to analysis knowledge point by teaching practice experience. But, the student did not understand background and enterprise development environment so that difficult to comprehend practice experience and knowledge.

### *49.2.2 Strong-Comprehensive, Case and Project Practice Absence*

The software modeling is a strong comprehensive and systematic engineering which demand developer to have Object Orientated programming fundamental and be able to put software modeling method and technical into software develop process system. Actually, the student is exactly lack of this ability. The traditional practice teaching also emphasizes apply various UML model technique during the software development, but doesn't receive good teaching effect because it can not effectively combine with the actual case and the item.

### *49.2.3 Student is Independent and is Not Good at Team Cooperation and Communication*

Foundation courses, such as object orientated programming and data structure…etc. mainly is to improve the independence logic thinking and programming ability of students. But, it neglects a communication, interaction and cooperation with each other in team. The former practical teaching links also emphasizes a team development, but it is formalistic because lacks of the cooperation soft hardware environment and the necessary evaluation mechanism. The student's feedback shows that many students did not apply UML model technique into the item development in the process of graduation thesis.

In summary, teaching of UML course must be adjusted under guidance of the outstanding engineer plan. The paper illuminates advanced engineering education cultivation model with a view that our students must be talents who are able to combine theory with practice, doing practical work, competitive in the marker.

## 49.3 LBL and PBL Double Track Teaching Method

### *49.3.1 LBL Teaching Method*

Lecture based learning (LBL) is the teaching method that mainly by teacher teaching [3]. LBL is a famous traditional teaching method. It takes academics as foundation and takes student as center. Its main form is teaching. Its characteristic is systematically teaching basic theory, large capacity information, and progress speed quick, systematic strong, the tuition of the basic concept more deep and overall, the student easily comprehends to the content of course. The LBL teaching method is the essential teaching form for long time because it is deeply accepted by many teachers and students. This kind of teaching mode also has problem, for example lesson is little but the task is heavy, pay attention to knowledge induction but neglect ability development, pay attention to system integrity of respectively academics knowledge but neglect integrating of related academics, pay attention to teacher tuition in the class but neglect student's participation. Under this kind of teaching mode, the student only learn theory, but lack of knowledge application, combine theory with practice and the development of innovation thinking ability.

### *49.3.2 PBL Teaching Method*

Problem based learnig (PBL) is the problem-based teaching method. It comes from the study theory of basis construction by the Memaster University in Canada at 60's in twentieth century [4]. The PBL teaching method takes problem as foundation and takes student as principal. It adopts way of student self-learn and discussion under the teacher's guide. Under this kind of teaching mode, the student regard learn method as principle, which application knowledge work out an actual problem as purpose, improvement of synthesize ability as standard. The PBL teaching method not only effectively brings the initiative of student into study and promotes the development of the communication ability, but also profoundly changes the teacher's education idea. PBL has already been adopted by more and more colleges in the world now because of its advantage in the aspects of training a student to creativity thinking. However, the PBL teaching method also has its flaw such as lacking the theory knowledge system study like traditional teaching method. The contents included by PBL teaching will be less than the content of traditional course, the student is probably concentrated the attention in working out a problem, but neglects the farsighted target of study. If use the method to the beginner, may cause that student is hard to having knowledge foundation ability and self-learn ability because of lacking professional base, hard to seek particularly document or monograph because don't understand the characteristics and research category of each academics. At the same time, the request of this teaching method for hardware condition is also higher.

### *49.3.3 PBL + LBL Double Track Teaching Method*

Both of LBL and PBL all has his own virtues and defects. This is the so-called double track teaching method that both methods combined with each other and reasonably application in the teaching process. The typical teaching process model of double track teaching method is: the teacher teaches related theory foundation for actual problem. Student acquires related professional knowledge by search information in order to work out this practical problem. And then communicate mutually as group and discuss how to use knowledge acquisition to promote problem solution. In the process of discussing, in again and again carry on check to seek data-exchanges-discussion of process repeatedly, until the problem get solution. Fore theory study, the teacher aims at the special subject contents adoption LBL the teaching method teach theory knowledge, and then combining with the PBL. In the discussion of PBL, the student toughened to discover a problem and consider a problem and distinguish the ability of problem and promote student the development of thinking ability and can well transfer the interest that they investigate knowledge.

## 49.4 Double Track Teaching Methods Application in the UML Course of Outstanding Plan Major Class

During the teaching of outstanding plan major class unified model language, the theory system, basic concept and basic principle of modeling is introduced by the multi-media teaching to solute important and difficult knowledge point. The course provides student supporting point for the following self-educated discussion lessons. This is the LBL teaching process.

And then, PBL teaching is concrete implemented. Issue PBL teaching data (the teaching data comes from actual software development example) to all students first, request the student divide themselves into group and choose a topic, look for reference material by various resources (library, and network...etc.) at after-class time. Teacher carries on simple of guide before the lesson [5]. Students carry on discussion. Teacher guides and encourages student in time. The student freely speaks and explains in detail, discusses and brings up a new question in order to form the teachers and the students' effective interaction.

Student carry on an investigation to the system that wants to develop according to software engineering course (course beforehand) and write a statement of requirements, mainly is a function need. The statement of requirements is the pure text file now. In the class make the use case model (RUP development foundation) according to statement of requirements. Let the student measure the rationality of use case diagram from different angle (customer and development personnel), and make use of tool to carry out use case model. The next modeling work is done according to the achievement of ascend stage. For example after turning into an analytical stage, we will look for a class base on use case model born at requirement analysis produce and each file. The class only has name and parts of attributes at this time. They just carry on a concrete design including complete attribute and method after getting into to design stage. At analytical time, the new use case will be discovered. Student can learn iteration design process. In the learning process, the student improves system by own analysis and constantly. Student can continuously get a stage achievement, keep to till the last get a system model of integrity by practice. This kind of learning method raised the student's self-confident and developed the interest that they continue to study. The teaching practice can develop the function of the comprehensive experiment or group work and foster the student's engineering ability. At course study during, set up comprehensive experiment, and set out group work, combine with other methods such as group discussion and foster the student's engineering ability.

Finally, teacher evaluate each group of performance according to its performance and the report writing circumstance and issue the double track teaching method feedback questionnaire at the same time.

UML technique and application course has strong abstraction and practicalness. It demands student all have the lasting interest in studying the abstract idea, theory and model and also need the aid of the concretization example to match with an abstract theory to deepen the comprehension of the student to the course knowledge and promote actual knowledge usage ability.

Famous managerialist peter Sanchi elaborates the value of study type organization in his work ≪the fifth self-discipline≫. The study type organization is the process that the organization complete collective target by the cooperation. We led this thought into the teaching link of the UML course. The basic unit of teaching expands for the study group and imports the competition mechanism. Different role is assigned to each member in the team. This way can effectively stir up the student's interest in the study, creative power and cooperation spirit. Student can actively participate in the discussion of knowledge, discovery problem on one's own initiative, propose a solution and adequately communicate with other members in the team. The method not only promotes teaching effect of the UML course but also be of great benefit for the concrete work of students in the business enterprise or the organization in the future.

## 49.5 Summary

What higher engineering education accomplishes still isn't "outstanding engineer", but makes good foundation for an outstanding engineer. The outstanding of "Outstanding engineer" mainly means comprehensive character higher doesn't mean professional knowledge abundant or problem-solving ability stronger. So, to make today of university student in the future become "outstanding engineer" must be have comprehensive character foundation. The establishment and implementation of outstanding engineer plan make a new request to the choice of the teaching method for teacher. The contents of this paper develops new path for the educational reform pilot project reform and provides new way of thinking.

## References

1. Barrows H (1986) A taxonomy of problem-based learning methods. MedEduc 27:481–486
2. Chen H, Ming Z, Peng X-G (2010) Exploration on implementing the CDIO in the experimental teaching of UML Course. Comput Educ 11:125–128
3. Diana F (2003) ABC of learning and teaching in medicine problem based learning. BMJ 12:328–330
4. Edward FA, Johan M (2009) Rethinking engineering education: the CDIO approch, vol 33. High Education Press, Beijing, pp 367–370
5. Yu L-J, Zhang F-Z (2010) Application of PBL and LBL Double Track Teaching Model in Clinical Pharmacology Teaching. J Inner Mongolia Univ Nationalities 08:210–212

# Chapter 50
# On Teaching Reform of Life Science Public Elective Courses in Higher Education Institutions

**Wei Meng and Xiaochun Lai**

**Abstract** Humanity is in the face of a series of problems in life sciences such as resources, environment, population and mass extinction of biological species, so it is of urgent need to offer life science public elective courses in higher education institutions to improve the life scientific literacy of college students. After analyzing the course characteristics and student sources of whole-school life science public elective courses, the paper discusses about the appropriate teaching modes and methods, determines the principles of building life science public elective courses teaching system, and puts forward the strategies of carrying on life science public elective courses teaching, to promote the teaching effect and cultivate the comprehensive quality and creative thinking ability of college students.

**Keywords** Higher education institutions • Public elective courses • Life science

## 50.1 Introduction

In recent years, to get rid of traditional, closed and narrow disciplinary curriculum model, colleges and universities have set up public elective courses, the necessary component of university education, whose teaching content and mode have attracted more and more attention of education field. In this backdrop, a growing

W. Meng (✉)
College of Life Science, Jiangxi Science and Technology Normal University, Nanchang 330013, China
e-mail: mengweijx@163.com

X. Lai
Science and Technology Development Office of Jiangxi Provincial Higher Education Institutions, Nanchang 330038, China

W. Du (ed.), *Informatics and Management Science V*, Lecture Notes in Electrical Engineering 208, DOI: 10.1007/978-1-4471-4796-1_50,

number of colleges and universities take initiative to life science education for all as an important part of contemporary quality education, offering life science public elective courses to all college students. Life science is one of the oldest disciplines in the world [1]. In remote antiquity, people gradually accumulated a wealth of life science knowledge in the original agricultural production. From the mid-20th century, life science has developed by leaps and bounds [2]. It is of very important significance for students to learn some life science knowledge to promote the rationalization of knowledge structure and the knowledge merging of liberal arts and science. Students of science should learn some life science knowledge, so do students of liberal arts, to enrich knowledge structure, broaden vision and continuously improve humanities and scientific qualities [3]. Based on biological engineering and biological science in College of Life Science, Jiangxi Science and Technology Normal College offers to the whole-school students life science public elective courses such as neuroscience, introduction to life sciences, life science and biotechnology, which are welcome and highly appreciated [4].

## 50.2 Analyses on Characteristics of Life Science and Student Resources

### *50.2.1 Characteristics of Life Science*

The development of current disciplines is no longer independent, and the intersection between disciplines has become inevitable [5]. Life science public elective courses offered for the whole-school students came into being. As the science to study the nature, characteristics and the law of occurrence and development as well as the interrelation among various creatures and between creatures and environment, life science is applied to effectively control life activities, dynamically transform biosphere and benefit for mankind [6]. Closely related to human survival, people' health, economic construction and social development, life science is the basic natural science most talked about on the global scale. In the advanced countries of Europe and America, life science and material science, including physics, chemistry and earth science, are the two branches of natural science. The status of life science in natural science in the world can be seen from the following facts. In Science Citation Index (SCI) published American Institute for Scientific Information, more than 8,000 scientific journals around the world are carried, among which the top 10 of the greatest impact, namely citation frequency, besides comprehensive periodicals Nature in Britain and Science in America, are all in the field of life science [7].

Life science is offered as the whole-school public elective courses, with following main issues under research. What is the chemical nature of the biological material? How do chemicals in the body mutually convert and demonstrate life characteristics? What are the composition and structure of biological macromolecules like? How do cells work? How do various cells accomplish a wide variety of functions? How do genes as the genetic material work? What mechanisms promote cell replication? [8] How does a fertilized egg cell use its genetic information in the development of a singular process by a number of very different types of cells constitutes highly differentiate multicultural organisms? How do a variety of cells combine to form organs and tissues? How was species formed? What factors caused evolution? Are mankinds still in evolution? What are the relationships between species in a specific ecological niche? What factors dominate the number of each species in this life environment? What is the physiological basis of animal behavior? How is memory formed? Where is memory stored? What factors can affect learning and memory? Where does intelligence come? Are there other intelligent organisms in the universe? How did life come from? Meanwhile, life science as the whole-school public elective courses is good for students to analyze problems in the perspective of different disciplines and improve the comprehensive ability and innovative ability to solve problems [9].

### *50.2.2 Student Resources of Life Science Public Elective Courses*

Specialist students of whole-school public elective courses are with certain requirements, generally from the first grade next semester or second-year elective public elective courses in the allocation of resources permit, any elective can be used as public elective courses to school open to students across colleges, multi-disciplinary elective. According to the calendar year elective life science students to look at where the professional part of the data (Fig. 50.1), of the course students from 18 colleges of the school of all elective School of Chemistry and Chemical Engineering, School of Pharmacy, School of Physical Education and School of History and Culture, a high proportion of four schools accounted for 53.17 %, and 14 other college students accounted for 46.83 % from the proportion of non-bio major and the majority of students from the student's grade distribution, from and third grade students in the majority, accounting for 83.66 % (Fig. 50.2).

Note: School of Chinese Language and Literature (SCLL), School of History and Culture (SHC), School of Foreign Studies (SFS), School of Economics and Management (SEM), School of Mathematics and Computer Science (SMCS), School of Communication and Electronic (SCE), School of Chemistry and Chemical Engineering (SCCE), School of Civil Engineering and Architecture (SCEA), School of Art Design (SAD), School of Music (SM), School of Physical Education (SFE), School of Life Science (SLC), School of Pharmacy (SP), School of Material Science and Engineering (SMSE), School of Law (SLC), School of Education (SE), School of Vocational (SV), School of Tourism (ST).

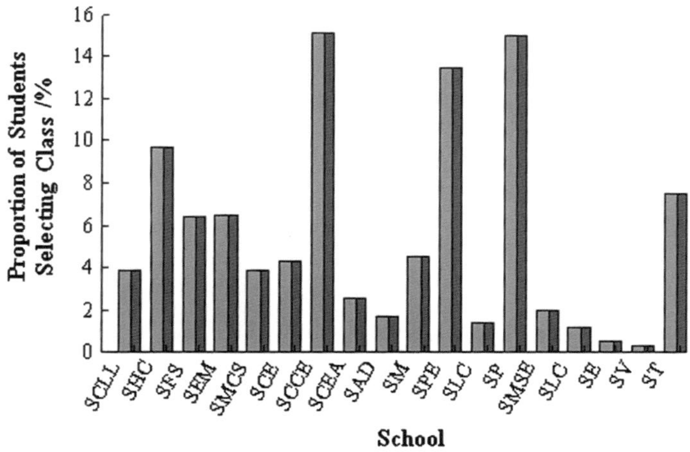


**Fig. 50.1** Majors distribution of students in selective life science

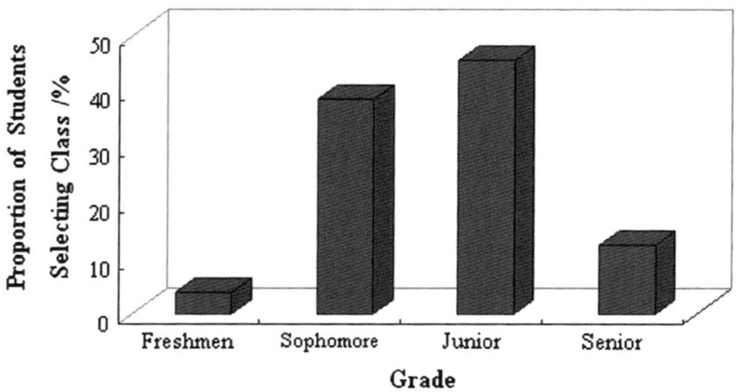


**Fig. 50.2** Grade distribution of students in selective life science

## 50.3 Establishing Teaching System of Life Science Public Elective Courses

Offered as optional courses, life science usually has 32 teaching hours in teaching programs of most universities. Since life science is a subject of extremely complex content, there are more than 260 teaching hours in the curriculum programs of professions related with life science in general institutions [10]. Therefore, how limited hours of life the essence of the scientific knowledge to the students is a challenging problem. As an optional course, on the one hand, can expand students' knowledge, basic knowledge of the life sciences can be learned in a very short period of time so that students are interested in life science; on the other hand, vivid knowledge of life sciences interesting is required, so that students want to hear interest in the life sciences, as a good this course is an important task. The following principles should be followed in establishing the curriculum system [11].

### 50.3.1 Combining Theoretical Knowledge and Practice Content

The teaching of life science, a very practical subject, should follow the principle of combining theories and practice. Theory teaching should focus on the thorough understanding of basic knowledge. The practice should be combined with reality, increase students' interest to help them understand biological knowledge involved in all aspects of daily life and cultivate students' observation of nature through various practical activities. Through field practice, site observation, record and analysis of a wide range and mix of animals, students can learn more about the morphological structure, ecological habits and dynamic number of creatures, thus recognize the diversity and complexity of biological organisms, and unity with the environment, to further consolidate, verify and enrich the content of classroom teaching, to deepen the understanding of theoretical knowledge; the same time, field practice, and further train students to engage in the actual operation of the animal field work and independent ability to work and train hard-working, consciously abide by habits of organization and discipline and the spirit of collectivism.

### 50.3.2 Systematically Explaining Basic Knowledge of Life Science

To reflect subject characteristics, the teaching arrangement of life science should be guided by the basic theories and follow the principles of combining science, system and practice, carefully design curriculum system and teaching content, so that knowledge can be gradually carried out from shallow to deep. The choice of classic teaching content should be carefully studied, boundaries of compartmentalization broken, and life science studied as a whole, so that students can have a systematic, comprehensive and accurate understanding of the concept of life science, fundamental features of life as well as the origin and development of biosphere.

### 50.3.3 Introducing Discipline Developments to Enhance Students' Interest in Learning

Some significant findings of life science, a rapidly developing discipline, may give the production of human life to a series of major changes. Should strengthen disciplines developments in the while explain disciplines basic knowledge, teach, attention interspersed with the latest scientific research of life sciences and domestic and international dynamics. For example, the annual Nobel Prize for Medicine and Physiology award-winning achievements, major scientific progress of the annual global life science research to address the progress of the current development of human society is facing a major crisis, such as population explosion, food shortage, energy shortage, environmental damage, species extinction, ecological balance disorders, so that students understand the latest developments of modern life sciences and application prospects, and mobilize them to learn positive life sciences, a sense of crisis and urgency, and inspire a sense of mission and responsibility.

### 50.3.4 Based on Students' Current Knowledge Level

Teaching activities carried out should take full account of the life sciences knowledge has been mastered by the students at the secondary level, to ensure that the curriculum system, the advanced nature of the reasonableness of the premise, build a scientific and rational knowledge framework. In view of the life sciences are generally set up for non-biology majors, taking into account the student's ability to accept and the level of understanding, a bold choice for a number of highly specialized subject knowledge. Specific explanation for some knowledge already learned in middle school, just by asking questions and assignments to review. Less involved in the theoretical, the students' daily life and appropriately reduced.

### 50.3.5 Strengthening the Explanation of History of Science, Reflecting Humanistic Connotation of Science

In the process of life science development, a batch of famous scientists emerged, who treat the spirit of the scientific research we learn from, but also a valuable asset in the course of development of human science. Addition to the exposition of the basic knowledge in the life sciences to explain the process, selected some of the history of science can not only enhance students' interest in learning, but also reflect an important way for scientists of the scientific spirit. Such as British life scientist Charles Darwin, despite his father's opposition, the courage to give up paid pastor career, interested in natural science research, and finally write "Origin of Species". Learned dedication towards science in his diary, to treat the nature of objective truth the spirit of science, has set an example for future generations.

Another example is the French entomologist Fabre, whose earlier life penniless, and the second is barely enough food and clothing, but he did not succumb to the "bias" and "poverty", decades on, the power of spending a lifetime in-depth the world of insects in the natural insects in the environment, observation and experiment, recorded under the instincts and habits of the insect, real serious and detailed description of the insect instincts, habits, labor, marriage, reproduction, and made outstanding contributions.

## 50.4 Strategies of Improving Teaching Effect of Life Science Public Elective Courses

### *50.4.1 Carefully Selecting Multimedia Teaching Methods*

Life science is rich in content and a large amount of teaching, traditional teaching methods alone, it is difficult to complete the task of teaching. The use of multimedia tools can greatly enrich the teaching contents and improve the utilization of the classroom. At the same time, the life sciences have a lot of pictures, it is difficult with traditional teaching methods to give students a distinctive visual impression, the use of multimedia teaching methods, can greatly make up for this shortfall, from multi-angle, multi-faceted, comprehensive in-depth teaching content. Biological knowledge and an intuitive, vivid, concrete image is displayed in front of students, not only to improve teaching effectiveness, but also can effectively mobilize the enthusiasm of students.

### *50.4.2 Adjusting Assessment Methods to Stimulate Students' Innovation*

In traditional teaching methods, there were a variety of assessment methods of optional courses, such as close exam, open-book exam and homework test, which are relatively single. Students were easily limited to rote knowledge, so the features of life science optional courses cannot be reflected. In order to enhance students' practical ability to reform the requirements students must complete an extra-curricular practice report, such as the anatomy of the fish, and insect species survey of a region.

Life science teaching reform involves a wide range. How to reflect the unique characteristics of their elective courses is a more complicated problem that requires general educators reform explores two aspects of the teaching system and means, in practice, constantly sum up experience, commitment to reform, and innovation. To popularize the knowledge of life sciences in the teaching process should be based on training students to analyze problems, problem-solving ability, ability to cooperate and exchange, an independent working ability, the ability to reform and innovation, so that the overall quality of high application type.

### *50.4.3 Conducting Inquiry Teaching to Cultivate Students' Scientific Literacy*

Inquiry teaching students to use inquiry to learn under the guidance of the teachers take the initiative to obtain the knowledge, capacity development and to experience the practice. In the classroom, according to the content of the life sciences, combined with students' characteristics to create a scenario to guide students to identify problems, ask questions. Students based on existing knowledge and experience, bold hypothesis, and then experiment to verify. In the teaching process, appropriate inquiry learning can be adopted according to course content, so that students can enhance their observing ability, thinking skills and hands-on practical ability.

## 50.5 Conclusions

Life science is rapidly developing discipline. Offering public elective courses related to life science to non-biology major plays an important role in increasing students' understanding of the discipline, improving the quality of college education and enhancing students' quality. Introducing quality education into public elective courses helps not only break students' traditional concept of learning, but also cultivate talents of all-round development. Strengthening the construction of public elective courses is beneficial to optimizing students' knowledge structure, improving their overall quality and promoting education reform, but in this process some ideas of quality education must be combined with, truly introducing quality education into teaching. It is an important issue in the coming period to sum up experiences, identify problems, strengthen curriculum development, and continuously improve teaching methods to serve for promoting students' quality.

**Acknowledgments** The study on which this paper was based was supported by Jiangxi Provincial Educational Sciences "12th Five-Year" planning Project (10YB029). The financial support to this study is greatly appreciated. Authors gratefully thank the anonymous reviewers for their valuable comments on this manuscript.

## References


1. Macleish MY, William TA (2009) Sustainable space life sciences education: strategies for innovation and global engagement. In: 60th international astronautical congress IAC, vol 1. pp 398–408
2. Nancy PM, Roberts JK, Barbara ZT (2005) Increasing student learning through space life sciences education. Living in space: scientific, medical and cultural implications. In: 14th IAA humans in space symposium, vol 36. pp 9–12
3. MacLeish MY, Thomson WA, Moreno N, Gannon PJ, Smith RB, Houston CW, Coulter G, Vogt GL (2007) National space biomedical research institute education and public outreach program: education for the next generation of space explorers. Acta Astronaut 38(5):321–332
4. Yongjian B (2010) On the reform of teaching mode for the public elective course-conservation biology. J Yulin Norm Univ (Nat Sci Ed) 31(2):89–91
5. Siegel MA (2007) Striving for equitable classroom assessments for linguistic minorities: strategies for and effects of revising life science items. J Res Sci Teach 44(6):864–881
6. Life Science; Bioscience (2012) http://baike.baidu.com/view/937.htm document. Cited 10 Feb. 2012
7. Nie C, Yin Y, Yuan G (2011) Teaching reform of public elective courses of breeding of special economic animals. J Sci Teach Coll Univ 31(3):103–105
8. Siegel MA, Wissehr C (2011) Preparing for the plunge: preservice teachers' assessment literacy. J Sci Teach Educ 22(4):371–391
9. Halverson KL, Freyermuth SK, Siegel MA, Clark CG (2010) What undergraduates misunderstand about stem cell research. Int J Sci Educ 32(17):2253–2272
10. Ning X, Xun W (2003) The study on elective courses of environmental science major. J Guangdong Univ Technol (Soc Sci Ed) 3(9):108–109
11. Larkin DB, Seyforth SC, Lasky HJ (2009) Implementing and sustaining science curriculum reform: a study of leadership practices among teachers within a high school science department. J Res Sci Teach 46(7):813–835

# Chapter 51
# Study on Teaching Characteristics and Methods of Interior and Furniture Design History Course

**Jianhua Lv and Ming Chen**

**Abstract** The main focus of this paper is to extend the teaching reform results of interior and furniture design history course in China through analysis and research. To achiever this, we made the characteristic analysis and judgment of interior and furniture design history course, and then analyzed the existing problems and difficulties in the teaching process of interior and furniture design history course. At last, the reform methods of teaching means and forms in interior and furniture design history course were discussed and summarized.

**Keywords** Interior and furniture design history course · Teaching methods · Course characteristic · Innovation

## 51.1 Introduction

In 1998 the Chinese Education Ministry adjusted the national universities undergraduate major catalogue, the furniture and interior design major was cancelled, and changed to a part or field of industrial design major or wood science and engineering major [1], which has a strong interdisciplinary and comprehensive characteristic in teaching. Research and reform of teaching system should developed continuously under the ever-changing background of major setting and industry requirements, this major field's teaching system and curriculum is filled

J. Lv (✉) · M. Chen
Forestry College, Sichuan Agricultural University, Ya'an, China
e-mail: ljh@sicau.edu.cn

M. Chen
e-mail: chenming@sicau.edu.cn

W. Du (ed.), *Informatics and Management Science V*, Lecture Notes in Electrical Engineering 208, DOI: 10.1007/978-1-4471-4796-1_51,

with many matters need to be explored. The teaching goals, ways and means of curriculum are also need to further study. Interior and furniture history course is one of these curriculums. The following are the discussions and conclusions on this issue.

## 51.2 Interior and Furniture Design History Curriculum Orientation and Teaching Status Quo

### *51.2.1 Curriculum Orientation*

The curriculum of furniture and interior design major field can be divided into three large curriculum groups.

#### 51.2.1.1 Material Science Courses

The first one is material science courses. These courses tapes are including wood science, adhesives and coatings, interior decoration materials, and other courses relates to the furniture and interior design, manufacturing or construction process.

#### 51.2.1.2 Processing or Construction Technology (Structure) Courses

The second one is processing or construction technology (structure) courses. These courses tapes are including the interior decoration engineering, wood products technology, furniture structure design, non-wood furniture processing, wood-working machinery and other courses relates to interior construction, furniture manufacturing or the structure design.

#### 51.2.1.3 Design Theory and Methodology Courses

The third one is design theory and methodology courses, which can be divided into three main types: (a) design basics, including sketch, color, preliminary design, design expression skill and other foundation courses; (b) professional design, including different types design, such as design planning, creative, production and expression; (c) design theory and history, including design theory and history, such as furniture design outline, interior design principle, industrial design outline, design methodology, interior and furniture design history and so on.

Therefore, interior and furniture design history course belongs to the curriculum group of design theory, history and methodology, which is the foundation theory of design science.

### 51.2.2 Course Characteristics

The Interior and furniture design history course has an important position and role in curriculum system and professional education, it's very necessary to understand the major field correctly so that we can develop the course efficiently.

Firstly, as a design discipline, the furniture and interior design major requires the students be familiar with the design objects, and study the design history and theory completely by grasping the connotation, essence, styles and trends of interior and furniture design. Modern design discipline emerged at the end of 19th century. With the development of modern design, its core concept of humanism has become the consensus of people, a good design should be a "humanitarian design" and "humanism design", which not only focusing on human nature, but also paying attention to the history and traditional culture. It's necessary to learn the design history and theory systematically, if the designer want understand the humanism clearly and use the principles in design projects [2].

Secondly, furniture and interior design major field is a unique and comprehensive major field, which has different characteristics that are relation with industrial design, environmental art design and engineering science and technology, therefore the teaching methods should have strong comprehensive design course teaching characteristics. In addition to the above-mentioned design history and theory, design psychology, ergonomics, other human and natural sciences are also indispensable. Because, furniture and interior design is not a pure artistic, pure social or pure technical activity, it's a kind of combining theoretical and practical activity, a multi-interdisciplinary subject cross integrated discipline [3].

### 51.2.3 Teaching Status Quo

#### 51.2.3.1 Lack of Coherence and Awareness in Design History Teaching

In many colleges and universities, the design basic courses' hours such as sketch, color and design expression are much more than theory courses' like design theory, design history and design evaluation.

At the same time, in some courses the teachers overemphasizes the pure art education methods besides teach the close relationship between theory and practice courses and guide the students to correctly treat and learn professional theory foundation courses.

China Interior Decoration Association honorary president, Professor Zeng Jian said: The modern furniture has become one part of modern society from the beginning, and the development of modern furniture has kept pace with the modern science and technology, modern art, modern architecture development. But our nation's furniture development is not very close with science, art and architecture, which can be proved by backward technology, old form, and poor function of our furniture [4]. The interior and furniture design history course is necessary for furniture and interior design major students to expand the visions of design and set up right design concept.

#### 51.2.3.2 Pragmatism Prevailed

In the process of teaching, the pragmatism idea is too popular in colleges and universities to limit the development of student's design ideas.

In traditional design education, people put more attention in the ideas like "pragmatism" and "technology first, theory second", which reduce the students' innovative thinking and motivation ability.

On the other hand, the design industry needs graduate students became the experienced designers at graduation. So the students spend a lot of time on drawing, computer-aided design and so on. At the same time, A few teachers ignore the design theory and history courses because of their own reasons or industry requirements. All those reasons cause a vicious spiral: ignore design theory and history learning; the design level is difficult to improve; industry and society is dissatisfied with graduates' design level; students spend more time in learning skills; ignore design theory and history learning.

Design theory and history is one of important parts of design majors' education system, which links different courses. And for the same reason, interior and furniture design history is indispensable to design work, even plays a more important role. In many colleges and universities, the design basic courses' hours such as sketch, color and design expression are much more than theory courses' like design theory, design history and design evaluation.

## 51.3 Analysis and Discussion on Teaching Methods

### *51.3.1 Typical Image Method*

The carrier of the development of interior and furniture art and technology is countless interior environments and furniture objects. Many related content and context are reflected through these objects. In the teaching process, we need to fully understand this point, and using all the means and ways to reflect the objects, which is also an important method to teach and learn design history, such as architecture or industrial design history.

For example, when we study the British arts and crafts style furniture, it's better to use the many furniture images to analyze and evaluate the points, reasons of development and historical background of British arts and crafts style furniture. This is the so-called typical image teaching method.

Finally achieve the design history learning purpose by combining the abstract theory and concrete image and confirming each other.In the process of teaching, the pragmatism idea is too popular in colleges and universities to limit the development of student's design ideas.

### 51.3.2 Point-Line-Surface Comprehensive Method

In teaching process, students generally reflect a problem that the content is scattered and lack of obvious directly contact. This phenomenon is the result of interior and furniture design history course's characteristic. Because interior and furniture design history is full of style, history, natural science, humanities and other integrative content, which make it wide, complicated and all-around.

In view of this phenomenon, the point-line-surface method can deal with problem well. We take the specific image, important design idea or designer as a point, take history development as line, take the region and time period once the style or school was popular as the surface. Through this way, the important materials can be connected, and the learning effect can be better than usually.

For example, when we teach the western modern furniture history, we can treat the famous designers or theorists like John Ruskin, William Morris, Samuel Bing, Henry Van de Velde and other famous design products like crystal palace, red house, red and blue chair, the cowboy chair and others as the point, treat art and crafts movement, Art Nouveau, De Stijl as the line, finally the whole surface of western modern furniture history can be shaped by those points and lines.

### 51.3.3 Multimedia Teaching Method

Because the interior and furniture design history has a lot of information and content, the teaching method besides traditional ones will be helpful. One of these new methods and technologies is the multimedia teaching method.

During courseware manufacturing process, we choose more pictures than words. The picture and its accompanying essay are both excellent. We were trying to use more pictures, texts, audios and videos material in the courseware production, which can attract the attention of students, improve the efficiency and results of learning.

### 51.3.4 Fusion of Disciplines

Furniture and interior design history has comprehensive and complicated characteristics. The teaching process, the teaching general outline and the teaching contents should be satisfied with these characteristics.

For example, when we talk about the furniture styles changing, the progress of science, architecture, art, the evolution of the social system should be analyzed. In some cases, these elements could have decisive influence on the design history development.

Through the analysis of these different backgrounds, the essence of the development representation will be revealed, thus effectively strengthen the interior and furniture design history teaching purpose, and could achieve a better teaching effect.

## 51.4 Conclusion

This article has a systematic study on furniture and interior design history course, makes systematic analysis on the generation of curriculum orientation and teaching status quo, and explores some reforms on teaching methods of furniture and interior design history course.

The research results are as follows:

It researches on the curriculum orientation of furniture and interior design history course, elaborates on the teaching status quo of furniture and interior design history course, makes detailed introduction on development of furniture and interior design history course, and reveals the reasons why the course needs reform.

It analyzes the teaching methods of furniture and interior design history course, and analyzes the new teaching methods of furniture and interior design history course from the view of adaptability.

It studies the new teaching methods of furniture and interior design history course: typical image method, point-line-surface comprehensive method, multimedia teaching method, fusion of disciplines.

## References

1. Tang L, Liu W (2005) The research on furniture design education in forestry university of China. Furniture Inter Decoration 12:46–49
2. Xiaoding Z (2000) Discussion on art design history lesson. Explor Music 2:95–96
3. Zhu H (2001) Technology and design–design culture and design philosophy, vol 18. Henan Art Press, Zhengzhou, pp 37–43
4. Hu J, Fang H, Peng L (2005) 20th Century modern furniture, vol 12. Central Compilation and Translations Press, Beijing, pp 19–24

# Chapter 52
# Analysis of Social Anxiety Based on 1161 Cases Teaching

**Lixin Wang**

**Abstract** *Objective* Survey and analysis of social anxiety on college students to understand social anxiety conditions and influencing factors of college students, and proposed countermeasures to alleviate social anxiety. *Methods* Interaction Anxiety Scale (IAs) conducted a questionnaire survey and analysis of social anxiety status quo of 1,161 students. *Results* Students' social anxiety significantly, and the high degree of social anxiety grouping accounted for 13.3 %. The survey sample, the average anxiety level is very significantly higher than normal average, different gender, different students, whether the student leaders of the Students' degree of social anxiety there are significant differences. Different family parenting style, social anxiety level of parents closely related to the degree students exist very significant differences. *Conclusion* There should be targeted to strengthen the anti-social anxiety and other mental health education for students.

**Keywords** Survey and analysis · Social anxiety · Cases

## 52.1 Introduction

The term "social anxiety" (Social Anxiety) was first proposed by British psychiatrist Mark Gelder [1]. Social anxiety refers to anxiety in the process of socialization, because of their social roles and social behavior can not meet the predetermined expected to be objective, generated by [2]. The basic performance: the experience when interacting with others uncomfortable, fear, tension and

---

L. Wang (✉)
School of Education Science of Leshan Normal University Leshan,
Sichuan 614004, China
e-mail: scieiistp@foxmail.com

W. Du (ed.), *Informatics and Management Science V*, Lecture Notes in Electrical Engineering 208, DOI: 10.1007/978-1-4471-4796-1_52,

worry, fear with others, fear of being watched, fear of being misunderstood and others negative evaluation, fear of eating in front of the faces of the people, writing, resulting to avoid the withdrawal behavior.

Social anxiety college students more common type of mental disorder [3], which exist to varying degrees of social anxiety in college students, some more serious [4], it not only affects the social interactions of college students, but also affect their daily lives, learning, employment and so on. In recent years, the social anxiety of college students already have research mainly focuses on research: If the differences in the way of parenting exchanges anxiety of university students have a significant effect [5], anxiety and self-harmony there is a big [6] self-harmony various factors and social anxiety was positively correlated [7], the implicit social and explicit social comparison on the social anxiety of college students have different effects [8] has been the study of the status quo of social anxiety and specific analysis less data, this study attempts to Survey and analysis of large sample survey on the social anxiety of university students, in order to understand the situation of university students in social anxiety and its influencing factors, and to propose corresponding countermeasures, and to provide information support for mental health education.

## 52.2 Methods

Use prepared by Leary Interaction Anxiety Scale (Interaction Anxiousness Scale, referred to as IAS), [9] random sample survey of the seven universities in college students, combined with interviews. Issued a total of 1,450 questionnaires were electronic version of which 250 questionnaires were paper 1,200 questionnaires, 1,300 were recovered, the recovery rate of 89.66 %, of which 1,161 copies of valid questionnaires, the effective rate of 89.31 %.

Through statistical analysis, survey, social anxiety questionnaire Cronbach's Alpha coefficient of 0.727, standardized item Cronbach's Alpha coefficient of 0.739 and a half coefficient was 0.716, indicating that the results of this survey is authentic.

## 52.3 Results and Analysis

### *52.3.1 Overall Situation of Social Anxiety*

The results of this survey college students social anxiety of the (M 41.12, the SD is 7.54) with the United States three different University's 1,140 college students often die [10] (M to 38.9, the SD is 9.7) exist very significantly with the difference

($t = 10.014$, $P < 0.001$). The survey results show that the social anxiety of the students surveyed were significantly higher than American students are the norm.

From the frequency distribution of different degrees of social anxiety, the results of this survey: the social anxiety score lowest score of 19 points, the highest score of 71 points. Scale (norm $M = 38.9$, $SD = 9.7$) provides that: less than or equal to 29.2 divided into the low group, low level of social anxiety; 29.2–48.6 points for the middle group, the general level of social anxiety; greater than or equal to 48.6 high packet, the higher degree of social anxiety. The survey results shown in figure: low group 70, 6 % of the total; the middle group of 937 people, 80.7 % of the total number; high packet 154, 13.3 % of the total number. Visible, the survey sample social anxiety incidence of 13.3 %, the degree of social anxiety was normally distributed, high packet proportion is significantly higher than the low group, further to a higher degree of social anxiety description of the survey sample.

### *52.3.2 Students Gender Differences in Social Anxiety*

It can be seen from Table 52.1, the female students of social anxiety was significantly higher than boys.

### *52.3.3 Different Family Education Students Differences in Social Anxiety*

It can be seen from Table 52.2, family education on the degree of social anxiety in college students is extremely significant difference. Further post-mortem examinations showed that: the democratic type of family education of students social anxiety level was significantly lower than the severe type and laissez faire; social anxiety level of the laissez-faire type of family education of students significantly higher than the severe type.

**Table 52.1** Gender differences in social anxiety ($M \pm SD$)

| | N | M ± SD | T | P |
|---|---|---|---|---|
| Male | 521 | 40.59 ± 7.12 | −2.18* | 0.029 |
| Female | 640 | 41.55 ± 7.87 | | |

* Indicates the $P < 0.05$ ** $P < 0.01$ *** $P < 0.001$

**Table 52.2** Family education students differences of social anxiety (M ± SD)

| | N | M ± SD | F | P | For multiple comparisons(LSD) |
|---|---|---|---|---|---|
| 1. Democracy | 757 | 40.44 ± 7.41 | 11.38*** | 0.000 | 1 < 2**,1 < 3***,2 < 3* |
| 2. To crack | 282 | 41.83 ± 7.60 | | | |
| 3. Laissez-faire | 122 | 43.65 ± 7.56 | | | |

* Indicates the P < 0.05 ** P < 0.01 *** P < 0.001

**Table 52.3** College students with their parents closely related to the degree of social anxiety difference (M ± SD)

| | N | M ± SD | F | P | For multiple comparisons(LSD) |
|---|---|---|---|---|---|
| 1. Close | 768 | 40.46 ± 7.14 | 9.60*** | 0.000 | 1 < 2***,1 < 3**,2 < 3 |
| 2. Generally | 343 | 42.22 ± 7.91 | | | |
| 3. Alienate | 50 | 43.66 ± 9.43 | | | |

* Indicates the P < 0.05 ** P < 0.01 *** P < 0.001

### 52.3.4 *Different Students with Parents Closely Related to the Degree of Social Anxiety Differences*

It can be seen from Table 52.3, with parents closely related to the degree of social anxiety level of students there are significant differences. Drawn further post hoc test: there is no difference between the social anxiety level of students in general to the relationship with parents and the degree of social anxiety and estranged relationship with parents and students; but close to students and parents of the degree of social anxiety was significantly lower than with their parents social anxiety, the highest level of the estranged relationship of general college students; estranged relationship with parents and students of social anxiety level was significantly higher than with their parents close to college students, relationship with parents and students.

### 52.3.5 *Students' Social Anxiety Students to Differences*

Be seen from Table 52.4, the degree of social anxiety for rural students significantly higher than urban students.

### 52.3.6 *Whether as Student Leaders of Social Anxiety Differences*

It can be seen from Table 52.5, not as student leaders, college students, social anxiety level was significantly higher than that as student leaders, college students.

**Table 52.4** Students' social anxiety students to the differences (M ± SD)

| | N | M ± SD | T | P |
|---|---|---|---|---|
| Cities and towns | 521 | 40.59 ± 7.12 | −2.22* | 0.027 |
| Village | 640 | 41.55 ± 7.87 | | |

* Indicates the $P < 0.05$

**Table 52.5** As the student leaders of social anxiety difference (M ± SD)

| | N | M ± SD | T | P |
|---|---|---|---|---|
| As student leaders | 653 | 40.51 ± 7.38 | −3.11** | 0.002 |
| Not as student leaders | 508 | 41.89 ± 7.68 | | |

** Indicates the $P < 0.01$

## 52.4 Discussions

### *52.4.1 Analysis of Social Anxiety*

The survey found that social development, China's social anxiety is evident, the degree of social anxiety of college students significantly higher than the norm of American college students. Results of this investigation with the Social Anxiety Scale for measuring ethnic Chinese with English version and a white American college students social anxiety results are consistent (Chinese college students were significantly higher than the white American college students) [11], findings consistent with many domestic scholars. Abroad to infer the third world countries due to the relatively backward economic and social survival of a greater pressure on the incidence of social anxiety should be higher than the West. However, the results of this investigation Pengchun Zi (2003), Zhang Zhiyong (1998), [4] survey results vary, the Peng Chunzi survey shows that any smooth the level of social anxiety between the two countries were no significant differences. They believe that its findings can not rule out a different layout of the Social Anxiety Scale may be deviation of understanding of Chinese college students due to cultural influences, as well as the outstanding popularity of the concept of mental health and to cover up higher, that conclusion was established, this results the reason can be explained. Such findings inconsistent, may also result from the difference of the sample selection. Sample survey of the literature support the findings of this study for the 200–400 regions ranging from a few college students, the small sample size, it is difficult to achieve measurement samples stratified sampling, sample representativeness; the Peng Chunzai social anxiety Survey sample Hunan the seven colleges and universities (including the key universities of two, three universities in general, universities, 2) 241,439 students, the sample represents only Hunan Province, students of social anxiety situation; any cis element in the Higher Education Campus 5 University of random 300 copies of questionnaires to investigate and study, 273 low representativeness of the survey sample; The survey sample of 1,161 undergraduate students from seven universities, two of the "211 Project" key universities, four two institutions (including a provincial key universities), one of three independent schools, this sample can represent the students of social anxiety status quo. This sample of the students surveyed degree of social anxiety was significantly higher than Pengchun Zi (2003) ($t = 10.556$, $p < 0.001$), the Horizontal (2008) ($t = 3.236$, $p < 0.001$) findings.

Social anxiety status quo of college students today is more serious, to analyze the reasons are mainly two points: First, China is currently in transition, social, political, economic, cultural, is undergoing major changes. Second, the college is by the school to the period of preparation of the community college students lack of social experience, not enough social skills, social relations are relatively narrow, easily lead to social anxiety. Specifically: (1) As a developing country, China's socio-economic development is unbalanced, the east–west gap between rich and poor, the employment pressure on college enrollment directly increasing year by year, and other social factors affect the level of students' mental health. (2) Chinese students' learning career, mainly at school and learning, the education system of the examination-oriented education is the heavy burden of learning, academic pressure, thereby enabling the College students lack of experience in social interaction schemata and techniques, can not find the topic, I do not know how to socialize, resulting in different levels of social anxiety. (3) Chinese family education does not focus on fostering children's independence, excessive doting hinder their communication with the outside world, and now college students in the 1980s after too much protected by their parents, experience exchanges setback will easily escape retreat.

### *52.4.2 Students' Social Anxiety Demographic Variable Differences*

Combined with the existing literature, the social anxiety level of female students was significantly higher than boys on the reasons: First, the traditional concept requires the girls to show shy, have led to expectations of girl's ladies girls by the implied social cause varying degrees of social anxiety. Second, the equality between men and women in today's society girls are still not absolutely equal status, and rural patriarchal ideology still plays a dominant role, the traditional patriarchal ideological preconceptions affect people on the evaluation of girls. Be discriminated against girls in the life and work, lead to low self-esteem, not the good of self-harmony, which can affect their social interaction. Social requirements are usually boys should have the dedication, initiative, "to do" strong life and social development to dominate, more to reach the community, explore new issues, which makes them generally more than girls strong social skills. The traditional gender division of labor "men, women" requires women's family roles than their social roles and female college students reserved, restrained some, so the character of the female students may be generally more introverted, shy, but shy psychological affect a person's social skills. Fourth, studies have shown that personality and temperament is phlegmatic type, melancholic type or two types of hybrid, the nervous system is more fragile and prone to shy psychological social activities, these types of women than in men. Due to differences caused by the traditional culture, make the men have a greater sense of superiority, so that the higher their general self-efficacy, lower female, women's social anxiety than men. Sixth, Chen Xuefeng, who found that social anxiety partially mediated the relationship between the Students' source of stress and mental health since. Female students are more vulnerable to loss of pressure and interpersonal pressure. Seventh, Li Lingyu "social anxiety, attachment and Mental Health" draw female college students to avoid the new environment, fear of negative evaluation and social scores were significantly higher than male college students, the difference reached a significant level. More women than men to fear of negative evaluation in social, tend to avoid the new environment, and social focal status quo is more serious. Eighth, parents and teachers on the education of boys and girls different. For example, to encourage more male children to "break" to try to innovate, to take responsibility. Female college students in general, there is no male courage to try, the courage to contact the new environment. Nineth, girls enter puberty, self-evaluation will be reduced, are very aware of their own lovely, whether accepted by the others, and causing the anxiety of the girls on the "negative evaluation" is much higher than the boys. Tenth in the survey sample from rural college students (N = 640) than towns (N = 521), which can better explain this survey differs from previous studies gender difference, can be seen to be more number of survey data to support verification.

Combined with analysis of relevant literature, students and to differences of college students significantly because: (1) Study found that rural students and students of cities and counties under stress, greater use of self-blame, avoidance response of immature way [10], an inferiority complex in the performance of the rural students who are more prominent, which in life produce more avoidance behavior than urban students, lead to social anxiety (2) the large gap between urban and rural areas in China, the economic conditions of rural households than the town, education and living environment is not as good as the town, which could easily lead to university students from rural mental pressure and inferiority, shy away from dealing with people, the degree of social anxiety was significantly higher than urban students. Learned from the college students from rural, rural household income and family status affect their self-confidence and educational backwardness of rural households, the family contacts a narrow range, they took less than urban students contact people did not get the social exercise formed some erroneous perception, such as less fewer mistakes, and often take the form of withdrawal response to the social. Time and region from the point of view, and in recent years with urban planning into a large number of rural land acquisition, this series of factors gave college students from rural to cause a certain amount of psychological pressure can also affect the degree of social anxiety. Five city a good educational environment and culture will also reduce the social anxiety of the students. In the university environment gives rural students on the one hand exchanges desire is very strong, eager to learn about the inner world of others, and also want others to really understand what they, on the other hand they are relatively closed, closed with a strong interaction between the contradictions increase the difficulty of interpersonal relationships, the psychological is bound to bring a certain degree of anxiety and stress.

The level of social anxiety presents a very significant difference between student leaders and non-student leaders, college students, non-student leaders of the degree of social anxiety was significantly higher than the student leaders. The survey results Guo Xiaowei (2000), LI Zhao-xia, Liu Fang's findings. Studies have confirmed that college students serving as student leaders engaged in social work, and is conducive to the more widely with people in work, social knowledge, experience, skills and confidence have increased significantly. On the contrary, as student leaders, university students of social knowledge, experience, skills and confidence is poor, likely to cause social anxiety [4].

Excess of non-student leaders ($N = 508$) of the survey sample college students high school students ($N = 653$), we can see today's universities have paid attention to the cultivation of students' comprehensive quality, to encourage students as student leaders, the exercise of its organization and coordination skills and social skills. Sichuan University carried out the focus on community activities, and student leaders in the process of organizational activities has accumulated experience in social interaction, social competence has been the conscious exercise, and thus the degree of social anxiety than non-student leaders. Student leaders generally selecting the best and competition, they generally have strong organizational skills, the ability of social skills, get exercise and the work of student leaders, their sense of control and competence capacity is higher than the non-student leaders, students, and continuously improve self-confidence in social skills.

## 52.5 Conclusions

This survey shows that: social anxiety was significantly higher degree of social anxiety grouping accounted for 13.3 %. The survey sample, the average level of anxiety is extremely significantly higher than the norm of the average level, gender, urban and rural areas, whether the student leaders on the degree of social anxiety of college students there are significant differences. Family upbringing with parents closely related to the degree of social anxiety level of students exist very significant differences.

## References

1. Liu L (2007) Social anxiety. Educ Theory Pract 27(4):37–39
2. Zhang C (1991) Psychology dictionary Shanghai, vol 5. Lexicographical Publishing House, Shanghai, pp 608–612

3. Shi Xun Shen (2001) Social anxiety disorder. J Clin Psychiatry 11(4):243
4. Smooth A (2007) Investigation and analysis of social anxiety. J Yangzhou Univ (Higher Educ Study Ed) 11(6):75–78
5. Wang W (2006) Students self-harmonious situation and anxiety. Chin Ment Health J 15(5):403–404
6. Peng C (2004) Interaction anxiety scale reliability and validity and its applicability in the chinese college students. Chin Ment Health J 18(1):39–41
7. Lin L et al (2007) Social impact of social anxiety. Psychol Sci 30(5):1220–1228
8. Feng L, Hong ZK (2005) Urban and rural college students anxiety and social anxiety investigation. Shandong Arch Psychiatry 18(1):30–31
9. Xueli Y (2006) Foreign language college students coping style and social support. Chin J Rehabil Theory Pract 12(7):637–638
10. Qin H (2009) Luli normal college students of social anxiety and coping styles. Shanxi Med J 38(1):13–14
11. Jin Hong Z (2000) Female college students, social anxiety and family functioning relationship. Shandong Branch Chin Women's Univ 2:26–28

# Chapter 53
# Application of Advertising Aesthetic Appreciation in University Aesthetic Teaching

**Mingyong Zhou**

**Abstract** As a form of aesthetic activities, the development of aesthetic education has a close relationship with Chinese education and China's destiny. The university is the cradle training high-level talents, in current aesthetic education, there are still many imperfections. There are some university educators believe that the aesthetic education only refers to the art aesthetic education, which neglecting the natural aesthetic, craft aesthetic (technological aesthetic and practical aesthetic), social aesthetic. For these three artistic components, the educational content is single and boring and the educational form is inflexible, so that the aesthetic education is not so ideal. This article thoroughly understands and analyzes the status quo of university aesthetic education, proposes to apply the advertising aesthetic appreciation in the university aesthetic education, it regards the practicality of advertising aesthetic appreciation form as a brand new teaching resource of university aesthetic education, so as to realize the purposes of enrich the content and form of university aesthetic education.

**Keywords** Aesthetic education · Advertising aesthetic · Social aesthetic

## 53.1 Introduction

The high-speed development of science and technology and social science in the contemporary world appears the open trend of system, which is reflected in the emergence of interdisciplinary. In view of the continuous development of aesthetic,

M. Zhou (✉)
Design Department, Humanities College, Xi'an Shiyou University, Xi'an, China
e-mail: zhoumingyong1@163.com

W. Du (ed.), *Informatics and Management Science V*, Lecture Notes in Electrical Engineering 208, DOI: 10.1007/978-1-4471-4796-1_53,

the technological aesthetic, commodity aesthetic, design aesthetic, environmental aesthetic, clothing aesthetic, the interdisciplinary of practical art and aesthetic, etc. appear, and the advertising aesthetic is a practical aesthetic, i.e., the interdisciplinary of advertisement science and aesthetic, the advertising aesthetic is a practical aesthetic with social functions and market value, its research object is the aesthetic law and artistic characteristics in the advertisement, the advertising aesthetic applies the basic theory of aesthetic to solve the aesthetic issues in the advertisement, it combines the aesthetic theory and market to build a new platform for the creative thought and performance of advertisement design [1, 2].

The contemporary university aesthetic education integrates the advertising aesthetic, it firstly respects the modern social economic, discipline development and subdivides the reality situation, so as to open up a new teaching curriculum resource for current university aesthetic education; furthermore, for the character and moral value formation of a human, it can not only rely on the direct education, and the indirect influence and influence also plays an important role. Our aesthetic education has the same situation, the aesthetic on practical art is a kind of temper cultivation in the reasonable beauty form of the object, and it can further transform into emotion and consciousness, and finally reach the aesthetic feeling [3, 4]. The traditional aesthetic belongs to the artistic of pure artistry, this kind of artistic is a higher level one, for non-arts major students, they may know little about it; while the practical artistic, the general public and students may more familiar with it and this kind of artistic may be experienced by themselves. Advertisement is a very popular issue for all of us, combining the advertising aesthetic, a kind of practical aesthetic, with aesthetic education, the teaching can base on the experience and feelings of the students, which is exactly what the American writer Pascin Bullock said in Aesthetic and Art Education, “Aesthetic can not be regarded as a fragmented curriculum to be learned by the students”[5]. The paper seizes a subject to analyze the status of our university aesthetic education, and proposes the specific teaching model and implementation method of combining the advertising aesthetic appreciation and the university aesthetic education, and cultivates the students with scientific thinking.

## 53.2 Concept of Aesthetic Education and Status of University Aesthetic Education

The aesthetic education, as the name implies, is an activity to educate human with aesthetic. Generally, it is an appreciation and creation applying the natural beauty, social beauty, especially the art beauty, which makes the educated person unconsciously influenced and edified, so as to enhance the appreciation and creation ability for beauty, cultivate the ideal of beauty, improve the moral quality, and make it the educational activity of the new socialist human with all-round development of morality, intelligence physique and aesthetic. Since the aesthetic education is a kind of educational activities, it is also an organized system activity of teachers, learner and educational media. And its difference from other educational form is that the aesthetic education is the teachers planned uses the artistic media in accordance with a certain artistic concept to trigger a certain artistic effect of the learners (train and shape the artistic ability and artistic boundary of the learners). As the ideal educational activities, the implementation of the aesthetic education is not an isolated process, both the teachers and learners should reasonably plan and adjust a variety of artistic behaviors according to the requirements of their social and educational system (the current political, economic, cultural and other factors), in order to achieve the desired educational results.

The aesthetic education thought in China has a long history; however, regarding the aesthetic education as a theory to study was happened after the twentieth century. The concept of "aesthetic education" in China was firstly and explicitly put forward by Wang Guowei, subsequently, the well-known educator Cai Yuanpei also strongly advocated the aesthetic education. As important positions of training senior specialized personnel, strengthening aesthetic education is particularly important to colleges and universities. Since the new era, the aesthetic education of university gradually enters into a healthy development with vitality, which has many accomplishments, but generally, the implementation of university aesthetic education work is unbalanced, there are also some weak links, and it is difficult to adapt to the needs of vigorous higher education development.

The aesthetic education is a new initiative taken under the situation of reform and opening up in China, in order to improve the quality of university students and provide the talent for social development. China's past education focuses on the cultivation of students' natural science knowledge with little training and edification of the human spirit, which results in a neglected state of the cultivation of artistic qualities.

The aesthetic education was misunderstood as art aesthetic education only, a considerable number of university educators believe that the art aesthetic education is the only way of university aesthetic education, therefore, the university opened a lot of art elective courses, such as music appreciation, sculpture appreciation, Chinese and Western art history, painting art, art introduction and various other art courses. They misunderstood and thought that the aesthetic education in university means setting some art aesthetic curriculums and making the students to appreciate some artworks. However, they do not know that the art aesthetic should naturally occupy a certain position in higher education, but it is not the only content of university aesthetic education. As China's aesthetician Xue Fuxing said: "the development of human aesthetic activities generally experience: natural aesthetic, craft aesthetic (technology artistic, practical artistic), social aesthetic, art aesthetic. The art aesthetic is the concentrated expression of artistic consciousness, is the advanced form in aesthetic activities, also the most complex and refined artistic creating form, it is based on specialized skill, the art aesthetic can only be achieved upon professional training, which is not conducive to the link of the aesthetic and the artistic reality of current university students." Indeed, beauty things are everywhere in our life, but the most closest form making the university students accept is not art aesthetic, it should be natural aesthetic, craft aesthetic and social aesthetic. The art aesthetic is in the advanced development stage of aesthetic activities, which is based on specialized skill, so it is not universal viewed from the artistic point. The art major students may easily comprehend these courses. But for non-major students, it is difficult for them to grasp and comprehension these courses in a short period of time due to the lack of art knowledge, so they can not sense the implied meaning of artworks. In addition, most of the university art courses are electives (except art major students), the course arrangement is relatively less, coupled with the limited art sensing ability of the students, it results in the actual effect of this teaching is far from ideal.

We can see that the current university aesthetic education skips the first three basic artistic processes and directly enters into the advanced stage of artistic; therefore, the effect of aesthetic education can be imagined. The aesthetic education focused on the art aesthetic alienates the relationship between artistic and real life, it will hardly catch the interest of the university students, it's more difficult for university students to master, and thus it's difficult to realize the effect of aesthetic education.

## 53.3 Advertising Aesthetic Appreciation and Aesthetic Education

We currently analyze the four elements throughout the aesthetic psychical elements in advertising aesthetic appreciation; they are perception, imagination, emotion and comprehension, respectively. These elements mutually penetrate and integrate in the entire process of advertising aesthetic appreciation psychology, and thus create the human advertising aesthetic appreciation activities. In the past aesthetic education, the teachers have not recognized the importance of students' rationality towards advertising works. The advertising aesthetic appreciation can not be grasped simply relying on the perceptual awareness, in the process of aesthetic education, we should let the students has a clear understanding that the external performance form of advertisement, or the intrinsic information, culture philosophy, affective communication have guidance of mental process, and then they can deeply experience the essence of advertising aesthetic appreciation. In this paper, the author addresses this question with various practice teaching processes in the practice teaching in College of Engineering and Technology of Chengdu University of Technology.

People are living in this boundless universe, we have direct and perceptual contact with all things in the world, with these direct and perceptual contacts, we have a direct sense to things, for example, we can feel the sweetness of chocolate, the fragrance of lily, the silky of silk, these direct senses are the basic elements constituting our imagination, emotion and comprehension. The reason why a variety of pure colors, sounds and other pure feeling elements symbolize different meanings in different historical periods and cultural background and express different meanings, first of all, it is this primary sense that gives a psychological pleasure, it is the foundation for more advanced mental activity. Max Dessoir appropriately called this physiological feeling as “aesthetic reflection”, he thought this is a response almost physiologically; it’s the same as human scream when stabbed and could not help, or feeling warm in the sunshine.

The second important element constituting the aesthetic activities is the imagination. Artistic imagination can be roughly divided into two kinds of consciousnesses, i.e., consciousness imagination and creation imagination. The consciousness imagination in the artistic process is basically transforming and recombining the thing images gathered in former artistic perception link and schema already existing in the minds and smelting into new images.

The cultivation of consciousness imagination should pay more attention to understanding the “prestructure” of students, select suitable teaching material and flexible and vivid teaching medium regarding the “prestructure” of students as a very important consideration, so as to stimulate the students’ interest in learning, and then expose them to more artistic information. Here, we’d take outstanding advertising works as a medium of aesthetic education, while appreciating the new creation, expression methods and visual effects in the advertisement works, the students constantly breakthrough their own habits, adjust their own artistic visions, and accept the things in the works with an open attitude, and these things may be different from their former artistic vision, may even the opposite. In frequent contact with advertising works, people gradually recognize more new information, new thought ideas and new artistic standards, and they would gradually comprehend and accept these new things. In this step-by-step process, the “prestructure” of students has been updated and expanded.

In addition to the raw materials of “prestructure”, the emotion is also an indispensable important element for the development of imagination, it’s just the same as having soup, rich and nutritious ingredients are not enough, and these ingredients must be cooked by proper stove fire to create delicious soup. Similarly, both the excellent advertising works and other artworks, the thing really touched the heart of people is emotion, the artistic imagination can really occur with abundant intrinsic element plus the effect of feelings. Emotion, the third artistic element, is the thing we tend to discuss now, and the emotion can be divided into consciousness emotion and artistic emotion.

As the artistic element, “comprehension” contains three levels; firstly, the most basic comprehension is the comprehension to “reality” and the “fiction” in artwork, in other words, to distinguish the event and emotion in real life and art creation. On this comprehension level, the university students can distinguish the real life and the art performance in advertising aesthetic appreciation.

## 53.4 Advertising Creative and Culture Method of University Students' Scientific Thinking

How to adapt to the talent requirements of university students and further strengthen, improve and innovate the university art thinking education through advertising creative? The author believes that in addition to lectures by teachers and advertisement analysis in teaching, it also requires a combination of advertising aesthetic appreciation to train the specific art thinking, so as to combine the knowledge teaching and students' application ability.

In the aesthetic expression of modern advertisement, it can be roughly divided into two kinds of aesthetic expression, i.e., perceptible advertisement and intentional advertisement. The artistic expression form of artistic conception is an intentional one, while the compounded value judgments in advertising creative are often achieved through the artistic conception of artistic expression.

The artistic conception means that the advertising image refers to an artistic level or artistic standard inducing the sense and feeling. This kind of advertisement is different from the pursuit of perceptible advertisement. The latter expresses the subject of goods, the former focuses on the performance of emotion, or the environment and atmosphere construction, and then the "artistic conception" would naturally transfer to the goods.

In this type of advertisement, the appealed commodity and its trade mark, signs, etc., in the advertisement almost do not occupy a prominent position, it trends to show the beautiful picture and charming artistic conception; or guide the audience to get close to some feeling, experience same mood, taste some life, so as to convey the advertisement creative and advertising thought through elegant artistic conception or emotional appeal.

We have already mentioned that the aesthetic observation of advertising creative is the first step of aesthetic activities, "Beauty is everywhere, for our eyes, it lacks discovery rather than beauty". This means that cultivating a careful aesthetic observation habits should be the first issue of aesthetic education, apply the "holographic" observation methods in the advertising creative, grasp the various features from the exterior to the interior of object, thorough understand the artistic object in the analysis, the imagination, emotion and comprehension may arise only on the basis of understanding. As we have analyzed above, in the human artistic elements, the perception must form prior to any other elements.

Feuerbach said: "enthusiasm and inspiration are not mastered by will, or adjusted by time, and will not come out in accordance with the scheduled day and hour." However, this does not mean that it is an elusive and impossibly grasped thing. Many successful creative examples show that there is also law and trajectory for generation or the emergence of inspiration. Good and in-depth observation habits will give students more inspiration. Chernyshevsky said: "inspiration is a guest does not like visiting idler." Careful aesthetic observation and pursuit is a prerequisite to get inspiration, as Lenin said, the inspiration is just a reward to tenacious working. For example, digging wells needs digging layer by layer. Only when reaching the proper depth, can the spring water be seen?

Both the creation of artist and the findings of scientists, the inspirations are obtained after a careful observation, in-depth analysis and arduous exploration, when the thought relaxes from a highly excited vortex, therefore, the inspiration always takes the opportunity.

Among the advertising creative, the creative thought is ever-changing, these thoughts in different advertising creative can get a lot of inspirations if breaks the previous thought in the thought training. Especially for polytechnic students, release the restriction of their logical thought and one-way inductive thought in the traditional education, it will encourage the divergent thought, which can effectively combine the logical thought and image thought, induce and deduce the combined creative thinking.

The divergent thought, also known as the spread thought, dispersion thought, fantastic thought, etc., refers to applying different thought directions, do not limited by existing knowledge range, do not follow the traditional fixation methods, and adopting an open and divergence method, so as to derive various possible answers or different solutions. The divergent thought gives full play to the imagination and association in the thought process, and diverges from one point to all directions, this method can find out more updated solution through the recombination of knowledge and ideas. Based on this thought, the United States Osborne created a "brainstorming" and thus laid the foundation of creative engineering. There are often such instances in our life, the laser was used for punching, and however, it was widely used in medical, engineering, defense, communications, photography, audio in a short time due to its particularity.

## 53.5 Conclusion

As a form of aesthetic activities, the development of aesthetic education has a close relationship with Chinese education and China's destiny. The university is the cradle training high-level talents, in current aesthetic education, there are still many imperfections. There are some university educators believe that the aesthetic education only refers to the art aesthetic education, which neglecting the natural aesthetic, craft aesthetic (technological aesthetic and practical aesthetic), social aesthetic. For these three artistic components, the educational content is single and boring and the educational form is inflexible, so that the aesthetic education is not so ideal. This article thoroughly understands and analyzes the status quo of university aesthetic education, proposes to apply the advertising aesthetic appreciation in the university aesthetic education, it regards the practicality of advertising aesthetic appreciation form as a brand new teaching resource of university aesthetic education, so as to realize the purposes of enrich the content and form of university aesthetic education.

## References

1. Quanmei H, Huang Q (2006) Reverse thinking in news and advertisement. J Shanxi Youth Manage Cadre's Coll 02:191–199
2. Wang J (2009) Between art thinking and scientific thinking—on the nature of creative thinking. J South-Cent Univ Nationalities (Issue Humanity Soc Sci) 12:113–118
3. Wang S (2006) Attention must be paid to the emotion education of university students. Soc Sci J Anqing Teach Coll 9:88–93
4. Cheng L (2009) On emotion crises of Contemporary College Students and aesthetic education resolving ways. J Shaanxi Univ Technol (Issue Soc Sci) 20:20–27
5. Liu C (2008) Intension of emotion education and investigation of present condition. View Educ 21:23–26

# Chapter 54
# Research on Design-Conscious Cultivation in Basic Art Education

**Wei Wei**

**Abstract** This paper, talking about the design concept, clearly indicates that the design in the field of art refers to a kind of creative activities such as vision, planning and programs etc. from the practical and aesthetic point of view to realize certain purpose in the creation process. It, from the educational value of traditional craftsmanship and modern design as well as the cultural point of view, explains the importance of cultivating design-conscious and describes its theories and methods by classroom DIY and design criticism and teaching methods, with the purpose of let others know more about the issues related to design-conscious cultivation in the basic art education.

**Keywords** Basic art education · Design-conscious · Cultivation

## 54.1 Introduction

Design education in China started developing just in nearly two decades and it mainly represents in universities and colleges' design professional field, but can't be applied in art education [1]. Though it is displayed in varying degrees in the past Syllabus of basic art education, it mainly involved patterns, handmade, and processes with the main teaching objectives of the formal beauty of the work or handicraft production processes and technical issues. The modern design doesn't enter classroom until the "Design/Application" learning area officially brought into the compulsory education stage by the nation. The new arts curriculum standards of compulsory education clearly points out that the main objectives of

W. Wei (✉)
Art Institute, Xinxiang University, Xinxiang, China
e-mail: weiwei1233211@163.com

W. Du (ed.), *Informatics and Management Science V*, Lecture Notes in Electrical Engineering 208, DOI: 10.1007/978-1-4471-4796-1_54,

offering the learning area of "Design and Application" in the compulsory education stage is to cultivate students to form the design awareness and practical ability; make them understand the design thought of "let all things serve their proper purpose" and the unification of beauty of art forms and its design function; develop their innovative and creative ability and abilities of aesthetic evaluation for living objects and their surrounding environment to stimulate their willingness of beautifying life, develop their behavior of prediction and making plan before practice and persistent, patient and careful working attitude. These educational objectives different from the past are both opportunities and challenges for art educators [2].

From the current teaching situation in China, it is not very satisfactory. The vast majority of art teachers themselves don't accept any design expertise trainings [3]. They are not certain about its teaching meanings and value, not able to select appropriate teaching content and in particular, they don't quite understand the meaning of "design-conscious" the word in the sentence of its main purpose that "cultivate students to form design-conscious and practice ability." However, the teaching must be more target-oriented. If the understanding of design teaching purposes is not accurate or clear, it is bound to reduce its effectiveness [4]. Therefore, this article would like to explain the meaning of the design-conscious and the relationship among the concepts such as design and technology, production and so on and by how to cultivate design-conscious in theory to let others know more about the issues related to design-conscious cultivation in the basic art education.

## 54.2 Cultivate Students' Design Interest

Design teaching requests to develop students' observe ability, but on its premise teachers' own capacity in this area should be stronger and more prominent. New art curriculum standards take full account of the subjectivity of students and teachers, and make general expressions to the learning contents and form under different conditions to widen the teacher's range of options and strengthen their independency. Teachers are requested to be good at observing life and discovering surrounding design issues even if it is a very minor, only from which can they understand what topics will attract students' eyes and simultaneously apply to the limited material and technical conditions. Besides, they are also requested to be good at observing students so that to find their interest, based on which to help them complete the design learning. Only in this way, design courses will not divorced from students' real life, as well as not contrary to their interest.

The design motivation develops based on problem awareness, which means that the design motivation naturally emerges when you are not satisfy with all the existing designs or considering creating new designs to meet your other purposes. Design is from the living demand. It always comes out with a new face or an improvement or perfection of the existing design. If you are not an observant and conscientious, it is hard to search different information with normal vision, and so it is difficult to continuously put forward doubts and new ideas. In the past, pupils' bags are basically messenger bags with a single color. There is no other choice, because the styles are similar. But look at now the students' bags, there are shoulder bags, even pushed box-type and for their patterns with multiple colors. If the designer is not a person good at observing life, how can he experience these things are needed and loved by students? The very thing that children are at this particular stage of physical development considered by the designer can he knows that the overweight schoolbags will badly influence their normal development, and to add elements children like in the bag will be more welcomed by them. Teachers should not only teach students these simple and convictive instances that close to children life, but also explain the influence of design in life, provoking their interest in design, and guide them to observe the life to find issues and create design motivation.

Modern society needs to be communicated. Everyone must interact with other members and perform emotional exchanges, thoughts and experiences to learn from each other for common development. If a person all day just stays in his own world, his thinking will always be limited in his own expertise, so that his experience will be reduced and be lack of creativity. Art teachers, especially the teachers to teach design classes, are requested to have inter-bank ability to think, because the design itself is a culture, which has certain relationship with politics, economy, science and technology, philosophy and psychology etc. When in teaching, teachers should not only stick to the design of its stature, color, texture and structure, but also should go beyond disciplines to help students understand related issues from a broader range to develop their thinking, but this development must be combined with students' comprehension and knowledge. For example, when teaching the production of paper bells for the first grade students, teachers only need to teach them the simplest method and treatment of colors and patterns. But for four or five grade students, in addition to that, they also should expand students' knowledge and skills to let students think about: what materials are also available except for using paper? When the paper bells originated? What is the main function? Should be added with other functions except for decoration? Theses questions will make students, through data collection, hands-on practice and other learning methods learn materials science, history, folklore, functional materials etc. Maybe the results are not quite good, but at least it reflects comprehensive learning concepts. Thus, to cultivate students comprehensive learning ability needs teachers with extensive knowledge and comprehensive thinking ability at first.

## 54.3 Diversification of Design Element and Principal Exercise

Teaching students to practice in multiple ways in design elements and principles purposefully can help students know about the concept and role of the design elements and develop their sensitive to the design products in the life to cultivate their design-conscious, and enhance the appreciate ability.

In this phase's exercise, teachers can let students look for shapes represent lines in nature, for example, teachers can bring students to the woods or the park and teach them to carefully observe the veins of the leaves, the texture of tree stumps and the trend of branches, and then ask students to talk about the rules of these lines making up veins, texture and trend of branches; whether their intricate changes have effected on the unification. Besides, teachers also can ask students to look up to observe the sky high voltage wires, and then let them make the deference between the wire and the lines; it is available if the wires are twisted like the branches or texture? Through these exercises, students learn the line knowledge and combination rules. This learning approach, learn from life practice, is very interesting, which can improve students' observed ability. Teachers could purposefully collect objects with line characteristics, such as paper clip, staple, knitting yarn, wire and toothpicks etc. and carry them to the class for students to look at, painting and compare, and then ask them to talk about their difference in aspects of shape, materials, and the role.

Try to find the types of lines from a magazine illustration and see how those lines reflect the character of illustrations, such as joy, sadness calm, and other feelings then think of the lines' changes, such as thickness, length, curvature and direction.

Represent by a variety of methods and materials. Use hand to present is freely, or use rulers adding with different forces to go try. Use a brush, you can also draw a thin line, thick line, sometimes with less ink, you also can draw the clean fly white lines. Change the method of holding brush and angle of drawing to try to draw curves, fold lines, indirect lines and spiral lines, and then switch to the pen to try the same and see the difference between the two then think about what time is suitable for drawing with pen or with brush.

Today, the development of science and technology makes the shape of the product become more and more miniaturized, because miniaturized products are less subject to the internal parts and also with low-cost, cheap, lightweight, fast and other advantages. The miniaturization of mobile phones and laptop computers is the best example of the modern science and technology's role on the products' functional morphology. For consumers, the lightweight and simplicity of the forms of products display a new science and technology beauty and fashion. The Development of materials science enables designers to convey more information about the Products through products' form design and guide consumers to operate in accordance with the designed forms of humane products.

With the continuous improvement of life quality, the development trend of the role of work environment intelligence, three-dimensional communication, leisure personalized, pleasant communication, sale by chain and entertainment diversification, by social integrated factors, brings more thought and a wider development space to the humane product' form design.

People realize that except for transforming nature, the more important thing for human beings is to learn how to live harmoniously with nature. Its core is to conserve natural resources and protect the ecological environment, to make human society integrate with nature, realizing the unification of humanization of nature and the nature of humanization and sustainable development of society. The design of public services and social-economic globalization bring people with more extensive contacts and common value. The global issues and sustainable development of human society are also the representation of the "people-oriented" design concept in the new era. Facing the increasingly serious ecological and environmental problems, more and more people recognize that if the design and production only emphasize "people-oriented" principle and solely pursue human needs and improvement of living standards, it is easy to bring about negative influences.

## 54.4 Exploration of Cultivation Method and Mode

Developing design-conscious needs to be targeted, that is to help students to gradually establish correct design thinking and design concepts. However, culture methods should be varied, which requires teachers constantly to attentively think in actual teaching with bold attempt, and then to sum up. After accumulating for a long time, they will master many methods and will teach skillfully. This study provides the following specific types of training methods for teachers' reference.

Classroom DIY is a method that under the guidance and assistance of teachers, students design by themselves, which is different from directly buying products with instructions in supermarket. Students must, in accordance with the requirements, design depending on their own imagination and thinking. If it is bought in store, it also needs students to add their own design elements, such as change the shape and color, improve function or add some new elements etc. in the production process. Classroom DIY is not simply a casual handiwork. Teachers have to select a representative subject and the teaching content based on teaching objectives in the semester, which best to be close to students' life. For example, teach them to design a reusable bag, decoration design for a corner of classroom or campus and design of multi-function stationery case etc. to learn knowledge related to design by personally look, listen, ask, think and hand make. In this way, they could memory deeper, but also know how to learn and in favor of applying their designs to daily life.

One of the main contributions of basic education to design education is to improve the design quality of the whole people. However, how tastes come from? This is through appreciation. I think that appreciation is a common and important way to cultivate students' design-conscious. Appreciation is essentially the design criticism teaching and design history teaching referred here. The "Criticism" in design criticism comes from the Greek, with the meaning of distinction or identification, which contains the meaning of judgment later. The former can be said the understanding of the design works, such as analysis of the form of work, interpretation of the meaning of a work, and the interpretation of the work function, and the influence to the audience etc. while the latter is a judgment of works' merits and value. Design criticism and design history overlap in some aspects, but also have differences. In short, design criticism discusses works' form, content, function, significance and value; with this, the history of design is a discussion of the relationship between the works and its social and historical background, and explains the origin and evolution of its style, giving it a historical position. As the primary and secondary students are lack of historical knowledge, the design criticism, in basic education stage is applied to their interest and knowledge degree.

The process of design criticism teaching is description, analysis, interpretation and judgment. Teachers in discussion on key point and aspect are better to use question and answer method to stimulate students to observe and discover by their own, and encourage them to feel, think, solve problem and make judgment. Usually the most intelligent teachers are good at using question-oriented ways to make students' observation and thinking way leading to clear direction. The inquisitive way and procedure created by Sweeter and Ross are useful in appreciation teaching. It can stimulate students to examine the works, find the aesthetic character and encourage them thinking and making judgment. The following tables are instructive inquisitive ways created by them. The former is suitable for pupils, and the latter is for high school students or adults.

Design includes a broad range of different categories. In this study, the researcher only provides very simple training ideas to arouse the attention of the teachers in actual teaching to build on the idea of others. These types of design are the protagonist of modern design and are close to our lives. Although in design teaching, there are all sorts of difficulties, such as the limitations of the material conditions, students' lack of knowledge preparation and teachers' lack of knowledge development etc., they also should overcome difficulties. The following patterns are for your reference:

Observation, investigation and experience—discover questions—analyses questions—solve the problem.

In the problem-solving stage, students mainly do it by drawing working sketch, representing text or simply designing work.

Plane, three-dimensional and color constitute and the basic patterns.

In this part' basic knowledge teaching, we have accumulated more experiences, so it is easy to implement. What should be noted is that it should not do the pattern just for the pattern, or do constitute for the composition. Teachers should train students' sensitivity through basic exercises. The following are learning models for reference:

Appreciate the excellent work (starting from the elements, principles, and the overall shape)-copying masterpieces—creation—applied to real life. In the stage of applied to real-life, teachers should ask students to protect production by molding except for making homework of beautifying.

Clay sculpture, metalwork, wood carving, paper art, bamboo arts and other folk arts.

Arts and crafts or traditional crafts should be integral teaching resources in basic education and Although not every kind of arts and crafts are required to enter the classroom, teachers by means of network and multimedia could introduce students more traditional crafts' process content, so that they understand that this is part of China. Areas with good facilitates are better to combine the local resources and select one or two arts and crafts to conduct specific teaching. The reference model is as follows:

On-site observation and experience (which can also watch the video to understand the simple history and material)—simple production (familiar with material performance and processes)—thinking (the combination points between traditional folk craft of modern life), within which, the most point is thinking for students. First, they should think the crafts' generation, features and value from the past point of view, and then standing in today's perspective to think their value. They also should think about how to protect traditions and how to draw from the traditional essence to integrate into modern life then to develop the traditional arts and crafts. Though such subject or thinking is more suitable for experts and scholars, it will allow students to realize that the traditional protection and the significance of development, which is benefit for establishing awareness of protection and development of whole people.

## 54.5 Conclusion

This paper, talking about the design concept, clearly indicates that the design in the field of art refers to a kind of creative activities such as vision, planning and programs etc. from the practical and aesthetic point of view to realize certain purpose in the creation process. It, from the educational value of traditional craftsmanship and modern design as well as the cultural point of view, explains the importance of cultivating design-conscious and describes its theories and methods by classroom DIY and design criticism and teaching methods, with the purpose of let others know more about the issues related to design-conscious cultivation in the basic art education.

## References

1. Yin S (2005) Art education: wondering among idea and reality, vol 12. High Education Press, Beijing, pp 46–53
2. Liu S (2006) Design art psychology, vol 113. Tsinghua University Press, Beijing, pp 34–40
3. Yuan Xi (2003) Research on the development of Chinese art design, vol 72. Institute of Technology Press, Beijing, pp 74–79
4. He X (2001) Modern ten design concepts-the future style design, vol 22. Jiangsu Art Publishing House, Nanjing, pp 41–45

# Chapter 55
# Study of Efficient Sustainable Development Scheme of Basketball Education

**Shen Fu**

**Abstract** In recent years, the Beijing Olympic Games success to the development of sports in China has brought unprecedented opportunities for development, but also to amateur sports, sports development of the school has entered an important stage. In this paper, literature research, surveys, statistics, logic, analysis and case survey method and other methods of Jiaozuo basketball school case studies on working conditions, lack of reserve personnel for basketball, athlete education, athlete whereabouts issues proposed to adjust the layout, improve the management system; expand financing channels, trying to diversify operation mode; strengthening scientific training, and recommend strategies for sustainable development.

**Keywords** Basketball school · Sustainable development · The status quo · Countermeasures

## 55.1 Introduction

Jiaozuo basketball school basketball as a back-up personnel training schools, professional sports, as a key development in Henan Province and the State reserve talent pool of basketball, basketball training reserve personnel bear the arduous task, but also provincial and even national team for the delivery important base for high-level basketball players. With the social development and change, Jiaozuo City, basketball school in our sports talents in the form of changes in the next birth was born and in recent years to enter a broad space for development. Jiaozuo City

S. Fu (✉)
Basketball Department, Wuhan Sports University, Wuhan 430079, China
e-mail: 2430031317@qq.com

W. Du (ed.), *Informatics and Management Science V*, Lecture Notes in Electrical Engineering 208, DOI: 10.1007/978-1-4471-4796-1_55,

basketball school, on the one hand the importance of Henan Province in the country and, from their own reality, to carry out a rapid development of their own way, in the new era of the training of sports talents have made striking achievements. On the other hand, due to the transformation of social and economic transition, Jiaozuo City, basketball school is continually adapted to social changes, but also facing various problems and challenges. If you can not effectively address the development of Jiaozuo basketball school problems, it will stagnate, or even depressed. Therefore, Jiaozuo basketball school must take the road of sustainable development, we must reserve for basketball training for the existing management system innovation and development, the use of scientific concept of development, sociology theories, combined with back-up for basketball practice to the training Sustainable continue.

## 55.2 Research Methods and Study

### *55.2.1 The Study*

Jiaozuo City, Henan Province, the school basketball.

### *55.2.2 Research Methods*

Literature, expert interviews, case studies, questionnaire survey, logical analysis, and mathematical statistics.

## 55.3 Results and Analysis

### *55.3.1 Jiaozuo Basket Social Basis for the Development of the School Survey Analysis*

A society's emphasis on sports and sports atmosphere is its social basis of sport. "Tai Chi Holy Land," "National Basketball City", "National Advanced Unit of mass sports" Jiaozuo Sports has many honors [1–3]. In March 2004, Jiaozuo was officially named the State Sports General Administration of the country's second batch of Basketball City, a national basketball cities in the world 15 [4]. According to Jiaozuo Sports Bureau statistics show that: Jiaozuo City's total population 336.6 million people, regular participation in basketball activities, the population is 67.32 million, 20 % of the total population. Currently, Jiaozuo City, the district offices, large residential areas, towns, and villages have basketball court (including a small basketball board venue); rely on all levels Basketball Association, sports centers, youth homes, neighborhood primary organizations active in basketball activities. Jiaozuo City, enterprises, institutions, Hangyetixie, insist on carrying out trade union activities in basketball, the city's 100 or more units have basketball venue, and have set up a amateur basketball team, the industry within the system are organized basketball game each year (including 3-a basketball) 1 or more times [5, 6]. Jiaozuo basketball schools on this basis, not only back-up personnel in training basketball has made gratifying achievements, for a strong basketball atmosphere Jiaozuo, Jiaozuo basketball in the public schools have also made a great contribution, people establish themselves in influence, to create its own brand, vigorously develop the National Basketball reserve personnel, but also committed to universal access for the public and guiding the development of basketball and help, this is worthy of recognition Jiaozuo basketball school that is in the path of sustainable development a strong security on the road [7].

### *55.3.2 Jiaozuo Basket School Investigation and Analysis of the Development of Teaching Facilities*

Athletes training and teaching facilities are an important condition for learning, teaching facilities directly affect the quantity and quality of training athletes the quality of learning [8, 9].

Table 55.1 shows, the number from the basketball court and texture point of view, the early school Jiaozuo basket with four outdoor schools, an indoor basketball court of the land, including seven now has indoor and 4 plus the top 11, including block of plastic basketball court. The texture of the boards on wooden boards from the previous to the present glass rebounds in the quality has been greatly improved. Henan focus on training as a talent pool of basketball development, Jiaozuo has a strong basketball school teachers, adequate teaching facilities and training venues for athletes training and learning provides a strong

**Table 55.1** Jiaozuo basket survey the development of school educational facilities, now the early school teaching facilities

| Teaching facilities | School early | Now |
|---|---|---|
| Number of basketball courts | 5 (outdoor 4, 1 indoor) | 11 (seven indoor, plus the top 4) |
| Basketball field texture | Land, wood backboard | Plastic ground, tempered glass backboard |
| Athletic field | A cinder track and field | A plastic track and field |
| Dormitory | 5 | 70 |
| Canteen | 1 (only 4–5 table) | Two athletes canteens |
| Other facilities | Simple strength training equipment | A weight room, a new basketball hall |

guarantee. In such favorable conditions, the athlete's training initiative will greatly improve the level of exercise can also be the biggest upgrade, this back-up for the training of high-quality basketball talent has laid a good foundation for the basketball back-up personnel training for sustainable development provided a guarantee [10].

### *55.3.3 Jiaozuo Basket Survey on School Coaches*

Coaches in sports training activities play a leading role in the process of sports training trainers and organizers, athletes, educators and mentors. According to the survey (Table 55.2), the age structure of the coaches, school coaches basketball Jiaozuo, Henan Province, mainly young or middle age, age mainly 24–35 years, a total of 11 people who the prime of life and experience [11]. Young teacher, their energetic, pioneering and innovative spirit is very strong, is the force team coaches, basketball, Henan Province, shoulders the future of the traditional school development project.

On the coaching qualifications structure (Table 55.3), Jiaozuo school basketball coaches in preparing 11 employing 5 people, 14 were educated to undergraduate, graduate or postgraduate qualifications 2.

On the coaching structure of the title (Table 55.4), mainly in schools Jiaozuo school basketball coaches junior, accounting for 71.4 % of those surveyed, middle and senior professional titles account for 14.3 % of coaches, which shows that Jiaozuo basketball school teachers, the support structure should be optimized. Title structure should be consistent for all teachers and teacher ratio between the actual situation, meet the teachers, the quality and level of work in the capacity of the real differences, must be beneficial to all teachers to fully mobilize the initiative, enthusiasm, will help teachers at all levels of work inheritance and the rational flow of teachers, help to maintain the team's creativity and high efficiency [12].

In the coaches work attitude, according to the survey, Jiaozuo school basketball coaches have been to continue learning to enrich their theoretical and technical knowledge. It is suggested that the school's basketball coaches Jiaozuo great attention to improving the quality of their own coach, to further improve the quality of teaching.

Personal capacity in coaches, coaches focus on improving the level, often ask for high-level basketball experts, Jiaozuo basketball coaches to teach school, learn some new skills and tactical knowledge. At the same time send to the professional team coaches, basketball coaches course are advanced studies, achieve and

**Table 55.2** Structure table of Jiaozuo basketball school coach's age

| Ages | 20–35 years | 36–46 years | More than 46 years old |
|---|---|---|---|
| Number of people | 11 | 2 | 3 |

**Table 55.3** Jiaozuo school basketball coach title structure table

| Titles | Advanced | Intermediate | Primary |
|---|---|---|---|
| Number of people | 2 | 2 | 10 |
| Number of people (%) | 14.3 | 14.3 | 71.4 |

**Table 55.4** Jiaozuo school basketball coaches on the attitude indicator

| Index | Very challenging | More challenging | General | A little challenging | No challenge |
|---|---|---|---|---|---|
| The proportion | 49.1 | 38.7 | 13.2 | 0 | 0 |

domestic situation, the international basketball development situation of the times; Zhanping through the organization of lesson plans and teaching demonstration activities to improve the training of coaches and theoretical level.

High-quality coach's basketball team is Jiaozuo training school for the scientific management and the necessary conditions. Strengthen professional training and improve the quality of coaches, introduce a high level, highly educated teachers, the school is to ensure sustainable development of Jiaozuo basketball an important measure.

## *55.3.4 Jiaozuo Basket School Teaching, Investigation of the Game*

Teaching is a concrete manifestation of classroom activities, and teaching the game is a clear manifestation of the way to test the effectiveness of teaching. Therefore, to understand a school's teaching quality, teaching and competition analysis of the survey is necessary.

According to the survey results, the training school Jiaozuo basket after a reasonable time, planning, training day, each group kept at 4–5 h, respectively, for the two morning training, afternoon training twice, a morning or an afternoon training. Jiaozuo basket school athletes in the strict guidance of coaches, to develop a conscious, hard training, good practice, training in the next 8 pm, many athletes to increase their own practice until 21:00 or 10:00. However, while in training, athletes must learn cultural lessons, regular assessment, on completion of the training mission at the same time, pay attention to basketball school into a learning and training athletes to develop both the talent and adhere to both hands, one hand and training on the one hand learning, and strive to develop into athletes qualified personnel.

From the competition point of view, Jiaozuo basketball school every year to participate in municipal, provincial, and national competitions, from the survey results, Jiaozuo basketball game in municipal schools each year more than 2 times,

3 times to participate in the provincial competition, participate in national competition five times. For the contest, the Jiaozuo basketball schools participate in the competition more frequently, a higher level of competition, to achieve the competition for training purposes, in the race to improve their level.

### *55.3.5 Jiaozuo Basket Survey School Analysis of the Results Obtained*

Jiaozuo City, physical training classes and basketball schools in the grasping performance also attach great importance to talents, since 1995, and tertiary institutions to professional teams and sending more than 300 elite athletes. Jiaozuo familiar membership which many athletes, two-time Olympic champion Chen Zhong in the national team since 1995 has made the Olympics, World Cup, the championship of the "Grand Slam." There are service clubs Xue Yuyang Xinjiang Guanghui, 1994 years ago in amateur basketball training schools, and then transported to the Henan team, 2003 and 2004 twice named National Team, 2003 NBA draft, to participate in 2003 World University Games fourth. Wang Lei, physical training classes in 1999 for basketball training, selected country team in 2000, elected 2001 National Youth Team in 2002, transported to the Henan team, named the National Team in 2005, has represented the national youth team to the United States, Greece, Japan, Australia game in 2003, the Asian Youth Basketball Tournament was the first in 2006, Chinese Star team selected to participate in Korea CBA Stars Competition, the national team in 2007. PAN Li, Henan active team. Who have been selected for the National Team in 2004, the Asian Youth Basketball Championship in 2005, the Asian Women's Basketball Championship winner?

Active membership of Jiaozuo, Henan team of male and female basketball players are more than half. Membership in 2005, Jiaozuo, Henan basketball players as the main force of the Tenth National Games in fifth place finish made to achieve a historic breakthrough in Henan. In the CBA, NBA, WCBA arena, we can see, like Liang, Cui Lei, Wang Jing, Cai Jinchuan, a large number of outstanding athletes Jiaozuo membership.

With excellent performance, a large number of transport, Jiaozuo City in 1998, the State Sports General Administration of CBA, Chinese Basketball Association, Jiaozuo City, amateur basketball schools named as "National Youth Basketball Training of key units"; named in 2002 in Henan Province Sports Bureau Jiaozuo City, as a "basketball training base in Henan Province"; 2004, Jiaozuo City, State Sports General Administration named "National basketball city." Jiaozuo City in January 2009 school basketball officially approved by the State Sports General Administration of high-level sports reserve talents for the country base.

### *55.3.6 Jiaozuo Basket Survey on School Funding*

Basketball school construction and development is needed as a guarantee of adequate funding, Jiaozuo basketball school funding comes mainly from sports athletes, parents and the system of special grants to support other aspects.

Survey, Jiaozuo, Henan school basketball all the special section for the municipal finance. Jiaozuo City, the province's primary physical training class was the only one unit of professional basketball players, male and 24 women's basketball team prepare athletes, male and female 12, a full supply of financial funds from the government. City fiscal year allocated to physical training classes and the city amateur basketball schools in the city's normal business expenses amounted to more than 90 million. The first is to protect the school's daily operations and staff salaries sources. The athletes training funds is made by the Sports Council for the full allocation. Basketball Sports Bureau to give the school funding mainly from the sports lottery revenue, but because sports lottery revenue distribution and the amount of uncertainty, resulting in the amount of money are quite different, sometimes the implementation of this funding to allocated to be honored by sports equipment, so the sports system of funding is limited. Jiaozuo basket school athletes, parents bear the expense of the competition part of the cost, although only a small part of the funding, but it is also seeking funding school basketball on the road a try. In addition, the basketball schools can seek the help of some enterprises, access to corporate sponsorship funding, which is another source of funding sources.

Survey showed that although there is no school Jiaozuo basket there is a big lack of funding, but to ensure the sustainable development of the school basketball, it is necessary to protect the athlete's training and competition expenses, training, or not allow athletes to give up the game less to save money. Want to have adequate funding; adequate attention must rise to the leadership.

## 55.4 Conclusions

Jiaozuo City, Henan Province basketball school funding earmarked mostly based Sports Bureau, the proportion of social forces involved in sponsorship is too small. Jiaozuo school basketball schools should actively expand financing channels, collecting funds from society, to diversify operation mode, open up a new path of development.

Jiaozuo, Henan Province, the school basketball coaches of the age structure and job structure is more reasonable; teachers can be equipped with basic competence in peacetime training.

Jiaozuo school basketball coaches the educational structure is reasonable, beneficial to the school's teaching and research and sustainable development.

The use of existing national laws and policies, and actively seek government support to seriously implement the sports regulations, the training of sports talents to provide legal protection. To establish a large talent, firmly to the "body, and Education with" road; sports schools should actively explore the market economy, multi-channel form of education; gradually establish and improve the exchange of sports talents in the market; all levels of government should increase the body investment in schools, improve school software and hardware facilities, improve treatment of coaches, and further training of reserve personnel to mobilize local enthusiasm. Improve the training venues "appearance" and create a harmonious, beautiful and human environment, and dig and promote the unique charm and basketball value, enhance the affinity of basketball players. Improve the coaches job training system, deepen the coaches job training in order to "competency-based" as the value orientation training to effectively improve the practicality of using a variety of flexible forms of training to improve training, increase job training the policy of law enforcement work to optimize the structure title coaches, coaches to improve education, higher education is its own structure. Improve the athlete education system, strengthening the cultural education of athletes, young athletes to stimulate awareness of active learning, develop learning abilities of athletes strengthen the scientific selection and scientific training, training of sports talents to improve the quality and efficiency, scientific selection work to improve the training of sports talents in the success rate, ready to take sports development patterns and trends to develop in line with project development patterns and trends in back-up personnel training plan. Artist-centered, youth sports training to enhance the medical supervision.

## References


1. Lu Y (2000) China social physical education, vol 01. Beijing Sports University Press, Beijing, pp 89–92
2. Liu Y (1999) State administration of sports basketball management Center project the national basketball city, People's Sports Publishing, 22–28
3. Xu F et al (2006) "Chinese basketball city" status and management strategies for sustainable development. Fuzhou Univ 1:29–34
4. Jiaozuo physical culture "Jiaozuo sports records," Editorial Department. Jiaozuo Sports Chi (1902–1985). Zhengzhou: Henan People's Publishing House, 1992. [9] Cao Rongjun. Emotional teaching theory elective in college basketball Teaching in the application of .2008 Northeast Normal University, a master's degree thesis
5. Shu Z, Dong L, Li Y et al (2003) China's social transition. Training of sports talents and sustainable development of Beijing Sports University Press 10(1):78–85
6. Duan S (2005) Preparations for the 2008 Olympics. 2005 winter training-cum-mobilization speech to the congress 11(22):139–145
7. Xie X (2008) State sports general administration compiled the reform and opening up 30 years of Chinese sports. Beijing Sports University Press, 36–41
8. Shu Z, Dong L, Li Y et al (2003) The social transition. China's sports talents cultivation and sustainable development of Beijing Sports University Press 10(1):82–85

9. Yang H (1999) Students interested in sports and the way the basic principles. Chin Sch Sports 19:33–38
10. Jin Q School of physical education. Higher Education Press
11. Zhou YY (2012) Service quality measurement at university's libraries by AHP method. IEIT J Adapt Dyn Comput 4(2):1–4
12. Yu J, Zhao Y (2012) Weighted approximation of functions with singularity by q-Baskakov operators. IEIT J Adapt Dyn Comput 4(2):5–11

# Chapter 56
# Art Practice Research in Vocal Teaching

**Zhenli Shi**

**Abstract** This paper mainly discussed the role and significance of art practice in the vocal teaching, and from the testing and improving two aspects of artistic practice to vocal teaching, detail analyzed the necessity of the combination between the two, for the college vocal teaching further development of artistic practice to offer a few ideas.

**Keywords** Music · Vocal teaching · Art practice

## 56.1 Introduction

The music is one of the most beautiful human art, not only beautify the living, but also increase the level of the spiritual world, so, a long time, the music has always been the endless pursuit of human. Therefore, the music teaching has run through the human history, endless [1]. Today, with the improvement of living standard of people in China, the love to music is unabated [2]. With the portable music audio-visual equipment and the developed network, the phenomenon was stimulated, involved in more and more people in the music, and music teaching has received widespread attention of the whole society. In the composition of music education, vocal teaching occupies a very important position, after all, the voice is the most wonderful sound god has given people, and it's easier addictive to grasp and use of it. How to improve the level of vocal teaching? This article believed that artistic practice is very important, which a test and improvement to vocal is teaching [3].

Z. Shi (✉)
Langfang Normal College, Langfang, China
e-mail: zhenlishi123@126.com

W. Du (ed.), *Informatics and Management Science V*, Lecture Notes in Electrical Engineering 208, DOI: 10.1007/978-1-4471-4796-1_56,

## 56.2 Artistic Practice Test the Level of Vocal Teaching

Over the years, the old generation of vocal educators in college vocal teaching devoted a great deal of effort to try to turn our vocal education with international standard, to train a large number of specialized personnel to fundamentally promote the development of music teaching in China, under the guidance of such a great goal-based vocal teaching standard, our vocal teaching has indeed made great progress, many different kinds of teaching styles and methods are constantly emerging, the development of vocal teaching reform is still changing [4]. Almost every college in the domestic vocal teaching has very good experience and methods, but how the effect of these experience and methods are and whether students can achieve their desired goals, we need to practice [5]. Only experience the continuous test of time, students can find their true level, and thus continuously improve.

### *56.2.1 Reasonable Adjustment of Vocal Teaching Through the Practice*

All forms of music should be continuous improvement in practice, all artists are also grew up in constant hone, so we will find that in teaching this or that problems through the process of artistic practice, and then further adjust. Some teaching method problems can be corrected in the feedback of practice, some students problems need to discover problems and continuously self-improve through practice [6]. No practice, a lot of theories just stop at the written level, as an armchair strategist, the effect is not good, and students can not found their own problems through practice, has never been able to be excellent, after all, no one is perfect, to achieve the perfect level only continue to practice, otherwise never been able to become an artist [7]. The two are complementary; we can not deliberately attach importance to certain aspect. The way of their perfect unification is practice. Xiaoping Deng once said that the practice is the sole criterion for testing truth, so, too, in music teaching. Constant practice and exploration, to find the most suitable teaching method, to achieve individualized, is the ultimate goal of all teachers and students [8]. For example, if a student can not play well in practice because of tension, we can adjust his mentality, strengthen the number of tracks browsing, and increase the opportunity to practice to provide a more scientific way of teaching [9].

### *56.2.2 Stimulate Students' Musical Potential*

Potential is a very special phenomenon, human potential seems infinite, but not necessarily translate into ability. We recognize the ability, but not recognize the potential, so trying to translate potential into ability should be seriously faced by everyone. In the process of vocal teaching, trying to explore the potential of students in music, and enhancing the musical expression is a very important part of vocal teaching [10]. All beginners at the beginning can not maturely use skills, most people may even be struggling with it, such as breathing which is most common in vocal teaching, the location of the breathing is deep or shallow, and aspiratory intensity is large or small, beginners is difficult to accurately grasp. In the learning process, via teacher's guidance and help, many students are more intelligent, and they will master the skills faster, some slower, but slower students may not have small potential, just in practice, this advantage is built up, the potential will gradually form ability, and finally to improve students' confidence, their music performance ability will be played and the learning result will be very ideal naturally.

If this potential is divided, then there can be many different performances, such as the self-adjust ability in the big play. A lot of people paid enough effort in the process of learning vocal, every day, "u, a, u, a" vocal, this process is very boring, but after spend months and years of practice, the result can be good, but in the key play, often played disorder. In fact, vocal has high requirement to the control of the stage, because vocal is a kind of moment art, at this moment, to throw oneself into might be a timeless classic, if gingerly, then out of tune is very normal, which reflects the importance of practice. Practice more and become more familiar with this kind of atmosphere, capacity will be getting stronger. For example, the music is to be creative. Art comes from life but higher than life, music is an important form of art, it is indivisible with life. Only the vitality of the music coming from life will become lasting. Many well-known musical works were coming from life, such as the Yellow River Cantata, the Osaka City Girl, and so on. In the creation or discovery of these works, the artists experienced enough effort even suffering, finally the inspiration emerging in the process of art practice, and then created these popular works. If not his own painful experience, blind Bing could not produce such classic work the Reflection of the Moon. From music development history, through constant practice and improvement, some minor recited casually in primitive society have become many ethnic minority well-known folk songs today. Another example is the opera which is our quintessence, and which is Chinese opera in the eyes of foreigners. The constant practice of art gave wings to music to take off, and this creativity was across history and national boundaries. In the continuous development of modern society today, we continue to create more perfect music. Continuous exchange of domestic and foreign makes music has more elements, so the music has a broader space for development in the context of social progress.

Potential of music is infinite, and each person's potential is also infinite. Only constant practice can be as much as possible to dig such infinity. Music will develop, and each person will develop too. The workers engaged in music teaching should give students enough performance and creativity, and the best way is to do art practice constantly based on hard exercise, so that students can find out the most suitable way and method, and ultimately in this process to realize their own things, and music will have a new way and burst out its own unique artistic charm.

## 56.3 Art Practice is Benefit to the Improvement of Vocal Teaching

### *56.3.1 Art Practice is a Reflection of Students' Comprehensive Ability*

Vocal learning is a long insisted process, and after continuous efforts and dedication, students will get a variety of artistic practice opportunities. Constant art practice can change the students' past thinking inherent errors, and increase their interest in learning, as well as the ability to reflect the social and art, and also exercise the ability to cooperate with others. Continuous art practice makes students' self-orientation clearer, the gap with others also clear, and then makes students more proactive to learn and practice and the learning initiative of students can be enhanced. This is a kind of ability that contemporary college students generally lack. In practice, students also found that the strength of an individual is always limited, and only the cooperate with others can get the best result, such as a show, there is no common efforts it can not be done, the actors need cooperate with the props man, the host and other actor. A possible perfect show still needs good lighting, sound, stage, etc., to support, and in this process, the students' ability to adapt to social is exercised. Similarly, the students learned a lot of social knowledge in art practice, this knowledge is not in the books, helping them better understand the society and integrate into the society.

### *56.3.2 Art Practice Can Enhance the Level of the Vocal Teachers*

Classroom teaching alone is boring for students, teachers also need some new platforms to enrich the content of their teaching, and art practice is a very good one. Because of combining with theatrical performance, teachers can express ideas and content more specific, although they can not make every student success, but the students can make use of this platform to seek and find their own learning motivation. Conversely, there is a relatively high demand for professional teachers, and teachers must show their true ability to teach students, if lack adequate practice basis, students can not convince, which will directly impact teachers' evaluation and curriculum learning attitude. If the teachers perform well, in the learning process students will not only work hard to master the skills, but also inspire new motivation for learning with teachers as an example, then the quality of learning will increase. In this aspect Music College of Dalian University has done well. The university has a clear prescription that teachers must have excellent professional skills, sufficient academic research and performance practice experience, and have achieved certain results; such can only ask and guide students. From the students' point of view, in addition to normal school, students need regularly participate in observation of professional class and actively participate in a variety of art practice. By the stage practice, the students' understanding of musical expression is more deepened and they can find their own inadequacies from teachers, forming a good learning atmosphere.

### 56.3.3 Foster Students to Enhance the Stage Control Ability Constantly

There is a saying that "Knowledge comes from practice". The ultimate goal students receive education is facing the society and recognized by the society, so in schools students should actively cultivate their own ability to respond to a variety of emergency situations or unforeseen circumstances, which requires both solid basic skills as a foundation and constant practice to adjust their attitude. The ability to control the stage will have a direct impact on the evaluation given by others, so in order to have this ability; students must participate in art practice constantly. Of course, firstly the profession is basis, without a good professional basis it is difficult to have excellent stage performance, which has become a social consensus. To take singing as an example, only had a solid basic skill, we can sing the mind state of the song, and seize the heart of audience. If not, but only exaggerated surface, it is difficult to continue to develop on the stage in future. The second is a good attitude. Some people said that attitude decides everything, especially on the stage, only calm and natural we can better play our basic skills, so we must have good attitude, and diligent practice is necessary. From this aspect, outstanding vocal talent must have received a lot of social practice, and the fact is that. The growth of many artists started from the bottom of society, and a lot of artists had worked as a lounge singer. Of course, not to say that students must practice in society, and undoubtedly the effect is better in school.

### 56.3.4 Mutual Promotion of Vocal Teaching and Art Practice

Vocal teaching has many contents with entry, intermediate and advanced, the more advanced the higher perception requirement to students, and the perception needs practice participation. Vocal teaching can better guide the art practice and art practice can better promote the level of vocal teaching, so vocal teaching is inseparable from art practice, and art practice can not be separated from the vocal teaching, both mutual promotion and common for the students' development. In vocal teaching, vocal teachers also should give corresponding art practice guidance for different level students. After all, the comprehend realms of different level students are not the same, but students can refer and learn to each other under such an opportunity, getting some inspiration. Knowledge comes from practice. To increase the students' vocal learning interest in continuous art practice, to feel teaching content in practice, and to figure out new thinking constantly, reaching deep-seated art realm, which is absolutely difficult to obtain from the classroom instruction. It has a very good result to improve the students' initiative and to improve their self-confidence, better guiding the students the learning of vocal teaching courses.

## 56.4 Conclusion

In summary, the close relation of art practice and vocal teaching is indivisible. Art practice not only enriches the teaching method, but also gives students rare exercise opportunity, so that they can more clearly understand themselves, and understand their own deficiencies, and find the gap with others and make improvement, at the same time learn how to get along better with others, better integrating into the society. The particular value is that in the process of art practice many students self-reflect and summary, and eventually forms their own unique style, and has the potential to create a new art form, which is commendable. Hope that in future vocal teaching, including other music teaching can enhance students' art practice, and continue to enhance the musical career of the motherland.

## References

1. Guo J (1996) Vocal trunk lesson basic theory, art practice and total thinking of trinity. Musicol Chin S1:8–17
2. Xie L (2003) The shape elements of the music creation. J Gansu Norm Coll 01:20–27
3. Ma W (2007) Practice and thinking of music education art practice approach. Exam Wkl 47:46–53
4. Wang P (2007) The value of series Qin Hu in art practice. Jiaoxiang-J Xi'an Conservatory Music 04:120–127
5. Zeng X (2009) Diversity: vocal music teaching reform and exploration. J Chizhou Coll 01:24–32
6. Xue L (2009) Music professional piano accompaniment course of higher vocational college. Art Educ 10:140–147
7. Fang AL, Li S (2009) How to accurately grasp the singing style of works. Mod Sci 24:37–42
8. Zhang J (2009) Erhu teaching status and improvement. Vocat Education Forum S1:361–367
9. Zou L (2009) Vocational college's music performance profession art practice curriculum reform. Educ Vocat 21:37–45
10. Li M (2010) Kindergarten teachers music teaching reform exploration. J Liaoning Teacher Coll (Soc Sci Edn) 01:67–73

# Chapter 57
# Research on Graduate Ideological and Political Education

**Lijun Xie and Zhongzhi Han**

**Abstract** With the popularization of higher education, graduates become an important part of university. Moreover, strengthen the ideological and political education of graduate students will be the focus of the university ideological work. This paper finds an effective method to improve the graduate ideological and political cultivation by analyzing both positive and negative impacts of networks to the ideological and political education and the current problems. By using the actual situation of graduate students' network using and doing entire front online and offline ideological and political education, it makes the network more effectively serving the graduate students' ideological and political education system and becoming an indispensable part.

**Keywords** Graduate student · Ideological and political · Networking

## 57.1 Introduction

Since the beginning of the twenty first century, a number of college students increases every year in China [1, 2]. At the same time, China has stepped up the intensity of reform and opening-up, and foreign ideas flooded in. Under this background, to strengthen ideological and political education in schools has become an inevitable trend [3]. With the raising technological level, the networks and computers become daily essential learning tools, and students communicate with network more than teachers [4]. How to use the network to carry on

L. Xie (✉) · Z. Han
Agricultural University of Hebei, Baoding, China
e-mail: zhongzhihan123@163.com

W. Du (ed.), *Informatics and Management Science V*, Lecture Notes in Electrical Engineering 208, DOI: 10.1007/978-1-4471-4796-1_57,

ideological and political education of graduate students is the main point of ideological and political education innovation and reform.

## 57.2 Network Positive and Negative Impact to the Ideological and Political Education

At present, people are inseparable from the network everywhere in learning, working, and playing [5]. The network has become part of people's lives. To the students of universities, the network has influenced their sense of worth, philosophy, opinion of the world and the political orientation.

### *57.2.1 Network Positive Impact to the Ideological Education*

The biggest advantage of the network is the sharing of resources, leaping constraints of time and space, and passing people's minds equal and free. Moreover, making people's ideas and political concepts collide with each other, and come up the change, which brings China's ideological and political education new opportunities and challenges.

#### 57.2.1.1 Abundant Network Content

After network becoming our learning tool, where we now go most to find the information is no longer a library or reference room, but the Internet. Currently, we can use the network to know and master a variety of political events occurring throughout the world at the very first time [6, 7]. Moreover, it can supply many realistic materials for enriching and improving the Marxism-Leninism, Maoism, Deng Xiaoping Theory, Three Represents and harmonious society. The continue engines improvement of modern web search makes people find and search for information through the network more convenient. Moreover, the continued expanding network information is becoming our inexhaustible storehouse of knowledge.

#### 57.2.1.2 Internet Changes the Way of Education

The birth of the network makes people's lives change enormously, while education also occur a corresponding change. In basic courses and specialized courses teaching, the traditional spoon-fed, infusion education model changes into guiding, cooperation, excited-style education model. In the classroom, the simplest use of the blackboard and some of the mold show has changed in colourful multimedia courseware, animation show, video explaining, and various methods, which greatly stimulates the students' senses, and mobilize the enthusiasm of students. The ideological and political education also requires corresponding changes. It can not only simply read the documents and spirits conveyed from higher authorities as previously, but also use the network, and watch the country and the Party Central Committee meeting by the way of webcast to understand the spirit of the meeting.

#### 57.2.1.3 Network Can Broaden the Horizons of the Students

The network provides people a more free, open and equal communication platform. Students can speak out freely on the world stage, give the free rein to their imagination, and discuss their behaviour as their wishes, strengthen the exchange between worldwide students.

### *57.2.2 Network Negative Impact to the Ideological Education*

Everything has its own two sides. When we see its favourable side, we also found its adverse effects. The network has played a positive role in the ideological and political education, but at the same time, we must consider to its shortcomings and give students timely guidance. The negative effects are as follows.

#### 57.2.2.1 Network Shock the Value, Life and World View

The freedom of the network makes people's mind more open. The countries have their different cultures, regions, values, and religion, so they form into different ideology. We carry forward the traditional culture at the same time to inherit and carry forward of Marxism-Leninism. Meanwhile, some dross of the foreign culture was spread. Although the thinking of the graduate is relatively mature and has certain cultural foundations, the social experience relatively small and resistance is weaker, so they are susceptible to the different ideas and culture.

#### 57.2.2.2 Decadent Culture Invasion

Cause the social experience of the graduate students in the school is relatively small, through the network to communicate and exchange made students lack the sense of moral responsibility and historical mission. Some sites in order to earn more click and profits, they use the unhealthy information and junk culture to attract graduate students, and these are very unfavourable to the graduate students. Graduate students in the invasion of this decadent culture and ideas, it is easy to lose their selves. As time passed, addicted to the virtual networks cannot extricate themselves.

#### 57.2.2.3 Networks Affect the Students' Communicative Competence

In the past, communications between students through a variety of beneficial activities organized by the class or school. With the popularity of the network, more and more students make friends and communication through the internet, such anonymous exchanges, many of which are built based on deception. People hold a sense of playing games to begin the communication. If things continue in this way, forming the serious mental illness of apathy, numbness, but they do not know. Into the community, the normal communicative competence is not enough, not capable of the tasks entrusted by the enterprises.

## 57.3 The Problems of Networking Ideological and Political Education

### *57.3.1 Educator and the Environment Needs to be Improved*

At present, many colleges and universities offer online education model, but its website and pages making exists many problems. For example: less personalized service, rigid webpage making, and slow update and so on. For the graduate students in the school they have learned skills and ability, this kind of lack of affinity and do not attract online education is dismissive to cause the ideological and political education network useless. Meanwhile, underestimated for the security of the network environment provides chances for many unhealthy resources. Unhealthy resources take advantage of the vulnerability of online education platform for illegal transactions and publicity, and make the online education even worse.

Currently, many colleges and universities political and ideological educator do not pay enough attention on the network, and educators' self cultivation is also inadequate. They consider that the inner quality of graduate students is relatively high, and a mention is enough, so despising the ideological and political education. They latter know and latter sense on domestic and foreign affairs, can't correct problems appearing in students at the first time, and give the graduate a feeling that ideological and political education teacher is a dispensable role, which makes them despise ideological and political education in mind, while the educator's authority is lost either.

In the management of the school network, schools usually divided every department as a unit, the units act on its own will, and lack of communication between each other. On the network resources management, failed to reach a uniform standard, cause a great waste of resources. Inside the campus, most of the competent departments do not give enough importance to the software, more money invested into the hardware, the network of education to carry out the formation of significant constraints.

### 57.3.2 Online Education is Lack of Innovation

At present, many ideological and political education paper at the undergraduate and graduate students are published in each year. However, there has rarely a quality paper, especially the online education. This can be seen that the school's ideological and political educators lack of attention to the network.

The network is a big vat which inside have many colours. For the young man, the resistance of toxic information and culture is relatively poor. When they graduate arises ideological and political aspects of the deviation, it is important that how timely for educators to make the correction. At present, the public security organs through the "fight against pornography, illegal publications and piracy" fighting to turn off a large number of harmful sites, but this is far from sufficient to solve the problem. The current network is an open system and there are always a small number of people uses the network loopholes and poor supervision of illegal distribution of unhealthy content. This requires educators to ensure the safety of the campus network and student psychological counselling timely.

In the network mode education, refers to the tradition of instilling education, there is a lack of innovation. Not to reflect the breakthrough time and space characteristics of the network, just transferred materials from by writing on the blackboard and paper to the network and no effects are also evident. Ideological and political education is not a boring thing and it requires a variety of means shown. Therefore, graduate students at the same time obtain a new current affairs and concepts. Moreover, it has been able to interact through the network, to express their opinions and views, educators can through these published opinion to understand the graduate ideological and political trends, in order to further make sure the work direction.

In the online virtual world, all the kinds of information emerge in endlessly. Some media in order to earn click-through rate, exaggerated and fictitious facts. The young people are full of vigour and vitality and easily excited and they are lack of ability to divide the difference between authenticities of the information. Some people with ulterior take advantage for young people of the so-called justice, and cause some harm to society.

### 57.3.3 Online Education Research Results is Very Little

Network technology is constantly updated and the new network product emerges in endlessly. However, we can find out the research results for ideological and political education network, all these years does not appear. It is partly because of less attention of a combination of education and networking. On the other hand is the lack of the corresponding force. Networking, communication, education, psychology, ethics and other subjects cross together, but its scientific researching strength is lacking.

On the net, to the current commentaries on current affairs, although many colleges and universities have studied-teacher exchange platform, students leave messages and the teachers rarely provide the answers.

In the aspect of research outcome, some universities have also made a related study. Only focus on the result of a subject while the research and the results show too superficial. Its lack of depth and breadth in the study solves practical problems.

## 57.4 Solutions of Ideological and Political Education Networking

### *57.4.1 Amplify Campus Network*

For schools, it is the primary problem to ensure the safety and health of the campus network. From a technical perspective, installing firewall in the entrance of the campus network is a must, but network security is relative, and there is no absolute security network. Increasing the monitoring of the campus network, discovering problems in time, reporting and resolving them is what network management personnel should do every day. At the same time, China's relevant laws and regulations make the relevant provisions for the safety of the campus network, which provide a legal basis and guarantee for the campus network.

#### 57.4.1.1 Increasing the Network Capital Investment

Much of network intrusion is due to the aging of equipment and personnel management not in place. So increasing the capital investment in the campus network, updating the network equipment, training the network management can solve this problem. Especially to the campus ideological and political education website, it should pay special attention to the viruses or invasion of foreign illegal personnel and software development efforts must keep on pace either.

#### 57.4.1.2 Strengthen Internal and External Cooperation

To the internal colleges and the departments, they should strengthen communication of ideological and political work between each other and make plans to solve problems among students at a time commonly. For the special cases, nip an evil in the bud, stifling small problem, and small events at the beginning to stop them from magnifying. At the same time, convening the ideological and political education, exchange meeting on time among the universities to explore the social hot issue. By these, it can accurately position ideological and political education of graduate students.

### *57.4.2 Establishing Ideological and Political Education Network Platform*

Use the trait of students' long time net playing, by establishing ideological and political education network platform with its own unique style. Moreover, use the healthy, safe ideological and political concepts to guide the students, it can make students construct a correct life outlook, world outlook and sense of worth. At the same time, enrich the network services as far as possible on the platform. It can make students understand domestic and foreign affairs from different levels by playing real-time large-scale conferences, broadcasting expert seminars.

The final act of a person, in addition to the external forces guide, it also needs the inner self-cultivation. Making students find a moral foothold in the learning process, thus forming a strong inner self-discipline and sense of honour, it make students can discard the false and retain the true, distinguish right from wrong.

To strengthen ideological and political propaganda, especially to the postgraduate in the schools, it has not just to learn and improve their own knowledge. But more importantly pay attention to the overall quality and country's politicians. Since reform and opening-up, each thought continuously flood in, and graduate students as a higher intellectual should discern affair more. So using the Internet to increase guiding of students' thinking, and avoiding the invasion of foreign decadent ideas is essential.

### *57.4.3 Introduce the Course of Ideological and Political Education Network*

At present, specialized courses and theory lessons already have a large number of network courseware. There is almost no network tutorial for the ideological and political education except philosophy courses.

By network multimedia resources, it riches ideological and political education system, improves students' learning pathways, and makes students change from the previous spoon-fed learning into active and cooperative learning.

### *57.4.4 Strengthen the Building of the Education Team*

The network is a virtual world, but the students' education status is real. Society is constantly changing to meet the needs of the times, while it is a must to solve the problem of strengthening the construction of the education team. To educate on thought, it needs to make educators understand the importance of ideological and political education firstly, and then can truly serve the students.

## 57.5 Summary

This paper researches the graduates' ideological and political education in networking. First, analyzes the pros and cons of network to the ideological and political education. Secondly, focuses on the problems of the networking ideological and political education. At last, provides the solutions. Due to space constraints, the solution needs further explore and study. Hope this article can provide readers some help.

## References

1. Fengjin Z (2008) The thinking of ideological and political education that under the internet age. Sci Technol Inf 19:162–164
2. Jifeng W (2005) Analysis of the relationship between Ideological and political education and the reality education. Theory Monthly 23:12–14
3. Weiping Z (2007) Use the internet to do the university ideological and political work. J Agric Univ Hebei 3:43–44
4. Shoumeng W (2004) The plus-minus effect and strategy of networked to the ideological and political education in colleges and universities. Seeker 1:120–122
5. Xiaoxia W (2006) The existed work problems and strategy during the ideological and political education in colleges and universities. J Inner Mongolia Norm Univ (Educ Sci) 3: 45–47
6. Jintao H (2005) In the national work conference on strengthening and improving the ideological and political education of college students of speech. People's Daily 1:19
7. Qigui L (2006) New exploration on the moral education work in colleges and universities in the network era. Chin Adult Educ 2:56–58

# Chapter 58
# Research on Determination of Weaknesses of Research Quality Management in Universities

**Wang Weiguo, Wang Kai, Wang Shuai and Mou Pengbo**

**Abstract** At present, the research on college scientific quality management system was only limited to effectiveness evaluation, while the quantitative determination of the weaknesses was less considered. It would cause that the persuasion of research quality management was not strong. Based on this situation, the model determining weaknesses was established based on extendable optimal degree. Through the introduction of eligible degree, learn the principle determining the number of principal components in the principal component analysis, the formula for calculating the cumulative contribution rate of weak degree of the index was proposed. Numerical example shows that the weaknesses of the quality could be determined scientifically. Finally, the application prospects of the method were looked in related fields.

**Keywords** Research quality · Weaknesses · Extendable optimal degree · Weak degree

## 58.1 Introduction

The extendable optimal degree method, created in 1983, is an original theory which is jointed of many subjects and broad application fields [1]. It studies rules and methods to solving problems from the views of the quantization and the

W. Weiguo (✉) · W. Kai
Beijing University of Science and Technology, Beijing 100041, China
e-mail: chaoyang1974@126.com

W. Shuai · M. Pengbo
Shijiazhuang Mechanical Engineering College, Shijiazhuang 050003, China

W. Du (ed.), *Informatics and Management Science V*, Lecture Notes in Electrical Engineering 208, DOI: 10.1007/978-1-4471-4796-1_58,

quantification, the other said that is the qualitative method of matter-element extension and the quantitative methods by the correlation functions to calculating.

In extendable optimal degree theory, eligible degree is degree of the index factor to meet the requirements, and it joins up qualitative and quantitative index and different index metering ways by correlation functions to uniformly calculate. The main cases had been made brief introduction as the below [2].

Defining 1 (pitch), the pitch from any point x in real field $(-\infty, +\infty)$ to an interval $X_o = <a, b>$ is:

$$\rho(x, X_o) = \left|x - \frac{a+b}{2}\right| - \frac{1}{2}(b-a) \tag{58.1}$$

Defining 2 (place value): if $X_o = <a, b>$ , $X = <c, d>$ , and $X_o \subset X$, the place value about $x$ on $X_o$, $X$ is:

$$D(x, X_o, X) = \begin{cases} \rho(x, X) - \rho(x, X_o), & x \notin X_o \\ -1, & x \in X_o \end{cases} \tag{58.2}$$

Based on pitch and place value, this article put forward the establish way of correlation function using qualitative and quantitative index [3].

For quantitative index of optimization point in interval midpoint, then

$$k_i(x) = -\frac{\rho(x, X_{oi})}{|X_{oi}|} \tag{58.3}$$

where $k_i$ $(x)$ is eligible degree of the $i$th index, x is index value, $X_{oi} = (a,b)$ is optimal regional which is measurement of the $i$th index optimal value, $|X_{oi}|$ is the interval length [4].

For quantitative index of the optimal point in right endpoint of interval, then

$$k_i(x) = \begin{cases} \dfrac{\rho(x, x_o, X_{oi})}{D(x, X_{oi}, X_i)}, & D(x, X_{oi}, X_i) \neq 0 \\ -\rho(x, x_o, X_{oi}) - 1, & D(x, X_{oi}, X_i) = 0 \end{cases} \tag{58.4}$$

$$\rho_r(x, b, X_{oi}) = \begin{cases} x - b, & x > b \\ a - x, & x < b \\ a - b, & x = b \end{cases} \tag{58.5}$$

where $X_i = (c,d)$ is joint domain which is the whole value range of the $i$th index.

If $X = (x,y)$ is fuzzy quantitative interval of the qualitative index, $X_{\mathrm{oi}} = (a,b)$ is optimal regional, $X_i = (c,d)$ is joint domain, then

$$k_i(X) = \frac{\rho_{\mathrm{I}}(X, X_{oi})}{D_{\mathrm{I}}(X, X_{oi}, X_i)} \tag{58.6}$$

$$\rho_{\mathrm{I}}(X, X_{oi}) = \frac{1}{2}(\rho(x, X_{oi}) + \rho(y, X_{oi})) \tag{58.7}$$

$$D_{\mathrm{I}}(X, X_{oi}, X_i) = \begin{cases} -1, & X \subset X_i \\ \rho_{\mathrm{I}}(X, X_i) - \rho_{\mathrm{I}}(X, X_{oi}), & others \end{cases} \tag{58.8}$$

where $k_i$ $(X)$ is eligible degree of interval $X$ about interval $X_{\text{oi}}$ *and* $X_i$

## 58.2 Determinations of the Weaknesses

### 58.2.1 Weaknesses Analysis

Weight method and index space method were the main methods to determine weakness in the past [5]. In this article, we ensured weaknesses which involved weight method and index space method should be priority improved.

### 58.2.2 Weak Degree Determined

For the eligible degree $k$ of index factor optimal regional, this article made the following provisions: if $k$ was positive value, the actual value satisfied optimal regional, if $k$ was negative value, $|k|$ meant the length of actual value to optimal regional, greater the $|k|$, more improve the actual value.

Based the theory above, a formula had been put forward for calculating weakness $r_i$:

$$r_i = \begin{cases} \alpha_i \times |k_i| & k_i < 0 \\ 0 & k_i \geq 0 \end{cases} \quad i = 1, 2, \ldots, m \tag{58.9}$$

where, $\alpha_i$ was the weight of index $i$, $|k_i|$ was the absolute value of eligible degree of index $i$, m was index factor number.

The formula indicated that weakness involved weight and eligible degree, if the value was greater, index factor should be more improved.

### 58.2.3 Determined Weaknesses

Consulting the Ref. [6], this article put forward a method determined weak factor but first, we put forward a concept—weak degrees accumulation contribution.

$$\alpha_p = \sum_{i=1}^{p} r_i \Big/ \sum_{i=1}^{m} r_i \tag{58.10}$$

where, $m$ was index factor number, $p$ was weak factor number.

Index factor was ranked by the size of weak degrees: $r_1 \geq r_2 \geq$ \cdots $\geq r_{\mathrm{m}}$. If the first weak degrees contribution was greater than 80 %, only took the first as the weak factor, if the first weak degrees contribution was less than 80 %, according to accumulation contribution $\alpha_p$ size, took the first $p$ index factor when $\alpha_p$ was greater than 80 % as the weak factor.

## 58.3 Case Analyses

Taking the process of a domestic college research quality management system for example, using the theoretical approach in this article analyzed weakness.

### 58.3.1 Evaluation Index System Construction

Consult the Ref. [7], the indexes architecture can be obtained, and the results can be seen in Table 58.1.

### 58.3.2 Determined Index System Eligible Degree

Consulting the AHP method of the Ref. [8], index system had been empowered, asking experts scoring for each eligible degree between two factors by matching

**Table 58.1** Evaluation index system of college research quality management system validity

| Total objectives | Subgoals | Attribute objective |
|---|---|---|
| Quality management system validity $X$ | Management responsibility $X_1$ | Customer as focus $X_{11}$ |
| | | Scheme $X_{12}$ |
| | | Responsibility, power and communication $X_{13}$ |
| | | Quality policy $X_{14}$ |
| | Product realization $X_2$ | Supplier relationships $X_{21}$ |
| | | Design and development $X_{22}$ |
| | | Purchase $X_{23}$ |
| | | Production and service provided $X_{24}$ |
| | | Control detection device $X_{25}$ |
| | Resource management $X_3$ | Research human resources $X_{31}$ |
| | | Infrastructure $X_{32}$ |
| | | Work condition $X_{33}$ |
| | Measurement analysis and improvement $X_4$ | Detection $X_{41}$ |
| | | Control reject $X_{42}$ |
| | | Date analyse$X_{43}$ |
| | | Continuous improvement $X_{44}$ |

the 1–9 scales before, the judgment matrixes R of index factor on subsystem level and the judgment matrixes $R_i$ $(i = 1,2,3,4)$ of index factor on attribute target level had been calculated.

$$R = \begin{bmatrix} 1 & 1/2 & 2 & 1 \\ 2 & 1 & 4 & 3 \\ 1/2 & 1/4 & 1 & 1/2 \\ 1 & 1/3 & 2 & 1 \end{bmatrix} R_1 = \begin{bmatrix} 1 & 1/2 & 4 & 3 \\ 2 & 1 & 4 & 3 \\ 1/4 & 1/4 & 1 & 1/2 \\ 1/3 & 1/3 & 2 & 1 \end{bmatrix}$$

$$R_2 = \begin{bmatrix} 1 & 1/3 & 2 & 2 & 2 \\ 3 & 1 & 4 & 3 & 4 \\ 1/2 & 1/4 & 1 & 1/2 & 1 \\ 1/2 & 1/3 & 2 & 1 & 2 \\ 1/2 & 1/4 & 1 & 1/2 & 1 \end{bmatrix} R_3 = \begin{bmatrix} 1 & 3 & 3 \\ 1/3 & 1 & 2 \\ 1/3 & 1/2 & 1 \end{bmatrix}$$

$$R_4 = \begin{bmatrix} 1 & 1/3 & 1/2 & 1 \\ 3 & 1 & 2 & 2 \\ 2 & 1/2 & 1 & 2 \\ 1 & 1/2 & 1/2 & 1 \end{bmatrix}$$

Having calculated each weight vector $X$, $X_i$ $(i = 1, 2, 3, 4)$ by AHP theory, and through consistency check. Empowerment as follow:

$$X = (0.22, 0.48, 0.11, 0.19)\ X_1 = (0.32, 0.45, 0.09, 0.14)$$
$$X_2 = (0.20, 0.45, 0.1, 0.15, 0.1)$$
$$X_3 = (0.59, 0.25, 0.16)\ X_4 = (0.15, 0.42, 0.27, 0.16)$$

### *58.3.3 Determined Measurement Value of Index System*

Index value of quantitative index could be obtained by calculating directly, but qualitative index adopted Obscure Quantitative idea and divided into 9 levels:

worst—(0,0.125), worse—(0.125,0.25), bad—(0.25,0.375), little bad—(0.375,0.5), ordinary—(0.5,0.625), preferably—(0.625,0.75), well—(0.75,0.875), better—(0.875,1.0), best—1.0.

Consulting the Obscure Quantitative idea, we obtained Table 58.2 by analyzing validity evaluation index factor of the college research quality management system, and determined.

### 58.3.4 Determined Weak Link of the Index System

When measurement value of evaluation index factors had been obtained and optimal regional and joint domain had been determined in actual evaluation process, processing dates of Table 58.2 by weak link determination model which had been established earlier and eligible degree model in extendable optimal degree, consequence as Table 58.3.

**Table 58.2** Measurement value, optimal regional and joint domain of index factor

| Attribute target | Measurement value | Optimal regional | Joint domain |
|---|---|---|---|
| $X_{11}$ | (0.375,0.5) | (0.9,1) | (0,1) |
| $X_{12}$ | (0.5,0.625) | (0.9,1) | (0,1) |
| $X_{13}$ | (0.75,0.875) | (0.9,1) | (0,1) |
| $X_{14}$ | (0.75,0.875) | (0.9,1) | (0,1) |
| $X_{21}$ | (0.625,0.75) | (0.9,1) | (0,1) |
| $X_{22}$ | (0.875,1) | (0.9,1) | (0,1) |
| $X_{23}$ | (0.375,0.5) | (0.9,1) | (0,1) |
| $X_{24}$ | (0.25,0.375) | (0.9,1) | (0,1) |
| $X_{25}$ | (0.75,0.875) | (0.9,1) | (0,1) |
| $X_{31}$ | (0.875,1) | (0.9,1) | (0,1) |
| $X_{32}$ | (0.625, 0.75) | (0.9,1) | (0,1) |
| $X_{33}$ | (0.5,0.625) | (0.9,1) | (0,1) |
| $X_{41}$ | (0.75,0.875) | (0.9,1) | (0,1) |
| $X_{42}$ | (0.5,0.625) | (0.9,1) | (0,1) |
| $X_{43}$ | (0.625,0.75) | (0.9,1) | (0,1) |
| $X_{44}$ | (0.875,1) | (0.9,1) | (0,1) |

**Table 58.3** Optimal regional and weak degrees of index factors

| Attribute target | Weight | Eligible degree | Weakness |
|---|---|---|---|
| $X_{11}$ | 0.0704 | −0.5139 | 0.0362 |
| $X_{12}$ | 0.0990 | −0.4355 | 0.0431 |
| $X_{13}$ | 0.0198 | −0.3182 | 0.0063 |
| $X_{14}$ | 0.0308 | −0.3182 | 0.0098 |
| $X_{21}$ | 0.0960 | −0.4048 | 0.0389 |
| $X_{22}$ | 0.2160 | −0.1667 | 0.036 |
| $X_{23}$ | 0.0480 | −0.5139 | 0.0247 |
| $X_{24}$ | 0.0720 | −0.6528 | 0.047 |
| $X_{25}$ | 0.0480 | −0.3182 | 0.0153 |
| $X_{31}$ | 0.0649 | −0.1667 | 0.0108 |
| $X_{32}$ | 0.0275 | −0.4048 | 0.0111 |
| $X_{33}$ | 0.0176 | −0.4355 | 0.0077 |
| $X_{41}$ | 0.0285 | −0.3182 | 0.0091 |
| $X_{42}$ | 0.0798 | −0.4355 | 0.0348 |
| $X_{43}$ | 0.0513 | −0.4048 | 0.0208 |
| $X_{44}$ | 0.0304 | −0.1667 | 0.0051 |

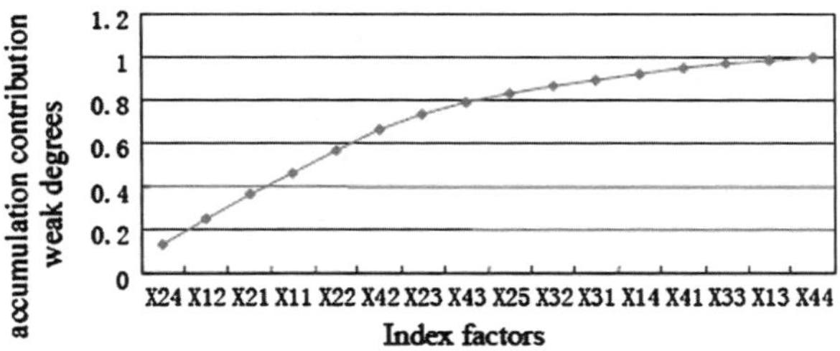


**Fig. 58.1** The determined figure of weak links

When weaknesses of index factors in Table 58.3 had been ranked in descending order, the determined figure of index system weak links was described by Excel, the figure also could be called line figure of index factors weak degrees accumulation contribution, as the Fig. 58.1 describe.

According to the determination method of weak link in established model above, when cumulative contribution rate of weaknesses was more than 80 %, these factors were weak link of the index system, the ranking the more front, the index the weaker, the way to improving the whole research management quality was priority improvement.

Analysed the improve order of index factors by different principles, the consequence as the Table 58.4.

Table 58.4 described that the determined improvement solution by the above three methods was different. Obviously, the improvement which based on the size of the weight and the size of the eligible degree |*k*| had some sidedness:

**Table 58.4** The improve order of index factors by different principles

| Attribute target | Weight order | Eligible degree \|*k*\| order | Weakness order | Weak link |
|---|---|---|---|---|
| $X_{11}$ | 6 | 2 | 4 | 4 |
| $X_{12}$ | 2 | 4 | 2 | 2 |
| $X_{13}$ | 15 | 10 | 15 | – |
| $X_{14}$ | 11 | 11 | 12 | – |
| $X_{21}$ | 3 | 7 | 3 | 3 |
| $X_{22}$ | 1 | 14 | 5 | 5 |
| $X_{23}$ | 9 | 3 | 7 | 7 |
| $X_{24}$ | 5 | 1 | 1 | 1 |
| $X_{25}$ | 10 | 12 | 9 | 9 |
| $X_{31}$ | 7 | 15 | 11 | – |
| $X_{32}$ | 14 | 8 | 10 | – |
| $X_{33}$ | 16 | 5 | 14 | – |
| $X_{41}$ | 13 | 13 | 13 | – |
| $X_{42}$ | 4 | 6 | 6 | 6 |
| $X_{43}$ | 8 | 9 | 8 | 8 |
| $X_{44}$ | 12 | 16 | 16 | – |

The more important index could be priority improvement by the size of the weight, but if the index had reached a certain level, we not only had little success, but also wasted funds and resource by blind improvement. Such as Purchase $X_{23}$, it was the first need improved link, but expert evaluation value was "better" level, it could not obtain obvious effect, even if positively improved.

The worst factor of the index value could be priority improvement by the size of the eligible degree $|k|$, but if the index factor had a little weight, or said its change had little influence for consequence, even though had a big improvement, the effectiveness of the university research quality management system had not significant improvements, such as Work condition $X_{33}$ in resource management, it was ranked 5 by eligible degree $|k|$ size, but it was ranked 14 by weakness size, it was not need to improve factors by the determined principle of the weak link.

In conclusion, the weakness method integrated index weight and index space value two factors, avoided sidedness, obtained more accurate index improve scheme, used the limited support resource in the most needs to improve index and avoided waste and unreasonable allocation of resources.

## 58.4 Conclusions

This method could be used in quantitative evaluation of having the interval index system and efficiently determining the system evaluated weakness, and it had important significance for optimizing system weakness and perfecting complex system.

## References

1. Wei M, Zhang S (2010) The effectiveness evaluation index system of colleges scientific QMS in military academy. J Sichuan Ordnance 92:133–134
2. Xia K (2007) The eligible degree evaluation of university students' business plan. Sci Technol Manage Res 24:14–15
3. Wan J, Zeng H, Zhu B (2009) Application of principal component analysis in evaluation of water quality of Lean River. Chin water wastewater 23:42–44
4. Bai X, Tao F, Jia C (2011) Research on ameliorated model of support capability weakness based on set pair analysis. Comput Digit Eng 23:432–441
5. Wei M, Zhang S (2010) Value engineering. Research on the validity evaluation of the colleges scientific QMS based on confidence. Ordnance Engineering College, Shijiazhuang, vol 34, pp 234–236
6. Shao C, Tang D (2009) An AHP-based on enterprise surplus quality evaluation 93:13–14
7. Xuan K (2008) Research evaluation index system of new Ordnance Equipment Support development. Master Dergee's Paper of Shijiazhuang Mechanical Engineering College, Shijiazhuang vol 25, pp 632–636
8. Zhang Z, Liu W, Meng Q (2007) Analyse and improve the weakness of the colleges scientific QMS. Sci Technol Manage Res 24:52–56

# Chapter 59
# Research of Psychology Teaching Based on the Information Technology

**Zhen Zhong**

**Abstract** Information technique and psychology teaching's integrating is a kind of information-based study method, its aim wants to develop learner to make use of an information technique efficiently of the target of completion course study in the information-based environment. Physically setting out from the teaching purpose and the reform in education, this paper mainly carried on information technique and psychology active, and promote psychology teaching to a layer, then set up technical according to the information teaching feedback model, finally researched how to raise efficiency and quality of teaching.

**Keywords** Information technology · Psychology teaching · Teaching feedback model

## 59.1 Introduction

To make the large teacher extensively and make use of an information technique to develop a teaching activity along with the quick development of modern information technique, in the reform in education, develop science and technology, concentration and convenience of information technique teaching thus. Therefore, blend modern information technique and psychology teaching mutually, have already become a psychology reform in education of a little bit hot and point.

Z. Zhong (✉)
Education Department, Wuzhou University, Wuzhou 543000, China
e-mail: zhongzhen21@yeah.net

W. Du (ed.), *Informatics and Management Science V*, Lecture Notes in Electrical Engineering 208, DOI: 10.1007/978-1-4471-4796-1_59,

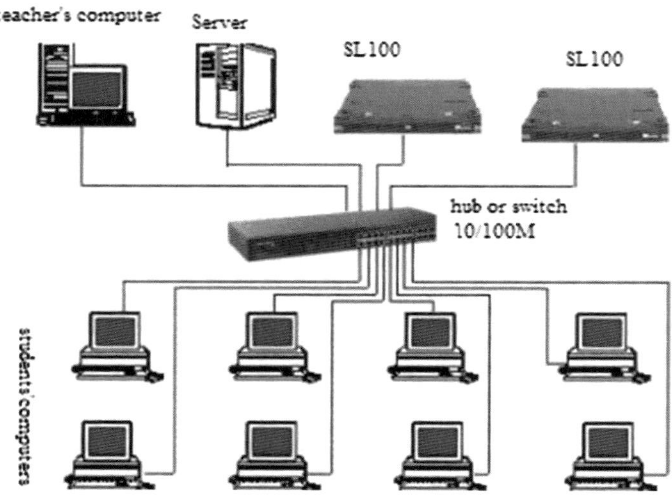


**Fig. 59.1** The schemes of the application of information technology in psychology teaching

Pass to develop both respectively of advantage; certainly will the meeting push psychology teaching toward a new development stage.

The psychology is a science that studies person and the animal mental phenomenon occurrence, development and movable regulation. The psychology is various foundations academics that the academics study, consequently the its key position is obvious. But, the teaching of the psychology exists a lot of irregularities currently, because of the psychology teaching adopt more the way of big class of teaching in the high school, having a class the number is numerous, but related concept and then unclear in meaning of psychology Be difficult to understand, therefore the teaching effect usually could not reach decent degree and influenced the student's study to the knowledge thus.

But the modern information technique is in the usage in the mental teaching and makes the psychology teaching follow an all new mode. In the information under the technical function, the psychology concept is no longer complicated and difficult to understand, the psychology phenomenon can dynamically play to show to listen to a lesson, the classroom teaching presents good interactive status and makes the student's innovation and quest get an exaltation in this kind of interactive teaching thus, therefore, can say that the information technique is a psychology teaching the catalyst of development. The information technique is applied to the sketch map of psychology teaching as shown in Fig. 59.1.

## 59.2 The Important Role of Information Technology in the Psychology Teaching

The fusion of information technique and psychology course is the process that is all-directions to combine. According to the student's knowledge need and accept degree, make use of an information technique in the psychology lectures, can from shallow go into deep of the rightness related knowledge carry on a demonstration; The teacher adopts a different information technique project to carry on a teaching according to the concrete teaching purpose, so the ability attain to have to the teaching process of put Shi; Educate governor to pass an usage information analysis technique, carry on a processing to the feedback information of teaching effect, do macroscopic analysis to the total development of psychology teaching thus. Information technology in psychology tutorial application process as shown in Fig. 59.2.

Information technology in the teach application psychology mainly embodied in the following three aspects.

The information technique is in the application within teaching of the psychology classroom.

The related concept in the psychology course is all abstract, cannot touch in the real life, therefore this brought course teaching a lot of difficulties. After leading into an information technique, can dynamically play to show a mental activity on the classroom of become because of and development step, make the student face lifeless psychology concept thus no longer indecisive. For example while explaining in detail the person's brain structure, the teacher can make use of Flash

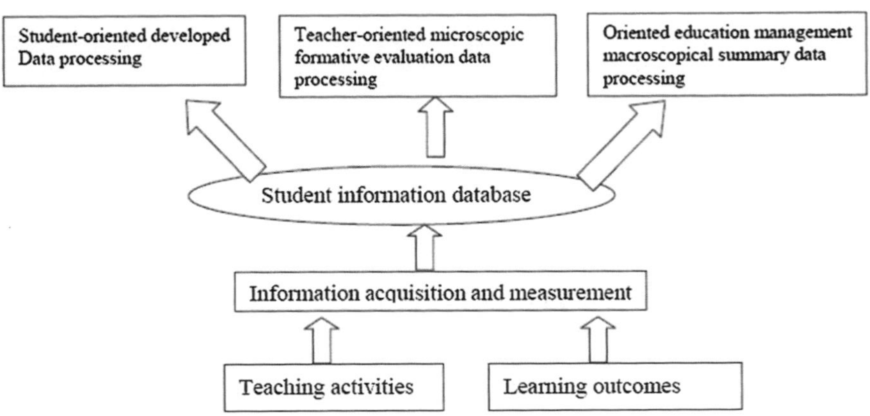


**Fig. 59.2** The process of information technology in psychology tutorial application

to play to show person while considering, the brain how carries on an operation, and brain the different part is concretely responsible for which abstract activities.

The information technique tests the application in the teaching in the psychology.

The psychology tests to adopt in the teaching with the calculator control system for the information technique of core, can avoid the irregularity of former big class of teaching, make the student able to carry on one by one and hand over with each other type study with teacher, raised teaching efficiency and quality thus, and make the mental state experiment more turn for the norm with scientific. At the same time, the usage information technique can carry on reviewing to measure to the student's study efficiency and then induce the factor of influence study efficiency.

The information technique is outside the psychology lesson the application in the classroom.

The psychology classroom outside the lesson the function is count for much in the teaching, and the teaching method is also abundant colorful. The student passes Internet and library resources outside the lesson, can carry on expanding and be thus advantageous to carrying on better absorption and digest to the knowledge in the classroom the knowledge studied. This has a very big function to the shortage and decrease for learning course that alleviates a psychology teacher. Consequently make use of an information technique the classroom study outside the lesson in, is a kind of form of good improvement psychology classroom teaching.

## 59.3 The Construction of Feedback Model of Psychology Teaching

The target of teaching is to train student's quest in doing not know knowledge ability and fulfillment ability. But in the traditional teaching, be subjected to restriction of the teaching place and teaching method, the effect of the psychology teaching always cans not gets a material exaltation, this causes student of actively active not ability drive good transfer, investigate desire to cannot get to build up, the thinking sloth lowers again and again. The information technique usage can make teaching form ever-changing at the teaching, and carry on a modification according to the teaching need, and the amount of information of the classroom became abundant, the teachers and the students' interaction raised, the operability is getting stronger, this consumedly stride up the student's study enthusiasm and investigation interest. But to the information technique in the analysis of the predominant specific weight in the mental teaching and the information technique to the positive effect that the student studies, can pass to build up teaching feedback model to carry on analysis and elaborate. Teach feedback model sketch map such as shown in Fig. 59.3.

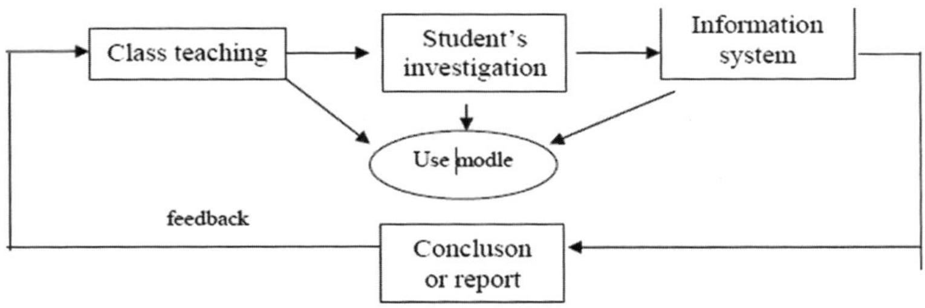


**Fig. 59.3** Schematic diagram of teaching feedback model

### *59.3.1 Class Teaching*

Understand the environment, condition of classroom teaching, actual student, and content of course etc., and carry on the information teaching before the lesson to program according to the above circumstance, for example prepare the information-based teaching in the classroom teaching to operate a step. Study difference according to the student's individual at the same time, if study degree and study method etc., pass to make use of the information system of the student whom the information technique sets up, will then give to match a student to the different student's adoption different teaching method individual development regulation of study mode. At the same time, make use of student information system, can choose that the data that the suitable student studies carries on teaching and makes the student attain to maximize to the accepting of knowledge degree thus. This kind of characteristic teaching, promotes to psychology teaching a new step, effectively changed the teaching method of tidy and uniform type.

### *59.3.2 Student's Investigation*

For carrying on an evaluation to the teaching quality, can carry on a questionnaire to the student, the feedback information of the student to the classroom effect directly influences a reform in education direction, therefore have to extensive but objective collections related information data, thus gather the original sex data that needs according to these information's.

### *59.3.3 Information System*

For collect of original sex data, also need to pass an information technique as to it's carry on Zhen do not, classification and calculation, and make use of related mathematics analysis theory according to the category and importance of the data, carry on settling sex and quantitative analysis. Thus carrying on a synthetic theory to the information is analytical.

### *59.3.4 Use Model*

According to the psychology teaching target, after carrying on statistics analysis to various information, can make use of model, comprehensive description influence degree of each factor to teaching effect, analyze information technique's influence specific weight for psychology teaching thus, thus system and thorough of analyze a traditional dissimilarity of teaching and information-based teaching.

## 59.4 Conclusion

Pass a classroom experiment research, can see the above-mentioned model having following function. Feedback function The model mainly used for the teacher carries on an usefulness programming to the lectures before the lesson and make the information technique can get an effective usage in the classroom, at the same time according to the macroscopic management need of teaching effect need and education of the student's personality need, teacher, course effect the feedback handle in the system to the information, thus to how raised teaching quality to provide concrete quantity to turn a data, but this has a decision function to the information-based reform of the psychology teaching.

Encourage function. The model can carry on keeping a view description to different teacher's teaching method, therefore can encourage the teacher carries on an improvement to own teaching method and excellent turn, attain to have the exaltation teaching that aims at sex thus of quality. At the same time, can also encourage the student whom the teacher aims at different personality, provide the teaching method of adequacy and reasonable, thus promote an information technique in the psychology classroom efficiently usage.

Compare function. The model can be used for some one teacher to different student's corpus and not and meantime the teaching of segment carry on a comparison, can also carry on in different teacher's of same profession interactive comparison, thus can the demonstration teacher completely at under various environment of teaching ability, provide the adjudicating of objectivity the standard thus. At the same time, the model can also more traditional teaching is with the good and bad of the information-based teaching.

Analytical function Pass to carry on a model experiment to the traditional teaching process and the information-based teaching process, according to the feedback function of the model, encourage function and comparison function, can synthesize to analyze two kinds of teaching models, get related conclusion.

**Table 59.1** The comparison of traditional teaching and experimental teaching

| Traditional teaching mode | Informatization teaching mode |
|---|---|
| Single subject teaching | Based on multi-discipline comprehensive teaching |
| Learning content stenosis | Learning content extensive |
| Take facts as the starting point of learning, teachers as teaching, teacher is the subject | Take the question as the starting point of learning to study, exploration, teacher as the auxiliary, the student is the main body |
| Didactic learning | Interactive learning |
| Learning has been organized information | Emphasis on discovery and innovative learning |
| Emphasis on learning results | Emphasizes the learning process |
| Subgroups by age and grade homogeneous | Subgroups by the wishes of heterogeneous |
| Evaluation on the specific knowledge and skills | Behavior based comprehensive evaluation |

Make use of teaching model to carry on the information technique review to measure in the function in the psychology teaching, isn't a simplicity to judge from the student's study result, but to influence a student to study of each link Be comprehensive to take into to consider. At the same time, can see from this model, the model set up can teaching information the feedback Islam learn classroom in, thus can effectively of carry on effective diagnosis towards the condition appeared in the teaching and put forward viable intervention measure, then raise the teaching quality of the teacher and the study efficiency of the student. Pass model analysis, can come to the conclusion such as shown in Table 59.1.

## 59.5 Conclusion

The usage of information technology make our teachings no longer tedious, but the information has great capacity, diverse form, accept one degree strong, its teaching effect is a traditional teaching to compare. However, face an information technique in the extensive application in the psychology teaching, the teacher can use flexibly a high-tech teaching equipments, can provide for student enrich colorful teaching information, can make the modern teaching process get away from the tie of traditional teaching, this is the problem that needs thinking urgently to reach agreement definitely, also the new challenge that the large teacher faces. What good teaching mode can be taught and study, but modernize a teaching demand is a kind of teaching with endless and abundant and colorful variety, therefore in the reform in education, cannot shine on again to move existing teaching mode. Ego exertion of the modern teaching demand teacher needs abundant teaching material and the forerunner's teaching principle and even needs a continuously developing of the information technique, is exactly at information technique of arouse under, then psychology teaching can at the new stage shining new source of vitality.

**Acknowledgments** This work belongs to the stage research performance of the key scientific research program of Wuzhou University, Research on university students' optimistic, thoughts of life and Dutch act (NO. [2010] No. 72). I would also like to thank all that help me finish the research.

## References

1. Wang L (2008) Applied psychology major experiment courses and teaching reform to explore. Chin Adult Educ 2:25–27
2. JieJiang W (2009) Network teaching environment of the teaching evaluation system. Chin West Educ 7:55–56
3. Jia Y (2006) Information technology and normal public psychology experiment course integration practice. Chin Audio-Vis Educ 4:35–38
4. Shengquan Y, Chen T (2010) Network teaching platform for the system structure and general design. Educ Technol Commun 7:58–60
5. Yang Y (2008) Mathematical model in the application of teaching efficiency evaluation. Pract Math 4:22–23
6. Ding X (2008) Distance education. Beijing normal university press 1:560–570
7. Liu J (2002) Education, Sichuan University Press, Chengdu, vol 1, pp 321–321

# Part IV
# Sports Management and Application II

# Chapter 60
# Experimental Study of Influence on College Students Psychological Health by Playing Football

**Limei Liu and Bin Yuan**

**Abstract** Sports have great influence on people's physical and mental health. At present, a great deal of college students has some mental health problems. Based on the documental accounts, questionnaires, sports experiments and data statistics, this article tries to find out the relationship between football-playing and students' psychological statement, and aims to explore the new model in college PE class.

**Keywords** Football · Influencing factors · Coupling mechanism · Psychology · Experimental psychology

## 60.1 Introduction

Mental health, from broad sense, it is a kind of high efficient and satisfaction and continuous psychological state. Speak from narrow sense, it is to point to the process, and that knowledge, the will, the social environment with the development. Claire and think: "mental health is a continuous psychological situation, a client in any particular environment can be good reflection, have and can get sufficient physical development; this is a rich. Not only is from just". Mailing geir thinks: it refers to people for each other and have the ability to adapt and happy [1–4]. Need is not only the efficiency, and not just the satisfaction, but to have

L. Liu (✉)
Institute of Educational Research, Handan College, Handan 056005, China
e-mail: limei_liu@yeah.net

B. Yuan
Department of Physical Education, Handan Polytechnic College, Handan 056005, China

W. Du (ed.), *Informatics and Management Science V*, Lecture Notes in Electrical Engineering 208, DOI: 10.1007/978-1-4471-4796-1_60,

between them. Mental health person should be able to keep calm mood, keen intelligence, is suitable for the social environment of behaviour and pleasant temperament. Contemporary college students are to belong to a higher quality and culture group is the hope of our country, has the country's mission, their physical and psychological quality related to the future of our motherland, related to the rise and fall of survival. But from a questionnaire, through investigation, statistics, at present our country college students' psychological health level quality is not optimistic [5–7]. Statistics show that 1.23 % of college students in our country are a different this or that kind of psychological problems. In 2007, a University of Hebei Mental Health Survey, (with a 90 scale SCL) 19.52 % of the college students' psychological barriers exist. These statistics results have shown that, China's college students' psychological health has come to the point of it. So, now we have to through the various technical means and education level to improve the psychological quality of college students in our country. Improve their psychological level, and healthy development of contribute to the society [8].

This article from the sport football and contemporary college students' psychological health in the starting point, through the experiment and research, it is a contemporary college students' psychological health related factors, and the impact factor analysis, this paper discusses after football sports of the new teaching mode, to college students' psychological problems of adjustment ability. While the traditional argues that, think it is just a kind of enhancing the body element is qualitative [9]. But the study found that the text, physical exercise on the treatment of depression and anxiety, fear, such as hostile to the positive role of the mood. This is the sports in the treatment of the university students' psychological health problems and provides a theory basis [10].

## 60.2 Research Method

### *60.2.1 The Research Object*

This paper is mainly through to Handan University of a professional level 10,200 students' quantitative selective examination, randomly assigned to two groups, each group of each 100 people. And for a six month experimental research and analysis, the final assessment test, draw the conclusion. A group of experimental research group, the football of the course to take teaching mode. A group is the contrast group, as compared with group for normal teaching mode. In order to ensure that the experimental results of rationality and authenticity, so that students in a relaxed state utterly test, all students are not receiving advance any relevant news suggests.

### 60.2.2 The Experiment Group

In order to make the results more significantly, randomly sampled in order to make the results more significantly, a random sample of Wuhan industry university 09, 10 level management professional students are divided into two groups as the experimental object, randomly selected from a class as the basketball training experiment group, another class for the control group. The experimental group used to understand the game teaching and based on the concept of physical education curriculum model of basketball teaching. The control group in addition to the daily school opened sports curriculum activities has not participated in other sports training. The 200 students were randomly assigned to two groups, namely the experimental group and contrast group. Each group of the 100 people. Group flow chart shown as shown in Fig. 60.1.

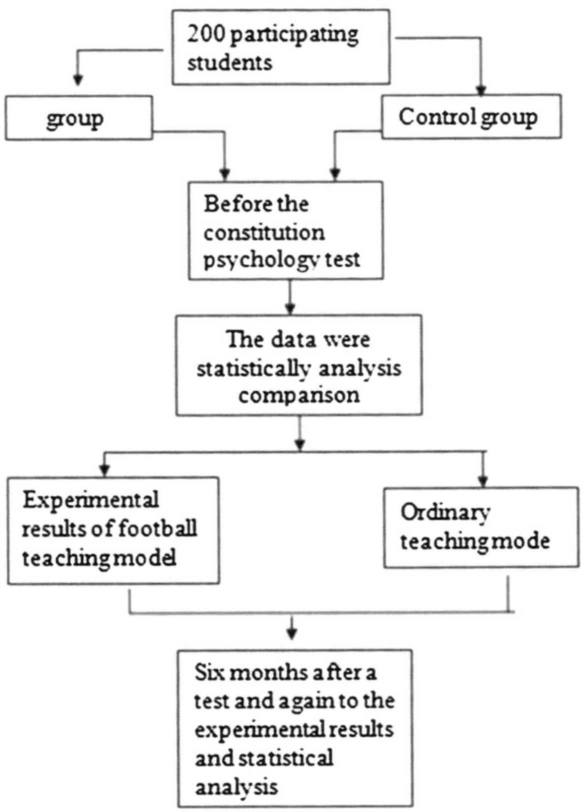


**Fig. 60.1** The flow chart of experimental group

**Table 60.1** The before and after experiment index of group and control group of psychological changes

| Factor project | Before test | | | | After test | | | |
|---|---|---|---|---|---|---|---|---|
| | Group | | Control group | | Group | | Control group | |
| | $\overline{X}$ | SD | $\overline{X}$ | SD | $\overline{X}$ | SD | $\overline{X}$ | SD |
| Body of health | 1.44 | 0.32 | 1.40 | 0.38 | 1.43 | 0.58 | 1.43 | 0.44 |
| Force | 2.06 | 0.56 | 1.96 | 0.52 | 1.82 | 0.58 | 1.95 | 0.53 |
| Interpersonal relationship | 1.96 | 0.54 | 1.99 | 0.54 | 1.84 | 1.60 | 1.97 | 1.55 |
| Depression | 1.82 | 0.54 | 1.74 | 0.49 | 1.70 | 0.64 | 1.75 | 0.50 |
| Anxiety | 1.76 | 0.53 | 1.71 | 0.46 | 1.54 | 0.46 | 1.69 | 0.54 |
| Hostile | 1.62 | 0.54 | 1.60 | 0.48 | 1.56 | 0.46 | 1.67 | 0.52 |
| Terrorist | 1.69 | 05.3 | 1.72 | 0.54 | 1.48 | 0.40 | 1.75 | 0.56 |
| Paranoid | 1.69 | 0.53 | 1.72 | 0.54 | 1.48 | 0.40 | 1.75 | 1.56 |
| Psychiatric | 1.57 | 0.49 | 1.54 | 0.48 | 1.51 | 0.43 | 1.58 | 0.59 |

### *60.2.3 The Data Processing and Analysis*

According to the application of statistical methods for all the SPSS1. Zero statistical data analysis of statistical software. Using t-test and average poor form side. The single factor analysis of variance concluded that the level of significance for Q = 0.05. Before and after the experiment analysis of results as is shown in Table 60.1.

The experimental group and control group psychological comparison abnormal situation incidence, as shown in Table 60.2.

## 60.3 Result Analysis

The data in the Table 60.1 can see, through experiment, the experimental group, all kinds of data than before the decline, of which the group, anxiety, depression, hostile, terror and have obvious drop dramatically. And, show ($P < 0.05$). Football as a, it reduced the tension between students dare and feelings, to strengthen the relationship of the exchange, enhance the mutual friendship. In addition, significantly reduce anxiety. The test results show that, the interpersonal relationship factors were significantly decreased. This shows that the collective exercise,

**Table 60.2** The table of mental abnormalities

| Group | Total number | Psychological anomalies know | Incidence of abnormal psychology (%) |
|---|---|---|---|
| Experimental group | 100 | 9 | 9 |
| Control group | 100 | 16 | 16 |

mandatory the ground increased their interaction opportunities, on the other hand exercise asked participants to physical and psychological activity in a certain degree to the excited state, which facilitates the exchange of feelings and to eliminate self claustrophobia, participate in regular exercise were more easily with others form intimate relationships, students can with the help of exercise way of understanding, adjust and change the psychological quality and the behavior way. Regular exercise can effectively improve college students' depression, reduce the degree of depression. Studies have shown that, exercise can increase the body endorphin content (whose role is to coordinate unified under stress state is the systemic function), content change in mood changes and interactions increases, so that depression loss, improve students' euphoria. Once upon a time table results also show, after the test of obsessive–compulsive symptom factor scores than the test before the fall, that through regular exercise after the forced to alleviate the symptoms, compulsive symptoms mainly refers to those who need not worry but could not get rid of the meaningless thought impulses and behavior, because exercise can be used as a vent, will all sorts of trouble, disturbing emotions, so that the psychology to balance.

The experimental study of identifying, collecting the relevant domestic university students' physical and mental quality status and basketball training on College Students' physical and mental quality of the data, analysis of basketball training course on College Students' physical and mental quality and comprehensive quality and academic performance of active meaning, combination of teaching evaluation and students' self exercise after evaluation of College Students' physical exercise interest, stimulate raise, train lifelong sports consciousness, through the basketball training "to help students build fitness", and then based on the results.

## 60.4 Conclusion

Through this paper, we can see, based on the experimental study of the psychology of football, the new teaching model can strengthen college students' psychological quality, in addition, sports and psychological education to coordinate with each other for teachers to provide a new method of teaching. But the study found that the text, physical exercise on the treatment of depression and anxiety, fear, such as hostile to the positive role of the mood. This is the sports in the treatment of the university students' psychological health problems and provides a theory basis.

## References

1. Yifang F (2008) Research on the analysis and diagnosis method of the competitive sports. J Phys Edu 9:12–13
2. Zhang L (2006) Sports psychology, vol 22. East China Normal University Press, pp 335–336

3. Sheng Z (2009) Health education guidelines of college students' psychology, vol 23. Zhejiang University Press, pp 315–317
4. Zhang Y, Hu J, Xie T (2009) Physical exercise—the effective way of psychological health education of college students. Hubei Sports Sci Technol 6:47–49
5. Qiwei M (2008) Physical and mental education of the students in the sports teaching. J Guangzhou Phys Educ 4:32–33
6. Pan D (2008) College football teaching for students' comprehensive ability training. Liaoning Coll Edu Adm J 12(8):11–13
7. Beili Z (1987) Sports psychology, vol 20. Higher Education Press, Beijing, pp 412–413
8. Zhong FS (2006) Mental health Chengdu, vol 34. University of Science and Technology Press, Chengdu, pp 445–448
9. Weizh iQu (2002) The information industry and China's economic and social development. Econ Res J 281:883–890
10. Meng X (2003) To cultivate the information culture industry. Inf Space 01:11–13

# Chapter 61
# Study on Efficient Training Mode of Youth Basketball

**Fei Zhang, Yonglin Zhao and Ping Huang**

**Abstract** This paper first introduces to juvenile basketball training in a number of common problems, and then set up a training process model and a performance evaluation model, introducing the relevant parameters, finally how to use the scientific concept to guide youth basketball training related issues were discussed and research, and for the subsequent juvenile basketball training work reference.

**Keywords** Teenager · Basketball training · Principle by science guides

## 61.1 Introduction

To whichever kind of sport in the nation, it is to investigate a national citizen's character and citizen incorporation, cooperation ability of a kind of outside performance. But want and want to thoroughly make the basketball business of our country and get a development, have to train business and start to grasp from the teenager's basketball. Therefore, how trained and then became a national development with the principle instruction basketball of teenager of science heavy medium of heavy [1, 2].

F. Zhang (✉) · Y. Zhao · P. Huang
Yangzhou Polytechnic Institute, Yangzhou 225127, China
e-mail: zhangf_ei@yeah.net

W. Du (ed.), *Informatics and Management Science V*, Lecture Notes in Electrical Engineering 208, DOI: 10.1007/978-1-4471-4796-1_61,

## 61.2 The Common Problem Appearing in Teenagers Basketball Trainings

The teenager trains at present medium pay attention to aggression, the thought lightly defending still exists [3]. To the basketball training of teenager in present, it is still to regard aggression training as principle, coach while carrying on basketball's training to the teenager, because of be subjected to individual personal performance desire of teenager of influence, make the process that it trains in basketball in, the aggression practice that valued its troops, make teenager of troops have to take the offensive consciousness, but caused personnel's degree in defending consciousness value of its oneself troops not enough, finally make oneself walk last the road of failure [4]. And, there is an individual coach member thinking, only take the offensive is the false information that bestly defends thought, make more the development pole of the whole troops unbalance, seriously influenced the whole development condition of troops.

The teenager's basketball trains, not the ability just only pays attention to the development of basketball training, but neglected the students' studies progress. Know according to the related survey of writer: Majority of basketballs train at present of students' all embracing this thought, basketball training and study are two kinds of different affairs, the training is a training, study and then study, two have no surely of practice [5, 6]. But is the fact really such? According to write the interview at the school know, such as if the teenager carried on basketball training of large numbers of quantity yesterday, it certainly would will cause next day's appearing in the classroom: Have a class to be late, poor mental state, truancy, leave early the occurrence of etc. phenomenon, but see at present, this kind of is heavy to train, the circumstance lightly studying usually takes place. This kind of vicious cycle makes more and more teenager's players start feeling that the study isn't the most important. And, at present the coach of numerous athletes sometimes for the sake of the result benefits of brief ball game, appear to make the student cut class the circumstance occurrence of carrying on the training, the occurrence of this circumstance, make the basketball player's understanding to the study of teenager be getting knotter enough and good enough more, finally caused cultural knowledge of athlete not enough, cannot pass examination smoothly the phenomenon occurrence of university.

The basketball training in present is still to regard vocational training as principle, to physical training still not. To basketball training now, it is mainly still is regard vocational training as principle. The coach member is in the teaching in present still just with shoot the basket technique, lead person's technique, defend technique and the luck ball technique teaching is lord, it to teenager of the physical training be placed in an entry-level stage of development, this kind of vicious cycle, make the athlete have a liking for go to basic achievement very firm, but in the middle of physically resisting appear physique bad, put together to rob the occurrence of appearing the problem. But cause this mistake's thoughting the essence reason of occurrence, be because of coach member the present education thought is still regard technique teaching as principle, it lacked the athlete's physical training. But the physique is in basketball training of the United States, it is highly valued. The basketball textbook in the United States writes such a word: "There is no healthy body character, cannot obtain the victory of end!" It is thus clear that the importance of the physical training.

## 61.3 Athletes Training Process Model

When the coach to athlete, in a certain role, will make the player state changes occur, if the motion toward the people desired direction, and exceed the coach acceptable level, it indicates a need to adjust the training scheme. Training model in Fig. 61.1 represents.

Each sign meaning in diagram:

T The radicle of time describes time of system variety to sit a mark;
X Input function, represent coach to the athlete's function;
Ω The importation segment gathers and describes to input mode inside a certain time partition, BEOf a statures gather;
Q The internal status gathers, is the core of internal system structure model;
$\delta$ The status transfers function (or call to deliver function), the definition system internal status is how to change;
$\lambda$ Output function, after expressing to accept an athlete to train an empress to train project of influence;
Y The exportation segment gathers, after expressing an athlete to train empress the amount of influence of environment in the exterior

The disaster characteristic in its variety, time and space of different category is different, the concrete importation function is also different. The disaster accepts disaster system after inputing how the status changes, transfer function with the status to describe, it embodied the weak characteristic or easy Sun characteristic of

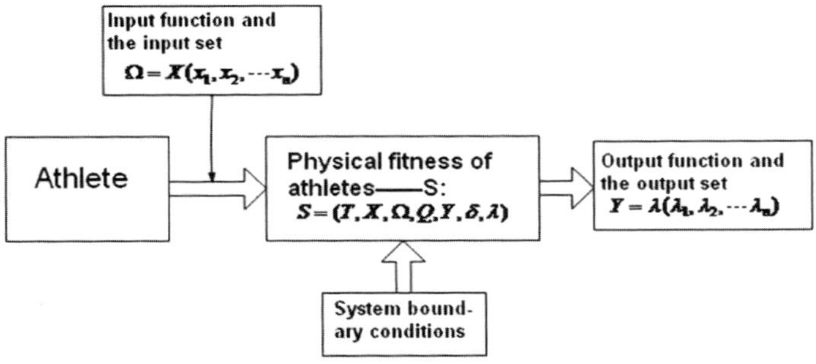


**Fig. 61.1** Athletes training process model

training the project. The variety of athlete's status outwardly outputs influence quantity by outputting function, it represents the serious degree of disaster result.

The training means as a result the indetermination of result, namely disaster source function $Y = \lambda(\lambda_1, \lambda_2, \ldots \lambda_n)$ after accepting disaster system the indetermination of result produced. From the Fig. 61.2 can see, train a result, then output the function $\Omega = X(x_1, x_2, \ldots x_n)$ be decided by to input function, basketball athlete's status of teenager and training system the boundary condition of place $S = (T, X, \Omega, Q, Y, \delta, \lambda)$, therefore, train result of the indetermination is to be decided by train oneself, but relate to the status, boundary condition of the athlete's effort, athlete. So the athlete's training level analysis and valuation want completely a characteristic of reflecting the athlete and train system at the certain boundary condition the characteristic is at his status variety, and train a property. Train effect valuation model is used descriptions. Among it, "I" means "interference", while "B" means "Boundary condition".

Figure 61.2 shows, the training effect depends on the training object, the athletes themselves and the quality of training $f(X, S, \lambda)$, i.e. Function $R = f(X, S, \lambda)$ of the specific expression, in addition to consider the whole system of all elements of the association properties, also considering the athletes on the examination results reflect the requirements (for performance analysis and evaluation is for the purpose of the performance management and control). Therefore, using elements of multiply structure model, i.e.:

$$R = X \times S \times \lambda \tag{471.1}$$

Type of representative athlete's physical characteristics, called the physical quality; psychological characteristics; for the athletes, training results for bearing properties.

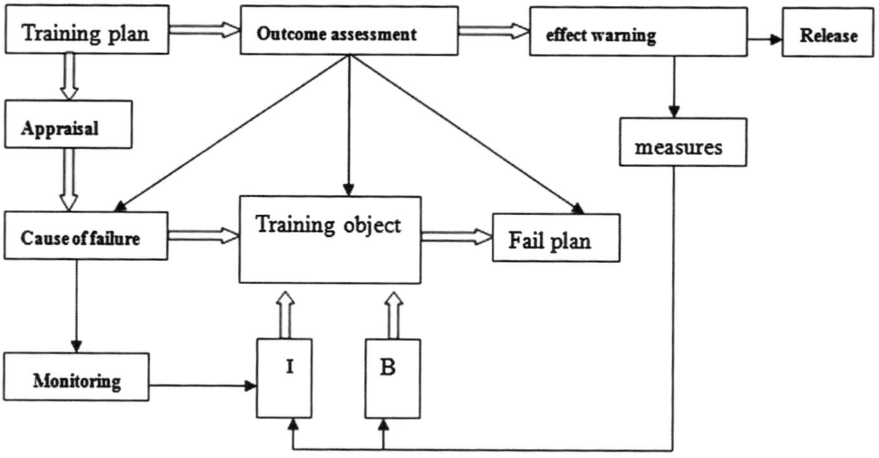


**Fig. 61.2** Trainning effect valuation model diagram

## 61.4 The Improvement Measure on How to Guide the Teenagers Basketball Training by Using the Scientific Principle

Teenager basketball that faces a such development trains of present condition, the teenager of the our country basketball train the education certainly will do adequacy of adjustment, so as to in essence and thoroughly guarantee the stability of the teenager's basketball training.

1. Coach member must the idea of the explicit basketball game of the biggest degree, the regular student's thought. Coach member in the process of carrying on basketball's training, must the biggest degree of carry on to the student systematize with whole turn of morals education, let the student have a good mindset to treat losing of each game and win, regular thought of the student him/herself of the biggest degree. On top of that, the coach member needs to train a basketball troops that offends and guards and has and borrows with the basketball troops that develops an omnipotent type, in the process of training, must paying attention to of the biggest degree scoop out the latent ability that each player has and fight for maximizing the special features that it has. When coach member still wanted in the training, defending individual consciousness level was the important sign that measures a basketball level height of athlete, let the athlete slowly changed to only take the offensive the thought of having the sense of achievement was just.
2. Coach member well develop the whole basketball brigade training education of at the same time, also pay attention to develop the cultural level of expecting the oneself. Coach member at to own student from the going training process, must paying attention to of the biggest degree observe the cultural classroom of the student to have a class a circumstance, utmost avoid the occurrence of two conflicts. Moreover, coach member must the thought of regular oneself and students, let it know to train surely important, but the teaching of cultural course is also important, coach the biggest degree don't want while training personal honor and train to practice, maximized the purpose of explicit training, borrow with developing of the biggest degree and encourage an athlete, attain to truly make the student able to obtain knowledge and technical ability on his/her own initiative, church they healthily develop to become an independence completely of integrity of person.
3. Basketball that the coach member wants to be utmost to strengthen a teenager train's consciousness stirs up interest in the training of student and raises whole consciousness of basketball. The coach must pay attention to express in the process of carrying on teaching's training to the student each the training exactly toughens of is what thought, what to toughen is that technical ability, make the student find out real training thought in the training thus, let it find out real fun in the training, let the student thoroughly fall in love with training, make it able to unconsciously raise basketball consciousness of oneself.

Meanwhile, the coach member still wants the biggest degree of set out from the student's interest, the biggest degree development student's this interest, borrow with athlete's interest in the training of stirring up of the biggest degree. Moreover, the coach member is in the disciplinal process, utmost avoid of carry on a great deal of repeated sex practice contents, make the athlete don't want to feel lifeless, make it train a creation treachery to basketball mental.

4. The coach member wants explicit basketball of the biggest degree to train the relation that physical training and technique train, then attains the physical training of enhancing of the biggest degree oneself and improves the ability of resisting of student. A basketball athlete has to have good physique level is one of the important guarantees of all athletics tournaments, particularly ball item, even request it to have good physique. But see to the development present condition of the basketball business of modern, it needs higher speed, strength, technique, and Gao Kong to resist to battle out. This was also the surface basketball business isn't to develop technique sport that train and then can win victory, but under the circumstance that needs physique and technique to develop together, so as to in essence and thoroughly raise the development of basketball business. So the athlete wants higher basketball technique level of acquiring of the biggest degree and obtains better sport result, have to also need to enlarge the physical training of training the oneself while training the technical ability of oneself. Therefore, while raising the skill military tactics training quality, enhance physical training, let each time particularly item physical training of eventually the class target become 1 kind to the ego to outstrip and make the abundant physique of oneself integrate into skill military tactics usage in, make own skill military tactics ability have vitality and fulfillment more, improve the ability of resisting thus.

## 61.5 Conclusion

In summary, for contemporary of basketball sport training business but speech, it isn't only a development engineering of long term, is still that one need to be had a virtuous completely, science ground training know thought, then thought in the training that well controls athlete's health, the decorum trains an attitude and borrows to train business by teenager's basketball of utmost development our country.

## References

1. Badri MA, Abdulla M, Kamali MA, Dodeen H (2006) Identifying potential biasing variables in student evaluation of teaching in a newly accredited business program in the UAE. Int J Edu Manage 20(1):192–195
2. Barth MM (2008) Deciphering student evaluations of teaching: a Factor Analysis Approach. J Edu Bus. 14(1):40–46

3. Boex LFJ (2000) Attributes of effective economics instructors: an analysis of student evaluations. J Econ Edu 31:211–227
4. Brightman HJ (2006) Mentoring faculty to improve teaching and student learning. Issues Acc Edu 21(2):09–15
5. Centra J (1982) Determining faculty effectiveness. Jossey-Bass, San Francisco
6. Chonko LB, Tanner JF, Davis R (2002) what are they thinking? Students' expectations and self-assessments. J Edu Bus 77:271–281

# Chapter 62
# Analysis of Short-term Physical Training Plan Influence on of Physical Quality of Female College Student

**Chunmei Peng and Wenping Ye**

**Abstract** This paper combines the literature material method and experimental method to discuss short-term physical training plan's influence on female college students' physical quality. Through short-term physical training of 120 female university students and physical quality condition records before and after the short-term training, this paper compares the training results before and after the training and discusses the physical quality of female college students. For these healthy female college students, we provided training of aerobic exercise, aerobic threshold and shoulders, strength training of abdomen and legs to muscle 2 h per week for 16 weeks. The results showed that their physical quality was improved after the physical training for 16 weeks. All in all, the above activity improved female university students' health. The experimental results showed that short-term physical training plan can improve female university students' physical quality, effectively improve female university students' physical form, strengthen the body function, and promote the physical health of female university students.

**Keywords** Female college students · Short-term physical training · Comparative analysis · Descriptive statistics analysis

## 62.1 Introduction

Sports education's goal is to develop knowledge, skills and confidence to impart physical education and enjoy lifelong healthy physical activity. Healthy habits, including healthy eating and physical activity, can reduce the risk of obesity and

C. Peng (✉) · W. Ye
Department of Physical Education, Huanggang Normal College, Huanggang 438000, China
e-mail: Chunmei_Peng@yeah.net

W. Du (ed.), *Informatics and Management Science V*, Lecture Notes in Electrical Engineering 208, DOI: 10.1007/978-1-4471-4796-1_62,

reduce the risk of the disease. School-based education has a lot of benefits, including increasing physical activity and improving physical and muscular endurance. Through the physical education increasing physical activity, they can effectively maintain good form. In addition, physical education can improve the students' health and improve their ability to learn [1].

With the speeding up of the pace of life, the increase of the pressure of social competition, people spend less time in sports, especially female college students who are faced with weaker physique, pressure, poor psychological bearing ability, and poor sleeping quality and so on, which lead to f present trend of gradual decline of female university students' physical quality. The more students are fit, the more they are willing to learn [2]. Physical education is a key factor of the students' physical qualities, health and learning, so we should strengthen the college sports education, and the state and society should increase college sports career support and investment to improve the quality of the education of college sports. According to the current situation of sports we discuss the influence of short-term training plan for female university students' physical qualities from the angle of sports in order to find a way to improve female college students to find the body of the quality of the way.

## 62.2 Research Methods

This paper uses the SPSS statistical methods, first through descriptive statistical analysis, including the concentration trend and discrete trend, and then making exploratory, comparative analysis of the distribution of frequencies [3].

$$Z_i = \frac{x_i - \bar{x}}{S} \tag{62.1}$$

We chose 120 female university students to research, and these female college students have never participated in regular physical activity before this investigation. According to the following formula and very good determination of the situation, there is no significant difference in the female university students' physical characteristics. See Table 62.1.

**Table 62.1** Female college students' physical characteristics Data: average (± SD)

| Items | N = 120 |
|---|---|
| Age (year) | 19.6 (0.7)–23.4 (0.9) |
| Weight (kg) | 45.6 (13.4)–48.5 (20.7) |
| Height (m) | 1.63 (0.10)–1.70 (0.10) |
| Weight index (kg m$^2$) | 21.5 (3.7)–22.6 (4.8) |

## 62.3 Research Steps

The testers involved in 1 or 2 h of training a week for 16 weeks. They first had 10 min warming up and carried out the main activities for 40 min (tensile, general and muscle strength training, stamina, agility and power). The recovery time lasted for 10 min. This training program started with low strength training, and then gradually continued high strength. In order to prevent injuries and be more familiar with test, they went through tests at the start before physical training program. The physical training program is shown in Table 62.2.

### *62.3.1 One Mile Walking and Running*

The program requires the tester to walk as fast as possible or running at a speed of 1,600 m. Time of walking or running test of time needed for the project is recorded.

### *62.3.2 Proneness Test*

Proneness test is a common test of flexibility, especially one way of measuring lower back and leg muscles flexibility. Subjects were bare feet and sit on the floor, with the foot flat against the table and leg unbend. They tried to get away from the edge of the table as much as possible and the distance from fingertips to the table is also recorded. The test project repeat two times, finally be used to calculate the average distance.

### *62.3.3 Sit-Ups Test*

Testers should lie low; bend their knees, hips about 25 cm heel distances, with arms bend, hand on head, feet on the floor by a peer will hold feet. When all those

**Table 62.2** Short-term physical stamina training plan project tables

| Number | Purpose of measurement | Training project |
|---|---|---|
| 1 | Cardiopulmonary health | 1,600 m walking and running |
| 2 | Flexibility | proneness test |
| 3 | Abdominal strength and endurance | Sit-ups |
| 4 | Shoulder strength and endurance | push-ups |
| 5 | Muscle strength | vertical jump |

testers are prepared, timers will start time for a total of 60 min. Testers are bending body, and their elbows touch the thigh, and keep hands to the chest or contact shoulder, then their elbows touch the thigh again. This is repeated. Either up or down players can have a rest. Sit-ups in one minute are correctly repeated as much as possible. Assistant supervisory participants have to ensure the correctness of the test form, at the same time, the number of execution and calculation of sit-ups. Posture of sit-ups doesn't count is it's not right.

### *62.3.4 Push-Ups*

The project requirement of the position that the hands are the same with broad shoulders, knee unbend, abdomen keep tightened, spine and keep in a middle position. And they bend their elbows, with their chest down to elbow 90°. They did push up, arms unbend, also their elbows can't be completely still. Total time is 1 min.

### *62.3.5 Longitudinal Jump*

The jump test includes measuring the difference between stand heights and vertical leap to the peak. The project makes the testers to jump and contact with these plastic rods as much as possible, and records the highest level of the high jumps and minuses the stand heights.

## 62.4 Research Results

Through the female university students' short-term training plan, the following are respectively standard deviation of conditions of the average physical fitness test before and after. We can clearly see corresponding change of female college students' quality before and after the test [4].

The traditional construction project management system can only be exchanged through the longitudinal way and thus limiting the data flow and accuracy on the working efficiency. Based on the technology of the Internet and the construction project management on all projects we can produce information points of the line centralized management. But for the managed project information we can provide an entrance so as to increase the efficiency of the engineering management and to facilitate various projects through the communication of the participants. The fusion diagram of the Internet technology and construction project management is shown in Table 62.4 (Fig. 62.1).

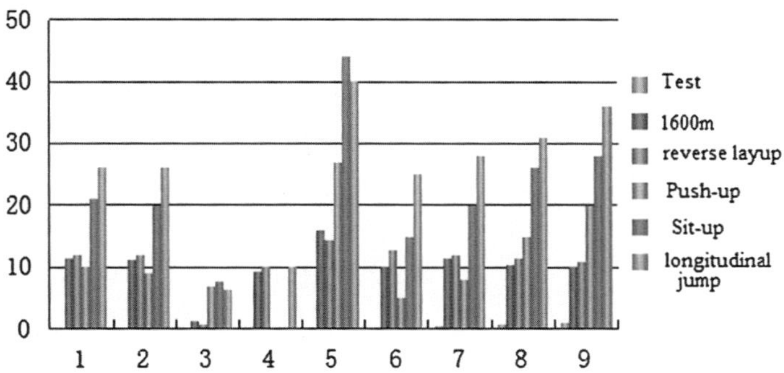


**Fig. 62.1** Physical parameters figure before the test

According to the test results, we found that 1,600 m walking and running, push-ups and sit-ups and longitudinal jump are different before and after test, this fully demonstrating that through the short-term physical training, female university students' cardiopulmonary health functions, flexibility, shoulders, abdomen and muscle strength and endurance has been effectively improved. The results of the test are shown in Table 62.3 and 62.4.

The test show that the before and after the test results of 1,600 m running (11/33, 8.87 min), the jump (27–31/6 cm), push-ups (10–18) and reverse layup (11/96–11/62 s). The improved tests found that comparing with the prediction of physical factors, the physical quality of the testers were increased by 25–50 % (Fig. 62.2).

**Table 62.3** Deviation percentages of the average and standard before the test

| Text | Average | Mode | Standard deviation | Minimum | 25 % | 50 % | 75 % | 90 % |
|---|---|---|---|---|---|---|---|---|
| 1,600 m | 11.34 | 11.28 | 1.18 | 9.21 | 10.1 | 11.45 | 10.37 | 10.2 |
| Reverse layup | 11.95 | 11.94 | 0.75 | 10.14 | 12.7 | 11.89 | 11.4 | 10.96 |
| Push-up | 10 | 9 | 6.98 | 0 | 5 | 8 | 15 | 20 |
| Sit-up | 21 | 20 | 7.74 | 0 | 15 | 20 | 26 | 28 |
| Longitudinal jump | 26 | 26 | 6.29 | 10 | 25 | 28 | 31 | 36 |

**Table 62.4** Deviation percentage of the average and standard after the test

| Test | Average | Mode | Standard deviation | Minimum | 25 % | 50 % | 75 % | 96 |
|---|---|---|---|---|---|---|---|---|
| 1,600 m | 8.86 | 8.57 | 1.03 | 7.13 | 9.37 | 8.42 | 8.30 | 7.53 |
| Reverse layup | 11.60 | 11.63 | 0.64 | 10.17 | 12.16 | 11.83 | 11.08 | 10.69 |
| Push-up | 19 | 19 | 7.40 | 0 | 15 | 19 | 24 | 29 |
| Sit-up | 20 | 31 | 8.01 | 7 | 26 | 31 | 36 | 46 |
| Longitudinal jump | 31.6 | 32 | 5.8 | 19 | 29 | 32 | 36 | 39 |

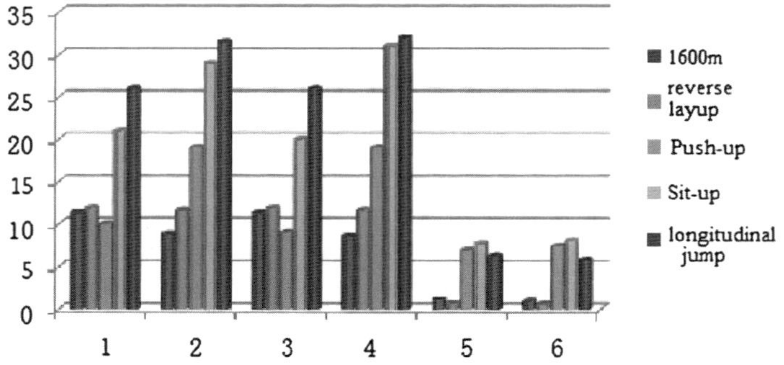


**Fig. 62.2** Results compare before and after short-term physical training

## 62.5 Conclusion

To keep healthy, prevent disease, we must establish healthy habits and form the good sports fitness habits. To keep a good health we must do the related fitness movement in one's sustainable atmosphere, thus not only help reduce the risk of disease, avoid injury, but also can enjoy physical activities besides work and have access to healthy body. This paper mainly selects five aspects of sports training factor to measure students' physical health and fitness function, and to test in the light of the five factors. And we also compare the standard test, which shows that the above factors for physical index reach average level, or even more than the average in the untrained projects. In short, short-term training plan have significantly change the female university students' constitution, and can enhance the physical health and quality of the female university students.

## References

1. Hang T (2010) Physical training's influence on female college students' physique. Sports theor 12:196–197
2. Guogui W, Yu F, Wang J (2010) Analysis of college extracurricular sports club exercise's impact on the students' physical quality. Sports BBS 9
3. Ying M, Zhang L (2011) Competitive aerobics training's influence on college students' sports fitness. J Tangshan teach coll 5:333–340
4. Guo J (2008) Content and classification of sports development movement. J hunan inst 3:87–95

# Chapter 63
# Analysis of College Students Weight Based on Physical Exercise and Dietary Pattern

**Chunmei Peng and Wenping Ye**

**Abstract** Weight gain and lifestyle behaviours during the university may cause overweight and obesity of the adults. The main purpose of this paper is to analyze the freshman and sophomore years of total 760 college students' (53 % women, 47 % men) movement and dietary patterns, and its influence on college students' physical and mental health. The study mainly involves measuring students' weight and height, and using the form of questionnaire survey as the measurement of their recent sports and diet mode. According to the results of the survey, 29 % of freshman students do not participate in sports, and 70 % of the people eat fruit and vegetables every day of less than 5; more than 50 % of people in at least three times a week before eat fried or high fat fast food. But by the end of the sophomore year, 70 % of 290 reopened the weight assessment, which shows the weight gain (4.1 +/− 3.6 kg, $P < 0.001$), but this has no clear link with physical exercise and dietary pattern. We need to make further analysis contribution of fat, muscle, and bone to weight gain, and determine the influence of physical exercise and dietary pattern on college students' physical weight.

**Keywords** Comparative analysis · Descriptive statistics analysis · Weight index · Diet · Exercise · Chi-square test

C. Peng (✉) · W. Ye
Department of Physical Education, Huanggang Normal College, Huanggang 438000, China
e-mail: Chunmei_Peng@yeah.net

W. Du (ed.), *Informatics and Management Science V*, Lecture Notes in Electrical Engineering 208, DOI: 10.1007/978-1-4471-4796-1_63,

## 63.1 Introduction

As people living standard improving, people pay more attention to diet and take relatively also too much of the fat intake. And people don't put enough importance for sports, which led to the popularity of overweight and obesity, including children and teenagers. According to the national health and nutrition survey, 64.5 % of adults are overweight, and this led to high morbidity rates. Results based on behaviour risk factors monitoring system shows that the group stages of 18 and 29 seems to be the most vulnerable to overweight and obesity [1–3]. The study results also show that the transition between adolescence and adulthood and college students of this age often come with dramatic and not proper weight gain.

In addition, survey data from 2005 college students' health risk behaviour shows that many college students diet and activity mode will give them future certain health problems. Teenagers early adulthood is in between overweight and obese, with habits of life like sedentary activities and eating too much calories greatly reducing their exercise [4–6]. The national health interview survey reveals that 10,645 investigators of the 12–21 year's old males and females and 43,732 of 18 year old males. The data show that the age of 12 and 21 step down in aerobic strengthen activity. Also, the national heart, lung and blood institute's growth and health research data show that, by 18–19 years of age physical activity began to significantly become less. Therefore, further study is needed to the person's constitution and weight to put forward a healthy mode of life [7].

## 63.2 Research of College Students' Physical Index

In order to research the mode of life beneficial to body health, and combined with the national health survey data, this paper chooses sport and diet as two aspects that are good way of life to the body. In order to better achieve the objective we choose a group whose living environment is relatively simple and way of life is easy to spot: freshmen and sophomore.

The survey found that there has been a lot of weight gain in the first year of university students. In order to understand the influence and trend of life of university from physical activities and body mass index (BMI), we focus on analyzing teenagers entry into university and second year university life's influence from their weight, sports and dietary pattern, etc., [8]. The purpose of this study is: (1) Assessment of the freshmen's weight, height, exercise and diet, etc.; (2) assessment of these changes in the first and second years.

We use the questionnaire survey to research people who participate in aerobic, strengthen, stretching self assessment. Exercise questionnaire is based on trans-theoretical model of behaviour change, and it is different from engagement in a particular behaviour preparation phase. The present study includes five stages: lack of deliberate practice, meditation (want to exercise), ready to (plan to exercise), action (present exercise) and maintain (continuous exercise). The first three stages mean that lack of exercise, and the latter two represent the current stage in the exercise. The specific standard is defined as three different sports: aerobic exercise, stretching and strength training. Aerobic exercise is a regularly plan to increase physical activity, for example, taking a walk, aerobic exercise, running, bike riding, swimming, boating, etc. [9, 10]. Conventional strength training is to use strong body for physical activity, for example, increasing or removing free weights, using weight or resistance training machine, etc. Regular stretching is any of the body's activities to enhance the flexibility, such as yoga, tai chi, etc. The results of the survey are shown in Table 63.1 (Fig. 63.1).

### *63.2.1 Survey Data Analysis*

In view of the investigation data, through the descriptive statistics calculation method [11], we make an analysis of the general features samples of the basic information. In addition, we have used continuous measurement on the sophomore, and the measurement results are such as BMI; and we use the Chi-square test results inspection to test the differences between men and women. We also examined results from freshman year to the second year to introduce the results with changes of time [12] (Fig. 63.2) (Table 63.2).

The results show that from beginning to the end of a sophomore year, 70 % of the 280 students' gains weight; 26 % has weight reduction, and only 3 % of the students' weight is remain the same. For those who gained weight, they had an average weight gain for 4.1 +/– 3.6 kg. And for the BMI index, 69 % of the students are higher.

**Table 63.1** Survey of students' body mass index parameters

| Factors | Female N = 410 | Male N = 350 | Total N = 760 |
|---|---|---|---|
| Human body measure | | | |
| Height (cm) | 163.2 ± 6.4 | 176.7 ± 7.2 | 169.5 ± 9.6 |
| Weight (kg) | 59.5 ± 10.7 | 72.5 ± 12.4 | 65.7 ± 13.2 |
| Body mass index (kg/m$^2$) | 22.3 ± 3.6 | 23.2 ± 3.4 | 22.7 ± 3.6 |
| Sports proportion (%) | | | |
| Aerobic exercise 3–5 day/week | 60 | 58 | 59 |
| Strength training 2–3 day/week | 35 | 55 | 45 |
| Stretching 2–3 day/week | 37 | 36 | 36 |
| At least one training | 66 | 75 | 70 |
| No exercise | 34 | 25 | 30 |
| Eating proportion (%) | | | |
| Five fruits and vegetables/d | 30 | 29 | 30 |
| > Three fried food | 51 | 29 | 41 |
| ≥ Three high fat fast food frequency | 56 | 35 | 46 |

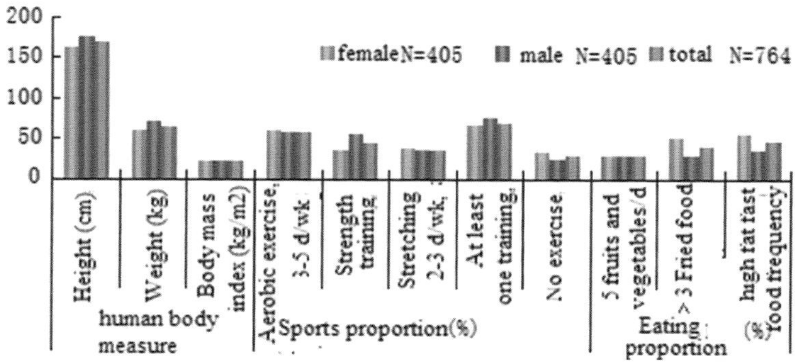


**Fig. 63.1** Comparison of the students' parameters

**Fig. 63.2** Change proportion of height, weight, and body mass index (BMI)

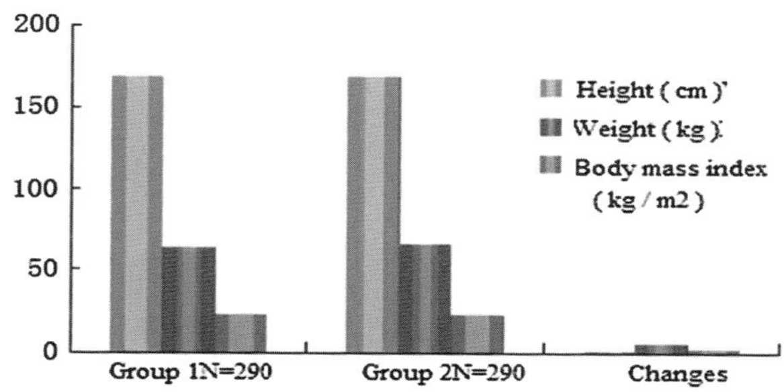


**Table 63.2** Change parameters of height, weight, and body mass index (BMI)

| Factors | Freshmen N = 280 | Sophomore N = 280 | Changes | p |
|---|---|---|---|---|
| Height (cm) | 168.5 ± 9.1 | 168.6 ± 9.1 | +0.1 ± 1.5 | NS |
| Weight (kg) | 64.3 ± 11.9 | 66.1 ± 12.8 | +1.8 ± 5.2 | $< 0.001$ |
| Body mass index (kg/m$^2$) | 22.6 ± 3.3 | 23.2 ± 3.5 | +0.6 ± 1.8 | $< 0.001$ |

And the data showed that students in the two years of participation in sports and dietary pattern situation are shown in Table 63.3. Data showed that the whole sports participation has not changed; aerobic exercise in number is in decline, but the number of stretching is on the increase. For freshman and sophomore students' consumption of fruits, vegetables and high fat fast food, the situation did not appear to change, and fried food consumption is on the decline. Through research and analysis and found no relationship between the changes of weight, body mass index (BMI), exercise and eating behaviour of the. Significant changes include: students in maintenance stages involved in aerobic exercise are reducing, the training of merger students involved in stretch action and maintenance stages increased, but the corresponding proportion is in the fall (Fig. 63.3).

**Table 63.3** Students' exercise and dietary pattern proportion in the first two years (%)

| Factors | Freshmen | Sophomore | p |
|---|---|---|---|
| Sports proportion (n = 290) | | | |
| Aerobic exercise 3–5 day/week | 63 | 54 | 0.039 |
| Strength training 2–3 day/week | 44 | 46 | NS |
| Stretching 2–3 day/week | 33 | 35 | 0.007 |
| At least one training | 72 | 74 | NS |
| No exercise | 33 | 26 | NS |
| Eating proportion (%) | | | |
| Five fruits and vegetables/d | 30 | 32 | NS |
| > Three fried food | 54 | 44 | 0.004 |
| ≥ Three high fat fast food frequency | 50 | 45 | NS |

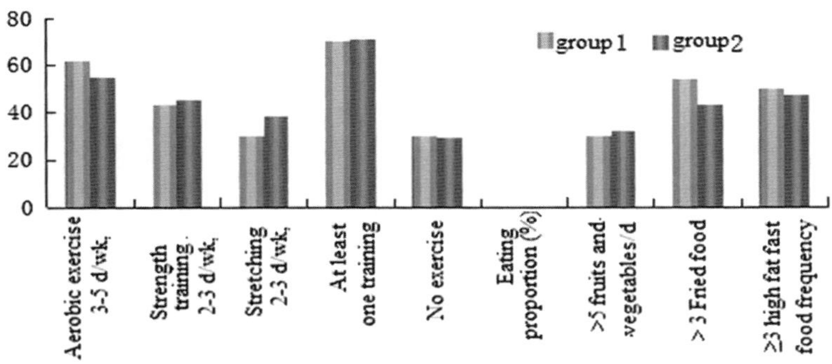


**Fig. 63.3** Proportion of students' participation in sports and diet mode

## 63.3 Conclusion

To keep healthy, prevent disease, we must establish healthy habits and form the good sports fitness habits. To keep a good health we must do the related fitness movement in one's sustainable atmosphere, thus not only help reduce the risk of disease, avoid injury, but also can enjoy physical activities besides work and have access to healthy body. This paper mainly selects five aspects of sports training factor to measure students' physical health and fitness function, and to test in the light of the five factors. And we also compare the standard test, which shows that the above factors for physical index reach average level, or even more than the average in the untrained projects. In short, short-term training plan have significantly change the female university students' constitution, and can enhance the physical health and quality of the female university students.

More than 12 million students are sent to the colleges and universities all over the country, and the vast majority of people in the 18–24 years old age. University students are a very appropriate group and their influence on evaluation and intervention of sports and eating behaviour's positive changes is very feasible and important. Prevention measures should be taken to reduce the incidence of overweight and to improve the health level of our country, and it's very important. In short, weight gain, lack of regular exercise and unhealthy eating pattern are the most common phenomenon of sophomore students in school. These findings highlight that the college students' potential bad behavior may cause overweight during the period of adults. With nearly two-thirds of adults are overweight, this will cause the attention of the society from all walks of life to fully understand the role of campus environment, and consider the system strategies as much as possible of to encourage students to pursue healthy exercise and diet, and to keep a healthy weight. And we need to further define the need to improve behavior model, which means to promote diet health and exercise behavior on college campus. We should advocate healthy life style as soon as possible during adulthood, so the college students will have long-term interest, and the ability to increase physical activity and reduce the overweight and obesity rates of the whole society, thus promoting the health level of the whole society.

## References

1. Qunling H (2004) Comparative analysis of student health standard and national sports exercise standard. J Beijing Sport Univ 9:33–35
2. Xiangchuan S, Yin H (2004) 2000–2003 Comparative analysis of adult's constitution monitoring results of Suzhou. Mountain east sports sci technol 26:90–92
3. Bo Xu (2007) Shanghai normal university students' physical health. East China Normal univ 4:14–16
4. Yu Kehong (2000) Research on the evaluation criteria of the national primary and middle school students' physical and health education. Zhejiang River University Press, Hangzhou
5. Qingzhu S, Shilin R (2001) Human physical measurement and evaluation, vol. 7 High Education Press, China. pp 90–94
6. Chaoqun Y, Yuhua K (1997) Research progress of immune influence of diet and exercise on fat people. J Beijing Sports Teach Coll 9(3):49–55
7. Zhang P (2007) Influence of exercise training on low weight female college students' heart rate and blood pressure. J Hunan Inst Sci Technol 28:9
8. He Zhongkai (2001) Physical and health relations theory empirical research, Ph.D. Thesis, vol 7, Beijing Sport University Beijing, pp 33–36
9. Xu H, Jiang W (2001) Research of BMI of Jiangsu adults sports and science, 6:39–43
10. Rong Shilin (2002) Mnistry of education and the state general administration of sports. About print and distribute "student physical health standard (trial implementation scheme)," and "the students' physical health standard (trial implementation scheme) executive way" notice. 207:225
11. Zhang L (2002) China office of international life science society obesity problem and China joint working group data collection analysis cooperation group. Recommendation introduction of China adult body mass index classification. J Chinese Preventive Med 5(5):349–350
12. Yang B, Gu X, Ni W (2005) Influence of BMI abnormal on college students' physical health. Chinese Sports Sci Technol 2:138–139

# Chapter 64
# Research on Nutrition Demand of Basketball Players

**Lijuan Hou and Sunnan Li**

**Abstract** As a basketball player, how vital of a role food and nutrition in the playing ability is critical. Somebody will ask "does it really matter what we eat be-for a game"? Optimum performance on the basketball court requires sound nutritional habits, as being adequately fueled directly affects your stamina and focus. You can get your daily requirement of nutrients and calories through everyday food. It is rare to need the use of supplements, with the exception being weight-gain shakes for those of you who have trouble consuming adequate calories to provide for muscle gain. You should most certainly steer clear of performance-enhancing supplements, such as creating and ephedrine, because of the possible side effects. Do not underestimate the role nutrition plays in acquiring maximum physical development. What you eat on a daily basis helps to determine body fat levels, as well as how much energy you will have for intense workouts and practices. This research was supported by the Fundamental Research Funds for the Central Universities.

**Keywords** Sports · Basketball · Nutritional habits

## 64.1 Introduction

As a basketball player, how vital of a role food and nutrition in the playing ability is critical. Somebody will ask "does it really matter what we eat before a game"? A professor compared our bodies to cars and the fuel that we put into them. If we put junk in our cars they don't drive well and the same is true of our bodies. This article will summarize his advice and will serve as a guide for basketball players

---

L. Hou (✉) · S. Li
Physical Education and Sports College, Beijing Normal University, Beijing 100875, China
e-mail: houlj@bnu.edu.cn

W. Du (ed.), *Informatics and Management Science V*, Lecture Notes in Electrical Engineering 208, DOI: 10.1007/978-1-4471-4796-1_64,

looking to improve their playing ability simply by being smart about what they eat and drink. The skiing players ought to have a special treatment to acquire strong body and in addition they should buy enough natural vitamins. There is a great way supposed to be getting shaping. Healthy nutrition inside fundamental action in the achievements any doctor athlete, and sometimes even amateur runners realize that will diet may have a major effect on their really perform.

At the professional level, most male players are above 1.91 m and most women above 1.70 m. Guards, for whom physical coordination and ball-handling skills are crucial, tend to be the smallest players. Almost all forwards in the men's pro leagues are 1.98 m or taller. Most centers are over 2.08 m tall. According to a survey given to all NBA teams, the average height of all NBA players is just less than 2.01 m, with the average weight being close to 101 kg.

The shortest player ever to play in the NBA is Muggsy Bogues at 1.60 m. Other short players have thrived at the pro level. Anthony "Spud" Webb was just 1.70 m tall, but had a 1.07 m vertical leap, giving him significant height when jumping. While shorter players are often not very good at defending against shooting, their ability to navigate quickly through crowded areas of the court and steal the ball by reaching low are strengths.

Fitness using for basketballs change from position set, but for the most part, the nutrient during knowledge periods are to be high about a nutritious exactly what foods, with moderate cholesterol protein and incline intake. Company, a substantial basketball's weight loss program is heavily according to bread, whole grains, and pasta and allows rice, all of us fresh vegetables and fruits, vegetables and lightweight dairy gain. Eating very same foods not merely allows players to completely prepare just because competition also helps their health to regenerate after goes along. Many guided on teams undergo matches as frequently as every a lot of different days; it's the same vital who players get trained properly. Through matches, players can easily drink isotonic drinks substitute carbohydrates and observe after their glucose levels and moisture levels master up. Your match, same drinks usually consumed as you restore carob levels the way they aid the back speedy upkeep of muscle incidents. Low-fat snacks in the likes of yoghurt's, cereal rungs and sandwiches may possibly be consumed, before a greater meal 2 or 3 h later repair their fortunes levels associated with important vitamins and minerals.

## 64.2 Nutrition for the Athlete

Sports nutrition is the study and practice of nutrition and diet as it relates to athletic performance. It is concerned with the type and quantity of fluid and food taken by an athlete, and deals with nutrients such as vitamins, minerals, supplements and organic substances such as carbohydrates, proteins and fats. Although an important part of many sports training regimens, it is most commonly considered in strength sports (such as weight lifting and bodybuilding) and endurance sports (for example cycling, running, swimming).

### 64.2.1 Carbohydrates

Athletes benefit the most from the amount of carbohydrates stored in the body. In the early stages of moderate exercise, carbohydrates provide 40–50 % of the energy requirement.

Carbohydrates may be classified as monosaccharides, disaccharides, or polysaccharides depending on the number of monomer units they contain. They constitute a large part of foods such as rice, noodles, bread, and other grain-based products [1]. Monosaccharides, disaccharides, and polysaccharides contain one, two, and three or more sugar units, respectively. Polysaccharides are often referred to as complex carbohydrates because they are typically long, multiple branched chains of sugar units.

Organisms typically cannot metabolize all types of carbohydrate to yield energy. Glucose is a nearly universal and accessible source of calories. Many organisms also have the ability to metabolize other monosaccharides and Disaccharides, though glucose is preferred. In Escherichia coli, for example, the lac operon will express enzymes for the digestion of lactose when it is present, but if both lactose and glucose are present the lac operon is repressed, resulting in the glucose being used first. Ruminants and termites, for example, use microorganisms to process cellulose.

During exercise, the glycogen is converted back to glucose and is used for energy. The body stores a limited amount of carbohydrate in the muscles and liver. If the event lasts for less than 90 min, the glycogen stored in the muscle is enough to supply the needed energy.

For events that require heavy work for more than 90 min, a high-carbohydrate diet eaten for 2–3 days before the event allows glycogen storage spaces to be filled. Long distance runners, cyclists, cross-country skiers, canoe racers, swimmers and soccer players report benefits from a precompetition diet where 70 % of the calories comes from carbohydrates [2, 3]. Extra carbohydrates will not help; any more than adding gas to a half-full tank will make the car go faster.

According to the Olympic Training Center in Colorado Springs, endurance athletes on a high-carbohydrate diet can exercise longer than athletes eating a low-carbohydrate, high-fat diet. Eating a high-carbohydrate diet constantly is not advised. This conditions the body to use only carbohydrates for fuel and not the fatty acids derived from fats.

### 64.2.2 Water

Water is excreted from the body in multiple forms. Therefore it is necessary to adequately rehydrate to replace lost fluids. To compensate for additional fluid output, breastfeeding women require an additional 700 ml/day above the recommended intake values for non-lactating women. It is dangerous to drink too little.

While overhydration is much less common than dehydration, it is also possible to drink far more water than necessary which can result in water intoxication, a serious and potentially fatal condition. In particular, large amounts of de-ionized water are dangerous.

Water is an important nutrient for the athlete. Athletes should start any event hydrated and replace as much lost fluid as possible by drinking chilled liquids at frequent intervals during the event [4]. Chilled fluids are absorbed faster and help lower body temperature.

### 64.2.3 Fats

Fat also provides body fuel. For moderate exercise, about half of the total energy expenditure is derived from free fatty acid metabolism. If the event lasts more than an hour, the body may use mostly fats for energy. Using fat as fuel depends on the event's duration and the athlete's condition. Trained athletes use fat for energy more quickly than untrained athletes [5]. Consumption of fat should not fall below 15 % of total energy intake because it may limit performance. Athletes who are under pressures to achieve or maintain a low body weight are susceptible to using fat restriction and should be told that this will hinder their performance.

Saturated fats (typically from animal sources) have been a staple in many world cultures for millennia. Unsaturated fats (e.g., vegetable oil) are considered healthier, while trans fats are to be avoided. Saturated and some trans fats are typically solid at room temperature, while unsaturated fats are typically liquids. Trans fats are very rare in nature, and have been shown to be highly detrimental to human health, but have properties useful in the food processing industry, such as rancidity resistance.

### 64.2.4 Protein

Protein is an important component of every cell in the body. Hair and nails are mostly made of protein. The body uses protein to build and repair tissues. In addition, protein is used to make hormones and other chemicals in the body. Protein is also an important building block of bones, muscles, cartilage, skin, and blood. Conversely, citing recent studies, recommends a minimum protein intake of 2.2 g/kg "for anyone involved in competitive or intense recreational sports who wants to maximize lean body mass but does not wish to gain weight".

Most authorities recommend that endurance athletes eat between 1.2 and 1.4 g protein per kg of body weight per day; resistance and strength-trained athletes may need as much as 1.6–1.7 g protein per kg of body weight (A kilogram equals 2.2 pounds.)

A varied diet will provide more than enough protein as caloric intake increases. Furthermore, Americans tend to eat more than the recommended amounts of protein. Excess protein can deprive the athlete of more efficient fuel and can lead to dehydration. High-protein diets increase the water requirement necessary to eliminate the nitrogen through the urine. Also, an increase in metabolic rate can occur and, therefore, increased oxygen consumption. Protein supplements are unnecessary and not recommended.

### *64.2.5 Vitamins and Minerals*

Vitamins and minerals make people's bodies work properly. Although you get vitamins and minerals from the foods you eat every day, some foods have more vitamins and minerals than others.

Minerals play an important role in performance. Heavy exercise affects the body's supply of sodium, potassium, iron and calcium. Sweating during exercise increases the concentration of salt in the body. Consuming salt tablets after competition and workouts is not advised as this will remove water from your cells, causing weak muscles. Good sodium guidelines are to: (1) avoid excessive amounts of sodium in the diet and (2) beverages containing sodium after endurance events may be helpful. Vitamins and minerals boost the immune system; support normal growth and development, and help cells and organs do their jobs. For example, you've probably heard that carrots are good for your eyes. It's true! Carrots are full of substances called arytenoids that your body converts into vitamin A, which helps prevent eye problems.

Calcium is an important nutrient for everyone as it is important in bone health and muscle function. Female athletes should have an adequate supply of calcium to avoid calcium loss from bones. Calcium loss may lead to osteoporosis later in life. Choosing low-fat dairy products provide the best source of calcium.

## 64.3 Basic Dietary Recommendations

Whether you are trying to gain muscle, reduce body fat, or maintain your current stature, it is very important to follow these basic dietary recommendations: A balanced diet consists of approximately 60 % carbohydrates, 20 % fat and 20 % protein; eat a variety of healthy foods. Limit the intake of fat, sugar, and sodium; Drink plenty of water. Many nutritionists recommend a minimum of 64 ounces of water per day; eat 5–7 small meals throughout the day. The size of the meal depends on the actual goal, as well as level of activity.

Weight Gain the Healthy Way: Most basketball players are tall and slender, and are looking to add muscular bodyweight. In order to gain weight, you must consume more calories than you expend on a daily basis. This means if you are looking to put on weight, you must eat, eat, and eat! Now for the select few looking to lose weight, they must do the opposite—consume fewer calories than they expend. This is done by controlling their portion sizes.

Below is just a very basic and general sample menu one can follow to get an idea of how much food he or she needs to consume on a daily basis to gain weight. A reasonable goal is to try and gain 1 pound per week for an 8–10 week stretch.

Example Menu: Breakfast: Orange juice, four pancakes w/syrup, and four scrambled eggs (Snack: one cup of low fat yogurt, granola bar, and a banana.) Lunch: two deli sandwiches on whole wheat bread, an apple, and a glass of milk (Snack: two peanut butter and jelly sandwiches and a glass of milk). Dinner: Steak, potatoes, steamed vegetables, and a roll (Snack: two sticks of string cheese and crackers).

## 64.4 Two Nutritional Habits to Incorporate with Exercise

Good health is not just the absence of an illness—it is the complete wellness of our physical and emotional selves. Our nutritional habits affect our health, happiness and our ability to get a good night's sleep. Besides helping you to effortlessly and naturally drop a few extra pounds, a healthy diet can aid you in improving every aspect of your life—including your stress and anxiety levels. The creation of a healthy lifestyle takes place within a multi-faceted approach that includes coping with daily stress, incorporating a positive outlook into life, and developing sound nutritional habits. Simply by paying attention to what you eat, you can uncover your habits and take a step towards changing them. Another key to good nutrition is becoming aware of what you eat and how it makes you feel. For example, if you eat a snack of fruit and cheese versus a chocolate bar mid-afternoon, do these foods make you feel any different later in your day? These are important things to take note of in your quest for better general health, which in turn leads to better life quality.

Take, for example, the car. You service your car—you put the correct fuel in it, check the oil, have it routinely examined for any potential problems, and you keep it clean. What do you think would happen to your car if you didn't care for it? It would break down and therefore you wouldn't be able to use it. Our bodies are the very same. If we don't do the regular maintenance, how can we expect to stay healthy? An important piece to our overall maintenance plan is good nutrition. In properly managing our body's "fuel" we need to include premium quality, as well as becoming mindful of when our "tanks" are running empty.

Forming good nutritional habits doesn't happen overnight. Following are two key points that can turn your new healthy lifestyle into a reality. Start slowly, and gradually build on these foundations.

## 64.5 Supplements Commonly Used in Basketball

Creatine is a supplement that can improve training ability and increase muscle power. One study showed that creatine improved jumping height. Creatine should not be used by athletes younger than 18 years because it is not known whether creatine is safe for people in this age group. If you use creatine, take 3–5 g per day. More than this is not needed.

Before your basketball game: Start hydrating 24 h prior to your game/training; 2–3 h before the game/training: Drink 400–600 ml of water or sports drink during active warm-up: Drink another 240 ml—about 8 gulps What and when to eat focus on carbs; 2–4 h before a game/training meals should be high-carb, moderate protein, low-fat, and low-fibre. If you have more time before a game/training (4 h) eat more; if you have less time (2 h) eat less to avoid stomach distress.

## References

1. Yasuharu N, Hirofumi I, Masami A (2011) Effects of jump and balance training on knee kinematics and electromyography of female basketball athletes during a single limb drop landing: pre-post intervention study. Sports Med Arthrosc Rehabil Ther Technol 3:1–14
2. Clark N, Polonchek J (1998) Outcomes of nutrition education for professional female basketball players. J Am Diet Assoc 98(9):A76
3. Sowell K, Ahmed SM, Warber JP (2006) The nutrition knowledge of collegiate division I male and female scholarship basketball players compared to a random sample of non-athletes at an urban university. J Am Diet Assoc 106(8):A67
4. Yoshinari T, Yasukazu Y, Yoshiki S (2010) Retear of anterior cruciate ligament grafts in female basketball players: a case series. Sports Med 2:1–7
5. Chen P (2003) Exercise and health promotion. Sports Sci Res 24(1):46–48

# Chapter 65
# Coupling Analysis of Social Background and School Health Sports Development

**Qingbo Kong and Lingyan Zhang**

**Abstract** By introducing social stratification theory and non-equilibrium development theory as the theoretical basis of the research, this paper focuses on the coupling mechanism between school health sports development and different social backgrounds. By establishing the fuzzy mathematical model, this paper then puts forward the constructing the school health sports non-equilibrium development mode: "upper-class school self-government development, middle-class school inside and outside development, lower-class school government support development".

**Keywords** Social stratification · School health sports · Fuzzy mathematics · Coupling mechanism research

## 65.1 Introduction

"Social stratification theory" is an important field of the study of the philosophy of science and technology, which means the possession of social resources of a country or society's individuals or group forming inequality between different levels and all levels. Grusky, Kerrbo, and Rothman summarized that this inequality is shown in economics, politics, culture, society, prestige, human resources and so on. Kingsley Davis and Wilbert Moore point out that social stratification is a kind of common phenomenon and also a kind of inevitable social

Q. Kong (✉) · L. Zhang
College of Physical Education, Chongqing University of Arts and Sciences,
Chongqing 402160, China
e-mail: Qingbo_Kong@yeah.net

W. Du (ed.), *Informatics and Management Science V*, Lecture Notes in Electrical Engineering 208, DOI: 10.1007/978-1-4471-4796-1_65,

phenomenon [1]. That is, social stratification is a kind of institutionalization of the social inequality system, and in this inequality system, some people or group can get more or grab more opportunities of social resources and even the public resources; other people or groups are in obvious disadvantage in the chance to get the resources [2]. Social stratification phenomenon is the background for present education development, and has laid the practice basis for the unbalanced development of school health sports [3].

"The unbalanced development theory" is an economic theory that was on the rise in the middle of the century; then economists expanded the theory from a single market economy to the market economy and planned economy [4]. The so-called non-equilibrium development usually refers to the imbalance or uneven phenomenon of the resource allocation, wealth accumulation, income, right application of different regions and industries, or the present more significant difference trend. From the world scale, social imbalance is a universal phenomenon, and is related to social development of all levels. The unbalanced development, from the effective distribution of resource, considers the economic development in the initial stage, and how to spend their limited resources distribution in a place of the most production potential. Balanced development and non-equilibrium development is the transformation of each other, and both are in a common among economic system; they develop in contradiction, and evolve to the senior equilibrium phase [5]. The purpose of school health sports non-equilibrium development model construction is the rational and effective distribution of sports resources and its use; in the whole development process, the "balanced development is the goal, the unbalanced development is the method"; finally it is to realize the higher level of balanced development of the school health sports [6].

## 65.2 Analysis of the Causes of School Health Sports Stratification

There are many factors supporting school health sports development, and its existing way can be material or spiritual, or as the human resources or wealth, or the target or content, inside and outside campus, the current or future, tangible or intangible and so on, as long as these factors can be used in school health of sports, and can effectively enhance students' physics and increase students' health. All can be listed in the category of school health sports resources. The universality of social stratification makes school health sports development certainly live with resources of inequality [7]. Referring to resources theory knowledge, and to the discipline of the "human, material and financial resources" as the foundation, from "the school subject, the gap between urban and rural areas, key levels and regional position" the above four dimensions to make descriptions on the differences of characteristics of school health sports development resources and lays the foundation for school health sports level classification [8].

1. Analyzing from running subject dimension

The school running subject is the investors and operators in the entire school development process, according to which the school are divided into three main types including public schools, private schools and privately managed schools. Among them, public schools' investment and operational subject are the state, and are under national direct management, support and protection; the school health sports development costs are all derived from the national or local financial tax, and they have enough resources guarantee. Public schools have high reputation and abundant teaching equipment and teachers. As a pioneer of quality education reform, it puts more attention on health sports.

2. Analyzing from the gap between urban and rural areas dimension

Along with the urban and rural stratified phenomenon in education, the gap between urban and rural school health sports is in a growing trend. In the school health sports resources on the gap, both human and material resources and financial resources, the rural school owns far less than the city schools. First of all, there is a big difference between urban and rural school physical education teachers' education background, professional title, and their numbers and so on. For a long time, the rural school education funds are too disperse and local financial raisers, together with the default intelligence education is the most important thing for education, and this is the tacit approval of the phenomenon that financial resources for urban and rural areas don't balance.

3. Analyzing from key levels

Key levels are a division made by education administrative department and the government. They compared the comprehensive strength and gap of the same category school according to the staff, the students' facilities, and other aspects by the end of last century. From the perspective of the law of social stratification, the division of the school is essentially the approval or consent to education resources' unbalanced distribution. Although on June 29, 2006, the standing committee of the National Peoples Congress passed the new compulsory education law article 22, which has already written: "the people's government at or above the county level and the education administrative department shall promote the balanced development of the school, narrow the gap between the school running conditions, must not divided them into the key schools and non-key schools", under traditional education factors' influence and restriction, there still exist between the key schools and the non-key schools, and this difference is obviously reflected in school health sports resources.

4. Analyzing from regional location dimension

In 1988, Deng Xiaoping put forward the regional strategic thinking of "two overall situations". According to this thought and the overall layout of the reform and open policy, in more than 20 years, coastal areas get fast development. In this era environment under the influence of trend, rich coastal regions and the inland impoverished areas of school health sports resources showed obvious difference and has confirmed a lot of research institute. Studies have pointed out that, because the development of the school sports work has a strong dependence on sports resources, PE teachers, location equipment such visible sports resources shortage of the poor areas make the rights and interests of students' accepting normal education sports pushed to the edge. In western minority areas poor sports teaching facilities widely exist: the PE teachers are not qualified; sports teaching are with less money; sports fitness ideas and exercise science weak method is the status quo. Compared with western poverty-stricken areas, the eastern coastal area school sports health resources are more abundant, which is mainly shown in the following aspects: leadership puts more importance on it; sports teachers' comprehensive quality and the facilities, etc. So it seems that the whole economic development in the western region relatively lags behind, and local public finance of the education has poor ability to pay are the main reason for causing poor area school health sports' difficult development.

## 65.3 Establishment of Coupling Mechanism Model

School health sports development under the background of different social stratifications is always our pursuit, but because of the constrained condition, the sports health in different parts can't have balanced development. For this we can analyze and expound their balance through the fuzzy mathematics model.

In this hypothesis, we accept evaluation objects are "n", namely n decision units (decision making units, including j = 1, 2…n); every decision unit has "m" types of the "on" and s types of "output", and the input and output are based on a negative value. $X_j$ Means the input of $DMU_j$ from the Ist; $Y_j$ means the output of $DMU_j$ from thirst. Therefore, the input and output of $DMU_j$ can be expressed as:

$$X_j = (x_{1j}, x_{2j}, \ldots, x_{mj})^T, \quad \mathrm{j} = 1, 2, \ldots, \mathrm{n}; \tag{65.1}$$

$$Y_j = (y_{1j}, y_{2j}, \ldots, y_{sj})^T, \quad \mathrm{j} = 1, 2, \ldots, \mathrm{n} \tag{65.2}$$

Set V = (,…) T, U = (,…) T, in which V and U are the input and output of m kind vector, and the selection of weight U and V makes efficiency evaluation index $h_j$ meets:

$$\mathrm{hj} = \frac{u^T Y_j}{v^T X_j} \leq 1, \quad \mathrm{j} = 1, 2, \ldots, \mathrm{n} \tag{65.3}$$

For in that weight coefficient v and u, $v^T X_j$ means input and $u^T Y_j$ means output, and this is the ratio of them.

If we take a DMU for the first evaluation, it is $DMU_0$, for its input is $X_0$, and its output is $Y_0$. Thus the first DMU relative efficiency evaluation model is:

$$\max E_0 = \frac{u^T Y_0}{v^T X_0}, \ u \geq 0, \ v \geq 0 \tag{65.4}$$

$$\sum_{i=1}^{s} u_i = 1, \sum_{i=1}^{m} v_i = 1 \tag{65.5}$$

$$\frac{u^T Y_0}{v^T X_0} \leq 1, \quad \mathrm{i} = 1, 2, \ldots, \mathrm{n}. \tag{65.6}$$

## 65.4 Theoretical Analysis of School Health Sports Development Model

Social stratification means that school health sports development will inevitably has many aspects of resources differences. In this context, classify all kinds of school according to the health of sports resources comprehensive possession ratio, as "upper, middle and lower" three levels. Based on the theory of the unbalanced development resources configuration request, that is, from the effective distribution resources angle, how to spend their limited resources distribution in the place of most productive potential, and make health sports development plans for school of different levels, and construct the school health sports non-equilibrium development mode of "upper-class school self-government development, middle-class school inside and outside development, lower-class school government support development". As is shown in Fig. 65.1:

1. upper-class school self-government development

Upper-class school self-government development means a management way that in light of their own characteristics and advantages of running, schools should fully mobilize various resources, and make scheme and implementing measures for school health sports development characteristics, and then realizing the

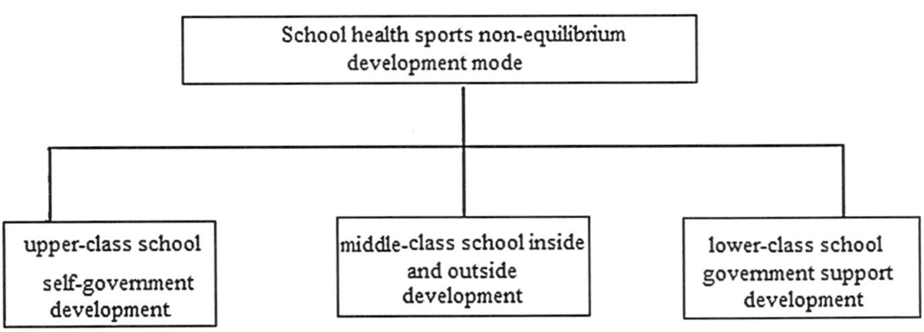


**Fig. 65.1** School health sports non-equilibrium development mode

development of effective school health sports. In the self-government development mode, the school physical education has autonomy in the specific affairs and party education department didn't directly involved in them. Party and government education department can only have indirect ways through all kinds of laws and regulations on school health sports implementation intervention, and its main function is to provide various standards or create goals for school health sports development, to provide the autonomous development of the system guaranteed, and to fully embodies the party and government education departments' top–down system planning and school health sports department's bottom–up autonomous management. As long as it does not affect the overall development of the region or country, every school meeting resource configuration condition can implement self-government development.

2. Middle-class school inside and outside development

Middle-class school inside and outside development is the symbiosis realized under the macro-control by the party and the government education department. In this kind of development mode, the school sports development intervention is more liberal with party and government education department of health, and its main function is planning and regulation, and to provide support and the help for the shortage of development of sports resources. The purpose is to promote good interaction middle-class school inside and outside development, which have three aspects need to be stressed. First, school health sports must be on the basis of complementary resources to achieve internal and external symbiosis development, such as school of human resources and enterprise, community of material and financial resources and complementary resources, etc.; Second, the education department of the party and government intervention is indispensable symbiotic in the development process of the school health sports inside and outside development, such as the extreme shortage of resources of school health sports individual should be funded; Third, the emphasis on various kinds of resources sharing association.

3. Lower-class school government support development

Lower-class school means school resources are scarcer or of scarcity, and the fact that school health sports has no independent operation. The development of the education department must be realized by party and government's direct and specific intervention. The characteristics of the government support are the close coordination of the government and school, which embodies strong administrative color and official color in management. For a long time, restricted by the economic conditions, human resources, education system, most school sports government were leading a model of development. Government support type development pattern is to explore on government "dominant" development pattern, and the emphasis is to make the model system get optimal allocation of a kind of operation ideas through the work of the education department of the party and the government behavior. In this process, the government behavior's main function is integrating the dispersed school health sports resources to achieve the balanced assignment of quality resources, and through the government power to make direct intervention and allocate reasonably the integration of the high quality health sports resources. And lower-class schools emphasize government equilibrium in the use of resources, play the advantage of government support to realize their own "haematopoietic function", and win the sustainable development, to have a gradual transition to middle-class schools.

## 65.5 Conclusion

Because the social stratification phenomenon exists generally, different schools in human, material and financial resources and other aspects of the resource allocation are different according to the development of the school health sports. In the uneven distribution of resources within the framework, school health sports adopting uniform development pattern is not suitable. From the point of view of future development strategies, the school health sports with different amount of sports resources should adopt different pattern of development, so that they can promote the school sports teaching quality health. The unbalanced development pattern is an economic theory proposed in the present situation of relatively scarce and uneven distribution of resources and its introduction has avoided school health sports development in trying to seek "keep in step", "one size fits all" management. The construction of school health sports development mode "upper-class school self-government development, middle-class school inside and outside development, lower-class school government support development" is based on the background of social stratification of the application and innovation of the unbalanced development theory.

## References


1. Xu J (2008) Public and private secondary vocational schools' enrollment game. Teach Manag 11:9–10
2. Guo J (2008). Research on the physical condition and some influencing factors of Henan ordinary private colleges schools. Beijing University of Physical Education, Beijing
3. Ma Z, Xing Y, Shi L (2009) "Class" integration of the teaching mode and the sustainable development of the school sports. Educ Art 11:188
4. Hu H (2009) Research on the development of school sports curriculum resources of Western minority areas in rural areas and heritage of national culture. Guizhou Ethnic Stud 29(4):159–162
5. Yan H (2009) Non-uniform distribution and its impact reviews to favorites of school sports resources in. Todays Wealth 11:141
6. Lin X, Yu Z, Yang Y (2006) China's sports industry development status and countermeasures. Sports Sci 26(2):3–10
7. Li N (2002) China's sports industry development and policy. Nanjing Inst Phys Educ 26(6):43–45
8. Liu M (2005) Sustainable development factor analysis of China's sports industry. Sports Sci 26(5):51–54

# Chapter 66
# Healthy Sports Situation for Different Social Background Based on Analytic Hierarchy Process

**Qingbo Kong and Lingyan Zhang**

**Abstract** Social stratification phenomenon universal existence, the school health education development also reflects the distribution of resources and the faculty and other factors influencing the prevalence differences. Based on the development of school physical education practice problems, the article tries to introduce the hierarchy analysis method the mathematical concepts, and the establishment of school physical education and healthy development of the mathematical model, through the index system weight coefficient, scientific and objective evaluation of different areas of the school sports development condition, have bigger use value.

**Keywords** Background · School health education · Analytic hierarchy process · Mathematical model

## 66.1 Introduction

On January 12, 2011, general office of the State Council issued the "on the national education system reform pilot notice" document, the private education in education field focus on the comprehensive reform of the pilot, which will lead to various levels of schools and state schools with graduate coexistence, and tends to diversify the direction of development [1, 2]. However, social stratification phenomenon exists; it will undoubtedly make teachers and teaching facilities such as the allocation of resources utilization differences. Especially in the secondary

Q. Kong (✉) · L. Zhang
College of Physical Education, Chongqing University of Arts and Sciences, Chongqing 402160, China
e-mail: Qingbo_Kong@yeah.net

W. Du (ed.), *Informatics and Management Science V*, Lecture Notes in Electrical Engineering 208, DOI: 10.1007/978-1-4471-4796-1_66,

school education, but also the existence of city and rural, key and general, inland and coastal difference in such areas, are resulting in the school education stratification. As an important part of school education, school health education development trend also affected by social stratification influences showed obvious difference, which makes the study of different social background of school health education is not balanced development, become the future of Chinese School Physical Education an unavoidable problem [3].

School health education from training target of physical education, based on "health first" as the core idea of the rebuild, refers to efforts to bring health into the teaching of physical education, change to sports teaching is the basic mode of physical education system, establish lifelong sports, health sports, happy sports, sports and other multi-level regression to nature, multiple forms of modern health sports teaching system, so that students from the physical education for health benefit, form the good habit of lifelong exercise [4]. It is the important part of sports in our country, directly related to China's young people's health, physical fitness and well-being of a happy life. Teenagers are the future of our motherland, and the hope of the nation, their physical and mental health is directly related to the country and the nation's thriving and prosperous, and also is a national society whether advantageous system and national civilization degree and important sign.

School health education study abroad compared to earlier, especially in the United States of America's school health education, both from theory and practice aspects of development, all countries in the world health sports reform offers many positive experiences. In 1980, the United States government in the "national goals: health promotion, disease prevention" reports on reiterated the school sports to improve the health of the people of the importance of. From Europe and the United States and Japan in recent years the research situation, using the theory of social action on the students' sports participation, health and sports organizations, sports health sports mode and method of more common. And home for school health education research after 10 years of development, but still limited to sport this single discipline with other disciplines, the correlation between the relative lack of still.

School health education stratification phenomenon is the main reason for the body dimensions, urban and rural difference dimensions, key hierarchy, regional location dimension difference caused by the imbalance of development. As shown in Fig. 66.1.

Compared with the extensive application and an early start in the study of basic theory, development study has certain depth. At present home about school health education basic theory research, more is to learn and use the Western health education theory research methods, mainly concentrated in the school health education concept, object, nature, etc. And combining the current situation of school education in our country is less, its breadth and depth is not enough, which is reflected in the research results on the low maturity. In the application of research, including health and sports management, health sport organizations,

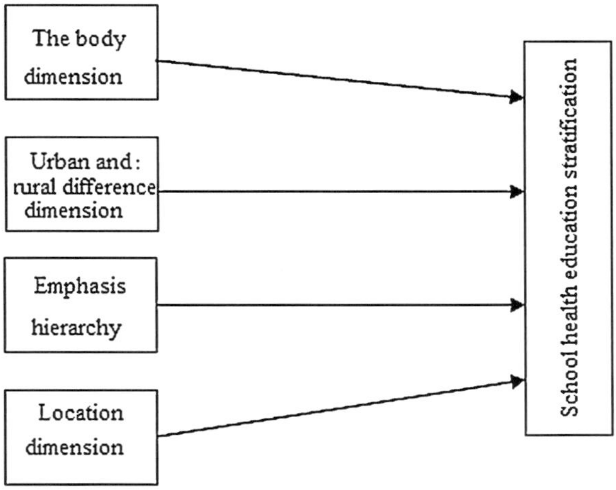


**Fig. 66.1** Stratification phenomenon causes

student health needs and perspectives of the research. Because only based on the actuality development countermeasure, the predictable is poor, cannot keep up with the pace of development of school health education [5]. Compared with foreign countries to start earlier, the mature development of philosophy of science and technology as the basis, the school health education for non equilibrium development of the positive effect of philosophy of science and technology, the domestic study on social stratification is relatively less, late start. But this also directly resulted in the context of social stratification, home for the school health education unbalanced development research is still almost blank state [6].

Because of this, the research on school health education development research as the basic orientation of philosophy of science and technology, the social stratification concepts for the practice of the background, the use of economics in the unbalanced development theory, to the present our country school health education development status differences were analyzed, in order to long-term since confined to physical education by the start of the school sports research limitation of breakthrough. On this basis, further links to Chinese education development problems, using dialectical thinking on social stratification in schools under the background of health sports development mode research [7–9]. At the same time, with the help of philosophy of science and technology in the social stratification theory, from the school health education development of human resources, material resources, financial resources and other aspects define the different school health education development of sex differences, and analyzes the reasons for the differences, to further clarify the current school health education in the context of social stratification in the development. Then on this basis, exploring the different levels of school health education development plan, to construct the social stratification in schools under the background of health sports development model.

## 66.2 Analysis Model

Due to the different social background have different social levels, different social levels and effects of different school sports development. Therefore, we use the hierarchical analysis method to study the different areas of the school sports development.

Analytic hierarchy process (AHP) is characterized by the complex decision problems in essence, its inherent relationship between factors such as based on the in-depth analysis, the use of less quantitative information to decision-making process of mathematical thinking, and thus more objective, criteria or more without the structural properties of the complex decision problems with simple decision method [10, 11].

AHP method is a combination of qualitative and quantitative methods, qualitative factors can make quantitative, the subjective judgment to a mathematical expression, and to some extent to inspection and reduced the subjective effects, make the assessment more scientific. The emergence of the level of analysis to decision makers to solve those difficult to quantitatively describe the decision problem bring great convenience, making it almost involved in any field of science application.

Analytic hierarchy process steps

1. The problem of. According to the problem gives the block diagram of the system structure, to explain the level of hierarchical structure and the relationship between subordinate relationships.
2. Constructing judgment matrix. According to the unified Saaty1–9 grade standard of judgment matrix table (see Table 66.1), for each element of two quantitative comparison, judgment matrix form.
3. Hierarchical ranking and consistency check. According to the structure of the matrix calculation of each element compared to a level of relevant factors important to quantify the extent value. This is a solution of rectangular array of maximum characteristic root and characteristic vector (x max) (W) calculation process. Consistency test is to verify whether the judgment matrix is consistency. Calculation formula for:

$$C_R = C_I / R_I \tag{66.1}$$

Type $C_R$ for the consistency test value, its value is small that judge the effect is better,

$$C_I = \frac{\lambda_{\max} - n}{n - 1} \text{ (N factor number)} \tag{66.2}$$

$R_I$ is for the average random consistency index, its value is determined by the provisions of the 1–9 order matrix table $R_I$ checking (Table 66.2). When

**Table 66.1** Saaty1–9 grade judgment matrix standard gauge

| Scaling | Content description |
|---|---|
| 1 | Factor 2, equally important properties |
| 3 | Factor 2, one important factor slightly |
| 5 | Factor 2, one factor was important |
| 7 | Factor 2, one factor strongly important |
| 9 | Factor 2, one factor is extremely important |
| 2,4,6,8 | The 2 scale adjacent judgment of intermediate value |
| Reciprocal | That factor I in comparison with J, get different values for Xij |
| | Then the factor j in comparison with I has different value of 1/Xij |

$C_R < 0.10$, the consistency test, can be considered judgment effect is good, and is not good, you should readjust the matrix xij,

4. The total sequencing. In the hierarchical ranking based on all factors, relative to the highest level of importance degree of total ranking order quantized values are calculated, and the total sequencing of the consistency check.

When $C_R < 0.10$ for the consistency test. Finally the problems related to the importance of each factor size accurate conclusion.

## 66.3 Research of Analysis Situation

According to the analytic hierarchy process to get the weight of each index, the formula $MWI_i = \sum_{j=1}^{h} X\prime_{ij} W_j$ to calculate the individual multiple weighted index. Type $X'_{ij}$ for the normalized indicators of individual data, Wj as indicators of relative weight, h index number, we get the corresponding weight values. We can see from the weight value of school health education stratification cause.

Effects of school health education carried out by many factors, its existence can be material or spiritual, human or financial resources, objective or content, inside or outside, the current or future, tangible or intangible, as long as these elements can be used in the school health education, and can effectively enhance the physical fitness of students, promoting student health, can be included in the school health education resource category. Social stratification universality that school health education development will inevitably exist resource inequality. Draw lessons from resource theory knowledge, the subject of "human, material and financial resources," the three major resources as the foundation, from the "main school, urban and rural difference, key hierarchy and regional location" of

**Table 66.2** 1–9 order matrix table

| Judgment matrix order | 1 | 2 | 3 | 4 | 5 | 6 | 7 | 8 | 9 |
|---|---|---|---|---|---|---|---|---|---|
| $R_I$ | 0.00 | 0.00 | 0.58 | 0.90 | 1.12 | 1.24 | 1.32 | 1.41 | 1.45 |

the four dimensions of school health education development of the differences in resources features described, for school health Education Level Division lay foundation.

Through to our country different years of school sports goal of comparison can be seen, along with the time development, the goal of our school physical education is expanding gradually, and toward independence, specific, clear, orderly and overall system of multi direction, consistent with the social development in the future. In order to adapt to the development and changes of the society, the function of education is constantly being developed, the school physical education as an important component, its function constantly expanding, with period and society, the development of education is consistent. Only on the basis of school physical education has the function of fixed target, it has direction, reality, will it be possible to achieve. So, to complete the school sports a variety of functions, it is necessary to establish the school sports "much polarization", multi direction and multi objective system.

## 66.4 Conclusion

Different social background caused the school sports and healthy development of the non equilibrium, the unbalanced development of different regions, industries, usually refers to the crowd in the development process of the resource allocation, the accumulation of wealth, income, rights of use and other aspects of the balance or imbalance, or present more and more significant differences in trends. From worldwide in light of, social imbalance is a universal phenomenon of coexisting, and relates to every aspect of social development. Non balanced development is from the effective configuration of resource perspective, considering the economic development in the initial stage, how to allocate the limited resources in the most productive potential. Balanced development and unbalanced development is mutually transforming, in a common economic system, in the contradictory motion of development, and to advanced equalization phase evolution. School health education is not balanced development mode aimed at health sports resource rational distribution and use, in the whole development process, "balanced development is a target, be not balanced development is a method", the final is to realize the school health education higher level and higher level of balanced development.

As long as we analysed seriously and think in different parts of the school physical education situation, study the different school sports development characteristic and law, strengthen the physical education teaching contents and process optimization, layer upon layer takes seriously, catch condominium together, can actually implement quality education, improve students' physical culture and accomplishment, enhance the body quality of students, thus facilitating culture teaching quality improved steadily, achieve the improvement of students' physical quality, promotes the student morality, intelligence, body, beauty, work full scale development, ultimately to achieve the purpose of educating people, rural school physical education will have a new situation, can truly nationwide to promote quality education, for the construction of a harmonious society to provide strong in support of.

## References


1. Huanchen Z (2006) Hierarchy analysis method—a simple new decision method. Science Press, Beijing, p 79
2. Dewei T, Huiling L (2007) Analytic hierarchy process in the evaluation of physical education teaching application. J Shandong Norm Univ Nat Sci Ed 12(2):223–226
3. Minghai L (2005) China's sports industry sustainable development factor analysis. J Sports Sci 26(5):51–54
4. Xianpeng L, Zhonggan Y, Yue Y (2006) The current development of sports industry in China and countermeasures. Sports Sci 26(2):3–10
5. Lining (2002) China's sports industry development and policy. J Nanjing Sport Inst 26(6):43–45
6. Hongjian L, Sun Celebrates, Rongbing T (2009) The symbiosis theory in China from the perspective of urban rural coordinated development of the mass sports. J Capital Inst Phys Educ 21(5):538–540
7. Cailing L (2010) Shaanxi ordinary high school sports and health courses analysis. Inner Mongolia Sports Sci Technol 23(3):122–123
8. Fang W (2007) Rural school physical education reform of philosophical reflection. Educ Explor 11:26
9. Naiwu H, Chen W (1994) Balanced and unbalanced development of basic industries: theoretical model and practical choice. Mod Econ Res 2:3–5
10. Weigang C (2010) Urban and rural education from the perspective of urban and rural school sports development of integration-to Chengdu, Chongqing overall urban rural education reform experimental area as an example. J Chengdu Sport Univ 36(9):80–83
11. Chunlin Q, Zhiwen M (2004) On non equilibrium development of Chinese mass sports. J Beijing Sport Univ 27(7):865–868

# Chapter 67
# Research on Outdoor Sports in Southern Sichuan

**Lin Liu, Chengxiang Liu and Xiaogang Li**

**Abstract** The investigation of the status quo in the outdoor sports of southern Sichuan, in-depth analysis of the major programs undertaken by the Kawaminami outdoor sports, participant age, gender, project selection tendentious and to participate in outdoor sports of the time, age, location, consumer, and forms of organization characteristics, and the corresponding countermeasures and suggestions to guide practice, and to provide reference for the better development of southern Sichuan outdoor sports.

**Keywords** Outdoor · Sports in southern · Survey

## 67.1 Introduction

With life increasingly fierce competition of modern life and the increasingly rapid pace, increasing pressure, has a sense of tired of the traditional competitive sports content and mode, the modern man needs a relaxed, spacious environment with different constraints space, to seek new thrills, relieve stress, and outdoor sports are able to meet people's needs [1].

Southern Sichuan refers to the Sichuan Basin, southern region, mainly refers to the four prefecture-level city of Zigong, Neijiang, Yibin, Luzhou [2, 3]. The superior geographical conditions, climate, resources and cultural resources to make southern Sichuan to become suitable for conducting outdoor sports one of the regions, the number of people involved in outdoor sports growing variety of

L. Liu (✉) · C. Liu · X. Li
Sichuan University of Science and Engineering, Zigong, China
e-mail: scieiistp@foxmail.com

W. Du (ed.), *Informatics and Management Science V*, Lecture Notes in Electrical Engineering 208, DOI: 10.1007/978-1-4471-4796-1_67,

outdoor sports growing prosperity [4]. This paper selects the southern Sichuan to carry out a good four outdoor sports outdoor sports clubs and club members as well as outdoor sports enthusiasts as the research object, the current situation of the surveys and studies carried out in southern outdoor sports [5, 6].

## 67.2 Objects of Study and Research Methods

To the Zigong pinnacle of outdoor clubs, Zigong salt Outdoor Club, Yibin city Ziyou Ren club, the Luzhou Lynx outdoor sports club members, spontaneous participation in outdoor sports crowd and outdoor sporting goods dealers conducted a survey for the study. In person by participating in club activities, outdoor sports facilities, a random sample of 336 participants conducted a questionnaire survey to participate in outdoor sports, and 287 valid questionnaires.

In this paper, literature, questionnaire, field trips, method of mathematical statistics and other research methods.

## 67.3 Survey Results and Analysis

### *67.3.1 Major Projects in Southern Outdoor Sports*

Southern Sichuan to carry out outdoor sports including mountaineering, hiking, camping, high altitude mountaineering, orienteering, picnic, caving, rafting, Skiing, and field, hiking, hanging down, rock climbing, to expand the project to carry out more but primary or secondary, is closely related to the participants in the selection of projects and environmental resources. Since this survey to the southern Sichuan, the investigation of natural selection of outdoor sports content by the impact of the geographical features of southern Sichuan and surrounding areas can be seen, according to the returned questionnaires, hiking (29.2 %) walk through (18.8 %), picnic (13.3 %) participants to choose the project, followed by high-altitude mountaineering (10.2 %), camping (7.3 %) and drift (5.5 %) and directional movement (4.0 %), other items of 11.7 %.

### *67.3.2 Selection Tendentious Analysis of Different-Sex Participants in Project*

Men like the excitement, adventure, fresh environment, able to satisfy their curiosity and challenge desire in order to get mental and physical relaxation, such as rock climbing, mountaineering, rafting and other projects, such projects with a strong challenge on the technical have certain requirements and the physical and men are often involved in the project more than women; women because of the week of strength, endurance, and less engaged in risky and violent project, often choose the less dangerous, easily, the action is relatively simple project. Such as: on foot through the directional movement, outings, both to shape the body, but also meets to breathe the fresh air, soothing mood in full compliance with the characteristics and needs of modern women.

### 67.3.3 Selection Tendentious Analysis of Age of Participants in the Project to Select the Orientation Analysis

From the age distribution of the number of 26–35-year-old age accounted for 44.8 %, followed by 18–25 years (27.8 %), again 36–45 years (17.4 %), over the age of 46–55 and 56 participating population only 10 % of the sum. The age distribution of outdoor sports to participate in groups, more concentrated, mostly for young people aged 26–35 such groups' work, the economy, a stable income, discretionary time is relatively large, relatively abundant physical, but at the same time working and living fast-paced, high pressure need to be able to vent and relax in the natural environment. Outdoor sports participation by the economic, physical, and social factors, the age of 18 and 46-year-old age groups are very low participation rate.

Different age groups in the project selection are also very different types of people with the characteristics of their own age, the purpose of participating in outdoor sports and form are different. Young people like to pursue novel and stimulating, easy to accept the forefront of information and things concerned about popular outdoor sports and related products, and gradually formed a unique insight on outdoor sports and outdoor products. Secondly, young people have better physical fitness and physical, after a certain amount of training that is able to master the basic skills of outdoor sports; the item difficulty constrained system is small. Economically powerful middle-aged, decompression, communication, health promotion is the basic purpose of their participation in outdoor sports activities can be exposed to different people, it's the development of mental and physical exchange of ideas of great help. Their choice of projects is the high consumption of the outdoor sports groups. Elderly people are chosen to fitness as the main purpose of the project, mainly excursions, mountaineering, etc., does not require specialized equipment and motor skills, place requirements are relatively low, less adventurous activities.

### 67.3.4 Educational Level and Occupational Status of Participants

Investigators, the number of secondary school or high school diploma, as well as secondary or high school degree or less the less, only 4.8, 71.7 % college or undergraduate degree, graduate or above accounted for 23.5 %, reflecting the outdoor sports to be compared with high education level of people awareness.

With the rapid development of socio-economic conditions, the spread of education in varying degrees to promote the development of new to outdoor sports, the more acceptable education, professional skills and knowledge and skills, the stronger the strength of social participation, while a wealth of knowledge is part of the indirect experience of the past, it is important to the transformation of past experience, the higher the education level, the more we can correctly apply the theoretical knowledge of the principles and methods as a guide, can greatly enhance the practical ability.

Survey showed that outdoor sports participants to type of occupation is primarily for technical personnel, management personnel, civil servants, including technical staff accounted for 43.0, 31.0 % of party and government organs enterprises personnel, business people, accounting for 9.4 %. Analysis because of technical personnel, civil servants relatively stable income, the income of sports activities play a role in safeguarding the human spirit, material and consumer awareness is awakened, when the reach a certain level of economic income, and decided its consumer orientation.

Participants in time, age and place of the outdoor sports.

Time choice, 43.4 % of people chose the weekend and other holiday's time, the weekend is fixed leisure time, most participants are working 5 days a week, less the whole time, before the weekend in advance some of the activities planned and arrangements to ensure the activities carried out smoothly. In addition, due to the adjustment of the national vacation time, the May Day holiday to 3 days, leading most people to reduce long-term plan for this period is relatively more reasonable on weekends or other holidays. About 16.8 % choose to participate in outdoor sports in the Chinese New Year, winter and summer vacations and other holidays, the kind of crowd to the teachers, civil servants. 31.4 % of people choose to avoid the peak period usually less busy time for outdoor sports, this part of the people involved in outdoor sports is more casual, higher flexibility, optional strong. And 8.4 % choose another time to participate in outdoor sports.

Respondents, 33.7 % of the exposure to outdoor sports less than a year, 2 years of exposure accounted for 20.4, 17.1 % in 3 years, 4 years accounted for 16.1 % of participants in more than 5 years was 12.7 %. One-third of the survey not long exposure to outdoor sports, reflecting the outdoor sports from the incoming to the popular universal speed is not fast; Another important reason is that the survey for the club member on a variety of activities and forms of unfamiliar novice outdoor sports club organization with more choices, so a larger proportion of novice. 53.6 % of the people involved in outdoor sports between 2 and 4 years, 12.7 % participate in activities in more than 5 years, indicating those outdoor sports more attractive, and the participants to choose the sport and to adhere to.

Participate in outdoor sports activity area has not fixed the active area of the respondents are more choices. 49.2 and 24.6 % of the outdoor sports in the nearby and other areas of the province, 17.5 % chose the neighboring provinces and cities, 8.7 % chose other areas. For participants in activities, transportation, climate and other factors, in southern outdoor sports venues are mostly concentrated in the suburbs and adjacent areas.

## *67.3.5 To Participate in Outdoor Sports Consumer Survey*

### 67.3.5.1 The Level of Income

There must be equipped with professional tools of the need for outdoor sports enthusiasts, while also taking into account the consumption of related industries such as transportation, accommodation and catering industry, so consumption is relatively high, participants must have a steady income. The survey shows that: income level accounted for 44.6 % in 2,000–4,000 Yuan/month; 31.6 % in 4,000–6,000 Yuan/month; 6,000–8,000 Yuan/month, 16.6 % higher than 8,000 Yuan, accounting for 7.2 %. Visible outdoor sports participants in the economic base are relatively strong.

### 67.3.5.2 The Consumption Expenses

Outdoor sports participants in south Sichuan on expenditures can be broadly divided into three categories: First, we need to pay the necessary consumption, the second category is the hats clothing or other necessities in life can also be frequently used, the purchase of large equipment (such as tents, bicycles, mountain climbing tools, etc.). 25 % of the population of the annual outdoor sports consumption 800–1,500 Yuan, 43.3 % of the population each year the proportion of outdoor sports consumption in more than 1,500 Yuan.

### 67.3.5.3 The Consumption Structure

Consumption of outdoor sports, including a variety of forms, in accordance with the purpose of outdoor equipment can be divided into infrastructure equipment and professional equipment, and basic equipment generally refers to most of the commonly used outdoor sports equipment, such as backpacks, sleeping bags, tents, outdoor clothing, compass, lamps, etc.; professional equipment is the need to use the equipment in special activities, such as seat belts in the high-altitude mountaineering, high altitude boots, ice ax, climbing, rock climbing shoes, and adventure activities in the maritime satellite telephone. In southern outdoor sports participants to have the sort of products and equipment, outdoor clothing (26.0 %), small tools (24.8 %), backpack (16.5 %), sleeping (13.7 %), tent (12.2 %), hiking shoes (6.8 %), etc.

In southern outdoor sports in the form of organization can be divided into three, one outdoor clubs in professional coach, manager, outdoor clubs, these factors have attracted most of the outdoor sports novice, you can quickly contact the love of outdoor sports to expand the outdoor circle of friends, enhance their outdoor capabilities and improve the quality of outdoor sports, but also attracted the participants to participate in outdoor sports in the difficult play to the strength of the team club such groups, to complete the more difficult challenges. Outdoor sports enthusiasts, self-organization, the Network convened this form is very popular, and participants through with the accumulation of experience in outdoor activities to participate in a certain period of time, and constantly sum up thus spontaneous activities of the Organization, this situation needs the organizers of outdoor sports experience and ability, participants need to because they have a strong degree of trust. Southern Sichuan outdoor sports through the network organized project QQ outdoor group activities through the organization; the Friends group has expanded rapidly, to carry out different outdoor sports. The third is a small part of the group's own activities, different personalities, who constitute the human society, like group activities, of course, experience alone will be like part of the population, their activities arbitrary stronger and more suitable for the modern era character.

## 67.4 The Conclusions and Recommendations

1. Survey carried out in southern outdoor sports, the gender differences in outdoor sports participants, the age distribution is more concentrated, to participate in the crowd getting younger and younger, highly educated, career stability, high-income and other characteristics, but to participate in group The universalization trend.
2. Participants choose to participate in the project, time, regional, consumer and other aspects of the impact of social and economic development in southern and restrictions, the outdoor Kawaminami Campaign still in the low phase. The relevant departments should actively encourage the public to participate in outdoor sports, to promote national fitness; related business units should step up publicity efforts to develop marketing strategies based on the actual situation, and guide people to participate in outdoor sports, and broaden participation crowd and expanding consumer market.
3. Reasonable development and utilization of a variety of outdoor sports, environmental resources and clubs, organizations and other market resources, make full use of resources, the development of the emerging outdoor sports, especially for women and older participants in the project; for students teenagers and office workers crowd, should focus on outdoor sports holiday period project activities, outdoor sports to participate in the population more widely.
4. Emphasis on outdoor training of professionals, all kinds of professionals (such as professional sports, tourism, professional, medical professionals, etc.) in southern colleges and universities training either as outdoor sports managers, coaches, athletes, research and medical personnel, but also can be used as outdoor sports publicity, development of outdoor sports will play an important role.

**Acknowledgments** The work is supported by Research Projects of Sichuan University of Science and Engineering under Grant Nos.2010XJKYW10.

## References

1. Lin L (1999) Historical materialism, leisure interests on the. Fudan University, Shanghai
2. Li H (2008) The theory and practice of outdoor sports, vol 12. Beijing Sports University Press, Beijing, pp 20–24
3. D Ma (2002) Outdoor sports accidentally pop up. Knowl Based Econ 9:18–26
4. Chen X (2003) Directional movement and field survival training, vol 19. Zhongshan University Press, Guangzhou, pp 128–134
5. Nan F (2005) Very promising new industries in the outdoor sporting goods. SME Technol 07:2–6
6. Luo D (2003) Western outdoor sports of. Chin Sports Sci 12:91–93

# Chapter 68
# Research on Humanity Spirit of Sports in Harmonious Society

**Nian Tang and Peng Li**

**Abstract** The unprecedented social transformation in China and the idea of building a harmonious society put forward by the CPC has brought a great shock to the social life, which make people's minds and spirit face enormous challenges, as well as the human-based sports with its core humanism be the first to be affected. This paper takes the humanistic spirit of sports as the research target, defining it and digging its value; raising the objective, principle, content and the available path of constructing humanistic spirit of sports in the harmonious society with the purpose of advancing it and developing it to play much greater role in the process of civilization of human life and society.

**Keywords** Harmonious society · Sport · Humanity spirit · Construction

## 68.1 Introduction

The rapid development of science and technology has largely developed people's living standard, so that to promote the life quality and pursue the value of life has been gradually becoming the utmost important for the contemporary people and the idea of "good health is over wealth" becomes the common point in twenty-first century [1, 2]. PE, an avant-garde subject be directly concerned with human quality, nation's power and the release one's potential, has become a hot subject nowadays [3]. Those ignorant thoughts such as "be strong-limbed but simple-minded" in history have gone forever. In twenty-first century, the development

N. Tang (✉) · P. Li
Sports Institute, Huzhou Teachers College, Huzhou, China
e-mail: tangnian1233@163.com

W. Du (ed.), *Informatics and Management Science V*, Lecture Notes in Electrical Engineering 208, DOI: 10.1007/978-1-4471-4796-1_68,

direction of science is integration science with humanities. When divorcing from humanities, science development will reach a dead end. Therefore, in the new century, the sports science as a complex cultural circumstance also will reach the opposite side of human development if it doesn't strengthen the new idea of "human-based sport"[4–7].

Humanities, acting as the soul of the development of human culture, are the core and foundation of ethical and cultural progress. Its essence in modern time is an action of development in all-around way to cherish freedom. In fact, sport humanities itself is a human phenomenon with the development of human social history, whose presence inevitably based on a certain human culture and ethical basis [8]. As a historical product, its deep spirit and rich contents are displayed in various physical activities in different forms. From the general trend of human culture, physical activities are absolutely impossible only with benefit purposes, but more the humanistic purposes.

Sports are a social and cultural activity with strong humanities spirit, which decides sports' social value, having common and long-term social needs, so it is quite important to keep and strengthen its humanities spirit. It is based on human beings and it grows up on the basis of the humanity constructed by social culture. It contains all the wisdom and grand ideals created by human beings, covering all meanings of people's life.

## 68.2 Core Principle of the Humanity Spirit of Sports

To highlight the development in all-around way and cultural features need to confirm the core idea of humanity of sport. I think the humanity of sport can be defined as "harmonious", because harmonious is China's traditional cultural essence; construction of "harmonious society" is an important objective of building a moderately prosperous society in all aspects; it is in line with the trend of peace and development and propagating harmonious idea could advance further development of humanity of sports.

Chinese culture, standing for the oriental culture, is epistemology and humanistic view influenced by harmonious ethics. It emphasizes the achievement of family, clansman and national interest, the responsibility and obligation of individual to the group. It takes individual as a member of the group, regarding them as interactive individual with the requirements of group living and ethics that every one's destiny is closely linked to the group [9, 10]. From the individual view, Chinese culture cultivates one's moral character as basis, taking the good, the true and the righteousness as the most important value, considering "every person can be a sage"; From the group view, it focused on relationship, relationship between people and society, ethics and harmony. In the relationship between people and nature, it emphasizes harmony and unity, but not takes conquest of nature as the start point. Chinese culture, featuring of entirety, balance and harmony, largely influences the historical development of Chinese.

Throughout the development processes of physical activities, some disharmonious events such as dug abuse, over commercialization, terrorist threat, sports politicization etc [11, 12]. This should be solved from sense and practice.

Harmony is a good solution to the co-existence for different cultures. Propagating "harmony" will reduce conflicts to get physical activities develop sounder. In all the processes of the sports, we should carry out harmonious principle to highlight China's bright human culture, making every participator fully feel enchantment of harmony, promoting complementary and communication between Chinese and Western cultures.

We should understand profoundly the basic intension of "humanity of sport", and always focus on it in physical activities to disseminate and carry forward the humanistic ideas of sport, displaying China's brilliant culture.

The humanity of sport is an important content of sport culture, which depends on the PE thought, sports ethic and especially the sports value. Its value has a rich intension with types of existence forms and representations. Thus its value is achieved through multi-layer, paths and ways such as its nature, function and cultural deposits etc. Man's all-round development is the topic in harmonious society, so it can be believed that the value of humanity of sports, in a ear of emphasizing people's development and converting the status of meeting people's material needs into that of meeting people's spiritual needs, will be rediscovered and developed. It will play more and more important role in human life and social civilization.

## 68.3 Promotion to the Humanity of Sports by Building a Harmonious Society

Sociologists repeatedly emphasize a theme that humanity is the epitome, focus and refraction of the society's intellectual civilization when doing research on humanity. They believe that sport, same as the traditional fields such as family, religion, politics and economy, is also one of the most extensive filed with common cultural and living style and basic social system in the industrial society.

The 16th Third Plenary Session advanced by the CPC raised scientific outlook on development: to adhere to the people-oriented, set up the comprehensive, coordinated and sustainable development, promote the economic, social and all-round development of human being, which predicts that the society will undergo substantial changes and reform, so as the sports in China will also develop under its guidance.

The scientific outlook on development is an overall development view. It tells us that anything is a system, containing many sub-systems which connect with each other, restrict each other and interact with each other, deciding the entire functionality. The overall development view can promote the sport and fitness program in the round, including the subsystem of humanity of the sports.

The scientific outlook on development is a coordinated development view. Dialectic method tells us that nothing stands alone, but is generally linked with other things. The humanity of sport is closely linked to the social development, degree of education and the development of sports and fitness, so it could move along well only inter-coordination with social culture, and economy.

The scientific outlook on development is a sustainable development view. Everything including humanity of sports must have sustainable development ability, which gives an important expression to the scientific development view. Sustainable development is to advance harmony between human beings and nature, realize economic development and inter-coordination among cultural, resource and environment. The sustainable development intension covered in the humanity of sport itself is consistent with the sustainable scientific development view.

The scientific outlook on development is a people-oriented development view. People-oriented is the essential starting point and the foothold. It brings the humanity of sport to a new height. Adhere to the people-oriented development view needs to aim at achieving man's all-round development, constantly meeting people's growing material and cultural needs, which provide good social environment for and point out scientific development direction in theory to the development of the humanity of sports, largely promoting the combination between human beings and sports. By means of sport to develop people's physical and psychological health is benefit for stimulating the masses' passion of participating physical exercises, which will strengthen their awareness of life-long sports to encourage the development of the humanity of sports.

In short, the scientific outlook on development is a new policy of managing state affairs and governing philosophy with an important guiding significance to the development of humanity of sports. Thoroughly implementation of it will propel the overall, coordinated and sustainable development between human beings and nature. Only adhere to the people-oriented development view can promote the in-round development of various relationships and survival modes including man's needs, quality and capability. The humanity of sport in harmonious society should represent and promote civilization, harmony, peace, progress and development of human society, strengthen the combination with environment protection; Physical activities should build individual robust, perfect human nature and harmonious development oneself; It must focus on the harmonious relationship with each other.

Characteristics of Internationalism and social communication for modern sports decide sport in a certain range, its technical characteristics, methods of operation, rules of the game must have a certain normative, binding and communicating with each other. Sports, only seriously treat and properly handle some technical, tactical, physical, psychological, ideological, moral, nutrition, diet, and a series of questions on the neighborhood and contingencies which has happened and will happen to ensure that sport smooth movement, and achieve good results. Any fraud, violation of sportsmanship behavior and actions will be opposed by the sports participants and sports viewer and also the pariah by the whole society.

Currently, the world has entered a rapid development of science and technology knowledge economy, science and technology has become the primary productive force. Scientific literacy is one of the important qualities which modern people must have; sports can not be replaced for developing the scientific literacy of people with other educational role. The moral value of sport lies in scientific literacy in the form and means through sport training the student to study science, are willing to sacrifice for the good moral science. Modern sociological research shows that: first of all moral education is a life science attitude, science awareness, the spirit of science education. In sports, which people must first abide by is the scientific attitude of sport, treat yourself scientifically, and treat competitors and also the surrounding natural environment, human environment, scientifically take nutritious material, in the training, scientifically impose physical and mental workload, only do like that, can obtain good results in sports, get beat after the happiness and joy.

Beliefs of perseverance and truth-seeking should be good quality which all scientific workers should have, also for them to search for truth and have a struggle all the life, or even the power sources of several generations struggling for. In the process of sports, as people can learn to deal with the problem, they must have the habit of thinking subject to the facts, Whether exercise technical and tactical use of athletes, emotions, emotional regulation, athletes individual skills mix, and reasonable to treat the media coverage and dissemination of methods, rational thinking should have a comprehensive, stand the test of practice, and in practice to distinguish the authenticity, in a courage to correct their mistakes. When the famous physiologist Pavlov concluded his successful experience, he said: "develop the habit of rigorous and patience, learn to do detail works, research the facts, compare facts, the accumulation of facts… … however, in research, experiment and observation, we should try not to stay on the surface of the facts." In sports, only to educate students from childhood to form the attitude of the realistic fact-based truth-seeking, so they can dare to face and explore the surrounding of the problems that arise, and to solve those issues one by one well-organized, so that they can pave the way for their lives to achieve the desired on the other side of the Strip.

## 68.4 Construction of the Humanity of Sports Under Harmonious Society

To bring the humanity of sports compatible with the development of harmonious society, it should be constructed to fit for the new historical time that pay equal attention to science and technology and humanity, cultivating talent with high humanity and cultural literacy.

The humanity of sport in China is facing hard situations that having human crisis, but China's spirit has its unique cultural value, integration and cohesion. The general theme of China's contemporary humanity of sports has formed, and increasingly represents keen people-oriented trend, but its ethnic cultural theme and time theme are still a historical tasks to be constructed and improved. China's sports have to be based on its social and cultural environment to build its spirit in line with time characteristics. It should hold high the greater banner of Deng Xiaoping theory, carry out seriously "3 delegates" thought and follow the harmonious society's basic line and program put forward by the CPC to represent the ongoing way of advanced culture and meet people's growing sport needs. Review the core value of China's contemporary sports to keep it in consistent with the overall development of the "harmonious society".

Democracy and nomocracy are not only the objective of humanity of sport in harmonious society, but also its basic guarantee. The essence of the legal spirit of sports is to pursue equality and justice and human rights and respect which is in line with the harmonious society's spiritual pursue. Only on the basis of democracy and nomocracy can the humanity of sports stick to the sports by law and administration by law to guarantee its healthy and orderly operation and citizens' basic rights of sport.

The humanity of China's sports has some discordant cacophony in the theme of social transformation, so it should seize the opportunity of the construction of harmonious society to improve it. For institutional innovation, theory is necessary, so it should start from aspects of nomocracy, rule of virtue and so on to convert its development mode into innovative direction during the reform of sports structure, bringing the humanity of sports compatible with the development environment of harmonious society and the needs of the masses.

Justice and equality are the basic spirit and essence of socialism and the basic principle and value standard of the harmonious society. The sport itself is equal. It denies inequality except for body, psychology and technique. Everyone has right to participate in physical activities and compete on an equal basis. The objective of equal humanity of sports will well harmonize various relationships in its development and guarantee citizens' sport right; further more, vulnerable groups' sport position and right will get attention. Therefore, it should make the justice and equality as objectives in the construction of the harmonious spirit of sports.

## 68.5 Conclusion

The unprecedented social transformation in China and the idea of building a harmonious society put forward by the CPC has brought a great shock to the social life, which make people's minds and spirit face enormous challenges, as well as the human-based sports with its core humanism be the first to be affected. This paper takes the humanistic spirit of sports as the research target, defining it and digging its value; raising the objective, principle, content and the available path of constructing humanistic spirit of sports in the harmonious society with the purpose of advancing it and developing it to play much greater role in the process of civilization of human life and society.

## References

1. Wang X (2006) Record of the search for the spirit of the humanities, vol 2. Wenhui Press, Beijing, pp 13–20
2. Xu S (2009) Humanity spirit theory, vol 36. Hubei People's Press, Wuhan, pp 256–257
3. Yang L, Zhang W (2007) Construction of current Chinese humanity spirit, vol 39. People's Press, Beijing, pp 24–26
4. Tong Z, Sun Q, Zhou N (2008) Humanism sports-culture of sport development, vol 81. China Customs Press, Beijing, pp 20–24
5. Xie Y (2009) Research on olympic, vol 04. Beijing Sport University Press, Beijing, pp 88–92
6. Jiang Z (2009) Build a well-off society in an all-round way and create a new situation in building socialism, vol 09. People's Press, Beijing, pp 87–91
7. Hu J (2005) What we will build a harmonious socialist society. Peoples Daily 219:45–50
8. Lu Y (2009) Sport Sociology, vol 10. Beijing Sport University Press, Beijing, pp 82–85
9. Qu Z et al (2005) Modern society and school physical, vol 84. People's Sports Publishing House, Beijing, pp 18–23
10. Sun Q (2010) Sport generality, people's sports publishing house. Natl Sports Inst Mater Committee 71:22–28
11. Yang W (2006) A number of opinions to further strengthen and improve ideological and moral construction. CPC Cent Committee 12:111–114
12. Mao Z (2003) New vision of physical education, vol 09. Beijing Sport University Press, Beijing, pp 77–82

# Chapter 69
# Research on Status of PE Teachers in Private Universities

**Guosheng Zhang**

**Abstract** Private universities are one of the important components in our country's higher education system. It has a great development in the number or on a scale. Through the current situation of research in Henan province private universities sports teachers. According to the analysis of existing problems, the author put forward the corresponding measures, only for the relevant part of the reference.

**Keywords** Private universities · PE teachers · Present situation

## 69.1 Introduction

September 1, 2003 in Private Education Promotion Law, "the private education has been clear on the legal aspects of the status of private universities, it is an important part of the socialist educational undertakings. Countries have private universities to implement" active encouragement, strong support, correct guidance and management according to law, the approach [1]. "CPC Central Committee on Education Reform and Development" clearly, the gradual establishment of a government school-based community schools system, the state also raised in the Ninth Five-Year Plan and the 2010 development plan, the basic form of public schools and private schools and common development. Under the relevant national policies to encourage private higher education in China has made considerable progress [2]. In short, the support and encouragement of the party and the state has adopted a series of policies and regulations, the private higher education is

G. Zhang (✉)
Zhengzhou Huxin Xueyuan, Henan 451150, China
e-mail: 2430031317@qq.com

W. Du (ed.), *Informatics and Management Science V*, Lecture Notes in Electrical Engineering 208, DOI: 10.1007/978-1-4471-4796-1_69,

becoming an important part of education in China, and it has entered a booming period [3]. The major part of physical education is also of higher education, sports education in private universities the normal smooth carried out, cannot be separated from the important role of physical education teachers. Only the construction of a number of high-quality teachers in order to ensure and improve the quality of private college physical education and scientific research level, in order to get the community more widely recognized and more government support. Therefore, the emphasis on private universities in PE-owned team building is to improve an important factor in personnel training of the private universities.

## 69.2 Object of Study and Research Methods

### *69.2.1 The Object of Study*

Physical education teachers teaching staff of 24 private universities in Henan Province, as the research object. This research refers to the object of study does not include all institutions of higher secondary School affiliated Institute. A random sample of Zhengzhou Institute of Technology, Yellow River Institute of Technology, Zhengzhou Sheng Da College of Business Administration, Nanyang Institute of Occupational, Songshan Shaolin Wushu Vocational College, the Xuchang ceramic Vocational College, Zhengzhou City Vocational College, Zhengzhou Vocational Technical College and other sports of the eight private universities teachers and the Department of Research and investigation.

### *69.2.2 Research Methods*

#### 69.2.2.1 Literature Data Method

View a large number of related papers on domestic and foreign private college physical education Teachers laid the foundation for the writing of this article in the paper writing.

#### 69.2.2.2 Questionnaire Method

Henan Province, eight private colleges' physical education teachers and the Department of Research and Director of the questionnaire issued in order to meet the needs of the research purposes. Questionnaires were distributed and 105 valid questionnaires 95, the effective rate of 90.5 %.

#### 69.2.2.3 Interview Method

Communication tools such as site visits and telephone contact-depth understanding of the status quo of private colleges physical education teachers.

## 69.3 Results and Analysis

### *69.3.1 Sources of Physical Education Teachers of Private Colleges in Henan Province*

As can be seen from Table 69.1, the private colleges of physical education teachers in Henan Province, the source of part-time mainly accounted for 57.9 %, accounted for 10.5 % of retired teachers. Society to recruit full-time staff accounted for 31.6 % seen the Henan Province of Teachers in the source channel is relatively wide, so to some extent, private universities in Henan Province on the management of teachers' sources not rigorous enough, in order to better development of the private universities.

### *69.3.2 Gender Analysis and Age of Physical Education Teachers of Private Colleges in Henan Province*

As can be seen from Table 69.2 Private University Teachers in Henan Province, the age of 30 years of age and 30–40 between the main, accounted for 73.7, 17.9 %, while 40–50 years old accounted for 5.3 %, accounted for more than 50 years old 3.1 %. The age distribution of teachers reflects the private universities in Henan Province with a young physical education teachers, they have much room for development; also reflects the private universities of Henan Province sponsoring a short time, lack of teaching experience in middle-aged and young teachers and the lack of disciplines to take the lead person in charge, to some extent to delay the normal development of the sports teaching in private universities in Henan Province and raise the level of sports research.

As can be seen from Table 69.3, private universities in Henan Province, the majority of male teachers, female teachers accounted for only a small part. The introduction of the new teachers in the past 2 years, in order to meet the

**Table 69.1** Sources of physical education teachers of private colleges in Henan province (N = 95)

| Category | Retired teacher | Part-time teachers | Social recruiting full-time teachers |
|---|---|---|---|
| Share of the number | 10 | 55 | 30 |
| The proportion (%) | 10.5 | 57.9 | 31.6 |

**Table 69.2** Distribution of teachers in the age (N = 95)

| Age | Less 30 | 30–40 | 40–50 | Above 50 |
|---|---|---|---|---|
| Guide | 70 | 17 | 5 | 3 |
| Proportion (%) | 73.7 | 17.9 | 5.3 | 3.1 |

**Table 69.3** Gender distribution of teachers (N = 95)

| Gender | Male | Female |
|---|---|---|
| Number | 70 | 25 |
| Proportion (%) | 73.7 | 26.3 |

**Table 69.4** Distribution of titles and degree (N = 95)

| | Titles | | | Degree | | |
|---|---|---|---|---|---|---|
| | Assistant | Lecture | Above lecturer | Specialist | Undergraduate | Above master |
| Number | 60 | 25 | 10 | 5 | 66 | 24 |
| Proportion (%) | 63.2 | 26.3 | 10.5 | 5.3 | 69.5 | 25.2 |

requirements of the institutions of higher learning school sports in the new era, the private universities in Henan province are starting to focus on the talents of female physical education teachers to introduce, such as dance, hip-hop, yoga full-time teachers, but the introduction of the was relatively small, the imbalance of male and female teachers the issues that must cause the attention of the relevant departments of private colleges.

### *69.3.3 Distribution Analysis of Titles and Degree of Physical Education Teachers of Private Colleges in Henan Province*

From Table 69.4 that, the physical education teachers of private colleges in Henan Province in the title, assistant accounting for 63.2 %, lecturers accounted for 26.3 % of lecturer or above accounted for 10.5 %. It can be seen that the distribution of the titles of the physical education teachers of private colleges in Henan Province is not reasonable, the small number of more than lecturers and lecturers with a wide gap between the vocational college personnel training requirements in the job evaluation Ordinance. Undergraduate degree in educational attainment are more accounted for 69.5 % of master's and above accounted for 25.2 %, accounted for 5.3 % of specialist. 2008 before the introduction of talent undergraduate degree in 2008, with the national emphasis on the Education, due to the severe employment situation, part of the master's degree can also choose to enter private universities, from this point can be seen in Henan Province Private College faculty teacher's far less than public universities. Therefore, to enhance the introduction of the teaching staff is an important factor for sustainable development of the private universities in Henan Province Sports.

**Table 69.5** Work intensity analysis (N = 95)

| Weeks the number | 10 h | 11–13 h | 14–16 h | 16 h or more |
|---|---|---|---|---|
| Proportion (%) | 10 | 11 | 14 | 65 |

**Table 69.6** Teachers salary (N = 95)

| Level | Less 1,500 | 1,500–2,000 | 2,000-2,500 | More 2,500 |
|---|---|---|---|---|
| Guide | 6 | 36 | 38 | 15 |
| Proportion (%) | 6.3 | 37.9 | 40 | 15.8 |

### 69.3.4 Work Intensity Analysis of Physical Education Teachers of Private Colleges in Henan Province

Can be seen from Table 69.5, the work intensity of the physical education teachers in private colleges in Henan Province to more than 16 h accounted for 65 % 14–16 h 14 % 11–13 h 11 %, accounted for 10 % and 10 h of week approved by the State Council promulgated the School Sports Regulations set forth in the colleges and vocational secondary school physical education teachers workload of 10–12 h [4], the workload of the visible private universities of hours of physical education teachers than the state regulations of hours workload beyond. Physical education teachers in addition to normal teaching activities, but also take on the extra-curricular training and group activities of the organization on campus, as well as the necessary scientific research and cultural knowledge learning and teaching of sports under the status of this overload of work, private universities in Henan Province quality can hardly be guaranteed. In recent years, increasing number of students admitted as private universities, the previous number of teachers did not receive a timely complement, resulting in a serious shortage situation of physical education teachers. In this grim situation faced by the private universities in Henan province caused by the intensity of physical education teacher's work career weary.

### 69.3.5 Salary of Physical Education Teachers of Private Colleges in Henan Province

As can be seen from Table 69.6, sports salaries of teachers of private colleges in 2,500 more than accounted for 15.8 % between 2,000 and 2,500 between accounted for 40 % between 1,500 and 2,000 37.9, 6.3 % below 1,500. To some extent, the wages of private colleges physical education teachers far less than they paid labor, a huge difference compared with the treatment of the public teachers, low-income young teachers will be difficult to feel at ease teaching, serious impact on sports passion of the work of teachers. Private Education Promotion Law

**Table 69.7** In-service training and research analysis

| Content | The content of training | Without training |
|---|---|---|
| Teachers | 23 | 72 |
| Proportion (%) | 24.2 | 75.8 |

Article 27 provides: "Private school teachers, the educated public school teachers, educated with the same legal status" [5]. The law enjoyed by private teachers and public teachers the same rights as there are specific provisions, which means, to improve the treatment of private teachers is an urgent need to address important issues. According to further in-depth investigation, the income of the private colleges of physical education teachers in Henan Province, and other subjects than the lowest, the welfare and security of teachers and inadequate incentives, resulting in a work of passion is not high, the decline in the quality of PE teaching, private schools sports development is relatively backward, the face of such questions school leaders should pay attention to.

### *69.3.6 In-Service Training and Research Analysis of Private Colleges in Henan Province*

Table 69.7 shows, sports teachers of private colleges in Henan Province, 23 were trained, while 72 people training accounted for 75.8 %, there are light weight technology theory This shows that physical education teachers of private colleges in Henan Province. Main reasons the following two points: First, private physical education teachers teaching task, although the Private Education Promotion Law provides the necessary professional training for teachers, but in the actual teaching process, due to the increase of students lead the class arrangements compared to full time training; private physical education teachers training funds are not in place; Finally, the support of the leadership is not enough.

In private universities in Henan Province Athletics Teachers' Academic field trips understands the research capacity of the private colleges of physical education teachers in Henan Province is relatively weak, mainly due to the following:

First, the larger the intensity of the work of private college's physical education teacher's heavier tasks is no extra time and effort to carry out scientific research. Second, most physical education teachers do not have the awareness of scientific research. Third, the lack of private universities research leaders of the sports disciplines. Fourth, the lack of incentives is also one of the reasons for restricting private college's physical education teachers to engage in research.

Analysis of the stability of the 3.7 Henan Province, private teacher of teachers.

Private Universities in Henan Province to maintain steady and rapid development cannot do without this backbone of the teachers. Physical education teachers in the private universities in Henan Province, young or middle age, so that the stability of the teachers of private colleges and universities in Henan Province is particularly important. According to field investigation learned that the stability of the factors affecting private college's sports teams in Henan Province is: First, private universities the development of the choice of teachers, in the highly competitive contemporary era, the private universities are also facing increasing competition. For example, low social recognition, social supervision of private universities is not enough, and the related benefits are not in place. Third, the teachers for their own reasons, such as the unreasonable teachers, teachers' mentality owed positive. These factors directly led to the greater mobility of the situation of teachers of private colleges.

## 69.4 Conclusions

(1) The sources of physical education teachers in private colleges in Henan Province wide, mainly part-time. Social recruiting full-time staff accounted for only a part of.
(2) Physical education teachers of private colleges in Henan province age to middle-aged and young teachers, in terms of gender, in men's favor.
(3) Titles and qualifications of the private colleges of physical education teachers in Henan Province vary with the requirements of Higher Vocational Colleges Assessment Ordinance.
(4) The intensity of the work of the physical education teachers of private colleges in Henan province accounted for 65 % of more than 16 h, resulting in the PE Teachers Professional weary of Private Universities in Henan Province.
(5) Private Universities in Henan Province PE teacher salary is not directly proportional to the intensity of their work; the welfare of teachers cannot be guaranteed and unsound incentives.
(6) Stability of Private Universities in Henan Province of PE teachers in three main areas. First, the schools own reasons. Second, social reasons. Third, the teachers' reasons.

## 69.5 Recommendations

In accordance with the relevant provisions of the state to address the welfare benefits of physical education teachers of private colleges in Henan Province, for the relevant insurance and public teachers to equal the legitimate rights and interests, thereby enhancing the work of physical education teachers.

Attach importance to the phenomenon of private universities in Henan Province sports teacher qualifications level, title, and enhance its research capacity, develop the capabilities of the overall quality of physical education teachers of private colleges in Henan Province to meet the needs of the New Age.

Any of the 23 Harvard University President Conant once said; "University of honor does not lie in its premises and the number, but rather it is the quality of the teachers [6] Burke, the first 25 Rector said;" to make our schools to the forefront, in the final analysis, is to have a good professor [7].

In the talent introduction, the main sports full-time teachers, and give the tilt on the policy of the female physical education teachers, to attract more female physical education teachers from teaching to improve the imbalance of male to female ratio.

Establish a reasonable evaluation system, to create a favorable environment to mobilize the enthusiasm of the teachers' work.

Increase of private colleges of education and propaganda, and the correct attitude towards the status of private higher education, to change the community lack of awareness of the private higher education and prejudice.

## References


1. Qiong L (2010) Private college teachers situation and development strategy—a case study of private universities of Hunan, vol 11, pp 50–51
2. http://www.studa.net/gaodeng/100421/14223832.html
3. Jianbo Z (2007) Private universities sustainable development, vol 3, National University of Defense Technology Press, pp 48–50
4. Yuemei H (2010) Jiangxi province, part of the private universities public physical education teaching situation and countermeasure, vol 12, pp 32–33
5. Qiong L (2010) Private college teachers situation and development strategy—Hunan private universities, for example, vol 11, pp 50–51
6. Guoqiao Y (2005) Private higher education in sustainable development research education finance and accounting, vol 6, pp 29–31
7. http://sdkdb.sdkd.net.cn/2006/580/1/tashan.html

# Chapter 70
# Research on Technical and Tactical Characteristics of Men's Professional Tennis Player

**Fusheng Jang and Ya Liu**

**Abstract** In this paper, literature and data, video observation, mathematical statistics on men's professional tennis players in singles matches in three different venues technical and tactical characteristics were compared. The results showed that: serve the data phase of success and failure is higher than the grass field site and the hard clay courts, a stalemate and hard clay than grass field site. More technical and tactical aspects of the use of lawn space and serve after serve sights, clay and hard to use space more technical and tactical stalemate.

**Keywords** Tennis tactics · Different venues · Singles match

## 70.1 Research Purposes

In this paper, the analysis of video game players to find men's professional tennis singles match in three different venues, status and success and failure in the technical, tactical choice law, to explore three different venues in the singles competition, which links well, using different techniques tactics to help players win [1].

F. Jang (✉) · Y. Liu
School of Sports Training, Shenyang Sport University, Shenyang 110102, China
e-mail: 2430031317@qq.com

W. Du (ed.), *Informatics and Management Science V*, Lecture Notes in Electrical Engineering 208, DOI: 10.1007/978-1-4471-4796-1_70,

## 70.2 Research Methods

In this paper, a video observation, literature and data, mathematical statistics and other research methods for men's professional tennis players in singles matches in three different venues in the technical and tactical characteristics of a comparative study [2].

## 70.3 Results

### *70.3.1 Men's Professional Tennis Players in Singles Matches in Three Different Venues Serve to Study the Technical and Tactical*

Men's professional tennis players in three different venues in both singles matches ACES ball scoring rate, or serve directly score scoring rate, serve in their own plays a decisive role [3].

From Tables 70.1, 70.2 and 70.3 show that men's professional tennis player tennis game in three different venues in ACES and serve directly serve the ability to score very high. The data from the table below can be seen, men's professional tennis player singles match in three different venues serve the ability to score very high, relatively speaking, hard grass field site and serve on the clay than in the more aggressive driving [4].

As can be seen from Table 70.4 in the men's professional tennis players in three different sites are sending and receiving winning number of opponents was hard to site 22, the grass field 20, clay 23. Men's professional tennis players serve in the opponent sending and receiving more in the serving side winning the second ball.

**Table 70.1** Men's professional tennis players in the field hard ball singles the ability to score table

| | ACE ball (162) | Serve straight (345) | ACE (507) |
|---|---|---|---|
| Total ball number (2514) | 6.44 % | 13.72 % | 20.16 |
| Serve the total score (1168) | 13.87 % | 29.54 % | 43.41 % |
| Total serve (300) | 0.54 | 1.15 | 1.69 |

**Table 70.2** Men's professional tennis players ace capacity of the grassland site singles match data sheet

| | ACE ball (270) | Serve straight (514) | ACE (784) |
|---|---|---|---|
| Total ball number (2938) | 9.19 % | 17.49 % | 26.68 % |
| Serve the total score (1566) | 9.19 % | 17.49 % | 26.68 % |
| Total serve (375) | 0.72 | 1.37 | 2.09 |

**Table 70.3** Men's professional tennis players in the singles match of the clay courts serves the ability to score data table

| | ACE ball (122) | Serve straight (326) | ACE (448) |
|---|---|---|---|
| Total ball number (2340) | 5.21 % | 13.93 % | 19.14 % |
| Serve the total score (1142) | 10.68 % | 28.55 % | 39.23 % |
| Total serve (285) | 0.43 | 1.14 | 1.57 |

**Table 70.4** Men's professional tennis players in three different venues singles match serve opponents sending and receiving the winning goal tables

| Site | Opponent sending and receiving winning number |
|---|---|
| Hard | 22 |
| Lawn | 20 |
| Red | 23 |

Sending and receiving the number of opponents from winning a reflection of the quality of the serving side serve, serve poor quality when it easier to attack the opponent serve receive opportunities to take the initiative, sending and receiving winning play ball. Men's professional tennis players in three different venues singles match victory by his opponent sending and receiving areas ($U = 1.43 < 1.96$, $P > 0.05$) was not significant [5].

With the development of modern tennis, players play, continuously updated, hitting the bottom line capacity, receiving more and more attention and ability to improve, to try to serve the end of phase one point the situation will be reduced, stalemate competition is particularly important (Table 70.5).

The presence of men's professional tennis players in three different venues in 24 games, men's professional tennis players serve a stalemate, hard venues serve the total score the winning goal stalemate serve the total score accounted for 15.41 %; serve to rally the grass field stage winning goal accounted for the total score serving 11.49 % of total score; clay courts serve the total score the winning goal stalemate serve the total score accounted for 11.65 %. Men's professional tennis players serve a stalemate in the bottom line in winning the ball scoring rate in the hard site venues and clay than grass winning the ball slightly higher than the bottom line, hard-ball field and grassland sites relatively fast clay courts more likely to hit the winning goal. Serve a stalemate in three men's professional tennis

**Table 70.5** Men's professional tennis players in the stalemate of the three different venues singles match winning ball scoring rate table

| Site | Bottom line of winning (%) | Midfield winning the ball (%) | Winning goal in the net (%) |
|---|---|---|---|
| Hard | 7.19 | 1.54 | 6.68 |
| Lawn | 5.87 | 1.60 | 5.30 |
| Red | 5.87 | 2.28 | 3.50 |

**Table 70.6** Men's professional tennis players in singles matches in three different venues serve a stalemate opponents mistakes by hitting ball before a sub-table conditions

| Site | Touchdown (%) | Midfielder score (%) | Net score (%) |
|---|---|---|---|
| Hard | 16.95 | 1.37 | 1.71 |
| Lawn | 11.05 | 1.34 | 1.40 |
| Red | 19.61 | 1.75 | 1.49 |

player on clay site midfield scoring the highest rate, because clay ball speed is slower players have more time to prepare to mobilize each other, forcing the other ball out there light more opportunities. Serve stalemate winning the ball in the net, men's professional tennis player in the hard grounds and grass sites serve stalemate net the winning goal scoring rate was significantly higher than the clay courts serve stalemate net score rate, because tennis courts to the different materials, resulting in the ball in space at different speeds, hard-ball field and grassland sites faster, hit a stalemate was to create opportunities for a higher risk of the net score.

According to men's professional tennis players in singles matches in three different venues serve a stalemate comparison table success and failure, can be found in men's professional tennis players serve a stalemate both in the hard site, grass or clay sites, serve and stand-off gains and losses [6]. The bottom line total score divided stalemate success and failure rates are higher than the midfield and the net, can reflect the men's professional tennis players serve a stalemate in the struggle more concentrated in the bottom range, while the midfield and front score also crucial.

From Tables 70.6 and 70.7 show that, in men's professional tennis players hard, clay lawn tennis singles opponents mistakes serve a stalemate situation of the previous sub-hitting, hard-to site a point in the opponents mistakes before hitting the bottom line score of the total score of 16.95 % rate, the grass field errors in the opponent scores first hit a sub-bottom line ratio of the total score of 11.05 %, clay mistakes in opponents batting score before a sub-bottom line ratio of the total score of 19.61 %, ($U = 2.39 > 1.96$, $P < 0.05$) significant difference, so the men's professional tennis players serve stalemate opponents mistakes by the bottom line before the ball hit a point score for the clay than hard space, space was hard to higher than the grass field. Men's professional tennis players in three different venues serve a stalemate singles match losing mistakes ratio of the total loss of points: Hard to site 35.87, 32.00 % grassland sites, 38.41 % clay, grass field men's professional tennis players serve Stalemate error rate is lower than the loss

**Table 70.7** Men's professional tennis players in singles matches in three different venues serve a stalemate error rate of loss of points table

| Site | The bottom lose points (%) | The midfielder lose points (%) | Mistakes in the net (%) |
|---|---|---|---|
| Hard | 35.87 | 1.90 | 3.81 |
| Lawn | 32.00 | 2.37 | 4.30 |
| Red | 38.41 | 2.64 | 3.42 |

of hard sub-sites and clay. As can be seen from the above study men's professional tennis player singles match in three different venues serve a stalemate and losing a scoring rate to be lower than the rate of the grass field site and the hard clay.

### 70.3.2 *Men's Professional Tennis Players in Singles Matches in Three Different Venues then Serve the Technical and Tactical Research*

In the men's professional tennis players pick three different venues singles match return of serve their technical and tactical stages of success and failure analysis of the characteristics.

Men's professional tennis player singles match in three different venues, sending and receiving phase rate of the total score of the total phase than sending and receiving much less loss of points, which is one of the three sites the same, which is due to the other's serve, and each began active the right hand side in the driving, driving after sending and receiving skills will be made tactical game tactical impact, thereby reducing the number of points then serve. Receiving Points and the sending and receiving sub-fault, non-fat grass field in the next stage should be higher than the hard grounds and clay; then winning the ball hard ball field, grass field, clay no significant difference; the opponents serve ACES in the grass ACES site to site and significantly higher than the hard clay; then made the second film score is higher than the hard turf field space and clay, in short men's professional tennis players in the singles match in three different venues Receiving Points fault, non-fat phase and phased return of serve, the grass field data more obvious competition, and hard clay courts of the data space and there was no significant difference.

## 70.4 Conclusions and Recommendations

### 70.4.1 *Conclusions*

Serve ACES scores of the total score and serve directly the ratio of grass and hard venue to venue than clay courts, clay courts serve as the grass field directly destructive and hard to site.

In the men's professional tennis players serve most points scored from the tee stage, grassland sites scored significantly higher than the hard-ball stage to venue and clay courts, grass visible on the site after the serve and serve relatively Attacking Tactics to take more.

In the serve and then serve stand-hard space and clay courts score is higher than the grass field, and hard clay can be seen in the technical and tactical venues serve to ensure the quality of the basis of rotation and the pursuit of point of view that serve the sights and stand-off after scoring basis.

### 70.4.2 Recommendations

To participate in the lawn tennis venue-based athletes, the training in peacetime to strengthen the training of three shot, serve to enhance the speed and strength, to develop the athlete's offensive capability.

To participate in the hard space based tennis serve and strengthen the first three athletes in the shot should also pay attention on the basis of a stalemate.

Participate in tennis clay-based athletes, training in peacetime should be added to the first tee and angle of rotation to increase the stability of the ball after ball sights and lay the foundation for a stalemate, stalemate and improve the athlete's defence capacity.

## References

1. Lin W (2000) Dry compilation of different venues that the impact of tennis tactics (continued). Tennis World Tennis 01:22–23
2. Waite R (1999) The way to play on clay. Turbo Tennis 08:25–29
3. Zhixiang T (2002) Who shot across the net-group men's single against the basic unit of competitive process project research. Dissertation, Beijing Sports University Ph.D
4. Xiaofeng S (2007) Roger Federer O hard tennis technical and tactical characteristics of the study, vol 5, Beijing Sport University (graduate) degree thesis, pp 17–23
5. Wu CC (2011) A Study of synchronous and bucket trading behavior of institutional investors. IEIT J Adapt Dyn Comput 4(2):14–25
6. Wu H, Xu JB, Zhang SF, Wen H (2011) GPU accelerated dissipative particle dynamics with parallel cell-list updating. IEIT J Adapt Dyn Comput 4(2):26–32

# Chapter 71
# Research on Reformation of PE Based on Culture Rebuilding for Industrial Technology

**Zhuang Liang**

**Abstract** The reconstruction of cultural significance and research culture of their own teaching and research are discussed. The paper discussed constituent elements of teaching and research culture and the relationship, and the main content of current teaching and research and teaching and research system, the establishment of the content system was analyzed, and the protection of research activities carried out by several measures.

**Keywords** Teaching and research culture · Reconstruction · Universities · Physical education curriculum

## 71.1 Significance Reconstruction of Teaching and Research Culture

Teaching and research in the field of sports colleges and universities has always been very great importance, it is teaching physical education curriculum reform, the driving force for sustainable development. Especially the Ministry of Education issued a new "PE Teaching Guidelines" in colleges and universities have implemented the occasion of the last 2 years, research summary, analysis, control of the current research activities carried out in basic conditions, it becomes is very important [1]. New "framework" in the nature of the curriculum, course objectives and the physical education curriculum time and space, type and form of

Z. Liang (✉)
Department of Physical Education, Chongqing Jiaotong University,
Chongqing 400074, China
e-mail: 287599341@qq.com

W. Du (ed.), *Informatics and Management Science V*, Lecture Notes in Electrical Engineering 208, DOI: 10.1007/978-1-4471-4796-1_71,

organization, resource use and development and evaluation, etc. are greatly broadened the horizons [2]. Thus, on the one hand to the physical education curriculum reform has provided a broad space and the stage, but also to study physical education a tremendous challenge, which requires us to carry out both teaching and research activities in content and form of organization must be new, but should be efficient [3]. However, the perspective of nearly a year of physical education teaching reform literature, we deeply appreciate: the content in teaching and research, most usually focus on the form or empty committed to the concept of physical education or mechanical changes to state all the sports curriculum, physical education methods and changes in teaching content; in carrying out the organizational form, the lack of dynamic innovation, the majority of teachers also used the original form of large and uniform teaching and research, showing all-acceptance, recognition of the lack of quality and innovation; timeliness of research activities is also unsatisfactory, the majority of teachers through hard work, we found the guidance of teaching practice significance. We are less than a fundamental reform of the new curriculum, "vertex" Perspective, to examine, study and research activity itself, research system and structure of teaching and research activities, research and research activities carried out to optimize the form of its maximum effect [4].

Thus, in the current study to establish a way to promote teacher autonomy, self-discipline, self-internalization, civilized, with inspiration, inspire, encourage research and cultural effects of the mechanism, it is very important. I proposed the establishment of such a culture of teaching and research, teaching and research culture of trying to "convert," "Confucianism" of the role into full play the initiative of teachers, creative, comprehensive and thorough reform of the curriculum development of a strong, inexhaustible driving force [5].

## 71.2 Reconstruction Interpretation of Research Culture

New "framework" for physical education were more new descriptions and requirements: (1) the nature of the re-orientation, physical education curriculum is "students as the main means of physical exercise, through reasonable exercise science and physical education process, to enhance the physical fitness, improve health and enhance the physical qualities as the main target of public compulsory courses "and" blending to promote the harmonious development of physical and mental, ideological and moral education, culture, science and education, life skills for physical activity and sports and combine the educational process "and" implementation of the comprehensive development of quality education and training of personnel an important way" [6]. To highlight the nature of the course is physically and mentally involved in the sports program, in practice and practical courses under exercise load characteristics, and other distinctive programs separate. (2) two levels of the curriculum objectives to achieve its objectives, namely, the basic goal of most students and established there for some of the school director and the school has spare capacity of students to establish development goals, the goal set is actually of the past "into the university cultured on a standard door", the negation of the mode of teaching, is a people-oriented, recognize and respect individual differences of students, differential treatment, individualized, and many new educational concept of praise, is a profound educational thinking, the concept of sublimation. (3) "Three self" of implementation of curriculum structure, is a real recognition of the status of the student body, is to create a vivid, lively, active learning atmosphere must. (4) the course of time and space, type and form of organization, resource use and development and evaluation of the system has also greatly broadened the horizons: time and space, providing a curricular, extracurricular, school, district, community, nature, combination. The course structure ideas and great vision; course type and organizational form, to promote the establishment of the "big sports course structure", do not advocate breaking the original system, the class re-composition classes to meet different levels, different levels, different interests of students needs; use of resources and curriculum development, full use of all available internal and external, social and natural environment of human, material and financial resources for the program services; program evaluation, and effectively change the "examination-oriented education" in the re-identification, screening, selection and evaluation guidelines, and establish a re-strengthen, encourage, develop evaluation guidelines, evaluation of the content from the past a simple process of re-re-turned to athletic performance, attitude heavy, heavy participation, with special emphasis on students' attitude and behavior, communication with the spirit of cooperation, affection and other aspects of performance evaluation, evaluation methods is also a separate evaluation by the teacher to student self-assessment, peer assessment and teacher assessment of three-pronged approach.

Obviously, a characteristic of the times, location science, must advance the guiding ideology of teaching throughout the new "framework" always. Physical Education research is open and unprecedented, which contains profound scientific and cultural connotations. We need to study the course; teachers need to study more, to understand the teachers and students. Science and physical education curriculum itself is the carrier of the human spirit. Teaching and research work should be able to effectively promote curriculum reform and construction: in teaching practice and teaching and research, identify problems, solve problems, promote the program goals; awaken awareness of teachers, curriculum resources, development, build rich, vivid, distinctive curriculum resources; encourage the professional development of teachers, listen, understand, discuss, encourage, recognize teachers in curriculum implementation in practice and create personalized; establish educational administrators and teachers, between teachers and teachers in collaborative interactive relationship of equality, in teacher self-reflection, access to self professional development. On the other hand, to make teaching and research work can be effective in promoting curriculum reform and construction, and promote professional development of teachers; we must study the teaching and research activity itself. Curriculum reform, "vertex" Perspective, to examine, study and research activity itself, research system and structure of teaching and research activities, research and research activities carried out to optimize the form of its maximum effect.

Reconstruction of teaching and research culture is the content of teaching and research to effectively address the problems in the implementation of curriculum reform, the construction of the curriculum to provide evidence; teaching and research mode, the operating mechanism should be conducive to experimentation and to promote curriculum reform, conducive to professional development of teachers, thereby promoting the development of students, but also to obtain their own professional development of teachers to form their own educational philosophy, the ideals of education. Is to establish a way to promote teacher autonomy, self-discipline, self-enlightenment, internalization, with the inspiration, inspire, and encourage research and cultural effects of the mechanism

## 71.3 Constituent Elements of Teaching and Research Culture

The constituent elements of teaching and research culture, including the content of teaching and research, teaching and research system of systems, methods of teaching and research work carried out in three areas. Teaching and research but also contains the contents of physical education teachers, students, physical education ideas and concepts, their physical education curriculum structure and organization of the implementation process, physical education, and many other aspects of the physical environment; research system is to effectively carry out research activities for the protection of the established rules and regulations; way to carry out teaching and research work is to carry out teaching and research activities of the methods used, the means and ways. Broadly speaking, the content of teaching and research work also includes teaching and research methods and research system to carry out two aspects, the content of teaching and research, teaching and research methods to carry out the work, teaching and research system is an organic complex of the three systems, each of the three characteristics of both, there are relatively stable between the coupling means between each other and produce an effect.

### *71.3.1 Main Content of Teaching and Research*

Physical education teachers are teaching and research the subject; students are the object of education is the main teaching objectives. We first study teachers, knowledge, understanding teachers, while students and teachers have to research a new relationship with students. The study was efficient for teachers to carry out teaching and research activities, the basic premise is to create ways to carry out teaching and research work, teaching and research system of source-based; study of students in education and teaching and research objectives to achieve the goals and whether or not the touchstone; on the relationship between teachers and students the quality of the educational needs of sports, is an important part of the new era of quality education and research. (1) Structural characteristics of professional activities of teachers and teacher growth, development are the main content of physical education teachers. The professional activities of physical education teachers structural features are: the current physical education teaching, real scientific research activities, the new era of interpersonal activities. Growth of teachers, development of a new century, the main process of growth characteristics of teachers and teacher development in the history of related issues such as teacher growth characteristics of the adaptation period, teacher burnout and so on. Research should focus on the characteristics of scientific research activities and teacher growth characteristics of the adaptation period. (2) New era of sports students learn the characteristics of the factors affecting the development of student creativity, creativity, measurement, and cultivation is the main problem of the students. Students learn how to sports knowledge and sports skills, learning, sports and learning capabilities of students to learn in sports emotions, attitudes and habits and so on also must be given attention. (3) Reconstruction of the relationship of physical education teachers and sports teachers and students interact with students in the psychological structure and characteristics of the analysis are to study the relationship between physical education teachers and students the importance of content.

Research guiding the current physical education, teaching theory and related conflicts of things very basic theory, study and solve physical education reform and development in fundamental and directional problems. Educational administrators, teachers must accurately grasp the fundamental "health first" and "lifelong sports", "quality education" and the relationship between physical education, to avoid the blindness in the work, agnostic, and then into their own efforts to integrate management and teaching practice driving force.

(1) Curriculum and structural reform. How to adapt to the new "Program" requirements, the establishment of a new curriculum; how compression required courses, elective courses to open, the proportion of compulsory and elective; theoretical courses and practical courses the ratio of how much is appropriate; curricular, extra-curricular problem how to combine. (2) The organizational form of teaching reform. New "framework" in teaching forms expressly provided for, to about 30 small classes taught. How to promote the advantages of small class teaching system, learn the strengths of their teaching organization the flexibility to implement the reform of teaching is an important part of teaching. At the same time, the system attempts to form a variety of teaching practice and exploration, but also must study the contents of the teaching reform. (3) The choice of physical education content and innovation. Includes two aspects: First, the choice of materials and processing, dilemma; Second, take advantage of existing physical education resources, combined with the geographical features of the traditional sports of national research and development and innovative mining, transformation of the original sports programs, the creation of new forms of exercise. Such as simplified sports dance, hip-hop after the introduction of change, the system of non-two sand volleyball. (4) Strengthen primary and secondary school and college

physical education research. Physical Education must strengthen the practice of physical education in primary and secondary research, the research for their PE curriculum and teaching content, methods and means of selection of great significance. The study includes, first, primary and secondary schools "Sports and Health" study the contents of text and practice, and university new "framework" of the spirit of the contact; second, to study how efficient and scientific students to adapt to the new Physical Education the actual. Such as understanding the new "outline" of the spirit and requirements, how to make their own rational choice to learn the sport, learning time, school teachers, etc.; Third, how to strive to avoid in the curriculum and teaching content, methods, and on the choice of means School of duplication of research. (5) How to create an efficient physical teaching and research incentives. How to carry out research on physical education curriculum reform itself is an important topic, establish a way to effectively promote teacher autonomy, self-discipline, self-enlightenment, internalization, an enlightened, inspiring, motivating effect mechanism research system is required. (6) To strengthen the management of computer research. Computer Management Physical Education is the future trend of development. At present most of the colleges and universities have begun to try in computer management, but the situation according to the information available, the development of application software in the operational stability, speed, convenience is also unsatisfactory. Computer systems should be strengthened research and elective "Student health standard" test management systems research, sports scores to students and teachers in teaching systematic study and research management systems research.

## 71.4 Building of Teaching and Research Institution

(1) Establishment of an expert support system. According to our hospital (Department of) senior teachers of research, academic attainments, teachers and academic research capacity, combined with the actual situation of the discipline of study, etc., distinguished schools, high academic standards of the discipline outside experts, professors, experts to establish a more stable support system for teachers to provide a variety of security research, teachers can easily find an interactive, for support of experts and scholars, focusing on enhancing teacher quality and level of scientific research. (2) The establishment of the school teacher with only system. Integration of the faculty of expert resources and establish clear objectives and responsibilities delineated, the teacher with appropriate rewards and punishments only system, key areas to improve teacher education, teaching practice. (3) The establishment of teachers' mutual aid system. Organization-related disciplines group, established among teachers to help each other, mutual promotion and common to improve the mutual aid system, to enhance teachers' team spirit, team and individual teachers to promote common development. The establishment of this system, trying to break the field of university teaching and research among teachers is relatively self-enclosed, rigid pattern of the status quo,

has important practical significance. (4) A group, teaching and research, exchange of teachers, teachers in three-dimensional multi-system. (5) The establishment of young teacher's self-improvement self-targets to achieve the target system. According to the case of young teachers, faculty proposed the development of more realistic goals, regulations and requirements of young teachers to develop self-improvement, goal achievement plan.

Department of Physical Education (Department) is director of the leading teaching and research, organizers, leaders, mentors, have really established Keyanxingjiao, Xing (Department) of the educational philosophy, a good grasp of teaching and research work, experts and scholars, the backbone of teaching and research, work together to establish a with academic criticism, emphasizing the scientific spirit, the spirit of democracy, unity and team spirit, have to create a realistic, practical, free and harmonious atmosphere of teaching and research, visionary, able to motivate teachers autonomy, self-discipline, self-improvement, self-innovative teaching and research planning.

The new curriculum reform requires that we must be open to teaching, teaching content, methods and means more diversity and creativity, teaching and research activities carried out will lead to change in the way, the original form of both large and uniform dull and lack of innovation, it is difficult to raise the overall educational administrators and teachers, management, and research capabilities. Carry out a variety and flexible forms of teaching and research is required. Of forms such as the current hot spots by reading professional books, study groups; teaching case analysis; research salon; teaching demonstration; inter-school exchanges; issues briefing; literature (information) exchange; expert reports and other means.

Teaching and research were carried out on the record, organize, preserve; describe the process of teacher growth and distribution of the content of teaching and research; focus on role models, publicly announce the award-winning information; teachers to encourage and protect foreign-oriented research in the teaching, promotion for the new research starting point platform.

We proposed the establishment of two quantitative assessment criteria incentive system, in which the establishment of large and broad inter-school incentive framework, the sports departments to establish a discipline with its own characteristics, detailed, quantitative and qualitative incentives. Faculty mechanism results found should include: published and unpublished teaching experiences, teaching summary, excellent lesson plans, courseware, academic seminars and other teaching practice and research all contribute to a valuable documentary material, should be rewarded.

## 71.5 Conclusions

Physical education and research programs of college sports an important part, is to reveal the laws of physical education, deepening educational reform, to improve the quality of teaching and teacher research literacy is an important way. Establish a way to promote teacher autonomy, self-discipline, self-internalization, civilized with inspiration, inspire, encourage research and cultural effects of the mechanism, both must pay attention to the issue of curriculum reform, it should make great efforts to study the issue.

## References


1. Huaen Z, Qinghe C, Shugang L, Xuyao Z (2008) Analysis on professional abilities and research on cultivation model of P.E Teachers. J Hebei Norm Univ Sci Technol Soc Sci 7(1):65–68
2. Feng Y, Xin-ying C, Jian-jun Z (2010) A study on the structure and evaluation system of P.E major's professional ability. J Wuhan Inst Phys Educ 2:71–73
3. Yangyang L, Xiaochun W (2011) Investigation on P.E teachers identities of profession competence under the background of new curriculum reform in middle schools. J Shenyang Sport Univ 26(1):80–91
4. Zhichao L (2008) College two sports curriculum guide for the similarities and differences. J Phys Educ 1:75–77
5. Ye YH, Liu WP, Dao B (2011) BIM-Based durability analysis for RC structures. IEIT J Adapt Dyn Comput 9(4):15–24
6. Li GF, Kong JY, Jiang GZ, Xie LX, Jiang ZG, Zhao G, Xu SQ (2011) Hybrid intelligent control of coke oven. IEIT J Adapt Dyn Comput 9(4):25–33

# Chapter 72
# Research on Serving Technique Material of Men's Tennis Player in Industrial Technology

**Fusheng Jang and Ya Liu**

**Abstract** An important technology exists in the tennis serve tennis. Documents, video observation research methods, analysis of study 2009–2011 third Tennis Masters Cup players in the game serving technology use, a more objective reveals the world's outstanding men's singles tennis in the use of the characteristics and inherent law, the establishment of a tennis serve mode of action, provide a theoretical basis for our tennis serve teaching and training.

**Keywords** Industrial technology · Tennis · Serve · Technical features

## 72.1 Introduction

The serve is one of the most important sign to evaluate the technical level in tennis, is a technology access to contest the victory score means, but also with the development of modern tennis and gradually developed the more difficult. The serve is the beginning of the attack, the level of a game outcome to service quality has a direct relationship. The first serve is an important weapon for the offensive attack and a good serve is the key to winning.

F. Jang (✉) · Y. Liu
School of Sports Training, Shenyang Sport University, Shenyang 110102, China
e-mail: 287599341@qq.com

W. Du (ed.), *Informatics and Management Science V*, Lecture Notes in Electrical Engineering 208, DOI: 10.1007/978-1-4471-4796-1_72,

## 72.2 Object of Study and Methods

### *72.2.1 The Object of Study*

2009, 2010 and 2011, three Tennis Masters Cup Finals men's singles players.

### *72.2.2 Research Methods*

Access to network resources via China Journal Net, China HowNet sports literature databases, retrieval and access to a large number of documents such as tennis, focus on reading and analysis of relevant data, sorting and summarized a comprehensive understanding and of this study the status and latest research results.

#### 72.2.2.1 Observation

Observed from 2009 to 2011 Tennis Masters Cup World excellent tennis player, the latest ball technology video to understand the world's most advanced driving techniques, analyzed the characteristics of the tennis serve.

## 72.3 Analysis of Characteristics of Serve Technology

### *72.3.1 Analysis of Characteristics of Grip*

Personal body structure and serve different grip methods differ, serving basic continental grip and eastern backhand grip [1].

Clearly see from Table 72.1, the players of each session of the Masters Cup participants, using the eastern backhand serve grip dominant, which accounted for 85.8 % of the 2009 Masters Cup Masters Cup in 2010 accounted for 85.8 % of the 2011 Masters Cup accounted for 100 %. Description Eastern backhand serve grip

**Table 72.1** Serve grip player list

| Events | Grip | Number of people |
|---|---|---|
| 2009 Masters cup | Continental | 1 |
| | Eastern backhand | 7 |
| 2010 Masters cup | Continental | 1 |
| | Eastern backhand | 7 |
| 2011 Masters cup | Continental | 0 |
| | Eastern backhand | 8 |

has become the mainstream trend of outstanding pro ball grip, and will continue to occupy the dominant position of the ball grip.

Eastern backhand grip become a mainstream phenomenon in the world's top players, there must be the nature of the technology advantage. Grip not only issued on topspin, the most important thing is that this grip compared with the Continental grip, the hitting surface of the film surface is more open racket late scratch back to avoid hitting surface open out and appear to pull the film, wrapped ball phenomenon. In addition, the state of nature, the Eastern backhand serve grip the racket face hitting surface and the forearm medial angle less than 90°, so when the ball, the racket head in the wrist and do not take the initiative under the control of natural first, more easily wrapped ball, to improve the success rate hitting. The racket head in the former can make more effective whip can be used to maximize the momentum transfer to the racket head, improve the speed of the racket head, will play a very significant role in improving the accuracy of placement and rotation strength.

## *72.3.2 Analysis of Characteristics of the Racket*

Serve began to beat down swing, this article will serve the racket into the hem of the bent lead shot (such as Roddick, etc.) and the hem whirl arm racket (such as Roger Federer, etc.) two. The statistical results (Table 72.2 show), 2009-2011 Masters Cup hem bent racket and the hem whirl arm racket ball players, respectively, 3, 4, 5 and 5, 4 3, respectively, 37.5 % of the total players, 50, 62.5 and 62.5 %, 50 and 37.5 %. It can be seen from the data, whirl arm hem racket mainly serve the racket, but the overall trend point of view for 3 years to whirl arm of the hem of the racket the number of stable and a slight decrease, no hem bent racket serve the number of steady increase. Instructions hem whirl arm racket contemporary tennis racket, but serve the racket has bent racket along the lead photographed by the whirl arm hem, from complex to simple this a direction of the trend, action, process the pursuit of more simple, to avoid the complexity of the action process.

Today's tennis racket divided into two types, a hem bent racket, and the racket and hem whirl arm. Whirl arm hem racket contemporary tennis racket, but serve the racket has bent along the lead photographed by the whirl arm hem racket from

**Table 72.2** Racket players list

| Events | Racket ways | Number of people |
|---|---|---|
| 2009 Masters cup | Not hem bent racket | 3 |
| | Hem bent racket | 5 |
| 2010 Masters cup | Not hem bent racket | 4 |
| | Hem bent racket | 4 |
| 2011 Masters cup | Not hem bent racket | 5 |
| | Hem bent racket | 3 |

the general to this direction. The trend of development, course of action is more the pursuit of simple, to avoid the complexity of the action process. As commonly used in the past, tennis pros serve the racket—there hem brandish arms racket, is still contemporary tennis racket the way dominant. The reason why it is enduring, its advantage is that the racket trajectory arc, the route is longer, movement stretch, the racket is not easy to interrupt to avoid the impact of swing speed, and natural, smooth action gives a sense of beauty.

Hem bent cited the film serve as a new ball technology in recent years is gradually more players to accept, has its unique advantages. Analysis from the perspective of Sports Biomechanics, hem bent lead shot serve the elbow-speed whirl arm higher than the hem of the racket to the front of the forearm muscle group stretch role, can be increased by hitting the shot speed. The action is simple, small amplitude hem bent racket larger than the magnitude of the hem whirl arm fixed shoulder of the racket can better [2]. In addition, the hem bent racket ball directly on the lift racket, pull shoulder followed by forward top of the swing the ball, the whip back concessions and forward above the waving of the movement trajectory is almost in the same plane this reduces the loss of strength and speed in the other direction, greater explosive.

### *72.3.3 Analysis of Characteristics of the Drop Way Analysis*

Statistical results found from a third Masters Cup, 15 players drop the racket in the order there are two types, first drop the racket and toss with the racket at the same time, did not find the first racket drop players (see Table 72.3), which dropped the racket frequency of simultaneous players is 13, well above the first drop the racket. Toss and racket order, drop, and the racket at the same time are the current world tennis men's singles outstanding tennis in the mainstream.

Ready to drop, the player holding and holding clapping separate holding player portrait on the movement, hold clapping remain low, synchronization, physical extension to the side, knee to maintain the peg-leg, relaxed posture. Drop from the moment of hand held shooting arm in the shoulder at the vertical direction of the body about an angle of 60–80° and the shoulder axis to extend the line into a 45° angle, elbow flexion angle of 150°, grip the palm of the hand, making surface down the racket head pointing to the ground, tossing hand move up at the same

**Table 72.3** Drop the racket order players list

| Events | Toss with lead shot order | Number of people |
|---|---|---|
| 2009 Masters cup | First drop the racket | 2 |
| | At the same time | 6 |
| 2010 Masters cup | First drop the racket | 0 |
| | At the same time | 8 |
| 2011 Masters cup | First drop the racket | 0 |
| | At the same time | 8 |

time, the focus of synchronous transfer to the rear. When the drop hands above shoulder height, centre of gravity began to move forward. Dropped from the moment of the hand, the centre of gravity about to fall on the feet intermediate trunk omitted flexion becomes erect. In the process of upward toss, trunk, lower extremity muscle groups relative relax to control the movement speed of the upper limb muscles around the shoulder joint, and when dropped from the hand, shoulder axis natural high to low profile. When dropped, the upper extremity of each joint is necessary to fully straighten the drop location is completely determined by the upper limb relative to the position of the shoulder axis and the torso surface. For information about the characteristics of the drop, the arm drop has been the main drop in the professional tennis players serve, the reason is very simple, only the arm case, in order to better ensure that the drop stability.

### *72.3.4 Analysis of Characteristics of Lead Shot Rhythm*

From third Masters Cup statistical results (Table 72.4) see racket process to the existence of two different rhythms, racket smooth type of apparent standstill and the racket obvious pause two, the former more frequent use of instructions, without pause. The racket is the rhythm of the mainstream of the current and coming period, the outstanding player in the racket.

Today's tennis racket process there are two different rhythms, racket obvious pause two apparent standstills and the racket. Pause the smooth type racket is the mainstream of the outstanding players in the current and future period rhythmic racket. This is due to the dynamics and anatomy of the advantages: smooth, middle pause the racket rhythm so that the racket forward movement of the top of the swing hitting fused, without pausing to gradually accelerate muscle elasticity potential energy of the play, to reduce a pause while the rate of loss [3]. Smooth racket rhythm so that the entire swing process before hitting the muscles are in a relatively relaxed state, elastic, which is conducive to the last beat action [4].

Serve the middle pause the racket rhythm, making it easy up the racket and forward the top of the swing out of line between hitting, it will cause muscle tension, the muscle elasticity, and the racket of the stretching of muscle elasticity due to the pause a result of the use of rate is not high, swing speed and significant loss due to a standstill [5].

**Table 72.4** Racket rhythm player list

| Events | The racket the rhythm | Number of people |
|---|---|---|
| 2009 Masters cup | No apparent pause | 7 |
| | Apparent pause | 1 |
| 2010 Masters cup | No apparent pause | 7 |
| | Apparent pause | 1 |
| 2011 Masters cup | No apparent pause | 7 |
| | Apparent pause | 1 |

**Table 72.5** Serve footwork player list

| Events | Footwork | Number of people |
|---|---|---|
| 2009 Masters Cup | Serve and step | 6 |
| | Not serve and step | 2 |
| 2010 Masters Cup | Serve and step | 6 |
| | Not serve and step | 2 |
| 2011 Masters Cup | Serve and step | 7 |
| | Not serve and step | 1 |

### *72.3.5 Analysis of Characteristics of Serve Footwork*

Table 72.5 shows nearly three Masters Cup, the world's outstanding men's singles players, most of them and step-serve footwork. Among them, the Masters Cup in 2009 and step-serve footwork players accounted for 75 % of the contestants. Accounted for 75 % in 2010 and 2011 accounted for 87.5 % of the contestants. The past 3 years, and step-serve players to maintain the development momentum for steadily. Instructions and step-serve footwork contemporary tennis serve, footwork, and may in a certain period of time to maintain a mainstream position.

Serve the transition stage adoption and step-style tee players to maintain the development momentum for steadily. And step serve as the rear foot forward of the foot to move closer to the body centre of gravity along with the full forward, hitting point will be more forward, to provide protection for the first click on the ball; In addition, the rear foot forward heel into the whole body in the same line so easy to form the back of the bow, "scratch back" more fully, the last whip formation, help to increase the strength of the tee. Therefore, balance and coordination and step-serve the players are higher and professional athletes and amateur high level players to use more.

Non-and step-serve, though not the mainstream of the tennis ball footwork and step of the non-serving footwork also has its own unique advantages. From the mechanical point of view, pedal to take off due to the non-and step-serve is a double support, good control of body balance to ensure that the ball has a better stability; hind legs support to take full advantage of the lower extremity strength, the highest point of the ball guaranteed. This serves footwork in addition to suitable for high level players, because the body balance is easier to control, more suitable for primary teaching and amateurs.

## 72.4 Conclusions and Recommendations

### *72.4.1 Conclusions*

Serve the technical aspects of the outstanding players of the man: the Eastern backhand grip is currently serving grip the mainstream; serve the racket the way along the hem cited photographed from the general to the hem of the racket.

This direction of the trend, action, process the pursuit of a more simple; drop the racket at the same time and pause smooth type serve the racket rhythm excellent tennis in the mainstream rhythm.

Tennis ball technology stations, grip the racket, drop, hitting, coherent with the play of six actions made. The tennis ball hitting the power system is composed of human multi-joint dynamic mechanical system, which includes the foot, ankle, knee, hip, torso, shoulder, elbow, wrist eight sessions.

### *72.4.2 Recommendations*

Note that the tennis ball during training stations, grip the racket drop, hit the ball, with the play's action technology, keeping in mind the factors that affect serve success rate, action technical coherence in order to ensure that serve success rate.

Strengthening physical training to improve the serve threatening to reduce the fat skills and tactical play and Serve receptor.

The negative impact of the factors. Especially in strength training for tennis serve, pay attention to the strength of the corresponding development of different muscle groups to focus on strength training held shooting arm on one side of the body, at the same time, to strengthen the non-shooting arm side of the force training. Pay attention to the training of the thoracic and abdominal muscles, should strengthen the training of the lower back muscles.

## References

1. Zhixiang T (2003) Tennis tutorial, vol 01, Higher Education Press, Beijing, p 105
2. Bin X (2009) Tennis, vol 10, Higher Education Press, Beijing, p 60
3. Plot L (2005) Graphic tennis tips, vol 5, Fujian Science and Technology Press, Fuzhou, p 51
4. Ming J (2003) Tennis serves analysis and training methods, vol 25(2), Liaoning Sport Science and Technology, pp 14–16
5. Wan-jun C (2009) Excellent tennis player serving the technical features. Hubei Sports Sci 4:22–25

# Chapter 73
# Study on Natural Ecological Environment of Interior Design for Gyms

**Xinhua Zhu**

**Abstract** In this paper, ecology an sports science theory are applied, the methods of literature, interviews and observation research are used, at the same time combining the gym interior architecture characteristics, puts forward the connotation of ecological environment and principle of the gym interior design. And try to discuss the application of ecological technology from space, color, light, sound, temperature, and the points for attention when designing the interior environment of gyres. The final purpose I want to reach is to make sure that every area of the gym shall accord with human body comfort and security and my viewpoint can provide reference basis for gym's interior design and adornment.

**Keywords** Gyms · Interior design · Natural ecological environment · Research

## 73.1 Introduction

In recent years, as the improvements of the quality of peoples' life and the paying more and more attention to health, gyms have been the first choice of leisure fitness at the rest time of working life, so that the fitness industry has been quickly development, kind of fitness centers come out like mushrooms after rain. At the same time, people has higher requirements for the environments of fitness, the fitness clubs with more characteristics, comfortable environments, complete functions will be more popular in consumers. For this affect, the interior design of modern gyms should to be safety and comfort with each area and each machine,

X. Zhu (✉)
Sport Institute of Jianghan University, Wuhan 430056 Hubei, China
e-mail: tigerandsheep@163.com

W. Du (ed.), *Informatics and Management Science V*, Lecture Notes in Electrical Engineering 208, DOI: 10.1007/978-1-4471-4796-1_73,

and then the machines and environments will to have better service for people, and comes to the optimization of people-machines-environment. So we should research for the natural ecological environment of interior design for gyms to have better service for consumers.

## 73.2 Connotation of Ecological Environment of Interior Design for Gyms

According to the related theory of ecology, as a dynamic open system, the sport ecological environment can be composed of many subsystems or elements. Based on different causes of formation, it can be divided to sport natural environmental system and sport social environmental system Among them, "the sport natural environmental system contains atmosphere environment, water environment (water, ice, snow), terrestrial environments, noise environment, biological environment" [1]. In the activities of sports, the natural environment offers the areas, supplies, and other conditions for the process and developments of sports, and try to be harmony with natural environment.

The interior design is that to create the interior environments with reasonable functions, comfort and beauty, satisfaction of people's physical and mental life, basing on the using property, location, and other standards of the building, using technologies and the theory of building beauty [2]. Depending on the ecology of interior design, it should be meet people's physical and mental need, and keep the fluxion of substance and energy, and make the interior environments in good cycle, and then to improve the quality of interior environments. According to it, this article has the view below, to achieve the ecological design of interior environment, ecological beauty should be a advocated; popularizing the using of "green" building materials; proposing saving and recycle; bringing the nature into indoor, developing the potential of new materials and new technologies, cooperating with other professional designers, etc. Only in this way can we create comfortable, energy saving and environmentally-friendly interior environment.

The interior space of gyms depends on the areas for people's activities, which will affect the interior layout and the display of machine. Reasonable layout and display not only give people to enough personal space, and keep people in complete and independent space, and also make the people's communication in the best situation. So that the gyms must determine the area of the gym and indoor different function areas to reflect the ecology of interior environment; using the dimension of fitness equipments and the areas of peoples' activities to layout a reasonable display; through different sense organs such as sight, hearing, skin sensation to make sure the light, sound, temperature is reasonable to peoples' adaptation. Through the survey, research, statistic, analysis people to collect more human's data to humanize the interior design for gyms.

## 73.3 Design Concept of Natural Ecological Environment of Interior Gyms

The design of fitness clubs which can attract people to doing exercise do not aim at luxury, but pay more attention on the reasonability of space layout and adaptation to peoples' physiological and psychological need, and emphasize the adaptation between design and the building environment, to show the design concept of energy-saving, environmentally friendly, low cost of maintenance so the fitness clubs with sense of sport, fashion, cultural care can be created.

### *73.3.1 Sense of Space*

The rationalization of the space and give people the feeling is design basic requirements open, narrow, low, special-shaped space will have different influence on people's feeling, so when to design, only exact sense of scale, proportion, distance can make the most ideal and comfortable feeling of space. The design and layout of modern gyms should under the need of the size of management place and actual requirement. It should guarantee some proportion and take more consideration to the convenience, functions, safety when used, and then distribute the gyms reasonably, so that the gym will play a good role.

The function area can divide into necessary function area and expansion function area. The necessary function area fall into; fitness machine area, which contains: aerobic area, anaerobic area and power area; independent exercise class area, which is separated from the public machine area, it contain: big gymnastics room, hot yoga room, dynamic bike room and so on; reception area, business area and office room, this kind of area can be designed depend on actual situation; sauna bath area, which include shower, sauna room (dry evaporation, wet evaporation), locker room, storage room, water massage room, SPA service, massage room and so on.

Expansion function area means some gyms increase some fitness service on basis of necessary fitness items. For example: swimming pool, taekwondo place, table-tennis hall, tennis court, Squash hall and so on. Expansion function area also contains leisure entertainment area, such as game centre, computer room. Nutrition restaurant is also an important item, although it is in expansion function area, it has been used in most of gyms. Some gyms has VIP area, it is used by some VIP, and offer special service.

### *73.3.2 Sense of Color*

Indoor color affect sight environment, and also affect the moods and mind of people directly, using color correctly is good to work and health [3]. The color in gym mainly be bright light color, which to have the feeling of broad, relaxed, cool,

Table 73.1 Psychological effects of colors

| Colors | Psychological effects |
|---|---|
| Red | Optimistic, warm, happy, dangerous |
| Orange | Active, cheerful, mild, romantic, mature, harvest |
| Yellow | Health, relaxed, bright, hopeful, sprightly, light, promising |
| Green | Quiet, peaceful, fresh, safe, young |
| Blue | Calm, quiet, cold, lone, open |
| Purple | Solemn, uneasy, mysterious, serious, noble |
| White | Pure, simplicity, fresh, cold |
| Grey | Ordinary, imperturbable, depressed, neutral |
| Black | Dark, serious, severe, uneasy, stressful |

tireless and unrestrained, and it is not good to use black or red, which will make people heaviness, uneasy, no distance and big irritation. On the other side, it can use color define functions, different color in different space make people marked notice to have clear division. For example, the keynote of exhibition area should be cool color to show dynamic design concept; sport area should be alternated with bright color to show fashion sports; the keynote of reception area should be worm color to emphasize cultural care.

From Table 73.1, red makes people anxious and stressful, which is easy to make people tired. So in rest place, red should be avoid using mostly; orange is good to recover and keep healthy, it should be applied in recreational place, it should not be used in quiet place; Golden decoration will make unstable and arbitrary behaviors, so rest place should be avoid golden furniture and other decorations; green can overcome fatigue and negative mood, it can properly used in rest place; blue environment will make people feel elegant and quiet, and it can remove nervous mood and relieve headache, fever, syncope, insomnia, it can be used in rest place [4].

### 73.3.3 Sense of Light

The light environment of the whole gym can be divided into natural light and indoor lighting. Combined with peoples love, most of the indoor environment use natural light, so that the location is very important to gyms, it should keep reasonable distance between buildings, choose adaptable daylight opening, using point-blank sunshine to lighten indoor, and make indoor atmosphere to ensure enough light and open, which will have a comfortable natural environment. Besides, in limit lighting conditions or at night, lamplight can render indoor environment, and build reasonable artificial light.

When design, artificial light can apply to different illumination and color of light to divide different function area and make kind, warm and friendly atmosphere, Lighting effects on people mainly intensity of illumination, color

**Table 73.2** Standard illumination of sport building from broadcast [7]

| Sport events | Reference plane and its height | Standard of illumination (LX) | |
|---|---|---|---|
| | | Train | Competition |
| Basketball, badminton, tennis | Ground | 300 | 750 |
| Bowling | Home bottles of area | 300 | 500 |
| Fencing | Table-board | 500 | 750 |
| Table-tennis | Table-board | 750 | 1,000 |
| Swimming | Water | 300 | 750 |
| Gym | Floor | 200 | – |

temperature and angle. Firstly, different situation should has different illuminations, and different illuminations will have different feelings in the same place (Table 73.2); secondly, the influence from color temperature of light, thirdly, the angle of light.

People for the light on the color temperature of subjective feeling has 3 kinds, one kind is partial cool color, one kind is partial warm color, there is the natural color [5]. For example, the environment of the rest area in gym should be filled with cool color, and it is good to reduce people's visual fatigue, reduce the anxiety when waiting and keep relaxed mood. Sport area should be filled with warm color, which can foil the atmosphere of sport, under the shine of worm light, the machine is more attractive. In addition, controlling of angle of light source can stand out the key of indoor design on artistry, which will make beautiful sense of vision and enjoy for people. Controlling the angle of light source can avoid of glare to meet the physical need of people, offer the best illumination can satisfy people's physical need.

### *73.3.4 Other Requirements of Design*

In the environment of gym, most of sound is audible, include of the noise of machine, voice of people, sound equipment and so on. Being a public place, gyms should be quiet, and avoid the pollution of noises. Because the noises have influence on people's life. Such as the discomfort in ears, notice of hearer, bother people, and reduce people's working efficiency. The location of gym should be far from traffic noises, and each room should be avoid to noises, sound should be not too higher, different rooms should be segregated.

When on the process of metabolism, human take in $O_2$, and take out much $CO_2$, so the content of $CO_2$ in the air increases, $O_2$ decreases. The quantity of $CO_2$ depends on the activities of human (Table 73.3). Which shows that, people exercise on the fitness centre will produce mount of $CO_2$, to ensure the ventilation of air indoor? Otherwise, much of $CO_2$ w will make people headache, dizzy brain,

Table 73.3 Different quantity of $CO_2$ in different activities [8]

| Working condition | $CO_2$ exhaled quantity [$m^3$/(h person)] | $CO_2$ exhaled quantity [g/(h person)] |
|---|---|---|
| Peaceful time | 0.013 | 19.5 |
| Lighter work | 0.022 | 33 |
| Light work | 0.03 | 45 |
| Middle work | 0.046 | 69 |
| Hard work | 0.074 | 111 |

shortness of breath and other bad reaction. The ventilation should be natural ways, not only save equipments and cost, but also be healthier.

The indoor temperature of air is very important to the environment. When the temperature of indoor in 20–25 °C, people in a comfortable situation can keep good mood and good to do all kinds of activities. So, gyms should keep stable temperature, when using the air-conditioner, it is not good to have very high or very low temperature, and so that will not make people in bad physical stress; when not using the air-conditioner, good condition of ventilation and stable temperature should be kept.

## 73.4 Announcements About the Process of Interior Design

Gym is a public place with special indoor space, as this kind of place would gather large quantity of people, so the fire safety is one of the important problem of interior design. There are kinds of causes to safety accidents, In short, it is the cause of human, so the source of danger should be controlled when design. The detail measures are below: equipped kinds fire-fighting equipment, fire alarm system and spraying system; there should be a first-aid case available, which to deal with medical negligence; escape way and emergency exit should be unobstructed, area of electronic machine should be set switch in the floor or leakage automatic disconnection system, front desk or reception room should be equipped with camera monitoring system to monitor accidents and stealing, smoking prohibited indoor, no wine be offered, safety notice and all emergency measures should be posted [6].

At the same time, special area should be specially dealt with to ensure the fitness club. Fox example aerobics room. The main room should be large enough to give people the opening feeling; layout should be avoid of uprights. The leading platform always be equipped with a big piece of glass mirror to enhance visual effect, it will make people to feel larger space in the aerobics room. If possible, the walls in the aerobics room can be separated by glass. Which will increase the fitness atmosphere, and keep close contact with machine area, the horizon affect will be improved.

The distribution density of machines. The distribution density machines should consider two aspects of problems: One is that whether the adjacent people will affect each other when exercise, the other is the safety problems. Such as the running machine, sometimes some accidents would be happened for reasons of wrong operation or other causes, the running frequency cannot follow the speed of continuous track, so people will fall down the running machine. For the effect of inertia, the people will have bigger step or pushed to one direction, if the distance between two running machine is too early, then two people nearly will be crashed, the bad result will be unpredictable.

Dynamic bike room. The decoration of dynamic bike room is the almost the same to aerobics room, which should be harmony to the whole gym. Dynamic biking is a special item with big attractive, it can make people to have the desire to fitness, so it should be set on the conspicuous place in gym.

The load-bearing of the floor. Load-bearing of machine area should be taken into consideration when zoning of the gym. Machines as running machines, dynamic bikes, Mountaineering machines, elliptical motion instruments are very heavy, so it has big pressure to the floor.

Water seepage. Bath facilities in the gyms should be used more than 10 h, the floor always full of water, so the decorate for bath center is the most important part in the whole gyms, 3 layers of waterproof should be set on the floor, water seepage should be test and checked for more than 24 h with strict supervision.

## 73.5 Conclusion

Through interviewing the different gyms, the effect of interior natural ecological environment on kinds of aspects, such as space, color, temperature, sound and layout and safety, all factors in the design for gym mainly be evaluated by people's feeling, and it is lack of data of quantitative criteria, so strengthening the research to the standard is to better effect of design. Only determine the detail standard can to evaluate with the level of design for gyms, which is to meet people's need of comfort and safety.

The concept of natural ecological environment of gyms combines the different subjects such as ecology, sport, culture, economy with the requirements and expectation to building design, which makes a organic integrating destination for the design of sport building, the designer can exceed the simple design of the space and style of building, wider field of vision to research the design and developing trend of sport building.

## References


1. Sun H, Cai M (2011) The theory research of the relations between sport ecological environment and all people's fitness. Sports 4:140–141
2. Liu N (2010) The brief analysis for situation, characteristics and developing trend of indoor design in China. Theory Mod Decoration 5:73
3. Ma L (2011) Discuss disciplines and performance methods of the indoor design of environmental art. Build Mater Liaoning 1:47–48
4. Li J (2007) Discuss the application of color in sport stadiums. Shandong Sport Inst J 23(1):69–71
5. Zhou N (2011) People-oriented interior design should serve to the feeling requirements of people. Chifeng Coll J 2:96–99
6. Li W, Zhu S (2002) Interior and furniture design human body engineering, vol 10, China Forestry Press, Beijing, pp 112–118
7. MOC of P. R. C (2004) Standards of illumination design for buildings, vol 12, p 1
8. Wu S, Xia Q (2003) Interior environment and equipments, vol 12, China Architecture and Building Press, Beijing, pp 3–7

# Chapter 74
# Researches of Effective Teaching in Physical Education Under New Curriculum Standards

**Xingdong Yang**

**Abstract** In order to train a large number of personnel adjusting to social and economic development with the progress of society, China's education is also undergoing tremendous changes. Education reform has already become a hot topic of the education sector and the academia. Physical Education Reform also needs to be reformed as a physical training course, besides reforms in the basic courses of the professional culture. In the entire reform process, all reforms are empty talk if their effectiveness can not be guaranteed as they have never be carried out, no matter they are changes on curriculum or teaching mode. China's economic is changing from the traditional intensive labor to high-tech labor since the beginning of the 21st century, which has a higher demand for personnel. As the cradle of personnel training, school plays a vital role on the development of the society. Therefore, it is to be considered and concerned for all schools that ensuring the effectiveness of their teaching under the new curriculum standards.

**Keywords** New curriculum · Physical education · Teaching

## 74.1 Introduction

For a long time, China's education is focused on imparting and receiving, which neglects of the exchanges between teachers and students [1]. Students receiving extremely limited after teachers' painstaking taught and this is a psychological

X. Yang (✉)
Physical Education Teaching Group, North China Electric Power University Science and Technology College, Beijing, China
e-mail: xingdongyang1@yeah.net

W. Du (ed.), *Informatics and Management Science V*, Lecture Notes in Electrical Engineering 208, DOI: 10.1007/978-1-4471-4796-1_74,

blow for both teachers and students. China has already made reform in physical education curriculum and framed new curriculum standards to make receiving physical knowledge more easily in recent years [2]. Though Physical Education Reform exists from primary school to university, there are still a considerable number of students who only exercise in gym classes, without formatting the sense of physical exercise. This is a problem in the validity of the physical education curriculum in the learning process, and Physical Education Reform is still a failure if this problem is not solved, even if amount of physical knowledge and physical skills are taught to students [3].

## 74.2 New Curriculum Standards and Teaching Concepts

### *74.2.1 New Aspects of the New Curriculum Standards*

With the development of the society, the advancement of technology, the constantly updating of knowledge, the basic ability of people to adapt to society is changing, too. New curriculum standards of schools are born in such environment. Physical teaching under the new curriculum standards reflects the new understanding of people in sports [4]. Compared with traditional physical education curriculum, new aspects of new curriculum standards mainly reflect in the following:

(1) Innovation of Concept

New teaching ideas highlight the Guiding Principles of health, which focuses on promoting the healthy growth of the students, cultivating students' awareness and consciousness in doing physical exercise by themselves. The center status of students is outstanding in the teaching process, which can give different education to different student based on their innate body status and so that each student can get benefit in the physical education curriculum.

(2) Flexible Goal

Excluding the uniform teaching methods, New Curriculum Standards effectively combine the health of students and the characteristics of the physical education curriculum, which can carve up the traditional learning programs and content by more detailed method as well as expanding the range of student learning. In the formulation of learning objectives, it uses the method with the health of students as the main line as while as the implementation of the physical skill as the auxiliary line. This can set physical education curriculum objectives of each student more specific and encourage the student to strive toward his/her goals. In the arrangement of teaching time and teaching content, it breaks the year constraints, which can maximize strengths and advantages the students.

(3) Diversified Evaluation

The evaluation criteria of the new physical education curriculum do not only focus on students, and it also pay attention to teachers, teaching content and schedule. It completely abandons the way of the past that has only a single examination, and divides the final quality assessment into gradually the amount of the evaluation, which pay attention to the exercise in students' physical health, as while as taking into account the ups and downs of the students' mental state. In the learning process, students are not only encouraged to publicize their personality, but also encouraged to promote team spirit. In the evaluation process, it is not just reflected between students and teachers, but also schools, parents, peers and social staff, to offer advice on curriculum reform, promoting its continuous improvement.

### *74.2.2 Teaching Concept*

The new curriculum standards also work out the relevant provisions for teachers. Compared with the traditional concept of teaching, new teaching concept no longer has knowledge as center, and the teaching concept changes of the physical education curriculum are mainly reflected in the following areas:

(1) Health First concept mixed into teaching

The concept instead of the content should be changed at first, if the reform is making. Only content which is state-of-the-art and right can work effectively to play a good role of guidance and education. In the new curriculum standards, Health First concept goes into the hearts of teachers through a series of process including learning, practicing, studying and re-practicing. In schools, physical education teaching is different from the training of competitive projects, for it is not intended to make sure every student to get the title, or becoming sports experts or scholars. It aims to give every student a healthy body. Therefore, in the process of physical teaching, it is necessary to let students exercise in pleasant and exciting movements, instead of pursuing the density and intensity of exercises or forcing students to skills training, in order to ensure that students can receive both mental and physical exercise.

(2) Student-centered

There are differences between each individual's physique, so it is extremely wrong and impossible to use a mold in the process of teaching as a standard that every student must achieve. Students in poor innate health to should be encouraged to exercise instead of being severely criticized. At the same time, it is important to see the progress of students, concerning about the psychological aspects of change and enhancing the potential of the learning.

(3) Integration into Society

The final attribution of the students is the society. In students' physical education process, it is important to continue to camp record this situation, in order to let students constantly exercise their communicative ability, team spirit and sense of competition in a harmonious learning exercise.

## 74.3 Implementation and Issues of Effective Teaching

### *74.3.1 Effective Teaching under New Curriculum Standards*

Since the promulgation of the new leaf curriculum standards, there have been two different voices: One is to overthrow all the traditional teaching, and then re-establish a new order of teaching; the other is to further develop and improve on the original basis. The latter one is recognized by the majority of the education sector and the staff for its inherited type of development. Although there are many deficiencies and shortcomings in the traditional education, its decades-long development has accumulated a lot of experience, of which there are many merits. In order to ensure the effectiveness teaching under the new curriculum standards, it should base on the original teaching, which mainly reflected in the following aspects:

(1) Classroom Rules

Nothing can be done well if there are no rules or discipline, and teaching goes in the same way. Classroom rule is a necessary condition to ensure the order of teaching being carried effectively. Although the new curriculum standards provide the students with main position and also points out physical education as the student development center, physical teaching is still a group campus activity instead of a one-on-one expert guidance. Therefore, it is necessary to enable students to comply with the course of discipline and rules, which is a positive effect to improve students' ability to adapt and exercise their self-discipline.

(2) Prompt Teaching

In the process of Physical Education Curriculum teaching, if a teacher explains and demonstrates to the students, as the accepting ability of the students is different, it is hard to achieve the scheduled teaching effect through the curriculum dozens of minutes. In the process of teaching, it is necessary give students the space and time to think for themselves as well as giving correct guidance to the students. It can stimulate students' positivity and initiative of learning better to allow students to use their own way of thinking to discover the correct understanding and master the correct motor skills.

(3) Control of Load

In the physical learning process, it is wrong to enable students to make the appropriate physical exercise desperately in order to improve students' physical fitness. Anything over limitation will be counterproductive. If the increasing load exceeds the affordability of the student, it may leads to muscle strain, bone dislocation or other visceral disease. Of course, the load can not be too small for being afraid of harm to students, which has no effect of exercise. None effects of training will be seen if the students are in the long-term low-load training, and it will produce the mood of weariness over a long time. For individuals, it will be a good choice to separately control the load of different students.

### *74.3.2 Problems in Effective Teaching Process*

After the new physical education curriculum standards were pulled out, not all staff are able to master the idea of the new curriculum and the problems are mainly the following aspects:

(1) Course Name can not Link Physical Education and Health

The original course name is gym class, while the new name is physical education and health after the announcement of the new curriculum standards. This makes most of the physical teachers puzzled, for they are thinking to give two courses: physical course and healthy course. Actually, though health is the first physical education guiding ideology, the content of teaching is exactly the same as before, with different names. This is because these teachers do not grasp the essence of the physical and health education, which should be effectively combined. They should enable students to understand the benefits on the body coming with physical exercise through the study of the theory, and verify the correctness of the theory through physical exercise, promote students self-awareness of exercises and develop lifelong healthy physical exercise habits.

(2) Confusion on Choices of Teaching Content

Original physical teaching has all identified the task of teaching, and teachers have very clear goals and adequate space in teaching. However, the new curriculum standards require students to master a variety of physics. This makes the teachers not know what to do, as they are used to group activities. Physical Education Teaching relates to mental health, sports participation, social adaptation aspects, but for competitive projects, it does not specifically pointed out if athletic does not need to learn and they only need to focus on training students to the theory and psychological exercise. This puzzles teachers a lot. It is shown if studies have a long-term exercise of a particular sport and ignore the other aspects of training, it will fail to achieve the purpose for the exercise of the internal organs. So it is necessary to equip students with a variety of sports instead of just learning a particular aspect in a semester. For strong school with lots of teachers, it means no problem to give cross teaching according to the case of students. However, for relatively weak school with few teachers, there are more problems which needs for further improvement.

(3) Lack of Syllabus

In our long-term of education process, the syllabus has played a pivotal role, which has a positive impact in the specification of our school education. But with the progress of society, the curriculum gradually exposes its inadequacies, which includes specific performances: without touch for the society, lack of modern content, limit of content definition, too large limitation from teachers to students, excessive pursuit of results, not perfect on evaluation system.

## 74.4 Principles and Analysis of Effective Teaching

### *74.4.1 Principles*

In the new curriculum standards, it is necessary to grasp the following principles for effective teaching:

(1) Optimized Principle

The optimization of the external environment is effective usage of school teacher, environmental, financial and equipment to make the best use to ensure the normal conduct of its teaching activities. The coordinated optimization is to ensure that students get both physical and mental health, as well as individual character development, which can make a perfect combination in the whole and part of the teaching. The optimization of teaching activities structure is to take full account of students' actual relationship between the content and teaching, and design scene games and carry out the project in a targeted manner to ensure students to improve comprehensive quality.

(2) Subjectivity Principle

For a long time, teachers are the center of the whole teaching and learning activities. With the development of education, student-centered was put up. Actually, during the entire teaching process, the teaching mode with teachers and students as two centers can ensure the effect of their teaching more effectively. Therefore, in the teaching activities, it is necessary to take full account of the relationship between the two to find its balance point, so that it can give full play to the role of teachers "teach" as well as arousing the students' enthusiasm and initiative to make students have their own thoughts and ideas into a useful talent of creativity.

(3) Development Principle

The development of everything is a process of changes in motion from small to large, from weak to strong, from lower to high-level. Physical teaching activities means teachers should consider students' knowledge before learning and ability, and then to get out the potential of students through training and get the goal of lifelong exercise.

(4) Health principle

Health is not merely external physical performance, and it also includes psychological factors. Through the learning of Physical and Health Education, students can understand and master the knowledge of nutrition, psychological, physical, social, security and environment better, so that they can form good health and exercise habits.

### 74.4.2 Effectiveness Analysis of Physical Education Teaching

Through the discourse of the problems and principles of physical effective teaching, the analysis of validity will be listed in the main following areas:

(1) Teaching Topic Foundation

By changing the model of teaching and learning, students' enthusiasm and initiative can be fully mobilized. The theme of teaching is fully integrating the learning objective with the actual situation together based on the student's own characteristics and the status of the body, to achieve physical and mental exercise and social adaptability. The theme is designed to combine with national, social, regional and school-related affairs, so that it can make easier the formation of resonance in the learning process to obtain better teaching results.

(2) Character Development Promotion

It is impossible to put all school students into a mold. The characteristics of students should be fully considered, to exuberant energy and creativity of students to get a good space for development.

(3) Teaching Diversification

In the process of teaching, every student should be treated fairly and equitably. Differences between different individuals should show a sober understanding of how to make the improvement of physical condition for different students, which is just what should be thought by the teachers. The actual exercise needs of students in different levels should be met through diverse group teaching model, in order to achieve the intended teaching objectives.

### 74.4.3 Recommendations on Effective Teaching

China is a populous country with hundreds of millions of students. If all these students have adopt a unified management model, there will be no doubt that the ingenuity of many students will be stifled. The main recommendations for physical effective teaching under the new curriculum standards are: taking full use of local resources on the actual situation of the school to give targeted teaching; fully learning from the foreign advanced teaching methods and means, while retaining the merits of the traditional teaching methods, so that to make the two fully integrated to identify a suitable path of development; recognizing the right from wrong to retain its truly effective content while abandoning the useless content, to ensure students to master as much useful information as possible in the effective time; treating differently on different student, and giving different teaching to different level students, to ensure that students get plenty of exercise on the physical education curriculum; finally, making breakthrough on innovation.

## 74.5 Summary

This paper made discussion on the effective physical teaching under the new curriculum standards. First of all, it made a detailed analysis of the new curriculum standards and teaching concept. Then, it discussed effective teaching as well as the problems in the effective teaching. Finally, it introduced principles of effective physical teaching, and analyzed the problems in the effective teaching. Due to space constraints, the recommendations of effective teaching are given at the end of the paper. This can give the reader some of the necessary reference, although the content is not much.

## References

1. Shenying G, Lianghua L (2004) Effective pedagogy, vol 8. Guangdong Education Publishing House, Guangdong, pp 28–34
2. Sun Y (2004) Researches on effectiveness criteria in classroom teaching. Ph.D. thesis in East China Normal University, vol 181. Wuhan, pp 32–27
3. Qiuqian S (2007) Concept and implementation of effective teaching strategies, vol 03. Zhejiang University Press, Zhejiang, pp 7–15
4. Wu XW (2005) Pay attention to students' subject status in classroom instruction of physical and healthy education. Educ Modernize Mag 11(62):66

# Chapter 75
# Research of Volleyball General Course in the Physical Education Major

**Wengang Ren**

**Abstract** As the international competition project, the majority of sports fans favor the volleyball. The volleyball project in our country is placed at the international advanced level. Especially the female volleyball has obtained the champion in the Olympic Games and the World Cup many times. It has the sufficient influence. In the physical education major of our colleges and universities, volleyball course is the required course that places the important position among the entire major education. Through the volleyball course learning, students can grasp the theory basement that relative with the volleyball. Moreover, they can obtain the enough practical abilities and establish the foundation of further employment. However, the present physical education major arrangement is not reasonable that cannot completely adapt to the social development requirement. Therefore, the course evaluation and reformation is necessary. This is the hot spot of volleyball researcher in the sporting world. It has the instructive significant for the volleyball development.

**Keywords** Volleyball course · College and universities · Sports

## 75.1 Introduction

In our country, the target of training the physical education major is satisfying the physical teacher requirement in middle and primary school. However, with the physical education reformation in middle and primary school, there need to do

W. Ren (✉)
Hebei Institute of Physical Education, Shijiazhuang, China
e-mail: wengangren@yeah.net

W. Du (ed.), *Informatics and Management Science V*, Lecture Notes in Electrical Engineering 208, DOI: 10.1007/978-1-4471-4796-1_75,

the relative transform of physical education major in colleges and universities [1]. Volleyball course is the regular program in middle and primary school. Therefore, in the physical education major, volleyball course is the required course for the students. Analyze and research the general course of volleyball has the significant meaning of promoting the volleyball development and increasing the education practice abilities [2].

## 75.2 The Reformation Overview of Volleyball Course in the Physical Education Major

The establishment of our physical education major follows the Soviet mode. The main target is letting students to grasp the theory basement during the training process and the motor skill of some sport events. It takes the low-level school as the principal thing. The undergraduates in our country are under the work distribution system during the long term. The teaching method and approach is singleness. At the end of the last century, our education system has transformed from the uniform work distribution into the self-employment. Students experience the survival of the fittest that influenced by the market economy law. For the further employment, students need to grasp the major theory and ability as far as possible during the school term. In our country, the middle and primary school provides the education idea of "lifelong sports" and "healthy first". It requires the physical education major to close up this theory during the target training and teaching activities in order to satisfy the present gym teacher standard of middle and primary school.

### *75.2.1 Target Research of Volleyball Course*

In recent years of our physical education major, there provides the different opinion and viewpoint of education target in the volleyball general course. Some opinions are placing extra emphasis on the motor skill practicing and training [3]. The key point is training the students follow the athlete direction that pays attention to the movement correctness and evaluation. Some viewpoints are placing extra emphasis on the theory learning that pays attention to train the academic coach and teachers without thinking highly of the practice. No matter what kind of standpoint, it is particular about the method of evolutionary and guide gradually.

Based on the summary and personal experience of the writer, here we set the general course target of volleyball is centered with major training, sufficient contact with the social practice, train and practice students' comprehensive qualities with pertinency, plan, and method [4].

### 75.2.2 Volleyball Teaching Content and Approach Research

Most part of the physical education students will go into the middle and primary school as the gym teacher. The present physical education in middle and primary school desalts the sports education. Each student can enjoy the physical course. It is not completely ignored the training of convention skill. This needs to popularize the hygienic knowledge, sports protection and abundant the teaching content for establishing basement for the further employment.

On the teaching method, we need to take entertainment and health building as the principal thing. Moreover, pay attention to the comprehensive ability training to meet the social education development.

### 75.2.3 The Textbook Research

In our country, there have so many textbooks for the volleyball course. Lots of colleges and universities have their own publishing teaching materials. In the summary, the most applied textbook is Ball Game—Volleyball, which is published by the Higher Education Press. In this textbook, it is mainly introduce the tactics, training management, judgment, theory, and method. The recent years of modification cannot change the content.

## 75.3 The Evaluation and Reformation of Volleyball General Course in the Physical Education Major

The course research of physical education in colleges and universities can be achieved through the following aspects.

### 75.3.1 Course Target

At present, our volleyball course target gives priority to the professional theory and tactics, which ignores the obvious attention to the training of students' value, view of life, worldview, and professional quality. Moreover, it completely neglects the creativity, only the perfect motion and tactics that follow the teacher standards will be better students. Otherwise, totally repudiate the students and look down upon the psychological states and mental attitude.

Except the requirement of professional theory, ability, and tactics, the course targets need to pay attention to training students the self-study ability, practice ability and creative ability. Under the training target of the physical education major, we establish the subject system with the social development requirement.

### 75.3.2 Students Source and Employment

Since the student increase enrollment, the students' number has increased with each passing day. The long-term higher education talent shortage and large population base, we obtain the optimistic enrollment about the physical education major. With the aging period coming, the student resource snatch is round the corner.

In the employment, we break the traditional distribution system and apply the two-way selection between student and enterprise that everything will follow the market standard. In recent years, the teachers' number is saturated in middle and primary school. The undergraduates of the physical education major have more and more pressure on the employment and the employment rate is decreasing year after year.

In order to ensure the enough student sources, we need to ensure the employment rate. The good employment rate will attract more students. Therefore, the employment rate is important. The employment arrangement cannot only pay attention to the industry of gym teachers in middle and primary school. We also need to focus on the position of coach, athletic director, and the sports health worker. The employment department needs to support the sufficient attention, select the employment information and feedback them to the students for expanding the employment rate as far as possible.

### 75.3.3 Teaching Tasks

Under the guide of education target, we need to detail each class and form the series of tasks. There need to grasp three aspects in the teaching tasks. First, students have grasped which kind of knowledge and ability; second, the required knowledge and ability for students; third, evaluate the teaching content and do the further arrangement.

In the present teaching activities, most teachers are following the outline process that rarely consider about the students.

Before each teaching activity, the teacher needs to have the primary understanding about the students' conditions. During the teaching, consider about the mentality variation, aiming at the different reception capacity to explain in the different level. The emphasis and difficult questions, the teacher needs to design the detailed plan. Careful classifies the learning result to ensure the different students can grasp the knowledge. This can promote the learning enthusiasm and increase the communication between teachers and students.

### 75.3.4 Teaching Content

Our teaching contents pay attention to the basic ability, theory and tactics exercise. It is directly decide the achievement of teaching target. Based on the teaching program, the teaching contents include the following.

(1) The volleyball introduction has the competitive method, character, origin, transmission, development, progress situation, and tendency; (2) Volleyball technology and tactics evaluation has the analysis of gesture and movement technology (half squat, combine steps move, cross step move, and step move), serve technology evaluation (under arm serve, overhead serve, overhead floating serve), technological evaluation of frontal forearm pass (frontal hitting, receive hitting), pass technology analysis (forward pass, the normal toss), spike evaluation (front spike), block analysis (one-man block, double block), battle array ("four-two" or "five-one"), position exchange (front row exchange), and combined rush tactics (pass attack), collective defense (five people receiving formation, one-man block defense, double block defense); (3) Volleyball competition and judgment has the competition (organization, Berger layout method, and scoring), and the judgment work (equipment, rule, and judgment method). Although each school has the different teaching content, there has no difference in the main details.

With the education reformation, we increase the volleyball theory acknowledge. The course content should add the mental health, social adjustment, and teamwork. At the same time, mix the relative volleyball entertainment activities.

### *75.3.5 Teaching Ability*

During the long-term education, our physical teaching activities mainly use the rough explanation, personal example to impart the knowledge and ability.

Develop the training of explanation, through the speaking teaching to pass on knowledge. This can stimulate students creative and imaginative.

### *75.3.6 Class Hour Distribution*

In the general college and university, the volleyball general course of the physical education major needs to complete in one semester with 64 class hours. Table 75.1 is the detailed arrangement about the volleyball course.

From Table 75.1 we can find out, during the whole semester learning, the theory learning reaches 10 % and most of the time spends on the practice. The Table 75.1 is "pay attention to technology and ignore the theory". This is the present question in most colleges and universities. On the class hour arrangement, there needs to appropriate add the class hour of theory class with the content of ideological education. In the practice part, increase the volleyball entertainment, decrease the class hour of technology, and wipe out the difficult and over numerous abilities.

**Table 75.1** Class hour arrangement of volleyball general course

| Classification | Teaching contents | Instructional mode | Class hour | Percentage (%) |
|---|---|---|---|---|
| Theory | Volleyball summarize | Theory | 2 | 11.7 |
| | Volleyball technological theory, tactics theory | teaching | 4 | |
| | Competition rules and judgment | | 2 | |
| | Hour subtotal | | 8 | |
| Practice | Volleyball technology | Practice | 36 | 88.3 |
| | Volleyball tactics cooperation | teaching | 12 | |
| | Judgment practice and teaching competition | | 8 | |
| | Hour subtotal | | 58 | |
| Total | | | 64 | |

## *75.3.7 Exam Evaluation*

The long-term test of sports major course is the once only marking about the students' abilities in the end of this semester. This method ignores the individual endeavor and development condition. It is not fair and attacks the positivity of learning volleyball.

Add the evaluation score as well as the usual performance and divide the ability in different levels will ensure the fair and scientific marking.

Combine with the years of work experience, the writer provides the relative evaluation solution for the reference. The total score is divided into two parts—usual performance and terminal performance. The usual performance is the attendance, attitude, innovation, effort level and so on. Table 75.2 has the details.

The detailed practice exam needs to have the detailed evaluation. Each student has the clear standard will not doubt the performance after the exam. At the same time, the student will acknowledge the independent disadvantages in order to correct them after class. The following will introduce the grading details of two abilities that often tested.

**Table 75.2** The marking scheme after reformation

| Classification | Exam content | Test type | Score |
|---|---|---|---|
| Usual performance | Attendance | Usual test | 40 |
| | Attitude | | |
| | Innovation | | |
| | Effort level | | |
| Terminal performance | Theory test | Written examination | 20 |
| | Practice test | Practice | 40 |
| Total | | | 100 |

**Table 75.3** Evaluation standard of frontal overhand serve

| Marking standard | Score |
|---|---|
| Drop ball steady, concerted power, consistent movement and bat the ball with the whole palm | 5 |
| Bad drop ball, simple movement, and bat the ball with the whole palm | 3–4 |
| Drop ball steady, consistent movement, and bat the ball without the whole palm | 2.5–3 |
| Bad drop ball, bad movement, bat the ball without the whole palm, and serve the score of 4 marks | 0-2 |

The first item is the overhead serve, with ten marks. The details are in the following:

(a) Method: 5 straight shot, 5 diagonal shot, 0.5 mark of each effective shot. 5 marks in total.
(b) Ability standard is Table 75.3.

## 75.4 Summary

This article starts the research of volleyball general course that aims at the physical education major in colleges and universities. In the first, the writer summarizes the course reformation. Evaluate the volleyball general course from course target, class hour, student enrollment and employment, teaching task, teaching content, and the evaluation standards. Moreover, the writer also discusses the disadvantages and provides the reformation plans and strategies. For the length limitation, this article still has some shortcomings, hope the interested reader can perfect and modify it on this basement.

## References

1. Zhang P (2002) The reformation research of deepen volleyball course system. J Shenyang sport Univ 12:444–449
2. Li H, Ge C (2005) The course test research and evaluation of physical education major. J Beijing Sport Univ 127:43–49
3. Huang HS (2005) Ball game–volleyball, vol 06. Higher Education Press, Beijing, pp 42–46
4. Hong W, Zhu H (1998) The application of students oriented method in the volleyball special class. J Sports Sci 38:10–15

# Chapter 76
# Research of After-Class Physical Training in Higher Learning

**Zhixin Sun**

**Abstract** With the rapid economic development, the education filed provides the series of reformation such as quality education, humanistic education, and lifelong sports for adapting the society development. Under the great base number of our population, the traditional physical education caters to the competition requirement. There have a few people practice the high target training after class. This condition will snag many questions along with the social development that under the largest population base in our country. School is the part of this society and it is the basement of training talents. The target is training the talents as much as possible. If we do not guide and focus on the students' after class training, it will decrease the training positivity. In recent years, our country has obtained good performance in the various national physical competitions. However, through the quality research about the teenagers, their physical quality is under the sustained decrease. The school has the shortage attention on the sports, especially the physical training after class. How to increase the students' physical quality and mobilize the physical training positivity among the whole students is the most considered question.

**Keywords** Colleges and universities · After class · Physical training

## 76.1 Introduction

At present, each college has opened the physical course with various contents that bring students more selections. However, it is impossible to improve the physical quality that depends on one or two classes. Only bring the training enthusiasm into

Z. Sun (✉)
P.E. Department, Hebei United University, Shijiazhuang, China
e-mail: sunzhixin11@yeah.net

W. Du (ed.), *Informatics and Management Science V*, Lecture Notes in Electrical Engineering 208, DOI: 10.1007/978-1-4471-4796-1_76,

the students' after class sports course, and let them join into the training of their own free will can obtain the unexpected effect. Therefore, the physical course in the colleges and universities need to include the after class physical training except the normal teaching contents [1]. Develop the after class physical training in colleges and universities can help schools to improve the abundant physical competition, which places the positive function to perfect campus culture establishment and promote the physical development.

## 76.2 After School Physical Training Theory

Through the 30 years investigation and research about the teenagers' constitution, the obesity rate and myopia rate is under the sustained increase. Although the students' height, bust, weight, and relative morphological parameters have reached the obvious improvement, the index of physical strength, speed training, and vital capacity is under the decrease [2]. Combine the recent years' practical condition, the education personage believe it is caused by the substantially decrease of sport training after class [3]. Our Sports Law has the specific formulation, the school should organize and develop various kinds of after class training and competition activities. The after class sport training has been the most important part of our physical education activities in the colleges and universities [4].

### *76.2.1 The Advantages of After Class Physical Training*

#### 76.2.1.1 Promote the Development of Groupment Activities

Each student has the different healthy condition. We cannot train all the students to be the sports major talents. However, encourage everyone join in the physical training can increase the understanding and comprehension among students. Moreover, it can improve students' communication ability, let them form the optimistic and intuitive spirit, and promotes the groupment project development. On the other hand, play the physical activities with other student, the sport specialty students can attract more students to join in the activities. This will guide students to form the excellent sports environment, improve the physical quality, and easily develop the team event.

#### 76.2.1.2 Promote the Increase of Competition Project Level

Through the after class physical training development, more students join in it. We can search some potential students at this event. The school will have greater chances to select the athlete through the increase trainers. Thereby, students can obtain the performance development of the sports event.

### *76.2.2 The Guide Thinking of After Class Physical Training*

In our country, we have the clearest guide thinking about the sports courses. However, there have no relative rules and papers to specify the after class physical training. The after class physical training in every school effectively supplements the course under the independent condition.

After the research in colleges and universities, 90 % people believe the development of after class physical training has the positive function for our physical industry development. Nearly 95 % people think the positive development of after class sports have the particular promotion to the school honor and influence. About two-thirds people consider the after class physical training will help students to improve the sports ability and competition level. Nearly four-fifths people think the after class activities can promote the interaction and communication among students, which has the positive function of the society.

Therefore, this article summarizes the guide thinking of after class-training development in colleges and universities in the following. Based on the education guide from the party and state, we should take full advantage of the school practical condition that under the huge environment of education reformation. Consider about the people first, encourage, and lead students to do the physical training. Moreover, market orientated establishes the after class sports physical training system, which is suitable for the society with sustainable development, just and fair.

### *76.2.3 The Training Target of After Class Sports*

The development of after class physical training caters to the requirement of developing physical events in colleges and universities. It is the most effective supplement of physics course. The physical quality increase cannot leave the after class physical training. It is the development requirement. The present college and university is the high openness system, it is very wrong to stand alone the school out of the society. The after class physical training is the requirement of physical industry development, which has the significant meaning to train the special people (new generation of people with lofty ideals, moral integrity, good education and a strong sense of discipline) and strong the national body. The training target of after class sports needs to based on the following principals.

(1) System principle

If we treat the after class physical training as one system, we need to take full the advantage of school, teacher, environment, resource and students. Moreover, it is necessary to connect all the above elements.

(2) Integration and diversity principle

The after class physical training needs to have the unified target. Aiming at the department, college, and research group even the teacher, there needs to have clearer target content. Without changing the unified target, different department and people will have the various events and targets.

(3) Operability principle
(4) No matter which kind of task, there needs to have the operability in the formation of target and plan. If the target is too indistinct, it is very hard to operate and everything will be the empty talk.

## 76.3 The Current Condition and Tendency of After Class Physical Training in Colleges and Universities

### *76.3.1 The Current Condition of After Class Physical Training in General Colleges*

At present, the physical training leaders in our physical courses are still the school managers and teachers. The acknowledge degree of the school directly decides the development condition of after class physical training. Based on the research, most colleges and universities have the positive function to develop the after class physical training for helping students to form the habit of physical training. However, they have the lack of communication function and education function that extended from the after class physical training.

For the current condition of after class physical training, we can research from the following aspects. The details are in the following.

(1) Participation condition

In recent years, our university games development become more and more mature. There have different levels from national grade, province grade, and serial colleges' competitions. Different from the old competition, the present students with no grade has the greatest number that nearly place half quota. This describe we have more and more common students join in the sports. On the other hand, the university games in recent years have many talents with high education background such as master and doctor. This means the physical training that after class has obtained the improvement. Most of the coach's age is between 30 and 45 years old. The new generation coach is growing up.

(2) Object development condition

In the university games, although the different places have the various geographic environments, cultural characters and object settings, the competition of traditional objects such as track and field, balls, martial arts, chess, and swim has obtained the development.

(3) Time and period arrangement

The universities that build the large-scale sports meets, half of them will arrange nearly half year or one year for the after class physical training. Some colleges even arrange over one year. This means the sports meeting can stimulate the after class physical training of school and student. Those colleges who do not have the competition event will have the unnecessary condition about the after class physical training. It is ordinary to arrange one time training in two or three days. It will not delay the learning and will ensure the condition promotion.

(4) Filed and funds situation

At present, the file and funds about the after class physical training is mostly provided by colleges and universities. Only few colleges that can attend the high-level competition will obtain the sponsor's support. The research shows, over three-fifths of our colleges have shown the disadvantages in the funds and field. This is the important element to restrain the development of our after class physical training.

(5) Equipment condition

For one college, the physical equipment provides the normal physical education activities. It has no more consideration about the after class physical training. If we increase the utilization rate of equipment during the after class physical training, it will cause the equipment wastage. The equipment complement behind the time will influence the after class training, even the general teaching activities.

### *76.3.2 Development Trend of After Class Physical Training in Colleges and Universities*

(1) Research the practical applicability and diversity on the function

In order to bring more students join in the after class physical training and develop the practice gradually, there needs to pay attention to the training function. Let the students to obtain the practical effect and joyful mood during the training.

(2) Research the system and basic on the content

The training objects of after class sports are different from the high-level athlete. They are the college students. Therefore, the training content needs to follow the method from simple to complicated, from easy to difficult. The content needs to be easy and joyful that can ensure the learning positivity. Moreover, it is necessary to effective combine the preparation and relaxation training to form the unified system.

(3) Research the competition and entertainment on the events

If only research for the entertainment, the sports event will lose the applied function. For example, if excessively research the competition, many students will lose the interest and patience to join the training in the long-term.

(4) Research universality on the objects

Attract students to join in the training as far as possible is the target to develop the activities.

(5) Research variation in the organization forms

More and more colleges bring the increase number of people and scale. Many colleges have expressed the combination of several areas. We can compete between campus to increase the communication and improve the students' organizing ability and creative ability.

## 76.4 Development Counterplan of After Class Physical Training

Integrate the writers years' of working experiences and the current situation and tendency evaluation, we improve the after class physical training in the following aspects.

### *76.4.1 Pay Attention to the After Class Physical Training in the Thinking*

We only pay attention to the cultural theory learning during the long time education. The sport is under the half neglecting. With the proposition of quality education, people focus on the sports courses but only express in the course arrangement. In most people's minds, it is just the assistant course. Even do not mention the after class sport training.

Aiming at this problem, the college manager and physical major teachers should arise acknowledge in the first place. Treat the training as the important part of quality education. In the second place, increase the students publicity, let them pay attention to the training in their own free will and understand the significant meaning of the after class sport training. In the last, coordinate the school, department, student union, youth league committee, and association to build the various sports activities. Let all the people believe the after class physical training is very important.

### 76.4.2 Define the Long-term Target and Short-term Target with the Detailed Implement

The after class physical training in the school needs to divide into several layers. Most students have the weak training foundation. The training will aim at the basic practice. With the students' level increase, the training content should change too. Each school needs to based on the individual condition to set the target and complement methods with pertinence. Otherwise, it will be the fantastic talk.

### 76.4.3 As Far as Possible to Bring the Various Training Contents

Different student has the various behaviors and the sports emphasis is different too. In order to let more students join in the after class physical training, there needs to take full advantage of the training content, satisfy the different interest, and enjoy the participation.

### 76.4.4 Improve the Manager Level

By the limited teacher team, the after class training will be replaced by the excellent students. However, the faultiness of quality and ability level, some technologies will not mature and even in the wrong way. The training on this kind of basement will makes students feel tired and produce the negative attitude.

### 76.4.5 Perfect the Sport Association in Colleges and Universities

In college, the student association is the effective supplement of campus administrative function. However, founds, filed, equipment usage obtains so many limitation. Many perfect thinking and ideas cannot develop effectively. The school should provide the sufficient support and supervise the rules perfection to work out the excellent working system.

## 76.5 Summary

This article starts the discussion of the after class physical training. In the first, introduce the theory basic of after class physical training and emphasis evaluate the advantage, guide thinking, and target. In the second, describe the current condition and tendency. In the last, provide some suggestions and strategies about the current questions during the after class physical training. For the length limitation, there still has some disadvantages about the solution during the after class physical training. Hope the readers can correct the problems and support the own strength to the physical industry.

## References


1. Tieli Y, Chen JZ (2003) The new view of after class training reformation in the school, vol 172. Beijing Sports University Press, pp 276–282
2. Zhang L (2009) Campus sports culture introduction, vol 12. Hunan University Press, Hu'nan, pp 3–39
3. Ministry of Education, Physical education curriculum teaching guidelines for common institutes of higher learning in China 2002
4. Chen J (2005) The establishment of sustainable development strategy about after class physical training in our colleges and universities. J Harbin Inst Phys Educ 71:372–375

# Chapter 77
# Innovative Research on Rating Quantification System of Tai Chi Courses Portfolio in Colleges

**Chao Zhao and Hu Ma**

**Abstract** This paper fully takes into account of the characteristics of tai chi course teaching in colleges, combines with the spirit guidelines of traditional rating rules of martial arts, uses mathematical and statistical functions of excel software, comprehensively applies research methods of consistency coefficient, range analysis and so on, through two major theme frameworks of tai chi exercise levels and difficulty coefficient, constructs a rating quantification system of tai chi courses portfolio in colleges, divides into six groups, three cycles of rating test, the test results show, the rating quantification innovative system of tai chi courses portfolio in colleges is reasonable, feasible and teaching directive. It recommends promotion in rating of martial arts courses in colleges, to improve the effectiveness and influence of Chinese traditional sports teaching.

**Keywords** Course teaching · Tai Chi · Rating system

## 77.1 Introduction

Tai Chi is a subjective rating project, in accordance with the content and spirit of related rules; it can achieve rule-based fair judge to a greater extent. Undeniably, there are differences in the aesthetic concept, the understanding and use of competition rules of rating performer, i.e. the referee, resulting in similarities and differences of the final performance ratings. Whether international events or

C. Zhao (✉)
Wuhan Institute of Physical Education, Wuhan 430079, China
e-mail: 27276287@qq.com

H. Ma
Hebei United University, Tangshan 063009, China

W. Du (ed.), *Informatics and Management Science V*, Lecture Notes in Electrical Engineering 208, DOI: 10.1007/978-1-4471-4796-1_77,

national events, or provincial, prefectural, municipal events, or internal competition in units, industry, institutions, there are varying degrees of judging controversy, which is a major feature of subjective rating project. This paper selects several rating test methods that are more operable and practical, provides data quantification analysis for reasonable, feasible and teaching directive characteristics of the rating quantification innovative system of tai chi courses portfolio in colleges, which is designed to improve the effectiveness and influence of Chinese traditional sports teaching [1].

## 77.2 Rating Quantification of Exercise Levels of Tai Chi Courses

Calculate the correlation coefficient of exercise referee and difficulty referee of top 8 to top 20 (see Table 77.1). Import in CORREL function statistics of Excel sorted by artistic points, respectively calculate the correlation coefficient of exercise referee and difficulty referee, and analyze the correlation between referee ratings and student ratings. Data analysis shows, according to exercise scores order of top 8, the consistency coefficient of exercise referee rating is higher than difficult referee rating, according to exercise scores order of top 9, the consistency coefficient of difficulty referee rating is higher than artistic referee rating, according to artistic scores order of

**Table 77.1** Exercise level scores in order of large to small

| Order | A1 | A2 | A3 | A4 | Exercise scores | B1 | B2 | B3 | B4 | Difficulty scores |
|---|---|---|---|---|---|---|---|---|---|---|
| 2 | 7.40 | 7.40 | 6.90 | 7.50 | 7.40 | 7.50 | 7.50 | 7.60 | 7.60 | 7.55 |
| 3 | 6.70 | 6.80 | 7.00 | 7.00 | 6.95 | 7.10 | 7.00 | 7.00 | 7.00 | 7.00 |
| 16 | 7.10 | 7.00 | 6.60 | 6.70 | 6.80 | 5.50 | 5.40 | 5.20 | 5.40 | 5.40 |
| 6 | 6.80 | 6.50 | 6.60 | 6.70 | 6.65 | 7.00 | 6.60 | 6.70 | 6.70 | 6.70 |
| 7 | 6.50 | 6.70 | 6.80 | 6.70 | 6.70 | 6.60 | 6.40 | 6.50 | 6.60 | 6.55 |
| 4 | 6.40 | 6.60 | 6.60 | 6.70 | 6.60 | 7.00 | 7.30 | 7.40 | 7.20 | 7.20 |
| 1 | 6.50 | 6.50 | 6.50 | 6.80 | 6.50 | 7.70 | 7.80 | 7.80 | 7.80 | 7.80 |
| 9 | 6.60 | 6.00 | 6.30 | 6.40 | 6.35 | 6.68 | 6.78 | 6.78 | 6.77 | 6.70 |
| 20 | 6.50 | 6.00 | 6.40 | 6.40 | 6.35 | 6.80 | 6.90 | 6.90 | 6.90 | 6.90 |
| 10 | 6.10 | 6.20 | 6.40 | 6.40 | 6.30 | 6.40 | 6.50 | 6.20 | 6.20 | 6.30 |
| 13 | 6.20 | 5.90 | 6.20 | 6.10 | 6.20 | 6.00 | 6.10 | 5.60 | 6.00 | 6.00 |
| 14 | 5.80 | 6.10 | 6.30 | 6.30 | 6.20 | 5.80 | 6.20 | 5.80 | 5.60 | 5.80 |
| 15 | 6.20 | 6.00 | 6.10 | 6.10 | 6.10 | 5.30 | 5.70 | 5.90 | 5.60 | 5.65 |
| 11 | 6.00 | 6.10 | 6.20 | 6.00 | 6.05 | 6.10 | 6.20 | 6.20 | 6.20 | 6.20 |
| 18 | 5.60 | 6.00 | 6.00 | 6.00 | 6.00 | 6.20 | 6.10 | 6.10 | 6.10 | 6.10 |
| 5 | 6.00 | 6.20 | 5.60 | 5.70 | 5.85 | 7.00 | 7.00 | 7.30 | 7.20 | 7.10 |
| 19 | 6.00 | 6.20 | 5.60 | 5.70 | 5.85 | 7.00 | 6.70 | 7.20 | 7.10 | 7.00 |
| 17 | 5.90 | 5.80 | 5.80 | 5.80 | 5.80 | 5.50 | 5.60 | 5.40 | 5.40 | 5.50 |
| 12 | 5.70 | 5.90 | 5.60 | 5.50 | 5.65 | 6.20 | 6.10 | 6.10 | 6.20 | 6.15 |
| 8 | 5.50 | 5.50 | 5.50 | 5.50 | 5.50 | 6.70 | 6.70 | 6.80 | 6.80 | 6.75 |

Table 77.2 Correlation analysis results between referee ratings and student scores

| | A1 | A2 | A3 | A4 | B1 | B2 | B3 | B4 |
|---|---|---|---|---|---|---|---|---|
| Top 8 | −0.35 | 0.02 | 0.09 | 0.37 | 0.97 | 0.96 | 0.97 | 0.97 |
| Top 9 | −0.18 | 0.10 | 0.16 | 0.41 | 0.96 | 0.97 | 0.96 | 0.96 |
| Top 10 | 0.052 | 0.330 | 0.388 | 0.563 | 0.970 | 0.958 | 0.972 | 0.961 |
| Top 11 | 0.268 | 0.416 | 0.476 | 0.628 | 0.975 | 0.949 | 0.972 | 0.967 |
| Top 12 | 0.338 | 0.498 | 0.585 | 0.699 | 0.977 | 0.958 | 0.959 | 0.971 |
| Top 13 | 0.367 | 0.515 | 0.597 | 0.688 | 0.977 | 0.959 | 0.959 | 0.972 |
| Top 14 | 0.496 | 0.565 | 0.674 | 0.738 | 0.932 | 0.935 | 0.922 | 0.935 |
| Top 15 | 0.427 | 0.522 | 0.398 | 0.536 | 0.934 | 0.937 | 0.924 | 0.937 |
| Top 16 | 0.476 | 0.579 | 0.481 | 0.595 | 0.943 | 0.945 | 0.933 | 0.944 |
| Top 17 | 0.502 | 0.597 | 0.505 | 0.605 | 0.940 | 0.946 | 0.933 | 0.941 |
| Top 18 | 0.436 | 0.508 | 0.420 | 0.525 | 0.941 | 0.946 | 0.933 | 0.941 |
| Top 19 | 0.442 | 0.505 | 0.420 | 0.515 | 0.939 | 0.944 | 0.923 | 0.940 |
| Top 20 | 0.441 | 0.502 | 0.411 | 0.515 | 0.932 | 0.943 | 0.920 | 0.938 |

top 10, the consistency coefficient of finish referee rating is higher. The overall data show little difference between ratings, indicating that the results of tai chi ratings in groups have high trustiness. Use CORREL function statistics of Excel, separately gather statistics correlation coefficient of exercise referee and difficulty referee, correlation analysis between referee ratings and student scores, quantitative analysis of the ratings is in Tables 77.2 and 77.3. As can be seen from Table 77.3, the overall evaluation of referees is still good, but specific items in the first round of test, such as group 3, the rating coefficients of two exercise referees are low, the ratings of two exercise referees are general, the reliability of rating of one artistic referee is medium.

## 77.3 Application of Range Analysis

Range refers to the difference between the maximum and minimum of the data sample, reflecting the discrete trend of samples, in following the first round of test, for example, rating results in group [2], use method of range analysis to analyze consistency degree of ratings of exercise referee and difficulty referee, as is shown in Table 77.4. In recent years, with the popularity and development of martial arts culture and people gradually began to realize the social and cultural values of martial arts, sports researchers begun to engage in research related to martial arts culture. Martial arts culture is the country's cultural treasure, in today's highly developed science and technology; it still has a high social and historical value. Therefore, the popularity of martial arts becomes the focus of the country's government departments and education sector, how to promote the cultural characteristics of China's martial arts, how to carry out martial arts education, which becomes one of the key issues of cultural studies [3]. In the majority of our institutions of higher learning, some PE educators begun to engage in martial arts

**Table 77.3** Range analysis of ratings of exercise referee and difficulty referee in group 3 first test

| Name order | A1 | A2 | A3 | A4 | Range | B1 | B2 | B3 | B4 | Range |
|---|---|---|---|---|---|---|---|---|---|---|
| 1 | 6.50 | 6.50 | 6.50 | 6.80 | 0.30 | 7.70 | 7.80 | 7.80 | 7.80 | 0.10 |
| 2 | 7.30 | 7.30 | 6.80 | 7.40 | 0.60 | 7.40 | 7.40 | 7.50 | 7.50 | 0.10 |
| 3 | 6.80 | 6.90 | 7.00 | 7.00 | 0.20 | 7.20 | 7.00 | 7.00 | 7.00 | 0.20 |
| 4 | 6.40 | 6.60 | 6.60 | 6.70 | 0.30 | 7.10 | 7.20 | 7.40 | 7.20 | 0.30 |
| 5 | 6.00 | 6.20 | 5.60 | 5.70 | 0.60 | 7.00 | 7.00 | 7.30 | 7.20 | 0.30 |
| 6 | 6.90 | 6.60 | 6.70 | 6.80 | 0.30 | 7.10 | 6.70 | 6.80 | 6.80 | 0.40 |
| 7 | 6.50 | 6.70 | 6.80 | 6.70 | 0.30 | 6.60 | 6.40 | 6.50 | 6.60 | 0.20 |
| 8 | 5.50 | 5.50 | 5.50 | 5.50 | 0.00 | 6.70 | 6.70 | 6.80 | 6.80 | 0.10 |
| 9 | 6.60 | 6.00 | 6.30 | 6.40 | 0.60 | 6.70 | 6.80 | 6.80 | 6.80 | 0.10 |
| 10 | 6.10 | 6.20 | 6.40 | 6.40 | 0.30 | 6.40 | 6.50 | 6.20 | 6.20 | 0.30 |
| 11 | 6.00 | 6.10 | 6.20 | 6.00 | 0.20 | 6.10 | 6.20 | 6.20 | 6.20 | 0.10 |
| 12 | 5.70 | 5.90 | 5.60 | 5.50 | 0.40 | 6.20 | 6.10 | 6.10 | 6.20 | 0.10 |
| 13 | 6.20 | 5.90 | 6.20 | 6.20 | 0.30 | 6.00 | 6.10 | 5.60 | 6.00 | 0.50 |
| 14 | 5.80 | 6.10 | 6.30 | 6.30 | 0.50 | 5.80 | 6.20 | 5.80 | 5.60 | 0.60 |
| 15 | 6.20 | 6.00 | 6.10 | 6.10 | 0.20 | 5.40 | 5.70 | 5.90 | 5.60 | 0.50 |
| 16 | 7.20 | 7.00 | 6.70 | 6.80 | 0.50 | 5.60 | 5.50 | 5.30 | 5.50 | 0.30 |
| 17 | 5.90 | 5.80 | 5.80 | 5.80 | 0.10 | 5.50 | 5.60 | 5.40 | 5.50 | 0.20 |
| 18 | 5.60 | 6.00 | 6.00 | 6.00 | 0.40 | 6.20 | 6.10 | 6.10 | 6.10 | 0.10 |
| 19 | 5.60 | 6.20 | 6.30 | 6.30 | 0.40 | 5.50 | 6.20 | 5.70 | 5.50 | 0.40 |
| 20 | 6.00 | 6.20 | 5.60 | 5.70 | 0.60 | 7.00 | 7.00 | 7.30 | 7.20 | 0.30 |

**Table 77.4** Table of correlation coefficients of referee ratings and student scores in all groups

| Group | Rating | Correlation coefficient | | | | Standard value |
|---|---|---|---|---|---|---|
| Group1 | Exercise referee A1/A2/A3/A4 | 0.503 | 0.516 | 0.602 | 0.637 | 0.05(16) = 0.468 |
| | Difficulty referee B1/B2/B3/B4 | 0.807 | 0.831 | 0.895 | 0.902 | 0.01(16) = 0.590 |
| Group2 | Exercise referee A1/A2/A3/A4 | 0.652 | 0.509 | 0.638 | 0.633 | 0.05(16) = 0.468 |
| | Difficulty referee B1/B2/B3/B4 | 0.837 | 0.879 | 0.835 | 0.867 | 0.01(16) = 0.590 |
| Group3 | Exercise referee A1/A2/A3/A4 | 0.830 | 0.651 | 0.749 | 0.748 | 0.05(15) = 0.482 |
| | Difficulty referee B1/B2/B3/B4 | 0.866 | 0.913 | 0.896 | 0.902 | 0.01(15) = 0.606 |
| Group4 | Exercise referee A1/A2/A3/A4 | 0.436 | 0.508 | 0.420 | 0.525 | 0.05(16) = 0.468 |
| | Difficulty referee B1/B2/B3/B4 | 0.941 | 0.946 | 0.933 | 0.941 | 0.01(16) = 0.590 |
| Group5 | Exercise referee A1/A2/A3/A4 | 0.498 | 0.316 | 0.616 | 0.571 | 0.05(16) = 0.468 |
| | Difficulty referee B1/B2/B3/B4 | 0.760 | 0.697 | 0.738 | 0.749 | 0.01(16) = 0.590 |
| Group6 | Exercise referee A1/A2/A3/A4 | 0.902 | 0.880 | 0.921 | 0.881 | 0.05(16) = 0.456 |
| | Difficulty referee B1/B2/B3/B4 | 0.888 | 0.885 | 0.886 | 0.917 | 0.01(16) = 0.575 |

research, but their research areas confined within the scope of PE. Currently, the martial arts became one of the important parts of college PE classes, but according to the relevant survey we found that in campus, the popularity of martial arts is far less than taekwondo, which is worthy of further reflection, to reflect China's purposes of current setting up of martial arts classes, to provide a way of entertainment or fitness only for college students, and cultivate in order to develop our heirs of martial arts culture, and develop China's martial arts culture, reflect the teaching methods and teachers of martial arts, etc., thinking about the above questions plays an important role in martial arts carried out in Chinese colleges in future, which will become an important part of martial arts culture studies.

At present, most physical education teachers in China's colleges possess the cultural knowledge, martial arts skills reserves and so on, mostly obtained in school before graduation. Because in the process of higher education, learning time is shorter, only to 54–72 h are used to study martial arts. Related research shows that in our teachers of martial arts option classes of higher education, teachers of martial arts major are only 55.7 %, they carried out system training of martial arts knowledge, skills in universities, have high level of theoretical knowledge and skills of martial arts, skilled grasp of teaching methods, teaching content of martial arts options classes.

However, the current colleges shift the martial arts option class as a separate sports elective course, students select a sports area from a number of elective courses, which will result in a relatively narrow knowledge range of college student, single physical exercise means as consequence, which is contrary to our advocacy with the "wide caliber" training mode, will ultimately affect enhancement of the overall quality of our college students. In addition, because our implementation of option class teaching in colleges started late, action system of option class is not perfect, the students during the elective courses, may be because the number of elective people, time, experience a variety of issues affecting the teaching quality of college option classes.

The majority of our colleges and universities in the setting of martial arts option classes, take the specialized courses as the core, first divide the martial arts option classes into different subjects, and then gradually refine teaching content in the subjects of martial arts option classes. This approach emphasizes the integrity and systemic feature of martial arts option classes, but there are some defects, hindering the students' process to enhance the overall quality of the option course. Specifically, there are several issues: first, although the martial arts option classes have better systemic feature and integrity, but in the learning process of martial arts option classes at different levels, there are more repeatable teaching content, which will lead to the low efficiency of teaching, result in wasted teaching hours; second, stratification between basic courses and professional courses of martial arts option classes is more obvious, less contact between the two, which is not conducive to improvement of martial arts knowledge, skills level of students, likely to influence students' interest in learning martial arts; third, the setting of martial arts option classes in some colleges and universities is not scientific enough, too much emphasis on theoretical knowledge, lack of relevance of skills development; martial arts knowledge and technology taught in some courses are relatively backward, it is difficult to adapt to current situation of the students novelty, which can't reflect the teaching trends of martial arts option classes in colleges, lack of practicality of the overall quality development; fourth, martial arts option classes are lack of necessary communication between semesters and between curriculums, in the four-year college career, students are difficult to obtain the systematical knowledge, skills learning of martial arts.

Teaching purpose of martial arts option classes in colleges, aims to better develop overall quality of students. Therefore, during the teaching of martial arts option classes in colleges and universities, pay full attention to the capabilities of students of martial arts knowledge, skills, colleges and universities are fully aware of the importance of martial arts development capabilities of university students, but focus on how to train their ability through martial arts option classes, make students develop their own quality through the martial arts option classes. Currently, most colleges and universities in two ways, cultivate the martial arts knowledge, skills and other capabilities: first, teachers of option classes teach a number of practical martial arts training methods and skills, such as tai chi, martial arts gymnastics; second, maximize time of student contact to martial arts. The above two methods, to a certain extent, enhanced the students habits engage in martial arts training. However, because of the study time of option classes [2] the system of elective course, the goal can't be well implemented, only can enable students to understand how to do the theoretical analysis, and their ability didn't gain the most direct improvement.

Teachers resources play an important role in major development, martial arts option classes teaching in China's colleges and universities, is comparatively lack of teachers, the professionalism level of many teachers can't reach the teaching requirements of martial arts option classes, which in a large extent hindered the development of China's martial arts option classes. The reason why are the martial arts option classes lack of teachers, mainly lies in the following two reasons: first, compared with other sports majors, martial arts major requires more systematic feature; second, there are significant differences of teaching contents in martial arts major and other sports majors, it is not possible to achieve resource sharing between teachers. In addition, [4] the theoretical study of martial arts major is relatively fixed, the content is relatively simple, lack of professional training materials and reference books, many colleges and universities tend to choose the adjacent teaching content as an alternative, these materials are lack of relevance, there is no authority, to a certain extent affected the university's development of martial arts option classes teaching.

Martial arts option classes in colleges and universities are just for the development of the curriculum, are not only lack of appropriate teaching experience, but also during the choice of college students courses, the purpose is less clear. At present, the talents training, the development of exam systems in our colleges and universities have a certain imperfection, guidelines for capacity development of college students are not sufficiently clear of the curriculum setting, which is one of the reasons that the choice of martial arts option classes of college students is more blind. Some students in martial arts option classes, after systematical learning of martial arts in a semester, few have access to more great progress, according to the relevant survey found that during the course choice of college students in martial arts option classes, students mostly held purpose in order to obtain credits.

In order to improve teaching effect of martial arts option classes in our colleges and universities, curriculum system must be first completed of martial arts option classes in our colleges and universities, according to the student's interest in martial arts, requirements of physical development, reform the current curriculum system of martial arts, set up curriculum adaptable to physical and mental development of students, practice module teaching, reform class structural of options classes. Scientific curriculum structure should be divided into two parts: the compulsory content and elective content, compulsory content includes fist martial arts and other basic quality practice, elective content according to student interest, provides rich martial arts projects, the content can be related to the general public fitness management, health fitness way to martial arts, leisure and recreation, etc. [5].

Teachers are the key power of the smooth development of martial arts option classes in our colleges and universities. In order to improve the teaching level of martial arts option classes in our colleges and universities, necessary continuing education and training of martial arts teachers should be strengthened, through effective and targeted business training, to improve the professional quality, teaching and research level of teachers of martial arts option classes; in addition, should continue to optimize the teacher resources settings of martial arts option classes, can hire experienced excellent martial arts teachers with professional skills from sports school in colleges, achieve improvement purpose of university teachers in martial arts option classes, strengthen the backbone teachers' training of the martial arts option classes, and promote the rapid growth of teachers in martial arts option classes. During the curriculum setting of martial arts option classes in colleges, through opening wealthy school-based courses of martial arts, improve richness and interesting of students' martial arts option classes, change the curriculum status of traditional focus on teaching, ignorance of school-based curriculum development, continuously build curriculum model with regional characteristics and civil resource curriculum. Improve characteristics of martial arts option classes in colleges, martial arts option classes in colleges and universities should abandon the monotony of teaching content setting conditions, by seeking meeting point of China's civil martial arts resources and school-based curriculum development, jointly promote richness and characteristics of teaching content of martial arts option classes in colleges.

## 77.4 Conclusion

This paper fully takes into account of the characteristics of tai chi course teaching in colleges, combines with the spirit program of traditional ratings rules of martial arts, uses mathematical and statistical functions of Excel software, comprehensively applies research methods of consistency coefficient, range analysis and so on, through two major theme frameworks of tai chi exercise levels and difficulty coefficient, constructs a rating quantification system of tai chi courses portfolio in colleges, divides into six groups, three cycles of rating test, the test results show, the rating quantification innovative system of tai chi courses portfolio in colleges is reasonable, feasible and teaching directive. It recommends promotion in rating of martial arts courses in colleges, to improve the effectiveness and influence of Chinese traditional sports teaching.

## References

1. Zhang L (2008) Sports science research methods, vol 71. Higher Education Press, Beijing, pp 398–402
2. Jing J, Qin S (2007) Situation and development strategies of teaching methods of martial arts in colleges under new situation. Inner Mongolia Sports Technol 25:467–473
3. Huang M, Chen Z, Ding H etc (2010) Sports experimental design and scientific quantitative methods, vol 19. Higher Education Press, Beijing, pp 38–44
4. Ma J (2009) On the selective course of Wushu in colleges and universities. Wushu Science 10:10–17
5. Haibiao F (2011) Study on reform of teaching model of Wushu selected course in ordinary university. J Xi'an Inst Phys Educ 42:12–15

# Chapter 78
# Brand Advantages and Media Image of Youth Olympic Games Based on Political Communication

**Feng Yang**

**Abstract** The Youth Olympic Games are global youth brand games based on element definitions of universal values, global justice embodiment, group identity, market potential and so on, games brand advantages are the global implementation of universal values and the construction of youth group identity. The worldwide media image of Youth Olympic Games constructs myth imagination focusing on the personality characteristics of young people, which fits for the pure myth tradition of the Olympic Games, commits to show the innocence and vitality of youth, there are three building models of media image, i.e. the Singapore mode of performance of cultural diversity and city vitality, the Nanjing mode of performance of new forces and China's rise, the western reshaping model of performance of fresh cool breeze and passion element. Understand the nature of cultural politics of Youth Olympic Games, serve for sports development in China and the media industry.

**Keywords** Political communication · Youth olympic games · Brand advantages · Media image

## 78.1 Introduction

Global commercial sports are booming, sporting events are becoming more competitive, the establishment of brand advantages and media image is very urgent. New founder of the brand event—Youth Olympic Games (YOG) market

F. Yang (✉)
Xi'an International University, Xi'an 710077, China
e-mail: yangyangdeyi1001@126.com

W. Du (ed.), *Informatics and Management Science V*, Lecture Notes in Electrical Engineering 208, DOI: 10.1007/978-1-4471-4796-1_78,

positioning and development, have research value of typical case analysis and events political communication [1]. 1Examine from two aspects of the inherent competitive advantage and the external development environment for sporting events, the Youth Olympic Games have unique core competencies and cultural style [2]. Inherent competitive advantages are cultural ideas, events positioning and brand advantages, external development environment that is global political economy and media environment. Here, in political communication research perspective, provide a theoretical explanation for internal and external competitive advantage building and interactive development of the Youth Olympic Games, to better understand the culture political nature of the Youth Olympic Games, serve for the Chinese sports construction and media industry development [3].

## 78.2 Worldwide Brand Advantages of Youth Olympic Games

Confrontation between East and West Camps after the Cold War lift the tension, the capitalist process of the second globalization is accelerated, the global sporting brand events rapid expanded and developed. By sports, comprehensive, regional, national and other elements of classification, formed a more mature brand competitions of global market, individual events such as football World Cup, integrated events such as summer Olympics, regional events such as Champions League football, the national leagues such as NBA, etc. Enter into the new century, global sports development system for the population stratification is not perfect, the world's sports market, build brand influence mainly around the center of top professional events, ignoring races brand development of the young, middle-aged and old groups, which is decided by the differences in spectators, economic benefits and social impact of population stratification target events. The global market positioning of Youth Olympic Games is committed to the universal values, and construction of youth group identity.

YOG linked with the global universal values, the main political motive is the western values system in the name of universality to defense for their own special road and special interests [4] the national development strategy needs, and this was in the same strain after the Cold War, the west sports organization searched for the international road, such as predecessor of International Fair Competition Committee is "Pierre de Coubertin Fair Play Award International Committee of the Organization" rooted in the European sports, the predecessor of the International Society of Sports Philosophy is "Sports Philosophy Research" rooted in American society. Youth Olympic Games aim to reflect the political will of Olympic Games, strong heritage of respect, friendship, equality, honesty, solidarity, tolerance, and other excellent spiritual content of fair competition, focus on display fresh style of youth vibrant, health and vitality, the name of the universal value, build the international impact in full use of cultural soft power, western culture apostle image of communication and the establishment of the Olympic Games systems. Groups of young people strongly demonstrate and carry the culture fashion spirit, are the major force leading the trend of urban development and the future social and cultural changes. Youth Olympic Games in cultural nature construct the world's memory consensus and history narrative, educate world young people of the universal values which is defined and rewrote. The universal values are justice embodiment and myth power in social reconstruction after the war, characterize themes of world order stability and peaceful and sustainable development. In an international community of diverse cultural values, together at the negotiating table is a global consensus, recognizing the value of human virtue is the most common global identity. Youth Olympic endowment embodiment of global justice and myth power (cultural diplomacy ability to persuade others) is a reasonable application of the western value system, as timely as the cultural vanguard for special interests defending.

YOG's seeking recognition of youth groups is their moral character and nature decided. From the point of view of consumerism economy, youth groups play an important role throughout the production and marketing strategies. Such as the meaning of Nike brand slogan "Just Do It" (just do it), as long as to keep up with the pace of young people and fashion, a consistent marketing strategy of young fashion, that is to gain the initiative of the consumer economy in the brutal competition. Kappa, Adidas and other brands have special strategic planning with young people, achieve the consumer performance at some stage of brand growth, catch the last bus of young fashion economy. Review the history of the modern Olympic Games, Olympic education strategy for young groups did not become a strategic axis of the Olympic Games early development, in the longer term as a slogan and ideal existence. Capitalist countries after World War II in economic and social reconstruction, in the consumption of the juvenile products, found high value-added features, the business widely recognized that the consumption potential of juvenile products, as well as the ability to drive product consumption for adults. Olympic education as a relatively systematic strategy plan and new concept produced in the 1970s, [5] became the turning point of the global Olympic education fully integrated into the market economy. Creation and development of Youth Olympic Games defined youth main body identity and temperament content, [6] reshaped the commercial potential of the Olympic education, filled the gaps in the brand of global youth sporting events. The first Singapore Youth Olympic Games cloned product marketing model of Olympic Games, exposed the real face of the YOG and Olympic education, existed as supporting value-added services in business system and strategic marketing relationship between the Olympic Games there. Consumerism and the visible business group are the main driving force behind achievements of YOG Olympic education with major contemporary influence.

The global market strategy of YOG starts along two main lines of value logic, brand advantages rapidly grow in the white heat competition. In the event brand positioning, there are business opportunities in young people consumer products, with high value-added industrial properties, the Youth Olympic Games take young people group identity as a strategic core, brand advantages not only in direct young consumption and indirect adult consumption, but also beyond the narrow framework of market economy, provide a real carrier for international expansion and value-added of the regional political, cultural and educational aspirations. In identity construction of the race legitimacy, YOG holds great banners of global universal values and human virtues, spread clearly the ideological barriers for the identity of local culture, political legitimacy, and promote international communication of events content, forms and values. Therefore, the Youth Olympic Games are global youth brand games based on element definitions of universal values, global justice embodiment, group identity, market potential and so on.

## 78.3 Global Media Image of YOG

In information network society, the social construction relations of the things development and the public perception must rely on the power of the media, put up a bridge between the two. To a large extent, the Youth Olympic Games image itself is the wonders world and the cognitive simulacra of the media construction, media plays an important role. Looking at the pattern of international diplomacy, any information content with a regional political and economic purpose will be in culture political resistance of others, let alone the information content can circulate in international community areas by imagination construction. Youth Olympic Games must seek legitimacy of global political economy to establish a good international image, which is to find values consensus, seize the cultural highlights, enlarge propaganda, shape the image of the tournament.

In recent years, the western political-led international organizations frequently expose scandals, which greatly relieved the organization's international image and trust, revive the image of the international organization becomes an urgent thing of the western political legitimacy strategy, the functional role of sports (Sports) of fair competition, hard work ahead, friendly exchanges have the use value for organizational development. For example, the United Nations in dealing with regional affairs show weak trend, which undoubtedly relieved the voice and legitimacy status United Nations as a "world government", the timely implementation of United Nations "International Year of Sport and Physical Education" (IYSPE 2005) activities, strengthened national culture interaction and political trust, maintained the United Nations as a good international image as the non-profit function development sectors. Again, FIFA actively carried out cooperation with the media, suppliers, started crisis public relations activities after a chain of corruption scandals, including global commercial film "Player" gained support in the name of FIFA, the film in the shape of the healthy development of international football ecology, restore a positive role in promoting a good media image of FIFA. Salt Lake City Winter Olympic Games after the bribery scandal, Olympic Committee faced a confidence crisis of the development of the organization. Timely innovate event concept, launch new brand competitions, improve the Olympic Games product system, take Summer Olympics and the Winter Olympics as quality products for the event, Special Olympics and YOG as

**Table 78.1** College students of different grades on mass media YOG's information needs (%)

| | Broadcast | | Newspaper | | Magazine | | Television | | Network | |
|---|---|---|---|---|---|---|---|---|---|---|
| | Low | High | Low | High | Low | High | Low | High | Low | High |
| YOG's news | 38.8 | 31.2 | 41.8 | 46.7 | 21.2 | 27.6 | 44.9 | 39.2 | 34.2 | 49.7 |
| YOG's event | 36.5 | 26.4 | 48.2 | 52.1 | 22.1 | 28.3 | 64.3 | 55.2 | 37.6 | 47.2 |
| YOG's knowledge | 8.3 | 6.8 | 23.9 | 17.9 | 27.4 | 19.7 | 11.5 | 5.9 | 25.2 | 39.4 |
| YOG's commentary | 18.7 | 13.4 | 32.4 | 27.8 | 28.6 | 20.5 | 24.3 | 18.7 | 18.7 | 37.6 |
| Entertainment | 31.8 | 37.6 | 49.6 | 44.2 | 47.9 | 44.2 | 49.3 | 37.4 | 45.1 | 50.9 |

secondary value-added services. Youth Olympic Games are type of product additional marketing, event viewing, social impact and the poor performance don't have the direct economic benefits, but will help to maintain a pure, non-utilitarian, the beautiful angel image of Olympics, brand advantages value-added benefits are obvious for quality products for the Olympic Games, reinforce the foundation of consumer products dependence and loyalty of Olympic Games in the future, achieve the sustainable development of the Olympic Games (Tables 78.1, 78.2).

The worldwide media image of Youth Olympic Games constructs myth imagination focusing on the personality characteristics of young people, which fits for the pure myth tradition of the Olympic Games, commits to show the innocence and vitality of youth, innocence refers to the non-utilitarian pure world and sincere young thought world, vitality refers to the social sustainable development with vitality. These two points are very conducive to the maintenance of the traditional myth of the Olympic Games, cover and defense for the Olympic Games increasingly blatant act of commercialism. The cultural highlight of first Singapore Youth Olympic Games is the expression of the concept of urban vitality, the full interpretation of the value proposition of the YOG. Singapore is an island Chinese, Malay, Indian and multi-ethnic coexistent from around the world, and filled with city vitality. Singapore livable urban environment and vitality of the media image, imaginatively conveyed cross-cultural communication and vibrant Olympic education and the value YOG proposition, city image and the image of the tournament match each other, so that the YOG innocence and vibrant global media image rooted in people's hearts. Compared to Singapore, stereotypes of the ancient city of history and culture of Nanjing, China, and deep-rooted Chinese cultural tradition, Nanjing can't make differences on the value clue of urban vitality. Nanjing

**Table 78.2** College students on YOG's media information overall visual (listen to) evaluation (%)

| | Newspaper | Magazine | Broadcast | Television | Network |
|---|---|---|---|---|---|
| Very good | 20.1 | 6.9 | 3.8 | 27.2 | 25.5 |
| Better | 29.3 | 30.2 | 19.3 | 29.5 | 29.4 |
| Ok | 28.9 | 30.1 | 31.8 | 10.2 | 17.3 |
| Bad | 9.4 | 13.4 | 21.4 | 21.2 | 11.8 |
| So bad | 3.2 | 6.8 | 7.3 | 3.2 | 2.9 |

should construct media image in city and national levels, first, from perspective to re-read the history and culture of Nanjing, reconstruct city taste and cultural character of Nanjing on history representation and the media narrative, [7] change the tragic Nanjing, short-lived dynasty historical impression, into the civilized, dynamic, modern image element (Wang Cheng's opinion). Second, from perspective to symbolize national development in Nanjing, endorsement Beijing, Xi'an, Hangzhou, on behalf of great changes of ancient history, from point of surface, display the process of China's reform and opening up, traditional and modern symbiosis face, symbolic implication of the rise of China and the newborn young people power combined with a heavy responsibility to build China's trustworthy (national and global), paid off, the image of big developing countries to catch up. While bundle the media image of the YOG for promotion, ignite the rising generation of young people, a symbol of the future change of the world.

## 78.4 Outlook of Political Communication Trends of YOG

Both Singapore YOG image of urban vitality, or symbol of the new force of the rising China in Nanjing YOG, can't conceal the economic logic nature of the Youth Olympic Games. YOG innocence and vibrant global media image hide more deeply conflicts and crises. The commercial economy nature of YOG and Olympic education, are buried by the media's young innocence and overwhelming creation of a dynamic image of the tournament. Similarly, some regional cultural values rooted in European and American touted by the media stereotypes, buried by the global universal values, these construction of regional and cultural identity in the name of competition in order to achieve global legitimacy for the local culture the emotion, and ultimately achieve the requirements in the name of universality to defense for regional, cultural and political demands.

YOG media image options of the future, besides the Singapore model (diversity and vitality) and Nanjing mode (newborn and rising), the remaining third possibility is "return to the west". That is anti-mainstream cultural movement system flourished in the process of post-war reconstruction of western society, extracted the contents of pioneering spirit in hippies, yuppies, hip-hop groups, Bobos and a series of cultural innovation power, changed the decadence and stick media image of old capitalist western countries, blew fresh cool breeze and injected passion elements into the lethargic western society, for the image instead by youth trying fresh and new way of life and passion, "return to the west" and "reshape the west" may well be a media image and brand building road with cultural endowment products competitiveness. YOG political communication after Singapore mode and Nanjing mode, whether is the interpretation of polyphony, or the formation of new culture, we'll wait and see.

## References


1. Xudong Z (2006) Cultural identity in the era of globalization: historical criticism of western general discourse, vol 53. Peking University Press, Beijing, pp 652–655
2. Devitt M (2010) The myth of olympic unity: the dilemma of diversity, olympic oppression, and the politics of difference, vol 74. University of Toronto (MA), Toronto, pp 45–49
3. Anthony T (2008) Reporting from the Beijing bureau. http://sports.espn.go.com/espnmag/story?id=3541475
4. Cyphers L (2008) China's hero, its DiMaggio, falls before race even begins. http://sports.espn.go.com/oly/summer08/trackandfield/columns/story?id=3540374
5. Genhong Z (2010) The youth olympic games and olympic concept of cultural transformation. J Nanjing Inst Phys Educ (Social Science) 24:26–28
6. School Sports Work Principles (2010) People's Republic of China State Education Commission 5:25–26
7. Cheng W (2009) On the theory and practice for 2014 youth olympic games. J Nanjing Inst Phys Educ (Social Science) 23:5–8

# Chapter 79
# Research on Essential Difference of E-Sport and Online Game

**Hu Ma, Yinbo Wu and Xinyu Wu**

**Abstract** After nearly a decade of development and evolution, online game is popular in the world with its entertainment, athletic, virtual, interactive, innovative features and so on, and in the ascendant trend. This paper questioned the current mainstream ideas of the difference between the two (e-sport and online game). The essential difference is the sports standardization of online games, a distinction between virtual and reality, and the resulting influence on different effects of things development. Online game marries to sport, for common development, which is inevitable in its development history. Online game needs the sport form to regulate itself, further develop and expand; include the online game (also can be e-sport) into the sports category, as a sport, which has positive role in promoting the expansion of sports content, reflecting the function of sports. So a win-win situation can be achieved. It concluded that the differences we are talking about between e-sport and online game, are historical point of view, from the development perspective, e-sport and online game are not completely separate, immutable.

**Keywords** E-sport · Online game · Internet

## 79.1 Question the Concept Identity of E-Sport and Online Game

Question one: "From the technical point of view, the two rely on different network environment or have different carriers", "online game is entirely based on the Internet", "and e-sport rely on LAN environment" [1]. By network-wide,

H. Ma (✉) · Y. Wu · X. Wu
Hebei United University, Tangshan 063009, China
e-mail: 27276287@qq.com

W. Du (ed.), *Informatics and Management Science V*, Lecture Notes in Electrical Engineering 208, DOI: 10.1007/978-1-4471-4796-1_79,

computer network can be divided into local area network (LAN), metropolitan area network (MAN) and wide area network (WAN). Internet (Internet) is international scope made by a number of LAN interconnection. LAN and the Internet have the same communication network topology, the same transmission medium, the same data communication technology, the same network connection hardware. Under certain conditions, the difference between the two is that TCP/IP protocol. It has its particularity. Intranet is the use of Internet technology; connect the basis of computer-based network architecture with individual users, to set up the internal network. Intranet has all the features of the Internet, internal use only, classified by network range, it is a local area network. It should be noted that: So, there is no practical significance for what type of way to divide the network, it is just one of different stand points in discussion. The difference between e-sport and online game is the difference between LAN and the Internet, e-sports are face to face sports, entertainment in reality, carry out in a relatively small space; and online game is entertainment, sports conducted in virtual, the space can be assumed to be infinite. But this is not absolute. Take CS (Counter Strike) game as an example; many gamers are in Internet cafes for the entertainment and athletics. Internet cafe as a LAN unit connects to the Internet, provides online game services to game fans, but in reality, because the Internet the game running speed of CS game will be influenced, and in the cafes of the LAN to "Battle", the game speed will not be affected substantially, most of the fans of online games take this approach. In fact, the fans during playing the online game are only in the LAN of cafe, but because the game fans do not know who "opponents" are, the results obtained with the game on the Internet are the same, both in the virtual infinite space. Thus, the reverse launch, use the LAN and the Internet environment as a standard to distinguish between e-sport and online game, then the gamers in the cafe within the LAN entertainment and sports activities can also be said to be an e-sport.

Question two: "E-sports are a sport, but its equipment, the game environment, etc., are achieved through information technology, which is its most essential difference between online games" [2]. Broadly speaking, sports (upper concept) also include recreational activities. Logical analysis obtained: If the online game is a sport, then e-sport belongs to sports areas, but online game in reality we are not talking about is not recognized as a sport area; if the online game is not a kind of sports, then e-sport should not be a sport, or a "special" sport. Because General Administration of Sports of China listed e-sports as the official launch of the 99th sport, we believe that e-sport is in sport area, and after determining the e-sports is a sports project, say the difference between e-sport and online game: "e-sport is a sport", but online games are not, the "convention" is suspected. Online game membership issue has been controversial, based on consideration of the environment, characterize e-sport as sport, is it reasonable? Competitive nature is not the unique characteristic of sports. The classic section of Darwin's "survival of the fittest, survival of the superiority", is the basic law of natural development. Competition (or the sports) is the basic characteristic of nature, and also the basic characteristic of human development. If online game is not in sports areas, then, with its development and growth, it will have its "rules", "law", "competition", holds international competition in the real world, makes sense, in line with the basic law of things development. In the process of finding relevant literature of this article, a strange phenomenon was found, most sports journals say "e-sport", and most of the computer magazines say "electronic sports"; sports field see e-sport as a sport, and other areas have their different views, and so forth. Is there a certain subjective factor at play? Like a nondescript "people", the Americans see it as the American, Russians see it as the Russian, seem to want it to be attributed to the command.

Question third: "E-sport is sport; online game is entertainment game, which is the essential difference between the two" [3]. The main line of the birth, development of online game is to bring people entertainment, but the competitive nature is also in online game. Fans during the online games, in order to upgrade (refer to "Legend" and other game species), win (referred to as "Counter Strike" and other game species) and have to against the "opponents" for competition, fans appreciate the fun from the "competitive" game. Everything has two sides, of which there are some game fans purely for entertainment and game in there is also a part so-called "professional" game fans, solely to sport, to win, fun, interest (such as "Legend" game, use "equipments" for trading in reality). Recreational and athletic features in online game are the simultaneous phenomenon exist, they complement each other, with mutual help and progress. On the other hand, e-sport, if players only for competitiveness, not for entertainment, and use the idea to participate in this e-sport, it is contrary to the original intention of the origin of the game—entertainment. It can be said, whether e-sport or online game, has fun and competitive nature, the ideal criteria, we can think, e-sport focuses on the competitiveness reflection, while online game focuses on entertaining reflection. Note that the specificity expression, as mentioned above the "professional" game enthusiasts in online games are in pursuit of the competitive nature; in reality, a small part of e-sport fans to participate in the game only for "participation", do not care the results of the competition, pursue e-sport entertainment.

Question four: "E-sport as a sport, can exercise and improve the participants' thinking ability, responsiveness, coordination, teamwork and perseverance, and the ability to adapt to the modern information society, thereby contributes to the participants' fully development". "Online game is simulation and role-playing based on the pursuit of experience purpose, the way of the game is entirely accumulated through time to upgrade and improve, the game skills are basically no, and no confrontation" [4]. Online game seems can't improve the participants' thinking skills, ability to respond … …how to explain this one, "The Internet is only used for e-sports as a way of training and entertainment only". E-sports participants who use the Internet (refer to online games) to training, to improve their competitive level, is this self-contradictory? In fact, currently, a large number of e-sport athletes are trained to use online games, improve their game level in online games, and then to participate in e-sports, to achieve self-expectations. Let us consider this problem from another angle. Gamers in online games for training, and participate in the online game competition with strict rules (or sports normalized), to achieve self-expectations.

Question five: "It is understood that e-sport masters in young people are mostly academic leaders, and independent thinking and self-reliance are stronger, which also illustrates from one aspect that the sport has a positive role for young people to develop in morality, intelligence and body building" [5]. Athletes through long-term hard training to achieve high level, chess players practice in long terms can achieve chess skills; this is the law of things development. E-sport athletes can "build Rome overnight"? In reality, e-sport players through long-term training to improve the level of intelligence, improve the "equipment" skilled operation, and thus achieve the purpose of improving the competitive level. And training means are through the Internet (online games) to achieve. Online game likes a "double-edged sword", has a positive side and negative side. Master of e-sport on the road of growth must pay efforts and sweat (including a lot of training time). They therefore have a high intelligence level (upper concept, refers to thinking, co-ordination, reaction capacity), which is why most outstanding teams of domestic CS (Counter Strike) games are in major colleges and universities, "the youth masters in electronic games are also mostly academic leaders". In the tower and the bottom of the "pyramid", how many games fans (or e-sport fans) are in the "pushing hard", and "getting fruition"? These are the real existence of the facts, but we did not put on to talk about it.

## 79.2 The Thinking of the Essential Difference between E-sport and Online Game

After nearly a decade of development and evolution, online game is popular in the world with its entertainment, athletic, virtual, interactive, innovative features and so on, and in the ascendant trend. According to the British "Economist" reports, the 2003 turnover of world computer game market will exceed $ 20 billion, the figure is just over the world film market in 2003 the total box office revenue. In 2003, online games created a total of 160 billion yuan output value for China's game industry, telecommunications, IT industry, media and publishing industry.

Highlights the drawbacks of online game, and also demonstrates its unique charm, which, for any country's government leaders, will "gain essence, abandon its dross", maximum inhibition of negative side, play its positive side. The virtual nature of online game brought many negative effects, it brought moral anomie to the game fans, the evils that game enthusiasts in online games play multiple roles in the virtual identity and so on. The standardization of online games in sports, can effectively inhibit the occurrence of such negative effects, standardize the daily conduct of game fans and ethics, enhance the competitive sense, collective pride and willpower. Online game marries to sport, for common development, which is inevitable in its development history. Online game needs the sport form to regulate itself, further develop and expand; include the online game into the sports category, as a sport, has positive role in promoting the expansion of sports content,

**Table 79.1** College students with different gender of e-sport information in mass media needs (%)

| | Broadcast | | Newspaper | | Magazine | | Television | |
|---|---|---|---|---|---|---|---|---|
| | Male | Female | Male | Female | Male | Female | Male | Female |
| Sports news | 49.1 | 15.4 | 58.8 | 28.4 | 39.2 | 9.3 | 58.5 | 18.5 |
| Sports event | 51.2 | 14.8 | 68.2 | 29.2 | 44.7 | 14.1 | 79.6 | 35.2 |
| Sports knowledge | 18.6 | 2.3 | 31.8 | 9.6 | 37.7 | 13.1 | 14.9 | 3.2 |
| Sport commentary | 27.8 | 3.2 | 49.4 | 10.8 | 36.3 | 14.9 | 34.4 | 11.9 |
| Entertainment | 16.2 | 14.9 | 49.1 | 47.4 | 59.6 | 33.8 | 57.7 | 36.4 |
| Hobby | 20.6 | 2.3 | 56.8 | 39.5 | 63.4 | 33.2 | 67.4 | 29.5 |
| Sports star charm | 7.4 | 7.2 | 21.8 | 20.6 | 27.4 | 21.1 | 38.7 | 18.9 |

reflecting the function of sports. So a win-win situation can be achieved. The essential difference between e-sport and online game is the sports standardization of online games. The difference between a virtual and reality, and the resulting different effects will influence on the things development. It can be bold ideas, in the growth of online game history, the negative reports of public opinion as one side are increasingly frequent, people in mind think that electronic games, online games are "bad things", and at this time, hold a "XX Cup Online Game Competition", it does not seem "authentic", is also very detrimental to sports standardization of online game development. Meanwhile, the e-sport and online game really have essential differences (virtual and real), create a new vocabulary in the 21st century to meet their urgent needs, make sense. In fact, before there is the word "e-sport", followed a variety of names in the events held: China Telecom "Broadband Abyss 2003—Chinese Interactive Game Center CS Competition" [6]; VIA AMD National Game Contest; 2003 Unicom-cup "Magic Book" Grand Prix and so on. Until the emergence of the "e-sports" and recognized and widely used by people, which mapping out the online game industry is gradually mature and standardized [7] (Table 79.1).

## 79.3 Conclusion

Salient feature of online game is that the reality expression of the audience about the virtual world. The virtual nature of online game brought many negative effects, it brings fans to the game of moral anomie, evils of game enthusiasts in online games play multiple roles in the virtual identity. For example, the audience in online games confront to virtual "object", when encounter some game content to stimulate their normal mental activities, they are likely to cause verbal gaffe, the specific performance of the "swearing dirty words" and even lead to the radical behavior in reality (such as breaking things for no reason, acts such as beat computers and ancillary equipment). In reality, online games competition held according to certain rules can effectively inhibit the occurrence of such events, this time, the salient features of online games (virtual nature) no longer exist, because the opponent of a game fan is no longer the virtual world "imaginary enemy", but the real competitors, the audience can follow the rules of a certain online game plus the rules restricted by the game organizers, contest race between people, which is equal to the same activities of chess, I-go, etc., reflects the same races focused on the conscious mind activities, but competition of online games replace board for the computer and ancillary equipment, rules are depending on the rules of a game (plus the rules that race organizers restricted) and perform as different standards for different competitions. For example, A company organized a CS game, it follows the CS rules of the game, while develop complementary rules (such as regulate the number of the rounds, times per game, etc.), the players follow the rules of the CS game and under the premise of the added rules, and race with the reality opponents, in the process, the audience must consider that what he face to is real people, words can't appear to be too radical, or even an unfriendly action, the reality environment let the network game dummy gone, the negative effects of online games have been avoided. Online game competitions hold meets the practical needs of those gamers, so that the gamer can temporarily out of virtual environment of online game, face to the reality of life. In the online game tournament held in South Korea founded in 2000 the most famous World Cyber Games (WCG), it gathered the world's best online game fans together for a common entertainment and cheered.

From the above analysis we can see, online game with the performance of the virtual world and online gaming competition with performance of real-world are two understanding areas, their essential difference is between the virtual and real, thus affecting the audience's attitude towards online games and lead to different behaviors of real activity. If understanding of network gaming competition is replaced by the concept of online gaming areas, it can't well reflect the characteristics of the benchmark of the reality of online game competitions, reference to the concept of electronic sports (Electronic Sports) can accurately express the essential difference between online games and online games competition (virtual and real), founded in 2003, Electronic Sports World Cup (ESWC) also reflects the concept changes of world online game.

"E-sport" concept and the "electronic sports" concept have the same meaning, but in the unique discussion domain environment are not equal to the same. "Modern Chinese Dictionary" has six interpretations about the "sport", which, as a noun "sport" has two interpretations, can be regarded as broad and narrow. Broad "sport" means: political, cultural, production and other aspects of organized, purposeful and momentum larger mass activities; narrow sense of "sport" means: physical activities. If the "e-sport" in the "sport" refers to "physical activity", then "e-sport" should cut on the content in the "electronic sports" connotation, so the "e-sport" (referring to the narrow meaning) reflects the "physical activities" content. It must be noted that, even if the "electronic sports" have been given a narrow meaning (sports activities "electronic sports"), are also limited to sports field, are "personality" understandings of the sports sector about the online game competitions (also electronic sports).

He Huixian in General Administration of Sports of China wrote: "General Administration of Sport of China list the e-sport as a national sport officially launched, the most important reason is the sport has the basic physical properties, and has developed more widely in China, there are tens of millions of fans, and also have a certain competitive level, it is listed as an official sport in order to adapt this project for further healthy development". Looking back at our country to carry out some of the new and fashionable sports development process, we can tease out the main line: the development system of Chinese sports had a positive effect on socialist construction, it directed the social activities which meet the physical content to the relatively mature and stable development of sports track, in order to facilitate their healthy development. Such as "hip-hop is popular in recent years, the emerging sport in the world, China's development is still immature, the introduction of the hip-hop in aerobics development track is stage development strategy to regulate their development directions in China, to guide and support healthy development in China, hip-hop temporarily classified as aerobics area, hip-hop in China's development is out of horizon, and then let out aerobics development model for extension and development of its personalization". In addition, many social activities in line with sports content are into the sports development track, China's mature sports development model provides the template for the social activities which meet sports content. There is no doubt that online game entered in the sports development track, the specific performance is the sports competition standardization development of online game.

## References


1. Moschella D (2002) The rights wave: the development of global information technology and prospects 1964–2010. Social Sciences Academic Press, Beijing
2. Sunping S (2008) E-sports competition forces, vol 11. China Internet Weekly, pp 25–28
3. Xi Q (2004) Online games, play out the wealth, vol 9. China e-Commerce Business, pp 16–24
4. Wuyang Z (2008) Online game industry research. J Jiangxi Univ Finan Econ 17:51–55
5. Yun L (2010) Online games included in the "863" science and technology development. Today's Technol 24:43–48
6. Dayu X (2011) Open the door to the treasure house of the online gaming industry. Digit Media 22:24
7. Zhang G (2003) Online game industry development research, vol 8. China e-Commerce Business, pp 112–114

# Chapter 80
# Study on Rhymnastics Gymnastics Teaching Method

**Miao Feng**

**Abstract** If we want to relieve rhymnastics gymnastics education from the pure physical training, the key is rhymnastics spread of culture. The realistic meaning and historical significance is the main purpose of the important change in gymnastics teaching rhymnastics gymnastics education, pay attention to the building rhymnastics gymnastics education theory system contains the humanities and science and training talents, has both the rhymnastics gymnastics humanity quality and scientific spirit.

**Keywords** Humanity · Science · Rhymnastics · Gymnastics education

## 80.1 Introduction

The main content of innovation also is the development of the society rhymnastics gymnastics is not only more and more bonus content of science and technology, and rhymnastics cultural innovation. At present, the main project rhymnastics research is still about gymnastics training and teaching to develop content including physical skills and physical exercise, healthy development. And the spirit and culture research quality content, this development, tolerance, respectation, integrity, sympathy and self sacrifice, few people here attention.

M. Feng (✉)
Cangzhou Normal University, Cangzhou 061001, China
e-mail: miaofeng11234@163.com

W. Du (ed.), *Informatics and Management Science V*, Lecture Notes in Electrical Engineering 208, DOI: 10.1007/978-1-4471-4796-1_80,

## 80.2 The Understanding of Rhymnastics Gymnastics Education

Rhymnastics gymnastics education of education process is the instrument of gymnastics rationality unity, value and practical rationality and infinite horizons. Their main purpose is the humanity to tell the students how to get the job, more important is how to control their life, this is a survival skill [1].

The essential difference rhymnastics gymnastics teaching and traditional and modern rhymnastics gymnastics education is the traditional gymnastics teaching rhymnastics stops in the sports skills and may cause the people lost their goals in the process, to lead the endless alienation. And rhymnastics gymnastics education teaching not only the skill and guide the thought and education method, human and science, it can recover lost in the main body of the process. Humans can sublimate from the education teaching.

## 80.3 The Formation of Rhymnastics Gymnastics Education

In the past few years, the rhymnastics gymnastics education seems to be a process in which the humanity impediment self-recovers them. From the macroscopic Angle, Rhymnastics gymnastics education is education, so Rhymnastics gymnastics education itself is also a kind of restoring physical defects. Desire like Socrates said: that is what do take care of health care love all body consciousness, all desires, want to emotions and feelings and longing. This is all things of the body. Western education followed his mission for many years. As a part of the education system, rhymnastics gymnastics is also a kind of exercise and development of human beings.

Scientific concept is more and more attention, this means that the position of the rational spirit will be final confirmation. Rhymnastics gymnastics education method also got in "the body of knowledge education" as the main projects. The purpose of rhymnastics gymnastics education in the school playground becomes more and more simply, purely for enough brain thinking. Rhymnastics gymnastics education of the education is for the regulator seemed knowledge. The main purpose rhymnastics gymnastics education changes, rhymnastics gymnastics become simply for success, kind of rigid standard education. As CAI yuanpei said in his speech, no matter it is dangerous. Once the movement as a mechanical work, it will lose its true value, even if there is no harm to the body. System, standardization and unity of the mechanical system, can be found in the nature of the knowledge system.

The humanities and science two wings let a person is perfect, for a country to, be strong, for the prosperity of sports culture. No human items, the development of science is anything but a bunch of things without soul. Both cultural and scientific culture, rhymnastics gymnastics can also have a good development. So in the future rhymnastics gymnastics education system of human is culture and science culture blend. Rhymnastics gymnastics may try to fitness and education other than the simple teaching techniques and skills.

To sum up, rhymnastics gymnastics education system is the combination of young blood sports activity, the profound connotation and colorful culture together. This is an education system, promote human common values. In this system, the rhymnastics gymnastics teaching education improves learning skills, in a simple inheritance culture. This rhymnastics gymnastics education should be considered as a kind of social activities can make the person perfect. The aim is to improve the speed of the human society needs and assembly of the social change.

## 80.4 The Innovation of Rhymnastics Gymnastics Culture

The Rhymnastics Gymnastics Culture was defined by three layers (Page2): the first layer is thoughts and culture which is the core of the Rhymnastics Gymnastics Culture. It is the mixture of humanity culture and science, The second is the style and culture, this is style, it was used in the teaching and the training process Rhymnastics gymnastics, the third is external layer called realize layer. This means that the visible technology, skills and achievement. We study Rhymnastics gymnastics comprehensive culture including culture and implementing the core layer of popular culture, mixed elite culture.

Rhymnastics gymnastics cultural innovation comes from the core, thought and culture (layer 1) the mixture of the humanities education and science to meet education concept, this is style layer (layer 2), along with teaching and training methods, it is the thought layer and tangible technology, skills and achievement. The three aspects of the whole culture mixing is the combination of the real ideological culture and implement layer.

### *80.4.1 Humanity Education Thoughts*

Human education thought rhymnastics gymnastics is promote students' humanistic quality. It is most important to develop the general value. It involved better quality, we will further. Or, no social benefit and economic benefit from the audience, who want to enjoy healthy and beautiful combination, rhymnastics gymnastics industry will not development. Rhymnastics gymnastics teaching the general value will is not a weak teachers' and students' emotional interaction between, or even a student do tools rhymnastics gymnastics.

#### 80.4.1.1 Education of Mental and Emotion

The psychological and emotional core is character and squirming understands emotion and will. The root of the spirit of truth to understanding, for good and feelings, because is spiritual and emotional pounds [2]. From the training and education of the theory and practice Rhymnastics gymnastics, more attention is strong moral will of the training, practicer consciousness. Moral cognition is a process of observation, is to understand the ethics common law knowledge and natural law. The moral feeling sensitive feelings are such as understanding and care for others [3]. Moral cognition and moral experience, moral, that is psychological law, is the ultimate aim is to cultivate spiritual and emotional. In the process of teaching, Rhymnastics gymnastics perfect and shy of virtue, the cognitive-the emotion motive is one of the key maturities.

#### 80.4.1.2 True Love and Temper

Emotional impulse temper is happy to love and to be loved 31 from. Everybody wants to be loved and respected and appreciated. If people love each other, all is love. There is a famous German psychologist said From art love: to improve your ability to be love, than to seek to be loved. Promotion ability be loved is a process of mature personality, including in the study, correct each other, to teacher, love to the stadium, etc. All these elements can promote the ability to love long-term rhymnastics gymnastics teaching and learning methods, and that is the true love and emotion, serious and responsible, altruism and keeps his promise.

#### 80.4.1.3 Standard Education

Education is a kind of education standards is to establish the right position, follow the laws and regulations and establish a good relationship with each other [4]. Including improved practice appliances, family love between practice, performers, and the friendship between the audience and the unique emotion and the student's teacher. This emotional form will lead good relations and bring the satisfaction.

#### 80.4.1.4 Supervisor Education

The competent education is a kind of education wisdom and ability [5]. In China, there are old said want to care and social commitment. A person can contribute to the world is by the district control. In practice, the training and competition, the knowledge, skills, management ability and leadership should be considered as the value of the education benefits the world scope than any other single person.

## *80.4.2 Scientific Education Concept*

Science education idea is to improve the level of thought and psychological quality along with health. Science education rhymnastics gymnastics is essential to study how physical and psychological adjustment, to psychology as the main body, the body as the research object, to complete the knowledge system and the mature, the unity of the personality of the physical and psychological. The old view is rhymnastics gymnastics is just an aerobic exercise. In fact, it is also a kind of sport can make the spirit and body coordinated development of each other. So we do research rhymnastics gymnastics of the combination of body, mind, and group. This aims to speed up to good direction rhymnastics gymnastics in the human will mix with science.

### 80.4.2.1 Physical Perfect

Rhymnastics physical education is not only to improve the perfect gymnastics skills and physical strength, more important is to penetrate tolerance, respect and honest. This rhymnastics gymnastics physical exercise psychology. It is the perfect action and perfect appearance, physical and mental.

### 80.4.2.2 Mentality Tempering

Rhymnastics gymnastics required graceful behavior and good temperament, the unique architectural style, time control achieved good effect. This means that the show from the heart.

### 80.4.2.3 Group Perfect

Rhymnastics gymnastics education is to promote the cultural quality and the thought movement process of level. Its purpose was to help athlete's rhymnastics in a miracle. The old view is, Rhymnastics gymnastics is a kind of physical exercise, this is a low level of sports. And modern education as Rhymnastics gymnastics not just the physical elements, more important is beautiful heart. Modern beauty is natural, accept and beyond life, it is a kind of ideological health and beauty. So in practice, technology and human exercise should note. What Rhymnastics gymnastics showed is life, not skill. Carry out the action, and the nation, the dream and the natural and spirit.

Rhymnastics gymnastics education of two parts: the basic theory and concept. Is the study of human legal below tip, skills and exercise and study two human layer and layer is the process of real science education. We tried to continue the whole culture thought and implementation and the combination of is hard and soft

Rhymnastics Gymnastics Teaching: P (person) -S (sports) pi (improved person)
(from P~S, Train the Skill of job hunting)
Rhymnastics Gymnastics Education: P (person) -S (Rhymnastics Gymnastics Culture) --- pi (improved person)
(from P~P', Cultivate the capacity to earn "person")

**Fig. 80.1** The difference in essence between rhymnastics gymnastics teaching and rhymnastics gymnastics education

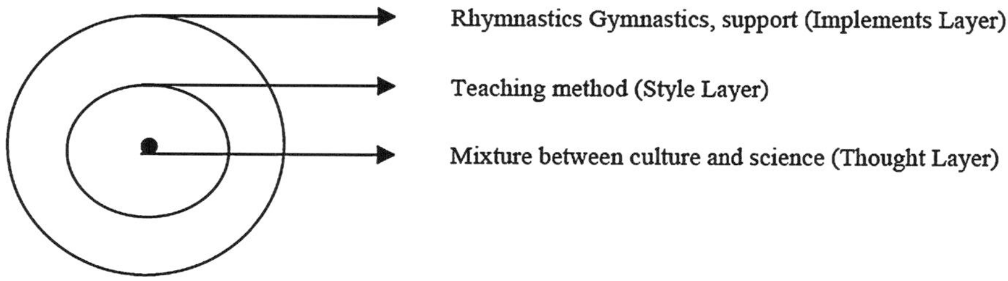


**Fig. 80.2** Structure profile of rhymnastics gymnastics culture

combining culture. Its purpose is to improve the all-round development of people rhymnastics gymnastics education. The new method is used for training and teaching rhymnastics gymnastics makes it become a gymnastics teaching process education. Nature is to meet the all-round development of people rhymnastics continue to gymnastics culture; this is a very important part of rhymnastics gymnastics cultural innovation. So the theoretical system of the future research rhymnastics gymnastics education of human culture and scientific principles mixed thought and culture structure (Figs. 80.1, 80.2).

## References

1. Jixin S (2003) Sports pedagogy, vol 79. People's Sports Publishing House, Beijing, pp 328–329
2. Bochun L (2003) Chinese culture and society of the 21 century-contem-porary scholars on the future of mankind trends, vol 12. Chinese Publishing House, Beijing, pp 357–365
3. Xue J, He Y, Jiang G (2004) The practice and study of building course program of quality education of physical education of unifying humanist and science. J Phys Educ 02:72–74
4. Zhao M, Xu G, Li C (2010) From education of body to liberation of body–prospects for P.E in 21 century. J Wu Han Inst Phys Educ 41:53–57
5. Jixin S, Juan G (2010) On sports integrated science and humanity—reforming of sports ideology theory and mode. J Phys Educ 11:7–10

# Chapter 81
# Research of Sports Teaching Network Support System

**Wengang Ren and Ligang Tian**

**Abstract** With the rapid development and popularization of network technology, network has been applied to all aspects of teaching and learning activities as an auxiliary tool. For the special subject of sports, shortcomings and deficiencies in the teaching activities can be compensated effectively through the application of network technology. Developing the corresponding network support system for physical education can break the time and space constraints of traditional physical education activities. It will lay a solid foundation for the popularization of the lifelong education concept. It is also an emphasis and a hot spot of the research for present sports workers.

**Keywords** Physical education · Network · Auxiliary system

## 81.1 Introduction

The support system's function of sports teaching network are: Playing students' leading role effectively, so that students make practices training purposefully according to their characteristics in their spare time [1]; making full use of the advantages of online tools to combine symbols, sounds, text, pictures, video and animation which are needed in the sports teaching process and form a vivid, three-dimensional process which can deepen students' learning impression; increasing the communication between teachers and students and breaking the

W. Ren (✉)
Hebei Institute of Physical Education, Shijiazhuang, China
e-mail: wengangren@yeah.net

L. Tian
Agricultural University of Hebei, Baoding, China

W. Du (ed.), *Informatics and Management Science V*, Lecture Notes in Electrical Engineering 208, DOI: 10.1007/978-1-4471-4796-1_81,

time and space constraints by the way of online Q & A and answering students' questions in the process of sports practice in real time [2, 3].

## 81.2 Requirements Analysis

Using the Internet is the key to sports teaching network support system; mainly all of the students can use the Internet at present [4]. Using a browser to interact between students and teachers through the Internet will save a lot of time. The overall structure of the system shown in Fig. 81.1.

As shown in Fig. 81.1, clients can be students and teachers and also can be system administrator. The entire system uses the Internet to access the server and fed back the results to the corresponding client. The system can be applied to each school independently. It also can be linked with and the school's original site effectively.

### *81.2.1 Business Analysis of System*

This system is the support system of physical education. Teaching activities centered in the whole business. It needs to follow the principle of sports instructional design. Give full consideration to the student's status and teaching objectives. The design of network courseware should give full consideration to students receiving skills and understanding to protect scientific and advanced of its system.

It uses modular teaching in the organization of the course content. The division of the curriculum form a separate module to ensure students' systematic and coherent in the learning process. On the basis of the syllabus, supplementing and perfecting the courses to expand the knowledge of students.

For the emphasis and hot spots of the course, it needs to strengthen exchanges and communications between teachers and students. Reproduce the teaching content by animation and video. It can reflect its practicality.

**Fig. 81.1** Overall structure schematic of the system

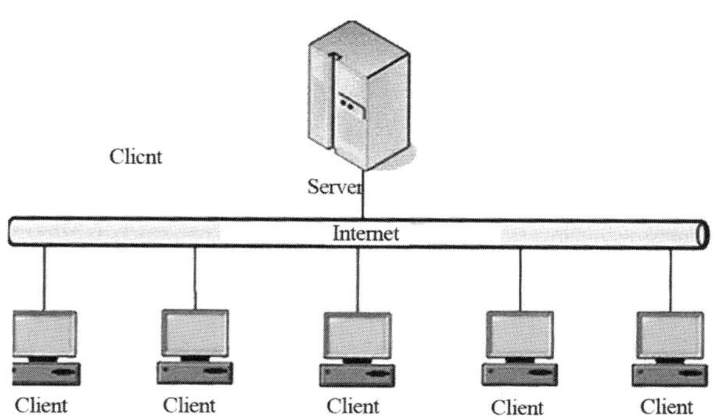

### 81.2.2 Principle of System

In order to protect the system's practicality, it needs to grasp the following principles in the process of building the whole system.

#### 81.2.2.1 The Subjectivity of Students

Regardless of classroom teaching and online teaching, the ultimate goal is to ensure that students can learn well. So as the body of teaching activities—students, play a very important role. The system's primary task is to enhance students' initiative. Stimulating students' enthusiasm through the establishment of the situational teaching, cooperative learning, etc. At the same time, for the arrangement of the content, you can't blindly pursue the syllabus and teaching programs. Only from the perspective of students and take full account of students' actual situation can make the arrangement of teaching content better.

#### 81.2.2.2 The Value of Teaching Content

In the arrangement of the content, it needs to combine the basics knowledge of teaching with inspired educational mode. Physical education can't be mastered in the way of spoon-feeding by students. If you want to make students to understand the essentials of the philosophy and actions, it is necessary to innovate in its approach to arouse the thinking capacity of students and stimulate the enthusiasm of students' learning. On the degree of difficulty, it needs to from easy to difficult which is a way of gradually promoting. We should mainly use the multimedia to show the contents of the teaching and attract students' attention as much as possible, so that students will focus on teaching content.

#### 81.2.2.3 The Interaction of Teachers and Students

After teaching through the network, there are fewer opportunities of face-to-face for teachers and students. This does not mean that the teacher's role is weaken, on the contrary, the teacher's role will become more clear. Within the stipulated time, tutoring the students in real time is an important part of the online teaching. At the same time, students seek the help of teachers through network message as a powerful complement in other time to increase the frequency of communication between teachers and students.

### *81.2.3 The Function of System*

Network Teaching System is a powerful complement to physical education It can make teachers and students to communicate through the network. Its major function is mainly reflected in: resource sharing, online Q & A, learning query, online theory test and system management and so on.

## 81.3 Summary Design of System

In order to reflect the advantages of network support system, it need to fully considered the "teaching" and "learning" unified contradiction in the development process. So it should consider the two sides of the teachers and students in the specific module design. The overall system consists of four major modules, namely: exam related modules, courses and function modules, online Q & A modules and teaching resources modules. The overall structure of the system shown in Fig. 81.2.

### *81.3.1 The Exam Related Modules*

There are mainly four scoring criterion in this module. They are exam results announced, the theory test, the standard of technical movements and project quality. Students can determine the incorrect or non-standard actions in their own training process by two scoring criterion to ensure their forward movement to

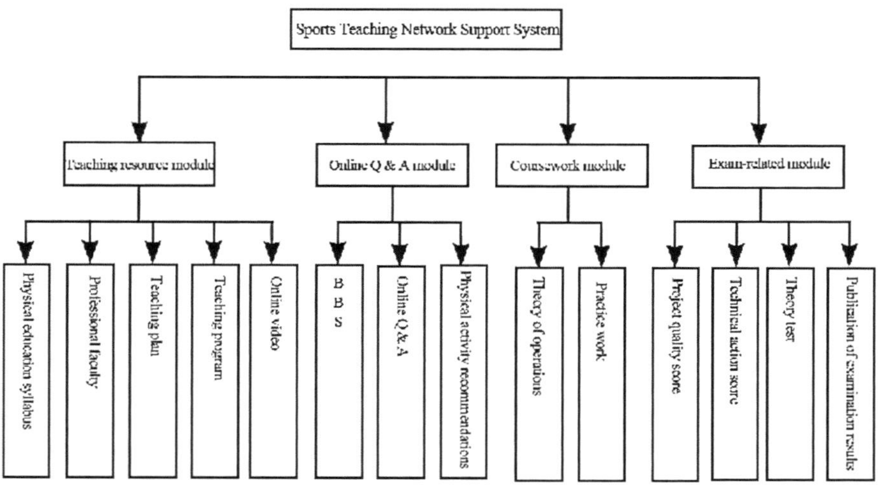


**Fig. 81.2** The overall system structure diagram

high-quality literacy. The theory test is an indispensable part of the physical education teaching activities. This part can proceeded through the online exam to ensure that the students improve the emphasis of the theory in the learning process to improve the professionalism of students. After submitted, the system will take the initiative to scoring for the objective questions. The final results will be announced on the Internet. It is facilitated to consult for students and can be saved through the EXCEL.

### *81.3.2 Online Q & A Module*

The module is primarily through the BBS, online Q & A and fitness advice to interact between teaches and students in the way of real-time online and the message. To solve the problems of students in the process of movement and the theory of learning and facilitate exchanges between students and students at the same time and form interest groups.

### *81.3.3 Teaching Resource Module*

The module mainly including online video, teaching programs, teaching plans, syllabus and Introduction of professional teachers. This module is mainly to make students understand the learning process and download some skills action video from the Internet. Strengthening their own learning through observations and getting rid to the restrictions of time and space in the learning process.

## 81.4 The Implementation of Specific Modules

### *81.4.1 The Design and Implementation of Online Q & A Module*

The subsystem mainly includes the following functional blocks: online message, message query and message viewing and so on. The Web subsystem which is a significant window of the service delivery is very important. This part provides a communication platform for readers and administrators to solve the various problems which are encountered by the user during use. It has used a FreeTextBox control to achieve the functions. FreeTextBox is the most common text editor. It is a ASP.NET open source server controls which is based on the MSHTML technology in the Internet. This is an excellent Free Software. This thesis can easily make it embed into Web Forms to achieve the on line editing of the HTML content. It can be used in the press release, blog writing, forums communities and other Web systems. The usage is as follows:

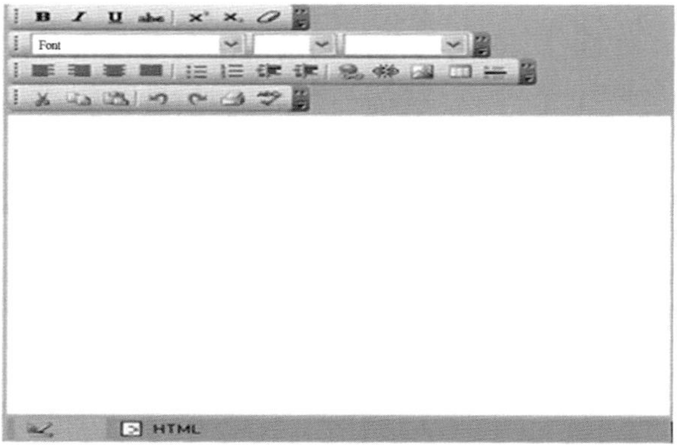


**Fig. 81.3** Online text editor

First, copy the FreeTextBox.dll file into the bin list of items. Then add a new reference to the project and select the items tab in the Add Reference dialog box and browse/choose your FreeTextBox.dl/Open/OK. The addition is completed.

To obtain the name space that which is used in the FreeTextBox.dll.
Using FreeTextBoxControls.Design;
Using FreeTextBoxControls.Common;
We could view the object browse of the referenced FreeTextBox.dll and see clearly by point the tree directory. All the above mentioned is name space.
Adding the freetextbox file to the aspx file
<%@Register TagPrefix = “ftb” Namespace = “FreeTextBoxControls”
Assembly = “FreeTextBox” %>
The cs file will have the following code:
Protected FreeTextBoxControls.FreeTextBox FreeTextBox1;
After added, the results will be seen (results shown in Fig. 81.3). We can also view the FreeTextBox property in the design pattern. The user can then see their messages and specifically view the inside information according to the title.

### *81.4.2 The Resource Management Module*

The system includes the following functional modules: user management, resource download management, teaching plans, syllabus and teaching content and so on. The system mainly introduces the resource management (reply, delete, view) and the management of user information (ads, delete, modify, and view). The specific code as follows:

String code = this.dg.DataKeys [dg.SelectedIndex].ToString ();
DateTime STR;
STR = System.DateTime.Now;
String content = Hcontent.Text;
String state1 = “Have been processed”;

```
SqlConnection conn = DB.GetConnection ();
conn.Open ();
string strSql = "update BBS set HContent = "' + content + "', DTime = "" +
STR + "', State = '" + state1 + "' where ArID = " + code;
SqlCommand comm = new SqlCommand (strSql, conn);
comm.ExecuteNonQuery ();
conn.Close ();
Hcontent.Text = "";
```

### 81.4.3 Bulletin Management Module

Bulletin management is to manage and set the bulletin issued which is issued by their teachers. Including add to a new bulletin information, modify existing announcement information and delete the old bulletin information. Part of the code are as follows:

```
Set rs = server.CreateObject("adodb.recordset")
sql = "select * from gonggo WHERE id ="&id
rs.open sql,conn,1,3
if not (rs.bof and rs.eof) then
rs("title") = title
rs("name") = name
rs.update ()
End if
```

## 81.5 Summary

This article studied the sports teaching network support system. It first analyzes the requirements analysis of network support system, then studies the system structure in general. Last it describes the resource management module of network support system, online Q & A module and bulletin management module. It also gives the core code and lays a solid foundation for the concrete realization of the reader system. It's hard to avoid shortages. Criticism and suggestions are welcome.

## References

1. Liu B (2006) Constructing the library's dynamic web site based on ASP. J Liaoning Adm Coll 9:213–214
2. Meng Y, He M (2000) Database programming based on web. Comput Sci 6:50–52

3. Liang J, Chen Y (2006) ASP program design, vol 63. China Water Conservancy and Electricity Press, Beijing, pp 255–262
4. Huang Z (2005) Computer teaching system design based on the campus network, vol 7. Chinese scientific & technical information, China, pp 196–197

# Chapter 82
# Study of Public Sports Service Demand on Human Movement Science Undergraduate Professional Training Programmers

**Jianyong Di, Yanmei Yang and Honghui Wang**

**Abstract** According to our country Human Movement Science of professional development, as well as the stage of public sports service demand, this paper aims to explore a new guidelines and implementation scheme based on the public sports service demand as human movement science professional training program.

**Keywords** Public sports service · Demand · Training plan

## 82.1 Introduction

Sports public service is a reflection of national sports level. Sports talent determines the level of sports public service. Our country is a big sports country, but definitely not a powerful sports country. A public sport service level cannot meet people's increasing by up fitness, leisure, entertainment; personnel training are the urgent needs to improve the level of sports public service. With the rapid development of the national economy, the supply mode of our country public service has changed significantly, a government-led, a variety of market main body actively involved in providing public service pattern is being formed stage by stage, under the new situation, how to cultivate the industry adapt to the development of talent becomes the internal demand of the public service industry.

J. Di (✉) · Y. Yang
College of Science, Heibei United University, Tangshan 063009, China
e-mail: dijianyong60@126.com

H. Wang
Qinggong College, Heibei United University, Tangshan 063009, China

W. Du (ed.), *Informatics and Management Science V*, Lecture Notes in Electrical Engineering 208, DOI: 10.1007/978-1-4471-4796-1_82,

## 82.2 The Current Situation of the Development of China Human Movement Science

Human movement science is the predecessor of sports health care and rehabilitation professionals, in 1989 at the National College of professional training directory in 1990, in Shanghai Sports University and capital Sports College officially launched. At that time in the professional training directory, the training specifications was identified as: enable the students master sports science, human science, sports medicine, rehabilitation medicine related basic theory, basic knowledge, basic skills, has engaged in physical health and physical rehabilitation clinical application, scientific research and teaching ability in practical work. In 1998 July the Ministry of education of physical education and Sports Science in the sports health care and rehabilitation specialty and sports biological science integration, and was renamed the "Human Movement Science", specialized training goal adjustment was shown as: we have the professional training of sports human science theory and research ability, can be in Colleges and schools, sports scientific research institutions, training base and rehabilitation of health and other departments, engaged in Human Movement Science of teaching, scientific research, sports and rehabilitation guidance of the senior specialized personnel. This adjustment fully embodies the education teaching reform should follow the deepening foundation, widening major caliber, cultivate the spirit of innovation, improve the comprehensive quality, and enhance the ability to adapt to the guiding ideology [1].

But this goal setting does not reflect the relationships between the market and social demand and the distinguishment of human sports science and sports with other professions, so that the Human Movement Science is neither accurate occupation should be relative, or alternative. Hongen passed in 2002 December held a "sports human science major construction seminar" on the investigation discovery, from personnel training direction of point of view, the start of human movement science professionals in Colleges and universities, the training program with the Ministry of Education promulgated 1998 professional directory was basically consistent eight colleges, one college which aims to cultivate talent for sports health care and rehabilitation the only goal of training, three emphasized in health and physical education teaching institutions; from the talent training level perspective, eight colleges to take sports human body science undergraduate professional training talents as "talents" colleges, four take sports human body science undergraduate professional training talents as "talents". The human science of exercise training objectives of the major limitations in the training of clinical physical rehabilitation professionals, graduates employment is restricted, the reason is the society some medical units in response to the needs of the establishment of specifications for upgrading hospital rehabilitation department, but not enough attention was paid to the corresponding rehabilitation medical professionals; some units for the reform of the personnel system, surely, posts and reason, arranged a number of other clinical departments because of age or health reasons but was not hired personnel into the rehabilitation office work and occupation personnel index, thus affecting the sports human science major students in hospital departments related to employment. In addition, because the society on human movement science professional knowledge and understanding is insufficient, and the educational system, curriculum, degree granting and people's ideology and other objective conditions, so that most regions of human movement science graduates in clinical rehabilitation work and on this post to their career development prospects are subject to certain restrictions. At present, many medical schools have begun to pay attention to clinical rehabilitation talent shortage, also is new or not preparing students for clinical rehabilitation therapy specialty, this situation will give the single training rehabilitation therapists sports colleges and universities bring new employment pressure. Hongen put forward the professional construction and development must adapt to the economic development and social needs for talent changes, pay attention to improve the quality of talent cultivation and adaptability, aims at the talent market demand, increase propagandist strength. Efforts to broad the professional training for students' employment caliber, extend a variety of channel. Human movement science specialty cultivation objective orientation should hold "body" features neither to medical school for seats, nor be equal simply at physical education, sports training, social sports specialty, but to highlight their own professional characteristics. As in the cultivation scheme, Zhan Changchun found that with the rapid development of Chinese economy and the deepening of the reform, some industry professionals' demand produces the change of structural sex, put forward that only to the market and social demand to talents training target direction for accurate positioning. Master the main pulse of the market is to formulate a scientific training scheme based. Till now, the specific research based on public sports service demand of the development of sports human body science undergraduate professional training are not found [2].

## 82.3 The Influence of New Social Environment on Human Movement Science Professional

At present, with China's rapid economic development, Beijing Olympic Games will be held in 2008, people's concept of life will change greatly, and people will pay more attention to a healthy lifestyle. In order to enhance physical fitness, health promotion for the main purpose of public sports service industry has been rapid and great development, in this context, as the route to the health of public sports service industry it will be taken to higher expectations and requirements. If human movement science can grasp the lifeline of the industry needs, and directly reflected it to the professional training program, the science professional development direction will be further clarified, and the undergraduate students' employment rate will be greatly improved. Study of public sports service market of sports human science of talent demand, as a basis of setting the training scheme, it will create the important application value for colleges and universities to improve teaching quality, as well as to improve the benefit of running a school.

It is the important content of the research for the undergraduate program and research of curriculum teaching reform in Colleges and universities, also a direct impact on the students' cultivation quality. The general cultivation scheme guided by three principles: the serious study of the Ministry of education sports professional guidance document, which is home to many experts collective embodiment of wisdom; the serious study of talent market demand, because this is the decision of its educational value foundation; the objective evaluation of the school's education resource advantage, because this is the professional running quality and features of the foundation. According to the above principle, in the medical colleges in our country have a large number of medical education resources and the health resources, in training medical talents and preventive health care personnel occupies absolute advantage. Teachers and students in Physical Education College's sports human body science talent into medical and health and public service industries have a competitive advantage. Current, professional sports mode of traditional obtain employment under a hitherto unknown pressure, on one hand, since the college enrollment influence, every year a large number of students graduated from the employment needs of physical education teachers, and occupation in a localized region contained limited staff; on the other hand, in recent years gradually new professional, such as athletic training, human movement, social physical education has not formed the independent employment direction, a large part of only for school employment, especially Human Movement Science to still do not have a corresponding specific occupation, before some colleges of sports human body science professional training target in a single training in clinical physical rehabilitation work of specialized personnel, due to the educational system, curriculum set, graduation degree conferred, the ideology of people all kinds of objective conditions, most parts of the country sports human science graduates in clinical rehabilitation work ability and career development prospects are subject to certain restrictions. Some schools will single training objective orientation in Sports Biological Sciences, graduate students engaged in sports biological science basic research and teaching work. However, students need to engage in teaching or research work on various aspects ability have bigger difference, cause the employment pressure, many students can choose to apply for graduate school, in order to further so as to adapt to the scientific research or teaching work need. Although such as rehabilitation medicine and other medical specialty is valued by more and more people, market is gradually increased, but many medical schools are actively building and preparation of relevant professional training, in recent years, the undergraduate medical students employment is facing more and more pressure. In this reality, human movement science professional training programmers to reflect their own characteristics, to meet the needs of society and the market has become a research topic of sports colleges brook no delay.

As the Sunny Sports is developing in our country, the government has paid attention to the health of people. So far our life has changed a lot. Now we have more time to do sport in the morning. It is reported that the health of young people in China, college students in particular, is not as good as is supposed to be. There are many reasons for this fact, but the main reason is that many people ignore the importance of physical exercises. Physical exercises help us to be successful in our lives. It is, therefore, strongly suggested that young people spare some time to take an active part in various kinds of physical exercises. Across the world many cultures have their own distinguishing games. Although the most of these cultures created the games themselves many of these games have similar attributes. One reason for this is that people everywhere find games entertaining. More importantly, however, games teach us about life m socialization, and about how to work as a team, how to win and lose gracefully. Everybody knows them without saying. Then, that is the sports' enchantment. Nevertheless, it also becomes one important part of life! Sports can make our health, someone said, health is the first wealth. Everyone has a good health body that is necessary. So sports play a important role of our lives. Sports also can teach us how to live and how to make a new life. Many popular sports involve a group of people. Being a team player means that one should work well with others in the face of adversity. Team sports are full of challenges and obstades that a team must overcome together. So more study about sports should be taken.

As to the system, from the founding of our country since execute is "the whole nation system", which means to develop sports by government, sports services has the same nature with public service. The current from the perspective of public sports services, sports public service cannot meet people's increasing by up to, become the main contradiction in sports. Liu Yanli, put forward that from the angle of economics, due to the incomplete information, the barriers to the implementation of government policies, political preferences and other factors, and the government to provide public goods process hard to avoid bilateral monopoly, budget maximization problems by the government, such as public sports service resource configuration of the only way, lack of economic system the mechanism of competition and cost benefit idea, lack of motivation mechanism and restriction mechanism. In addition, public sports service demand diversification, service object of civilians, sports resources and market price of its decision on the supply of public services tend to average people's needs and preferences, unable to respond to the needs of the new sport, "government failure" phenomenon. Many contradictions exist in urgent need of sports public service socialization, industrialization reform. Therefore, the public service of sports should be diversified operation, government, market, the third sector and diversified subjects in which to play the respective function. Yang Niansong divided three kinds: private goods, public goods and public goods, wherein the quasi sports public service is the typical form of sports service.

## 82.4 Summary

Public sports service market is a big market formed with breaking the original under the national sports system, sports administration monopolize situation. And it effects a dominant influence on other sports market. There is a certain dislocation between the demand of the talent majored in human body science and the current cultivation scheme. The major of sports human science should be located on the sports field, and avoid the overlap with other professional sports training direction. Human movement science professional training program should be based on the demand of public sports service market. Human movement science professional training should not only pay attention to the thick foundation and wide caliber, more attention should be paid to closely integrated with the market, the strong adaptability, and the market demand and anastomosis.

**Acknowledgments** This work was supported by Research project of science and technology for higher education institutions in Hebei Province of China (No. Z2011126 and No. SZ2011158) and the Startup Foundation for Docotors of Hebei Polytechnic University (No.35377601).

## References

1. Cui G (2006) In applied psychology for higher medical education research and practice on training project of higher medical education in China, vol 6. High education press, Beijing, pp 52–54
2. Hongen Y, Li Z, Jiang X, Zhou J (2006) Our country sports human body science undergraduate education present situation investigation and study of physical education. J Chongqing Phys Educ 13(1):82–84

# Chapter 83
# Analysis on Development of National Fitness Movement Based on College Sports Resource

**Bin Ding and Yan Ma**

**Abstract** Compared with social resources, the university sports resource is relatively perfect. But how to make use of the sports resources in colleges and universities to promote the nationwide fitness campaign, do not bring the burden for colleges and universities, and transform into a way that is conducive to the development of colleges and universities. This paper analyzes the college requirements of comprehensive sports, and designs a demand model. According to the demand, we find out it is necessary to introduce social capital. Then this paper analyzes the financing way of sports resources in Colleges and universities. From the University's social service function, this paper analysis on how colleges and universities attract the talents from the one participation in sports, absorb the human resources and financial resources for their own use, so as to achieve a win–win situation. After the final analysis, we draw some constructive suggestions.

**Keywords** Win–win cooperation · Civil construction · Sports resources · Social services · Demand model · Financing model

## 83.1 Introduction

Civil workout sport is an extension of athletics in the colleges and universities, the effective guarantee that carries out for life athletics [1], so need to adequately develop athletics resources in the colleges and universities and all the people to do exercise to keep fit to exercise to availably combine, at the same time make use of social capital to come to effective development combine enlargement strong

B. Ding (✉) · Y. Ma
Department of Military Physical Education, Heilongjiang Bayi Agricultural University, Daqing 163319, China
e-mail: Ding_Bin21@yeah.net

W. Du (ed.), *Informatics and Management Science V*, Lecture Notes in Electrical Engineering 208, DOI: 10.1007/978-1-4471-4796-1_83,

athletics resources in the colleges and universities. The colleges and universities collected an in great quantities precious resources in manpower, athletics infrastructure and information science and technology etc., currently, but the reality also exists a resources to effectively and reasonably install, the making use of resources efficiency is lowly and invest quantity to be partial to small and the outlet opposite one, waste of resources can not smoothly circulate etc., problem [2]. How well and availably make use of these a great deal of human resource, science and technology information and athletics facilities, urge the civil workout exercise business development originally the text exercise in the athletics of need model foundation up analytical civil need and present condition of workout sport, combine athletics resources in the colleges and universities to analytical all the people workout sport of hair development path and way, and combine society organization the group carry on the model of social capital margin to analyze the margin method of athletics in the colleges and universities and make athletics resources in the colleges and universities get full exertion of also get rapid development in the colleges and universities at the same time [3].

## 83.2 Sports Resources in the Colleges and Universities

Athletics resources in the colleges and universities' not only including existing athletics system in the colleges and universities but also including is treated to a development to make use of, have high level, forerunner can keep on the surroundings environment, athletics of talented person's resources of creating the value, infrastructure and athletics place in the colleges and universities education community and science and technology information.

Talented person's resources is colleges and universities' most basic resources, not only control athletics knowledge, but also have abundant athletics to practice experience in the athletics colleges and universities, for athletics education talented person as long as canning promise good work opportunity and environment, can create the glorious accomplishment of athletics business, will also carry out big economic efficiency and social efficiency [4, 5]. And colleges and universities can also through a full development make use of a talented person then can urge athletics business in the colleges and universities of whole keep on a development.

The infrastructure resources is athletics to exercise the most basic resources, at present China have 1,000 many colleges and universities, with greatly in city for mainly distribute ground, have the around 10,000 the athletics of various each kind sport building, at national field building number of have compare is 2.07 %. The field of the colleges and universities the building facilities is good and the device install opposite overall, but thus abundant of athletics facilities but used for peacetime of foundation teaching and training, most of facilities and place basically can be placed in idle status don't get the exploitation and development of full reasonable. If can adequately make use of these facilities and resources effectively to exercise at the civil workout, not only can raise all the people's body character, is carry out civil workout sport of the best path, and return ability full exertion resources advantage, effectively carry on the allocation of athletics resources, bring a larger social and economic effect, creation economy value.

The economic level, cultural environment and customs etc., of science and technology information and the region are vitally related, and people because of knowledge, age, love and occupation etc., of dissimilarity for workout exercise to have different viewpoint and understanding, at to workout way, workout contents in time place and workout facilities the device all have a different request.

## 83.3 The Current Situation of Nationwide Fitness Campaign

Outline of the nationwide body-building plan of the national promulgation strongly promoted the development of civil workout, currently; the present condition of civil workout sport is as follows:

Does exercise to keep fit the population circumstance of sport? Currently at China workout sport in the middle of the population the youth have compare be partial to small, age is 25–36 years old is the age layer of the proportion's least zone.

The way for training for physical fitness. According to the survey, people don't understand the method that the athletics toughens, didn't understand toughening of science way influence people carry on athletics workout to toughen, and this reason lines up in the sixth in 14 impact factors and also reflected people toughens to the athletics the imperfection of knowledge, also the feedback athletics educates to exist blemish.

Healthy condition. Mean according to the physical endowment monitor result of nation to people, all the people's physical endowment condition still needs a high value, particularly the teenager's healthy body and character problem, not only exist body character, function to bear descending of dint and speed, and tenacity etc., but also exist fat, be placed in status problems like "the second health", etc.

Does exercise to keep fit the understanding idea of sport? All the people has to compare biggest for Outline of the nationwide body-building plan of the national promulgation in order to have never heard, attained half of proportions. But only 5.01 % people know and understand a just-wanted contents, this means that the publicity that the China does exercise to keep fit to exercise in the athletics does still not enough.

The colleges and universities athletics resources make use of condition. Currently in athletics resources in the China colleges and universities although the instruction member has already strengthened to ten several more than 10,000 persons, the instruction member is more opposite than a flourishing nation, making use of resources condition serious shortage.

The workout exercise form and item. China for athletics workout sport with badminton, running and walk quickly, ping-pong, calisthenics, swimming and basket row football is lord. This means that everyone mostly chooses while choosing to do exercise to keep fit the way for exercising convenient, practical and stable opposite come to say a strength smaller traditional sport method.

## 83.4 Demand Model of National Fitness Sports in Colleges and Universities

Workout athletics need and modern fast-developing economy, society and science and technology of sport are vitally related. Economy is the economic foundation that urges the civil workout exercise and root, the society then does exercise to keep fit the power of sport, science and technology is assurance that civil workout sport can keep on a development. This text according to Ma Si Luo need layer theory combine athletics to learn, sport psychology and sociology of the knowledge exercise from the workout the cost and sport participation degree is two set up is civil workout athletics sport of need model the following diagram (Fig. 83.1).

From the last diagram, workout exercise cost and participation degree is with zero is point of departure, point of departure BE(A0,B0), the need of civil workout sport from satisfy physiology demanding sport is foundation beginning of, participant is be located into exercise to participate in of start most of entry-level stage; (A0–A1s) When passing by is horary of change, participate in canvasser to arrive (A1–A2s) of medium class stage, the canvasser gets contented foundation in the physiology need up, rises to social intercourse need of medium class sport need stage; But (A2–A3s) of deluxe stage, people then all get contented foundation in the physiology and the social intercourse needs up, have the condition the need for pursuing a self-actualization. The need of workout sport corresponds of will also be exercised by workout the influence of cost, the B1 A0 A3 Ds district is the physiology need that does exercise to keep fit to exercise stage, and at this time, the cost of civil workout sport is also is lowest cost (B0–B1s). The social

**Fig. 83.1** Exercise demand model

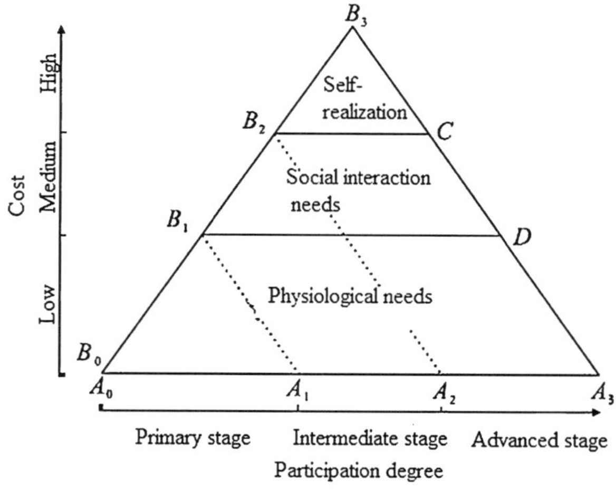

intercourse need of workout sport then is placed in the B1 B2 CDs district, the cost for corresponding is also medium cost (B1–B2s). The need of self-actualization of exercising then limits in the B2 B3 C district of the tallest layer. The need of 3 layers compares and talks, the need of self-actualization is high cost, is also sport need in the tallest layer.

Consequently aim at each layer need that the civil workout exercises, the athletics resources of colleges and universities should take into facilities to open exploitation and provide convenient place for all the people's workout sport, should also install at the same time homologous teams to guide civil workout sport and push forward the athletics of raising all the people workout business development at the same time.

## 83.5 The Current Situation of Colleges and Universities Sports Resource and Financing Model

Athletics in the colleges and universities is to educate unit business, is pushed forward by the nation and the government to develop forward, so the devotion of athletics funds mainly is come from a governmental supply. But the athletics corresponding belongs to a special industry, is the characteristics that takes to have social behavior, so will carry on collecting of funds from the methods like social organization, etc. (Fig. 83.2).

The governments provide public finance funds to give to support. Pass nation and government of unified operation, reasonably carry on public finance funds to orchestrate a programming, provide athletics appropriation to the colleges and universities and carry on each rules expenditure of athletics. And adequately develop the active of athletics and various communities and individual, carry on whole member participation and progressively operate under the market economy and lower expenditure to obtain repay and have already carried on a development to sex ground athletics sport that will have advantage.

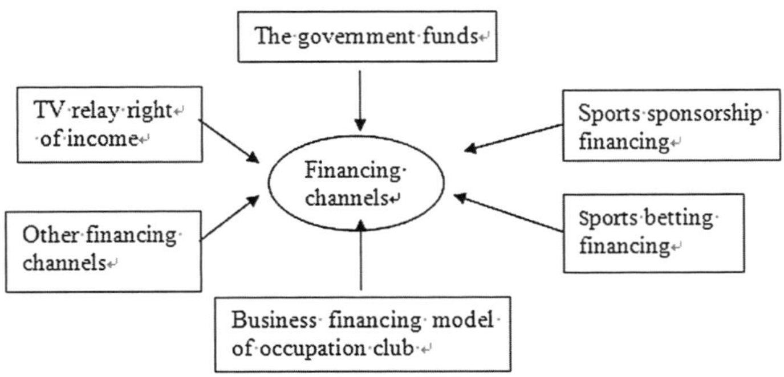


**Fig. 83.2** Financing model of fitness campaign

**Table 83.1** The total amount of incomes form of Olympic Games sponsor through the years

| Years | 1981 | 1985 | 1989 | 1892 | 1996 | 2000 | 2004 | 2008 |
|---|---|---|---|---|---|---|---|---|
| City | Moscow | Losangeles | Seoul | Barcelona | Atlanta | Sydney | Athens | Beijing |
| Sponsorship revenue (US $Billion) | 0.4 | 2.18 | 2.16 | 2.65 | 4.97 | 4.60 | 4.02 | 3.3 |

Provide the margin of athletics sponsor. The athletics sponsor is also a kind of very effective marketing method currently, not only can promote image, extend influence. Satisfy the demand of market. The athletics sponsor can prop up the common development of economy and athletics and carry out mutual benefit and common exaltation and progress. Data suggests, the athletics sponsors the income is in the gross earnings of have to compare to 34 %, Table 83.1 suggest, sponsoring income's tallest is Atlanta 96 Olympics, total amount is 497,000,000 USD.

Athletics Bo colorful margin. A lot of nations adequately make use of currently athletics sport lottery ticket collects funds for athletics sport, and this margin method has already become athletics workout to exercise colorful profession of Bo of main development outlet, and can draw on more social idle capitals to throw in workout business, for the country patrimony government increment revenue from tax.

Television transmission power income. Accompany with the rapid development of information science, medium and athletics of organic combine, the athletics workout exercises moreover an important margin outlet of business then will solid the television transmission power of athletics event carrying on face to television transmission organization and organization to carry on selling. Certainly Olympics' exercising so far single-item income's biggest funds source then is the marketing income of television transmission power that lies in Olympic Games event. Is relevant successive summer and the winter Olympic Games event of the income circumstance of television transmission power sees Table 83.2.

The management margin of professional club. This ways are seen to more the flourishing nation of west, the club is existed with the form of economic entity, so it obtains income through admission ticket, sponsor and various ways, such as advertisement and television transmission power...etc., and at the same time professional club not only the ability carry on promoting to the athletics tournament level, also can urge to push and universal and civil workout sport.

Other margin outlets. Mainly is pass a bank, investment the organization and organization, athletics consume community and social public of present to and investment.

**Table 83.2** Income statement of Olympic TV relay right

| Years | 1981 | 1985 | 1989 | 1992 | 1996 | 2000 | 2004 | 2008 |
|---|---|---|---|---|---|---|---|---|
| City | Moscow | Losangeles | Seoul | Barcelona | Atlanta | Sydney | Athens | Beijing |
| Television relay rights income (US $Billion) | 1.02 | 2.88 | 4.04 | 6.37 | 8.96 | 13.18 | 14.82 | 8 |

**Table 83.3** Financing amount and proportion of Tokyo, Atlanta, and Athens Olympics

| | Atlanta | | Tokyo | | Athens | |
|---|---|---|---|---|---|---|
| | Amount (US $Billion) | Proportion (%) | Amount (US $Billion) | Proportion (%) | Amount (US $Billion) | Proportion (%) |
| Government financial support | 0 | 0 | 54.7 | 68.40 | 508.5 | 24.30 |
| Commodity sales and service income | 0.83 | 2 | 9.72 | 12.10 | 233.3 | 11.10 |
| Sponsor income | 12.36 | 30 | 7.92 | 9.90 | 536.6 | 25.60 |
| Television relay rights income | 13.63 | 33 | 1.68 | 2.10 | 578.8 | 27.60 |
| Ticket income | 10.31 | 25 | 5 | 6.30 | 194.2 | 9.20 |
| Other income | 4.15 | 10 | 0.94 | 1.20 | 47 | 2.20 |
| Total | 41.28 | 100 | 79.96 | 100 | 2098.4 | 100 |

Descend a form reflection of respectively is at the 18th, 24 and the 3 greatest cities in 28 Olympic Games pass the amount of money that the different outlet carries on a margin and have a ratio (Table 83.3 and Fig. 83.3).

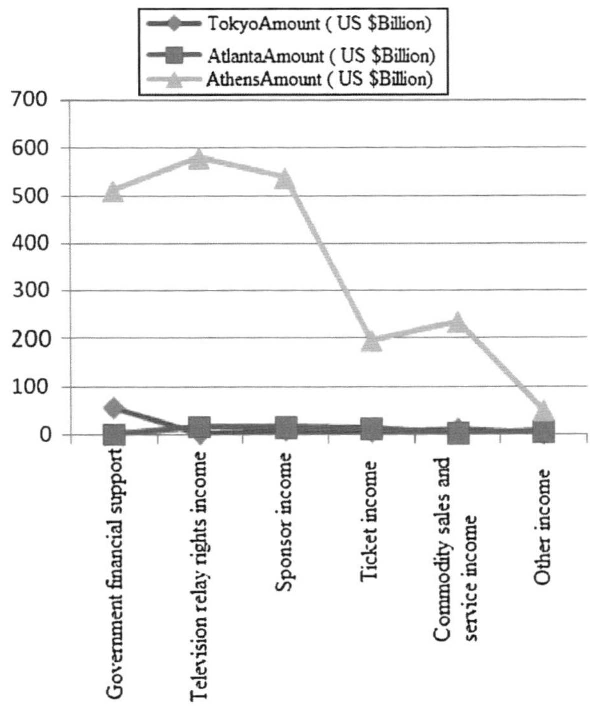


**Fig. 83.3** Financing amount and proportion of Tokyo, Atlanta, Athens Olympics

## 83.6 Conclusion

In view of human resources, information technology, sports facilities, this paper combines with the need model of exercise physiology, social and self actualization to collect some factors, such as government financial support, sports betting, sponsored financing, television rights income, occupation club management and other supports. And the integration of social organization, bank capital and the public power cause the nationwide fitness programs to develop rapidly, promote the sustainable development of physical education in Colleges and universities.

## References

1. Zhang S, Bojian X, Qi Y (2011) The countermeasures to promote the national fitness movement development by the sports resources in colleges and universities. J Chongqing Univ Sci Technol 5(1):190–192
2. Jiang T (2006) The comparison analysis on the financing present situation of China and America universities sports funds. J Cult 4(1):82–195
3. Zhang J, Sun Q, Mao L (2010) Research on financing status and strategy of high-level sports teams in universities. J Chengdu Sport Univ 21(5):6–9
4. Luo H, Tang Y (2011) Research on the financing mode of a large stadium. Sports Res 3(3):453–456
5. Hu J, Ji H (2010) The thoughts of the sports resources sharing model of College and society. Sports Ind 8(2):150–153

# Chapter 84
# Analysis on Development Mode of Island Sports Tourism Resource Based on Sustainable Development

**Lin Zuo**

**Abstract** Based on the measurement method of ecological footprint and the Security, judgment formula of Island Sports Tourism Ecological, Island Sports Tourism Resource mode is very dangerous state, which has appeared in Guangdong Province. On contrary to the strategy of sustainable development, four improved development model is proposed for Guangdong Province to give some advice for the sustainable development of island sports tourism resources.

**Keywords** Island · Sports tourism · Tourism resources · Sustainable development

## 84.1 Introduction

Guangdong, in the southern mainland of China, is a coastal province, located in the mountains to the south, the South China Sea, Guangdong Province has more than 700 in the area of 500 $m^2$ above the islands, the islands all over an area of 1,000 $km^2$, which inhabited only more than 40, province accounted for less than 5 % of all [the island1]. The more famous island of Nanaodao, East China Sea Island, Inc., Hailing Island, which is called "China fifth largest" Guangdong Province "largest" East Island is most famous, the East Pacific, South Asia, west hill to the southwest, China and India, the Pacific Ocean, the east the island, along

L. Zuo (✉)
Wuyi University, Jiangmen 529020, China
e-mail: lin_zuo88@163.com

W. Du (ed.), *Informatics and Management Science V*, Lecture Notes in Electrical Engineering 208, DOI: 10.1007/978-1-4471-4796-1_84,

the South Isles, national and European sea-land important points of intersection, the total area of 492 $km^2$, population 202,000 people; the main island of the East China Sea area of 401 $km^2$, it has a construction international is top-ranking big harbor of deep water conditions, depth 2,644 m from shore, the channel only 200,300 m, at the same time navigation in two to 300,000 t of class the above cargo and 500,000 t of oil tanker. Therefore, Guangdong Province Island is the national focus on sustainable development study.

## 84.2 Island Tourism Development in Guangdong Province

This paper from the tourism ecological footprint perspective, better show Guangdong Province Island tourism present situation, and the so-called ecological footprint refers to maintaining one, regional, national or global survival need and can absorb a human waste, with zoology productivity region area. In certain areas, on human activities and natural ecological effects of a measure [1]. By referring to "China Energy Statistical Yearbook", "2011 statistical bulletin" Guangdong Province tourism literature to draw the following related data:

1. Guangdong Province Island7 the tourism ecological footprint of a total of 103702.9909 $hm^2$, accounting for Guangdong Province Island National available ecological capacity in 15.72 %. Each tourist ecological footprint per capita average ecological footprint for 0.0189568 $hm^2$, 2.98827 $hm^2$, as shown in Table 84.1. Judge Island Sports Tourism Ecological Safe formula:

$$GESC_{ITD} = EF_{IC}/[(EF_{IRT} + EF_{IRB}) + (EF_{IET} + EF_{IEB})] \quad (84.1)$$

The above formula represents the island of rigid and elastic ecological footprint ecological footprint and ecological carrying capacity of rigid and elastic ecological carrying capacity (Fig. 84.1).

2. Various due to vehicles, aircraft, ships and other means of transport, they produce the energy consumption ecological footprint as shown in Fig. 84.2, the plane in all the statistics of traffic tools, the plane produces carbon dioxide up.

After calculation, Guangdong Province Island tourism safety coefficient is 75.26 %, in a safe and dangerous junction, at this time, the island sports tourism resources sustainable development becomes more important, if we do not take this development way, Guangdong Province Island Sports Tourism environmental carrying capacity will reach a critical value. Among them, the island tourism ecological environment and sustainable development of state division standard in Table 84.2.

**Table 84.1** Island tourism ecological footprint in Guangdong province

| Island tourism ecological footprint classification | Island tourism ecological footprint | Occupy the island tourism ecological footprint ratio | Island tourism rigid ecological footprint | Island tourism elastic ecological footprint |
|---|---|---|---|---|
| Island tourism traffic ecological footprint | 62408.2164 | 60.18 | 7997.2121 | 54411.0042 |
| Island tourism ecological footprint | 10823.9306 | 10.44 | 10823.9306 | 0 |
| Island tourist ecological footprint | 2143.3366 | 2.07 | 2143.3336 | 0 |
| Island tourism and recreational ecological footprint | 68.6286 | 0.07 | 68.6286 | 0 |
| Island tourism ecological footprint | 5334.5888 | 5.14 | 5334.5888 | 0 |
| Island catering tourism ecological footprint | 16628.2077 | 16.03 | 16628.2077 | 10050.638 |
| Island tourism waste ecological footprint | 6296.0852 | 6.07 | 6296.0852 | 0 |
| Total | 103702.990 | 100 | 103702.9909 | 64461.6422 |

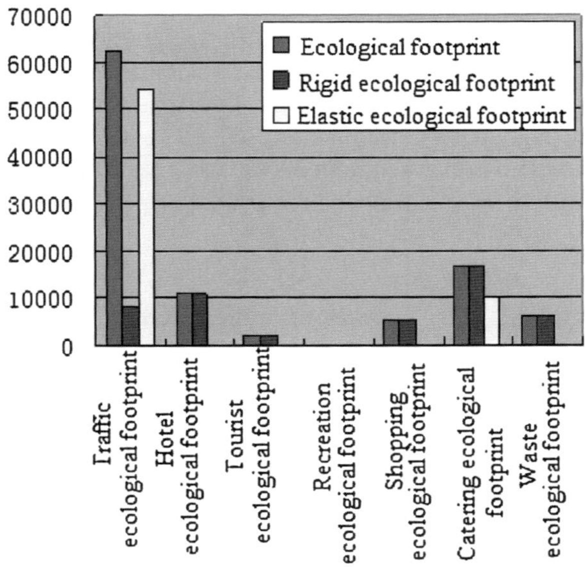


**Fig. 84.1** The classification of island tourism ecological footprint in Guangdong province

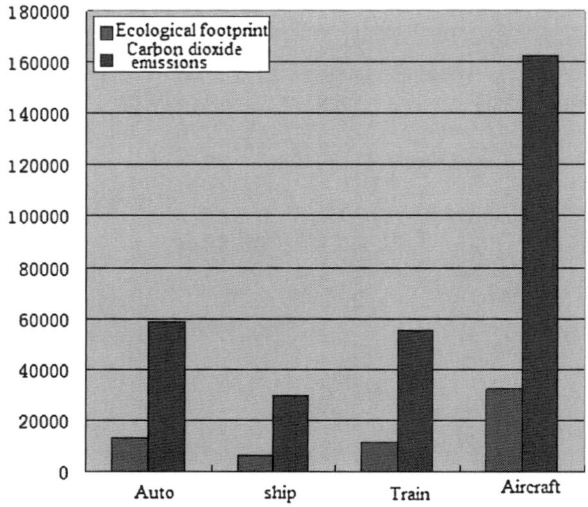


**Fig. 84.2** Transportation and carbon emissions in ecological footprint

# 84.3 Island Sports Tourism Resources in Guangdong Province

## *84.3.1 Sports Tourism Resources*

Sports tourism resources refers to: the nature or human society where to sports attract tourists, and can carry on the sports tourism, as tourism utilization and can produce economic, social, ecological benefit object. Sports tourism resources are the object of modern tourism, sports tourism resources with the existence of natural attribute, attraction is the evaluation of sports tourism resources the fundamental scale, he has some special characteristics, such as: regional, seasonal, diversity and the changes of Regional mainly in the tourism resources of the local and national characteristics. Tourism resource seasonality is by latitude and topography, climate, the sun, the motion of the moon and other factors. And its variations, is when it produces is tourism attribute, then for some reason, make it has undergone a qualitative change, a tourism resources.

## *84.3.2 The Development of Sports Tourism Resources*

Sports tourism resources development is refers to the people in order to develop, improve and enhance the attractiveness of sports tourism resources in the development and construction activities. The development of sports tourism resources aims to utilize tourism resources at the service of humanity, should turn into resource advantage product advantage, not to protect the resources used up, effective protection is not limited protection, effective protection and reasonable

**Table 84.2** The island tourism ecological environment and sustainable development of state division standard

| Ecological safety type | Safety | Threat to the security | Unsafe | The serious unsafe |
|---|---|---|---|---|
| GESC | GESC $\geq 1$ | $0.75 \leq$ GESC $\leq 1$ | $0.5 \leq$ GESC $\leq 0.75$ | $0 \leq$ GESC $\leq 0.5$ |
| SESC | SESC $\geq 1$ | $0.95 \leq$ SESC $\leq 1$ | $0.9 \leq$ SESC $\leq 0.95$ | $0 \leq$ SESC $\leq 0.9$ |
| Sustainable development | Sustainable development | Threats to the sustainable development | Sustainable development | Serious sustainable development |

development and utilization should be reciprocal causation, the key is to improve the planning of scientific and prospective. Strict protection, rational development, can make the resources to maximize efficiency, may use the life cycle is longer, achieve sustainable utilization, reasonable use can also play a role in the protection of resources. Sports tourism resources development and utilization of the contents include: attractions or visit the specific planning and design, traffic, communication can be entered, sports facilities and tourist facilities and training to provide professional service personnel.

Sports tourism exploitation principle shall include the following:

1. Protection development and utilization principle
2. Economy and efficiency principle
3. Planning principles
4. Highlight the uniqueness principle.

### 84.3.3 The Damage Sports Tourism Resource

Sports tourism resources are gradually suffered damage, so that our tourism industry can not follow the path of sustainable development to develop, main reason depends on natural fracture and damage, but the damage is cause tourism resources is the main reason, including the construction of damage and destruction of tourists, under the table according to the Guangdong Provincial Tourism Bureau of statistics data show that last year, Guangdong Province of tourism activities on the island caused by environmental load (Table 84.3) [2].

**Table 84.3** The tourism activities on the island caused by environmental load

| Category | Index | Environmental load/second | Environmental load/day | Environmental load |
|---|---|---|---|---|
| Resources | Energy consumption | 1828 | 585.8 | 8.18 * 108 |
| | Electric energy consumption | 39.57 | 15.73 | 2.07 * 107 |
| | Water resources consumption | 654.32 | 195.35 | 2.97 * 108 |
| Air pollution | $CO_2$ emissions | 125635.54 | 36210.24 | 5.87 * 1010 |
| | Carbon monoxide | 2798.62 | 875.16 | 13.01 * 108 |
| | Sulfureted hydrogen | 626.15 | 199.25 | 3.52 * 108 |
| | Nitrogen oxides | 78.57 | 29.54 | 35.17 * 106 |
| Water pollution | Waste water discharge | 498.93 | 135.24 | 2.04 * 108 |
| | Biochemical oxygen demand | 92.43 | 26.32 | 4.96*107 |
| Solid waste | Waste discharge | 2.08 | 0.82 | 9.52 * 105 |

### *84.3.4 The Sports Tourism Resources Protection*

According to the national sustainable development strategy, to the protection of tourism resources is imminent, the author from the following four points of Guangdong Province after the island sports tourism resources development model [3].

#### 84.3.4.1 Strengthen the Regional Tourism Planning

Tourism planning tourism planning is a strategic work, on the future of regional tourism development, development and management plays an important role in. General to be characteristic of tourism resources, tourism environment, market demand of tourists, tourism enterprises, workers and peasants are the location cost and industrial layout, and the road conditions, the influence and restriction of urban system construction. Therefore, tourism products and tourism industry space layout must consider tourism resources, tourism environment, tourism market, tourism traffic, industry factors such as system plan, comprehensive thinking, adhere to the "and the development of regional economy, social development and construction of town planning", "landscape boundary divided on the basis of scenic spots, to administrative boundaries based on division of functional areas", "to the highway as the clue of the cloth", "each regional tourism function complementary and interactive" in order to highlight the characteristics, forming a complementary, promote interaction. Detailed planning as follows:

1. Specifies the planning within the scope of various nature land boundaries, provides various types of land suitable for building construction, or conditional allows construction of building types.
2. Divides into the block, the block building height regulations, building density, the volume rate, ratio of green space and control index, and according to the scenic spot nature add other necessary control index.
3. Provides transportation import and export orientation, parking entrance range, parking, building, building back line spacing.
4. Put on the regional building volume, size, color, style and other requirements.
5. To determine the levels of road red line position, the control point coordinates and elevation.

#### 84.3.4.2 Preventing Vandalism

Environmental pollution is the main reason caused by man-made factors. Usually people in the production, life, discharge a large amount of "three waste" and certain industrial, living facilities for the unexpected accidents, as well as hospital untreated waste can cause environmental pollution, serious when can cause harm. The war because of the extensive use of various weapons to the residents of the killer and in residential area of damage can also cause environmental pollution and destruction. For example, the city's air pollution caused by dirty air, increase in the incidence of people and so on; pollution to water quality deterioration, a general decline in the quality of drinking water, threats to human health, causing prematurity or fetal malformation. Serious pollution incidents caused not only health problems, but also create social problems, so be sure to stop sabotage.

#### 84.3.4.3 Vigorously Carrying Out the Research and Personnel Training to Protect Tourist Resources

In order to better the implementation of the sustainable development of tourism resources, China must increase the protection of tourism resources in research and personnel training. In order to reach the target, the country should encourage more students to apply to school of tourism speciality; secondly, good planning postgraduate courses, so that these people can get better education, Fig. 84.2 is a tourism professional students on the current school training program satisfaction, these data are based on Guangdong Province several creation of Tourism Specialty school survey, has the following schools: Guangdong Polytechnic Normal University, Guangdong University of Business, Guangzhou University, Guangdong University of Technology, Shaoguan University, Huizhou University, Hanshan Normal University, Zhaoqing University, Jiaying University. Through the investigation, we can find that more than 80 % of the students in the school training program now satisfied, but there are still 20 % of the students to specialize more confused, we can't ignore the students, we should go to optimize our education programme, to truly achieve cultivate these talents, efforts for the sports of our country the sustainable development of tourism resources to make contribution (Fig. 84.3) [4].

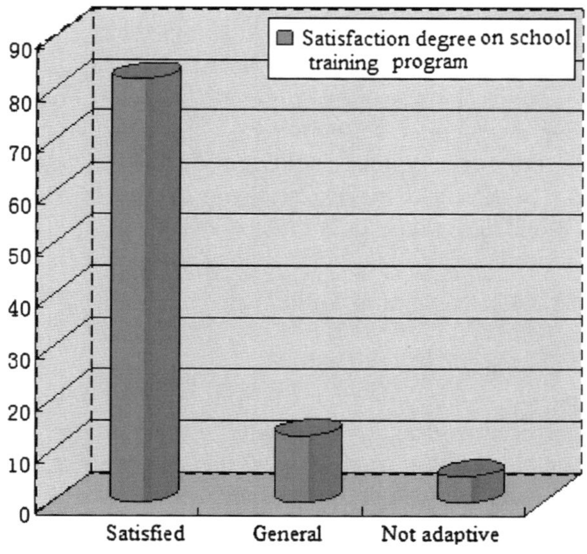


**Fig. 84.3** Satisfaction degree on the school's training program

#### 84.3.4.4 Making Legal System and Regulations Sound

Although in recent years our country eco tourism has made certain achievements, have issued many laws and regulations in tourism resources, but also the presence of bull management, environment and resources are destroyed, the tourism infrastructure ecological compatibility is poor, environmental protection laws and regulations related with environmental management measures do not place, as well as the tourist environmental consciousness is weak etc. We must perfect the legal laws and regulations, comprehensive use of various means and forms, strengthen the management of ecotourism, the healthy and sustainable development of Chinese ecological tourism, and strive to achieve tourism development and environmental protection, the person and the nature harmonious win–win goal.

## 84.4 Conclusion

In Guangdong Province, island sports tourism resources were analyzed, and the present situation is a dangerous state in the sustainable development and the sustainable development of the junction. So Guangdong provincial government should do regional tourism planning, and perfect relevant laws and strengthen the personnel training. Then increasing the tourist management, can promote the development of Island Sports Tourism Resource.

## References

1. Xiao J, Yu Q, Liu K (2011) Island tourist ecological security and sustainable development evaluation of Zhoushan as an example. Acta Geogr Sinica 66(6):842–852
2. Zhang Y, Li Y, Cao W (2011) Based on low carbon in China from the perspective of island tourism development. Land Nat Res Study 45(6):48–50
3. Zhao Y (2011) Island tourism sustainable development research. Ocean Dev Manage 1(1):79–82
4. Liu J, Li J, Zhou D (2007) Guangdong sports tourism resources development model. Sports Sci J 14(5):49–51

# Part V
# Multimedia Technology and Application

# Chapter 85
# A Novel Switching Vector Median Filter for Color Image Filtering

**Xiaohe Zhang**

**Abstract** Color impulse noise removing is a hot research problem in digital image processing, vector median filter (VMF) is a kind of traditional highly effective vector filter for color impulse noise removing. But it fails to distinguish details edges from impulsive noise, and usually filter them out. A novel vector median filter is proposed in this paper. This method uses quaternion rotation theory and the principle of Laplacian operators to construct the new switching condition. Combining this color impulse detector with the traditional VMF and some of the representative vector filters and recently developed vector filters, the proposed filters not only effectively preserves the edge details, but also provides better filtering performance.

**Keywords** Color noise removing · Quaternion rotation · Vector median filter (VMF) · Image details

## 85.1 Introduction

Image denoising is an eternal task of image processing; the most common image processing tasks is image denoising and image enhancement. The aim of color image filtering and denoising is to suppress noise while keeping the tone and preserving the edges or details [1, 2]. The filtering technology of color image has gone through the development process of filtering from scalar to vector filtering

---

X. Zhang (✉)
Modern Educational Technology and Network Management Center,
Gansu University of Traditional Chinese Medicine, Lanzhou 730000, China
e-mail: zyzxh@gszy.edu.cn

W. Du (ed.), *Informatics and Management Science V*, Lecture Notes in Electrical Engineering 208, DOI: 10.1007/978-1-4471-4796-1_85,

method. Vector filtering as a nonlinear filtering method based on statistical sorting is robust in the impulsive noise removing, keep the tone and protect the edges and details.

In recent years, a lot of vector filtering techniques are proposed [3, 4]. Among them, vector median filter (VMF) is a basic, classic and efficient vector filter; it can effectively suppress the impulsive noise. But basic vector directional filter (BVDF) and distance directional filter (DDF) are two kinds of vector filters with representativeness and high efficient. For BVDF, it can suppress noise efficiently while keeping the image tone; DDF is the combination of VMF and BVDF, it also can remove noise while keeping the image tone. In addition, adaptive nearest neighbor filter (ANNF), fuzzy vector directional filter (FDVF), adaptive hybrid directional filter (AHDF), quaternion representation and directional vector order-statistics VMF (QDVMF) are proposed in Ref. [5, 6]. AHDF takes into account the median value of output and mean vale of input; it is able to suppress impulsive noise and additive noise.

However, the vector filters mentioned above, which has low performance for noise removing and details preserving. In this paper, a novel vector median filter based on quaternion rotation is proposed, the new filters using quaternion rotation theory, similarly with the Laplacian operators. The experimental results show that, comparing with the existing vector median filters, the new vector median filter has better filtering effect and lower complexity.

## 85.2 Quaternion and its Properties

In 1843, British mathematician Hamilton presented quaternion theory, he extend complexity number from from 2-dimensional space to four-dimensional space. *A* Quaternion $q$ consists of a scalar $a$ and three imaginary part $b$, $c$, $d$, it can be written as [7]:

$$q = a + bi + cj + dk \tag{85.1}$$

where $a$, $b$, $c$ and $d$ is scalar, $i, j$ and $k$ are three imaginary, they have the relation as follows:

$$\begin{cases} i^2 = j^2 = k^2 = ijk = -1 \\ ij = k, jk = i, ki = j \\ ji = -k, kj = -i, ik = -j \end{cases} \tag{85.2}$$

When scalar part $a = 0$, we called $q$ is pure imaginary.

The module and conjugate of quaternion is defined as follows:

$$|q| = \sqrt{a^2 + b^2 + c^2 + d^2} \tag{85.3}$$

$$q^* = a - bi - cj - dk \tag{85.4}$$

The multiply of two quaternion can be denoted by $q_1 \times q_2$, the special representation as follows:

Let $q_1 = a_1 + b_1 i + c_1 j + d_1 k$ and $q_2 = a_2 + b_2 i + c_2 j + d_2 k$, then

$$\begin{aligned} q_1 \times q_2 = &(a_1 a_2 - b_1 b_2 - c_1 c_2 - d_1 d_2) \\ &+ (a_1 b_2 + b_1 a_2 + c_1 d_2 - d_1 c_2) i \\ &+ (a_1 c_2 - b_1 d_2 + c_1 a_2 + d_1 b_2) j \\ &+ (a_1 d_2 + b_1 c_2 - c_1 b_2 + d_1 a_2) k \end{aligned} \tag{85.5}$$

The polar coordinate representation

$$q = \|q\| e^{\mu\theta} = \|q\| (\cos\theta + \mu \sin\theta) \tag{85.6}$$

where $0 \le \theta \le \pi$, and

$$\mu = \frac{1}{\sqrt{b^2 + c^2 + d^2}} (bi + cj + dk) \tag{85.7}$$

$$\theta = \begin{cases} \mu \tan\tan \frac{\sqrt{b^2+c^2+d^2}}{a} & a \neq 0 \\ \pi/2 & a = 0 \end{cases} \tag{85.8}$$

$\mu$ is a unit pure imaginary, $\theta$ is the eigenangle.

## 85.3 Quaternion-Based Vector Median Filter

### *85.3.1 Quaternion Rotation and Chromatic Aberration*

Quaternion theory as early as 1843 has been proposed, but until recently was only began to be applied to image processing, such as it was applied to color image filtering and edge detection in Ref. and a description of the color edge detector based on quaternion also was proposed in Ref. [7, 8]. In this paper, the quaternion rotation theory is applied to design a high-performance switching vector median filter. The following is a brief introduction to quaternion rotation theory, and then explain it in this article.

In three dimension space, let $P = e^{\mu\theta}$, then $PXP^*$ represents a three dimension vector $x$ is rotate an angle $2\theta$ around $\mu$ axies.

In this application, let $\mu = (i + j + k)/\sqrt{3}$, then $\mu$ represents the grayline in RGB color space, therefore, the three components of the pixels at the grayline have the same value. Furthermore, let $\theta = \pi/2$, and $P = e^{\mu\theta} = (i + j + k)/\sqrt{3}$, then $PXP^*$ denotes $X$ is rotate 180° around $\mu$ axies. Hence, $X + PXP^*$ should be at the grayline.

Let $q_1 = r_1 i + g_1 j + b_1 k$ and $q_2 = r_2 i + g_2 j + b_2 k$, then the scalar part of $Pq_2P^*$ is zero, then we can get $q_3 = q_1 + Pq_2P^* = r_3 i + g_3 j + b_3 k$. If the hue of $q_1$ is similar to $q_2$, then $q_3 = q_1 + Pq_2P^*$ should near the grayline. In order to estimate the chromatic aberration between $q_1$ and $q_2$, a quaternion which represents the chromatic aberration is defined as below:

$$D_{pixel}(q_1 + Pq_2P^*) = D_{pixel}(q_3) = \left(r_3 - \frac{r_3 + g_3 + b_3}{3}\right)i + \left(g_3 - \frac{r_3 + g_3 + b_3}{3}\right)j + \left(b_3 - \frac{r_3 + g_3 + b_3}{3}\right)k \quad (85.9)$$

Therefore, when the hue of the $q_1$ is close to $q_2$, $\|D_{pixel}(q_1 + Pq_2P^*)\|$ should be relatively small; if the hue of the $q_1$ is the same as $q_2$, then $\|D_{pixel}(q_1 + Pq_2P^*)\| = 0$; and D, when the hue vary greatly between $q_1$ and $q_2$, then $\|D_{pixel}(q_1 + Pq_2P^*)\|$ is large, the switching vector median filter (SVMF) is designed based on this principle at the next section.

## 85.3.2 Design and Implement of the New SVMF

The morphological processing of hue is problematic. One cannot order colors from largest to smallest. As the hue is defined on the unit circle, the values wrap around, and we cannot directly build a lattice for the hue values. A method for ordering the hue values which requires the choice of an origin has been proposed by Aptoula [8]. Consider some arbitrary points $h_i$ distributed on the unit circle with centre $o$. After choosing an origin $h_0$, the distance from each point $h_i$ to the origin is defined as the acute angle $D_i(h_i, h_0)$.

For each pixel in the image, we first determine whether the pixel is noise, if it is noise, we can use the standard VMF to removing noise, filtering; otherwise, we do not take any action on this pixel.

Using quaternion notation, VMF can be written as:

$$q_{i,j}^{VMF} = \arg\left(\min_{q_k \in \{q_1, q_2, \ldots, q_N\}} \left\{\sum_{l=1}^{N} \|q_l - q_k\|\right\}\right) \quad (85.10)$$

where $N$ is the size of filter window (length * width), $q_1, q_2, \ldots, q_N$ are all the pixels within the filter window, $q_{i,j}^{VMF}$ represents the output pixel of $q_{i,j}$ located in the center of the filter window.

Therefore, the output of SVMF can be defined as follows:

$$q_{i,j}^{SVMF} = \left[1 - f_{noise}(q_{i,j})\right] * q_{i,j} + f_{noise}(q_{i,j}) * q_{i,j}^{SVMF} \quad (85.11)$$

where, $f_{noise}(q_{i,j})$ is the function which is used to judge whether $q_{i,j}$ noise is. If $q_{i,j}$ is noise, then $f_{noise}(q_{i,j}) = 1$, otherwise, $f_{noise}(q_{i,j}) = 0$.

In order to determine whether a pixel is the noise imitable the Laplacian operator, defined four quaternion operators:

$$\begin{pmatrix} P & P & 1 & P & P \end{pmatrix}\begin{pmatrix} P \\ P \\ 1 \\ P \\ P \end{pmatrix}\begin{pmatrix} P & & & & \\ & P & & & \\ & & 1 & & \\ & & & P & \\ & & & & P \end{pmatrix}\begin{pmatrix} & & & & P \\ & & & P & \\ & & 1 & & \\ & P & & & \\ P & & & & \end{pmatrix} \quad (85.12)$$

Let $\mathrm{M}_k(k = 1, 2, 3, 4)$ be the above four operators, and non-zero elements in the set corresponding to the image pixels $q_{i,j}^{k,1}$, $q_{i,j}^{k,2}$, $q_{i,j}^{k,3}$ and $q_{i,j}^{k,4}$, respectively, where {g, qi, j, g, g}, where $q_{i,j}$ is the pixel locate in the center of the filter window. Constructing the following formula to detect noise pixel:

$$D_{Line}\left(q_{i,j}^{k}\right) = \frac{1}{4}\sum_{l=1}^{N}\left\|D_{pixel}\left(q_{i,j} + Pq_{i,j}^{k,1}P^{*}\right)\right\| \quad (85.13)$$

$$D_{color}\left(q_{i,j}\right) = \min\left\{D_{Line}\left(q_{i,j}^{1}\right), D_{Line}\left(q_{i,j}^{2}\right), D_{Line}\left(q_{i,j}^{3}\right), D_{Line}\left(q_{i,j}^{4}\right)\right\} \quad (85.14)$$

$$f_{noise}\left(q_{i,j}\right) = \begin{cases} 1 & D_{color}\left(q_{i,j}\right) \geq T \\ 0 & otherwise \end{cases} \quad (85.15)$$

Using Eqs. (85.12–85.14), we can detect noise pixel in color image corrupted by impulse noise, and we also can use switching theory and VMF to smoothing image.

## 85.4 Experimental Results

The "Monkey" image(512 * 512, 24 Bit) as the test images. In the source image, add 0–30 % of impulse noise. Then, some classical vector median filters, and this article proposed filter are used to compare filtering performance. Figure 85.1a is an original color image; Fig. 85.1b, c is noise images (15 % and 20 % impulse noise). Figure 85.2 presents the filtered images using the various vector median filters. We can see that SVMF has good filtering performance, it has better effect on color noise removing and details protection (3 * 3 filter window).

The objective evaluation criteria used color peak signal to noise ratio (CPSNR), color mean square error (CMSE) and normalized color difference (NCD).

$$CPSNR = 10\log_{10}\frac{3 \times M \times N \times 255^{2}}{\sum_{i=1}^{M}\sum_{j=1}^{N}\|f(i,j) - g(i,j)\|^{2}} \quad (85.16)$$

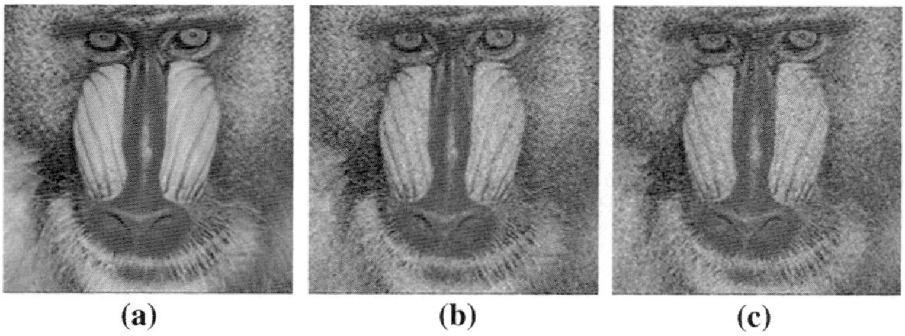

Fig. 85.1 Original image and noise image. a Orginal image. b Noise image impulse (15 % impulse noise). c Noise image impulse (20 % impulse noise)

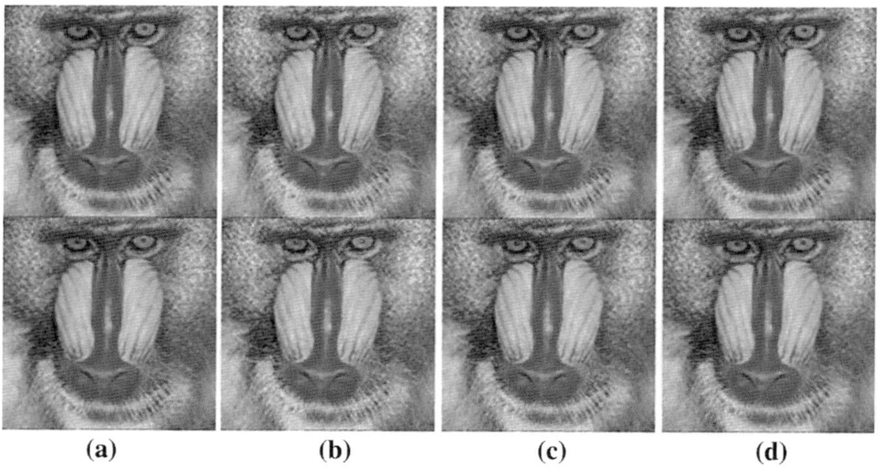

Fig. 85.2 The comparing results of filtering using various vector median filters (The first row is 15 % impulse noise, and the second row is 20 % impulse noise)

$$CMSE = \frac{1}{3M \times N} \sum_{i=1}^{M} \sum_{j=1}^{N} \| f(i,j) - g(i,j) \|^2 \tag{85.17}$$

$$NCD = \left\{ \sum_{i=1}^{M} \sum_{j=1}^{N} \begin{matrix} \left[L_f(i,j) - L_g(i,j)\right]^2 + \\ \left[a_f(i,j) - a_g(i,j)\right]^2 + \\ \left[L_b(i,j) - L_b(i,j)\right]^2 \end{matrix} \right\} / \left\{ \sum_{i=1}^{M} \sum_{j=1}^{N} \left[L_f(i,j)\right]^2 + \left[a_f(i,j)\right]^2 + \left[L_b(i,j)\right]^2 \right\} \tag{85.18}$$

**Table 85.1** Comparative results of CPSNR of different vector median filters for various percentages of impulse noise (salt and pepper) for the image "Monkey"

| Noise level | 5 % | 10 % | 15 % | 20 % | 25 % | 30 % |
|---|---|---|---|---|---|---|
| VMF | 22.4702 | 22.0070 | 21.4284 | 20.6056 | 19.5527 | 18.2372 |
| BVDM | 19.6943 | 19.4850 | 18.8667 | 17.8423 | 16.3566 | 14.8202 |
| DDF | 21.6875 | 21.2938 | 20.6256 | 19.5334 | 18.1253 | 16.5813 |
| SVMF | 22.5092 | 22.0999 | 21.6431 | 20.9340 | 20.0846 | 19.0552 |

**Table 85.2** Comparative results of CMSE of different vector median filters for various percentages of impulse noise (salt and pepper) for the image "Monkey"

| Noise level | 5 % | 10 % | 15 % | 20 % | 25 % | 30 % |
|---|---|---|---|---|---|---|
| VMF | 368.178 | 409.6169 | 467.9897 | 565.6205 | 2720.7973 | 975.789 |
| BVDM | 697.671 | 732.1132 | 844.1353 | 1068.73 | 1504.64 | 1504.66 |
| DDF | 440.887 | 482.7220 | 563.0147 | 724.0042 | 1001.37 | 1001.39 |
| SVMF | 364.8893 | 400.9478 | 445.4179 | 524.4266 | 637.7023 | 808.2844 |

**Table 85.3** Comparative results of NCD of different vector median filters for various percentages of impulse noise (salt and pepper) for the image "Monkey"

| Noise level | 5 % | 10 % | 15 % | 20 % | 25 % | 30 % |
|---|---|---|---|---|---|---|
| VMF | 0.1308 | 0.1382 | 0.1474 | 0.1608 | 0.1816 | 0.2138 |
| BVDM | 0.1469 | 0.1543 | 0.1689 | 0.1946 | 0.2379 | 0.2965 |
| DDF | 0.1320 | 0.1397 | 0.1517 | 0.1717 | 0.2042 | 0.2507 |
| SVMF | 0.1301 | 0.1367 | 0.1439 | 0.1554 | 0.1704 | 0.1926 |

where $f$ is original image, $g$ is filtered image, $M, N$ denote the width and height of image, respectively.

Tables 85.1, 85.2, 85.3 show the experimental data CPSNR, CMSE and NCD for Fig. 85.1b using VMF, BVDF, DDF and the proposed SVMF. Obviously, the results show that SVMF has better effect on color noise removing and details protection than other algorithms.

## 85.5 Conclusions

This paper proposed a novel switching vector median filtering based on noise detection. In this algorithm, the scalar filtering and vector detection are composed to remove color noise, and only those detected noise pixels are smoothed, this algorithm can effectively improve the accuracy of noise detection. The new filters can better maintain the image detail and edge, and do not introduce new colors.

## References

1. Andreadis I, Louverdis G, Chatzianagnostou S (2004) New fuzzy color median filter. J Intell Rob Syst 41:315–330
2. Jin LH, Li H, Song EM, Xu XY (2010) Impulsive noise removal using switching scheme and adaptive weighted median filters. Opt Eng 49(1):1–7
3. Plataniotis KN, Androutsos D, Venetsanopoulos ANN et al (1996) An adaptive nearest neighbor multichannel filter. IEEE Trans Circuits Syst Video Technol 6(6):699–703
4. Lukac R (2003) Adaptive vector median filtering. Pattern Recogn Lett 24(12):1889–1899
5. Jin LH, Li DH (2007) An efficient color impulse detector and its application to color images. IEEE Signal Process Lett 14(6):397–400
6. Jin LH, Li H, Xu XY, Song EM (2010) Quaternion-based color image filtering for impulsive noise suppression. J Electron Imaging 19(4):1–12
7. Sangwine SJ (2000) Colour image filters based on hypercomplex convolution. IEEE Proc Vision Image Signal Process 147(2):89–93
8. Sangwine SJ (1998) Colour image edge detector based on quaternion convolution. Electron Lett 34(10):969–971

# Chapter 86
# Color Management System Based on Spectral Image

**Songhua He and Bo Li**

**Abstract** To solve the defects of existing color management system which is colormetric match, a new mixed color management system architecture based on spectral image and color appearance attributes is proposed, its profile connected space may be spectral reflectance, spectral transmittance, spectral radiance or color appearance values, so its data processing workflow is discussed, and a low-dimensional interim connection space is introduced to reduce cost of calculation.

**Keywords** Color management · Spectral image · Color appearance

## 86.1 Introduction

In recent years, cross-media color reproduction has become a trend with the rapid development and wide application of digital imaging devices, such as digital cameras, monitors, printers, etc. But different devices describe color in different ways [1, 2]. In order to exchange color data between cross-media devices, ICC color management system is built based on device-independent color spaces. Although it is better to solve the problem of colormetric match, but causes metamerism that colormetirc match doesn't exist if illuminant and the observers are changed. So it cannot meet requirements of the high-fidelity printing and cross-media publishing. In order to solve the problems, Spectral image are introduced to

S. He (✉) · B. Li
Institute of Media and Communication Shenzhen Polytechnic, Shenzhen 518055, China
e-mail: hdh1818@szpt.edu.cnl

B. Li
e-mail: 330815866@qq.com

W. Du (ed.), *Informatics and Management Science V*, Lecture Notes in Electrical Engineering 208, DOI: 10.1007/978-1-4471-4796-1_86,

color management. This paper analyzed the lack of existing color management system, and proposed a new color management system architecture based on spectrum image and color appearance attributes, and discussed its data processing workflow.

## 86.2 ICC Color Management System

The existing color management model is proposed by International Color Consortium (ICC), which is based on the XYZ device-independent color space.

ICC color management process is divided into three stages: image acquisition, image processing, image output [3]. The most important stage is image processing, which is called characteristics of input and output device. Transformations are fulfilled from a device-dependent color space to device-independent color space again to the device-dependent color space. The process is shown in Fig. 86.1.

There are two research directions to solve the color reproduction deficiencies in cross-media color management system, one is spectral match, the other is color appearance match. Because color appearance model needs more parameters, and computational processes are very complex, and application conditions are very strict. The calculation of color management based on spectral match is relatively simple, and it will be the focus of the study of next-generation color management system.

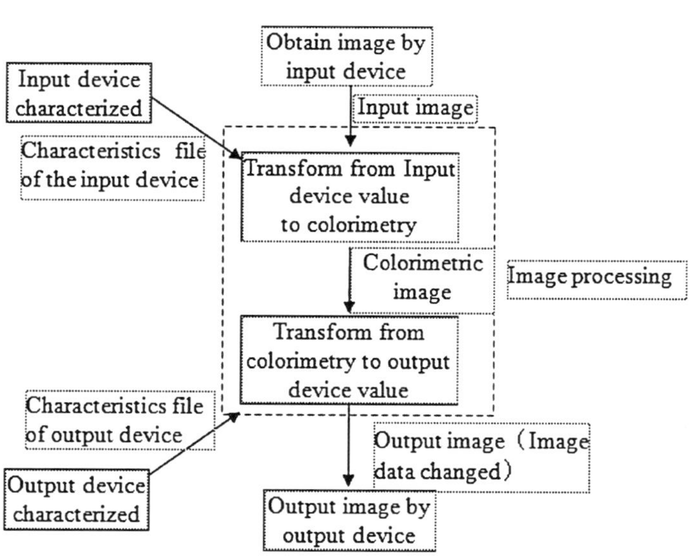


Fig. 86.1 Image processing workflow of ICC color management system

## 86.3 Color Management System Based on Spectrum

The color management system based on spectrum is to achieve a spectral match, it can ensure spectrum of the original color and spectrum of the reproduced color are exactly same. So it can reproduce color under any light source [4].

## 86.4 Workflow of Spectral Color Management System

If a pure spectral color management system only consider supporting spectral data and spectral characteristics of the devices, and the spectrum-based profile connected space (SPCS) is composed of the spectral reflectance, spectral transmittance, spectral radiance and other spectral values, then it will only realized spectral match [5], and has the similar structures shown in Fig. 86.2. However, considering the compatibility of current color management system, the next-generation color management system must support both spectral match and colormetric match. PCS may be one of CIEXYZ, spectral reflectance, spectral radiance. It can realize respectively color match, spectral match and spectral radiance match (color appearance match). If PCS is CIEXYZ, it will achieve a colormetric match under constrained condition and can't solve metamerism. If PCS is spectral reflectance, it can be realized colormetric match in arbitrary light and viewing conditions theoretically. If PCS is spectral radiance, it can reproduce color appearance of objects under a certain illuminant, so it is color appearance match (For a simple spectral color management system does not consider influences of surrounding environment such as surface properties, viewing distance and other color appearance attributes, it only consider the changes of lighting source).

What kind of data can be as the connection space data if there exist spectral data and color data at the same time in a color management system? The spectrum can be directly converted to colormetry, but the inversion conversion is not easy to achieve, so colormetry is the best choice for connecting space. In actual application, mostly spectrum is obtained using multi-spectral device and output using device with colormetric characterization, but rarely acquired colormetry was converted into spectrum and then output. You can directly obtain spectrum by multi-spectral device and output.

The color management system shown in Fig. 86.3 does not take into account color spacial relationships and influences of surrounding environment, it belongs

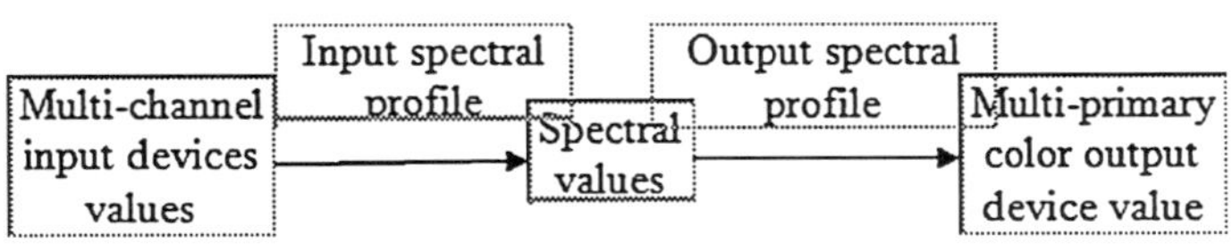


**Fig. 86.2** Color management system structure based on spectrum

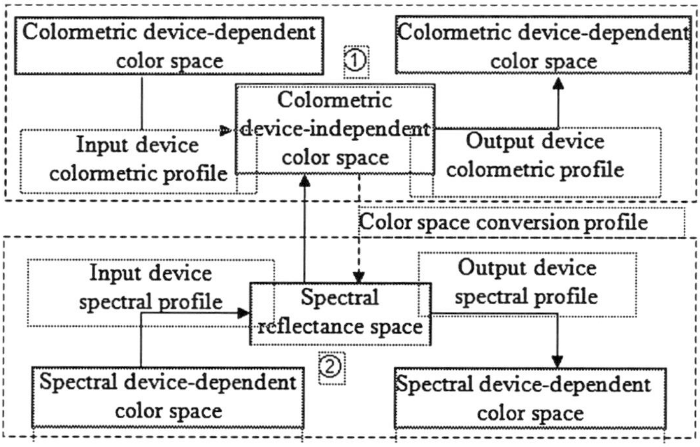


**Fig. 86.3** Mixed color management system based on spectrum (without considered color appearance factor) ① Color management system based on colormetry ② Color management system based on spectrum

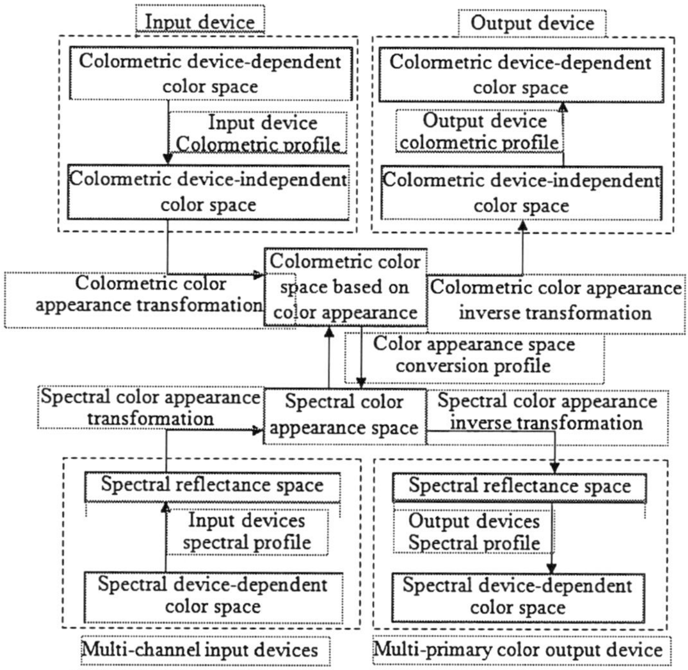


**Fig. 86.4** Mixed color management system considered color appearance factor

to constrained color match that two color stimuli realize color match which Must be subject to strict conditions: the same environment, background, size, shape, surface properties, illumination and other conditions. The two color stimuli may not match in color if any of these restrictions to be changed. In practice, these restrictions are difficult to achieve, so color match needs consider the impact of these variables, the color appearance model is put forward [6]. The new spectral color management system may take into account the color appearance characteristics. If spectral profile connection spaces is spectral radiance, it will be a simply color appearance match which only considered illuminant. Figure 86.3 is mixed color management system considering color appearance characteristics (Fig. 86.4).

## 86.5 Data Conversion of Spectral-Based Color Management System

Compared with ICC color management, Spectral-based Color management system is not simple to replace simply colormetric values with spectral values in PCS. It must meet the following requirements.

### *86.5.1 Input Profiles Can Describe the Following Conversion*

Input device values to reflectance

Input device values to radiance

Input device values to colorimetry (compatibility with ICC color management system)

### *86.5.2 Color Space Conversion Profile Can Describe the Following Conversion*

Reflectance to radiance by multiplying illuminant radiance power

Radiance to reflectance by dividing illuminant radiance power

Reflectance to colorimetry by illuminant radiance power, multiplying by color matching functions

Radiance to colorimetry by multiplying by color matching functions and then integrating

### *86.5.3 Output Profile Can Describe the Following Conversion*

Reflectance to output device values

Radiance to output device values

Colorimetry to output device values

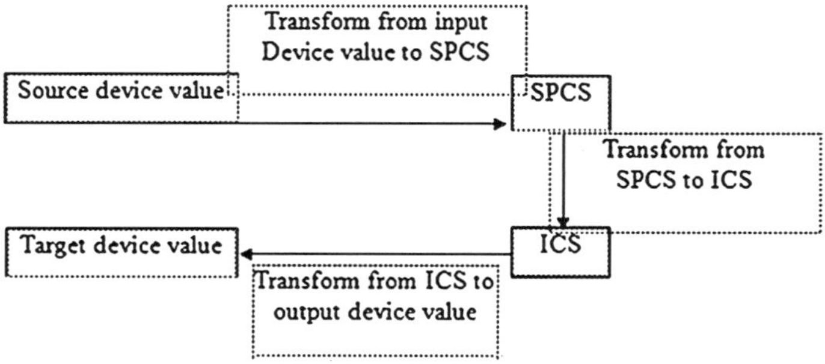


**Fig. 86.5** Processing workflow of spectral-based color management system introduced IC

Spectral reflectance is a 31-dimensional space, it is a high-dimensional space compared to CIEXYZ. The computation will be very great and its practicality is not high if using directly the 31-dimensional reflectance space as PCS, it is difficult to directly calculate the M-dimensional data of multi-color printer from the 31-dimensional spectral data (If the printer is a four-color printer, then M is four), algorithm is not easy to realize and cost of calculation is great. So multi-dimensional look-up table is generally used.

The specific solution to this problem is to introduce a low-dimensional interim connection space (abbreviated ICS), and then to realize the conversion from ICS to the printer device value by look-up table. ICS is relatively low in dimensions and is situated between a high-dimensional spectral profile connection space and output units [7]. An appropriate ICS must meet two conditions: conversion must be calculated quickly between ICS and spectral reflectance; ICS must be a low-dimensional space, so the size of lookup table established from the ICS to the printer device was small enough. The conversions between SPCS and output device values have two-step after introduced ICS: from SPCS to the ICS; from the ICS to output device values. Multi-dimensional look-up table was used in the second step of conversion (Fig. 86.5).

## 86.6 Summary

Spectral-based color management system reconstruct spectrum of objects from input device digital values by mathematic algorithms, and use spectral data as the interim connection space to achieve spectral match. The relationship between the source device and target device is no longer chromatic mapping but spectral mapping. Color under arbitrary illuminant can be reproduced by reconstructed spectrum, therefore, spectral-based color management system can include color appearance characteristics of human visual system, and can combine color appearance model to realize color appearance match. For printing output, ICS is a bridge between spectral space and device colormetric space; it can solve effectively problems which caused by high-dimensional of spectral space in spectral-based color management system.

## References

1. Yang WP, Zhao DZ, Fan QM, Huang QM (2005) Cross-media Color reproduction based on color appearance. Opt Tech 31(1):101–103
2. Hill B (2000) Color capture: color management and the problem of metamerism: does multispectral imaging offer the solution, color imaging: device-independent color, color hardcopy, and graphic arts. In: Eschbach VR, Marcu GG (eds) Proceeding of SPIE, vol 188(3963). Bellingham, pp 2–14
3. Novati G, Pellegri P, Schettini R, Zuffi S (2005) S-CMS: towards the definition of a spectral color management system. Colore e Colorimetria: Contributi Multidisciplinari, Collana Quaderni di Ottica e Fotonica 19(13):9–14
4. Rosen M (2003) Navigating the roadblocks to spectral color reproduction: data-efficient multi-channel imaging and spectral color management, vol 27(19). In: Phd thesis, Rochester Institute of Technology, Rochester, pp 24–28
5. Rosen M, Imai F, Jiang XY, Ohta N (2001) Spectral reproduction from scene to hardcopy II: image processing. Proc SPIE 133(4300):33–41
6. He SH, Liu Z (2009) The relationship between CIELAB color space and color appearance model. Packing Eng 30(5):68–70
7. Rosen M, Ohta N (2002). Spectral color processing using an interim connection space, vol 134(13). In: Proceedings of IS and T/SID eleventh color imaging conference, pp 187–192

# Chapter 87
# Efficient Approach of Image Quality Evaluation Based on Fuzzy Logic

**Qinghua Ji and Baojing Chen**

**Abstract** The image quality evaluation is an important subject in image processing techniques. Aiming to improve the existing methods of image quality measurement, a new fuzzy logic-based image quality evaluation approach is presented in this paper. It is well known that an image is composed of details (edge, texture) and smooth regions, according to several fuzzy inference rules, three different quality factors are defined and applied to their respective regions. The proposed approach separates the distorted effects from the original edges in the image, and avoids the edges being mistaken as distorted effects. Experimental results show that the proposed method is robust for various image, in addition, it has the general performance of image quality evaluation approaches.

**Keywords** Image quality evaluation · Fuzzy logic · Smooth regions

## 87.1 Introduction

Image processing technology has wide application in military, medicine, and daily life, including the image denoising, watermark technology, image reconstruction and pattern recognition technology. All the techniques affect the image quality,

Q. Ji (✉)
Mechanical and Electrical Engineering, Xinxiang University, Xinxiang 453000, China
e-mail: qinghua_ji@163.com

B. Chen
Information Communication Center, Gansu Province Power Company, Lanzhou 730050, China
e-mail: cbj_gstx@tom.com

W. Du (ed.), *Informatics and Management Science V*, Lecture Notes in Electrical Engineering 208, DOI: 10.1007/978-1-4471-4796-1_87,

thus we require a reasonable image quality assessment methods to estimate the image quality [1].

The method of image quality assessment can be divided into the subjective evaluation and objective assessment. The subjective evaluation means a group of observers scoring the same image at different scales, and then a value weighted size by value to evaluate the image quality [2, 3]. Although the method can reflect the situation of image quality, it is not practical in that the computation complexity. At present, the objective assessment methods are widely applied, such as peak signal-to-noise ratio (PSNR), mean square error (MSE), Mean Square Deviation of Slope (MSDS), etc. Recently George extended MSDS by increasing the measurement of the angular direction on block effect. MSDS value reflects block effect. PSNR and MSDS are the assessment methods of ignoring the image content. In fact, an image is composed by the smooth region, edge region and texture region, and the different regions should adopt a different approach to estimate the image quality [4]. Suthaharan presented PS-BIM standard, which use several different visual sensing regions to evaluate the block effect, this approach although use some characteristics of the human visual system model. However, PS-BIM has not well-defined mathematical theory framework. The simulation also shows that there is a contradictory between the theory and the results [5, 6].

In this paper, firstly, we define three suitable evaluation factors for smooth region, edge region and texture region, separately. Then, as the conclusion of the fuzzy inference rules, a Takagi–Sugeno fuzzy system is constructed to evaluate the effect of image quality. In Sect. 87.2, we gives three image area measuring factor which are suitable for the image quality assessment. A new method of fuzzy block effect for image quality assessment is proposed in Sect. 87.3. Sections 87.4 and 87.5 are the simulation and conclusions.

## 87.2 Measurement Factors

Image signal instability leading to its quality is the spatial variation, and it is therefore more suitable for the measuring the visual characteristics of the image by using local method. Here, we will use the sliding window method to measure the local image quality, one of the sliding window size is assumed to be $(2n + 1) \times (2n + 1)$. Now, three image quality factors, which are suitable for are smoothing area, edge area and texture area, are defined as follows.

### 87.2.1 Smoothness-Based Measurement Standard $Q_1$

Smoothness is an important indicator which can be used to reflect block effect. Here, the local mean is defined based on the smoothness of measurement standards, to evaluate the image quality of the smooth region.

Let $X = \{x(i,j)|i = 1,2,\ldots,M,j = 1,2,\ldots,N\}$, $Y = \{y(i,j)|i = 1,2,\ldots,M, j = 1,2,\ldots,N\}$, $X$ and $Y$ represent original image and block effect image, respectively. $A_{ij}$ and $B_{ij}$ are sliding window of center $x(i,j)$ and $y(i,j)$, then $Q_1(i,j)$ can be defined as below:

$$Q_1(i,j) = 1/(\theta_1|\bar{A}(i,j) - \bar{B}(i,j)| + 1) \tag{87.1}$$

where $\theta_1$ is a constant, and

$$\bar{A}(i,j) = \left[\sum_{k=-n}^{n}\sum_{l=-n}^{n} x(i+k,j+l)\right]/[(2n+1)\times(2n+1)] \tag{87.2}$$

$$\bar{B}(i,j) = \left[\sum_{k=-n}^{n}\sum_{l=-n}^{n} y(i+k,j+l)\right]/[(2n+1)\times(2n+1)] \tag{87.3}$$

### 87.2.2 Edge-Based Quality Measurement Standard $Q_2$

Because the human visual system is very sensitive to the structural information derived from edge, and therefore the edge is generally considered the most important features of the image, it plays a very important role in the image quality assessment. However, the existing method of block effect assessment does not distinguish between the edge and the block effect, and it will lead to inaccurate assessments. Here, the edge strength, as quality assessment parameters, is used to ensure a consistent of image structure and visual quality.

In order to compute the strength of edge, four edge detection operators are chosen, include 0°, 90°, +45° and −45°, as shown in Fig. 87.1, and represented by $ES_H(i,j)$, $ES_V(i,j)$, $ES_{NE}(i,j)$, $ES_{NW}(i,j)$, which can be computed according to the weighed sum of pixels in the window. Furthermore, we define the maximum of four respond value, and represented by $ME(i,j)$.

$$ME(i,j) = \max\{|ES_H(i,j)|, |ES_V(i,j)|, |ES_{NE}(i,j)|, |ES_{NW}(i,j)|\} \tag{87.4}$$

| | | |
|---|---|---|
| −1 | −2 | −1 |
| 2 | 4 | 2 |
| −1 | −2 | −1 |

**(a)**0°

| | | |
|---|---|---|
| −1 | 2 | −1 |
| −2 | 4 | −2 |
| −1 | 2 | −1 |

**(b)**90°

| | | |
|---|---|---|
| −2 | −1 | 2 |
| −1 | 4 | −1 |
| 2 | −1 | −2 |

**(c)**45°

| | | |
|---|---|---|
| 2 | −1 | −2 |
| −1 | 4 | −1 |
| −2 | −1 | 2 |

**(d)**-45°

**Fig. 87.1** Edge detection operators

### *87.2.3 Contrast-Based Quality Measurement Standard $Q_3$*

The response of the human visual system depends on the local disturbance of the absolute brightness relative to the surrounding environment, but not the absolute brightness itself. This property is well-known Weber-Fechner law. Here, the definition of $Q_3(i,j)$ is shown as below:

$$Q_3(i,j) = 1/(\theta_3|C_A(i,j) - C_B(i,j)| + 1) \tag{87.5}$$

where $\theta_3$ is a constant, $C_A(i,j)$ and $C_B(i,j)$ is defined as follows:

$$C_A(i,j) = |x(i,j) - \bar{A}(i,j)|/(\bar{A}(i,j) + a) \tag{87.6}$$

$$C_B(i,j) = |y(i,j) - \bar{B}(i,j)|/(\bar{B}(i,j) + a) \tag{87.7}$$

$a$ is an arbitrary small positive number.

## 87.3 Fuzzy Image Evaluation

The definition of the three image quality evaluation factors which are suitable to the smooth region, edge region and texture region, respectively. The next section discusses how to construct a fuzzy system for the assessment of image quality.

### *87.3.1 The Selection of Fuzzy Variant and Fuzzy Set*

The studies for the human visual system have shown that an image should include there important activities areas, i.e, smooth region, edge region and texture region. Therefore, in order to evaluate image quality, the proposed quality assessment model should give full consideration to the different regions of the distortion. However, how to identify the region of the current pixel? This is an openness problem. Here, we can use parameters $ME_X(i,j)$, $C_A(i,j)$, and the current pixel brightness values to compute their local mean absolute difference, denoted by $D_A(i,j)$, which can be used to identify the location of the current pixel in the original image, in other words, the three parameters are the input variants of the fuzzy system. Obviously, if the $ME_X(i,j)$ is large, then $C_A(i,j)$ is large, and $D_A(i,j)$ is also large, then the possibility of the given pixel located in the edge region is large; similarly, if the $ME_X(i,j)$ is small, then $C_A(i,j)$ is small, and $D_A(i,j)$ is also small, then the possibility of the given pixel located in the edge region is small; otherwise, we believe that the given pixel located in texture region. As for the output variants of the fuzzy system, $Q(i,j) \in [0\,1]$.

In order to fuzzy the actual input image, two fuzzy sets are defined, namely $S$ (small) and $L$ (large). $ME_X(i,j)$, $C_A(i,j)$ and $D_A(i,j)$, corresponding to the fuzzy set membership functions, where the parameters $a$, $b$, $c$, $d$, $e$ and play a very important role $f$ in the fuzzy system, and can be determined based on experience or statistical.

### 87.3.2 Fuzzy Principle Design

In order to reduce computation complexity, three fuzzy principles are constructed according to the format if-else-then:

Principle 1 if $ME_X(i,j)$ is small, $C_A(i,j)$ is small and $D_A(i,j)$ is small, then $Q(i,j) = Q_1(i,j)$.

Principle 1 if $ME_X(i,j)$ is large, $C_A(i,j)$ is large and $D_A(i,j)$ is large, then $Q(i,j) = Q_2(i,j)$.

Principle 1 else $Q(i,j) = Q_3(i,j)$.

According to the above rule, if a given pixel locates in the smooth regions probability, local image quality assessment based on the smoothing method is implemented; similarly, if a given pixel locates in the edge region probability, local image quality assessment method based on edge strength is implemented; otherwise, the local image quality assessment contrast-based approach is adopted.

### 87.3.3 Deblurring

For accurately computing the quality of estimated image, three equations are proposed as follows:

$$\lambda_1 = \min\{\mu_S(ME_A), \mu_S(C_A), \mu_S(D_A)\} \tag{87.8}$$

$$\lambda_2 = \min\{\mu_L(ME_A), \mu_L(C_A), \mu_L(D_A)\} \tag{87.9}$$

$$\lambda_3 = \max\{0, 1 - \lambda_1 - \lambda_2\} \tag{87.10}$$

where $\lambda_1$, $\lambda_2$, and $\lambda_3$, denotes activation degrees of fuzzy rules 1, 2, and 3, respectively, $\mu_S$, $\mu_L$, represent fuzzy variable "Small" and "Large" membership functions respectively. Then, the output of the fuzzy system is proposed according to weighted mean method:

$$Q(i,j) = [\lambda_1 Q_1(i,j) + \lambda_2 Q_2(i,j)\lambda_3 Q_3(i,j)]/(\lambda_1 + \lambda_2 + \lambda_3) \tag{87.11}$$

Finally, through computing the mean value of all $Q(i,j)$, we can get the final image quality $q$

$$q = \left(\sum_{i=1}^{N}\sum_{j=1}^{M} Q(i,j)\right)/(M \times N) \tag{87.12}$$

Table 87.1 Comparative results of different estimate standard

| | | PSNR | MSDS | PS-BIM | The proposed standard |
|---|---|---|---|---|---|
| Lena | Original image | | 1639.7 | 0.6358 | 0.9431 |
| 0.216b/p | JPEG decoding image | 29.92 | 5693.5 | 0.4658 | 0.9083 |
| Peppers | Original image | | 1881.4 | 0.5947 | 0.9642 |
| 0.207b/p | JPEG decoding image | 30.22 | 4666.6 | 0.4774 | 0.9425 |
| Fishing-boat | Original image | | 4760.7 | 0.7025 | 0.8738 |
| 0.244b/p | JPEG decoding image | 27.74 | 9262.9 | 0.5064 | 0.7455 |
| Test image | Original image | | 4326.5 | 1.8933 | 0.7838 |
| 0.08b/p | JPEG decoding image | 30.28 | 3686.9 | 1.2361 | 0.7568 |

**Fig. 87.2** Test image (0.08b/p)

## 87.4 Experimental Results

First of all, the relation between the size of the evaluation standard and the block effection is explained. According to the Ref. [7] and [8], expand MSDS and PS-BIM are not consistent with PSNR, the greater of their value, the quality of the images is more poor, the same as the fuzzy evaluation standard. In the design process of the proposed fuzzy system, let the parameters $a$, $b$, $c$, $d$, $e$ and $f$, be 50, 100 15, 75, 15 and 50 respectively.

Table 87.1 shows the results of different image for PSNR, expansion MSDS [7], PS-BIM [8] and the proposed standards. Because of the compression bit rate is low, JPEG decoding image performances more serious block effection than the original image. Table 87.1 shows the proposed fuzzy estimate standard is robustness for various images.

In theory, when the bit rate is high, the block effect is not obvious in images. Generally speaking, with the increasing of bit rate, fuzzy evaluation standard value is also increasing. Therefore, the proposed fuzzy evaluation standard has standards performance of the image quality assessment. Specially, PS-BIM simulation results are opposite with the theory, and so the standard of performance is questionable (Fig. 87.2).

## 87.5 Conclusions

We defined three local image quality estimation factors, which are suitable for smooth area, the edge area and the texture area respectively, and a new based on fuzzy inference rules image quality estimation method was proposed. This method can not only flexible use the human visual system knowledge, but also easy to implement. This method separates the block effect from the original image edges, which prevents the edge can be mistaken for a piece of effect and influence the results of evaluation. The simulation results show that the proposed method has standard image quality assessment performance, and can not be affected by edge on the block effect to get accurate evaluation result.

## References

1. Park HJ, Har DH (2011) Subjective image quality assessment based on objective image quality measurement factors. IEEE Trans Consum Electron 57(3):1176–1184
2. Li Q, Wang Z (2009) Reduced-reference image quality assessment using divisive normalization-based image representation. IEEE J Sel Top Sig Process 3(2):202–211
3. Sheikh HR, Sabir MF, Bovik AC (2006) A statistical evaluation of recent full reference image quality assessment algorithms. IEEE Trans Image Process 15(11):3440–3451
4. Narwaria M, Lin W (2010) Objective image quality assessment based on support vector regression. IEEE Trans Neural Networks 21(3):515–519
5. Suthaharan S (2003) Perceptual quality metric for digital video coding. Electron Lett 39(5):431–433
6. Kolaman A, Yadid-Pecht O (2012) Quaternion structural similarity: a new quality index for color images. IEEE Trans Image Process 21(4):1526–1536
7. Karunaseker SA, Kingsbury NG (1995) A distortion measure for blocking artifacts in image based on human visual sensitivity. IEEE Trans Image Process 4(6):713–724
8. Minami S, Zakhor A (1995) An optimization approach for removing blocking effects in transform coding. IEEE Trans Circuits Syst Video Technology 5(2):74–82

# Chapter 88
# Research on Medical Image Processing Method Based on the Matlab

**Shirui Gao**

**Abstract** This paper emphasizes the MATLAB-based medical image processing tools. It includes the theoretical background and examples. Through Matlab this paper make the introduction of the post-imaging quality in medical technology and medical imaging. It also introduces the medical image processing technology and describes the image processing and processing technologies, including the organ contours, interpolation, filtering, and segmentation techniques. In medicine, the DICOM image data processing using MATLAB is also widely used in this type of image processing.

**Keywords** MATLAB · Medical imaging · Data processing

## 88.1 Introduction

Advanced image processing and analysis technology are increasingly used in medicine. In medical applications, image data is used to collect the details of the imaging of the patient's process, whether it is a disease process or physiological process [1]. The information provided in medical imaging has become an important part of today's patient care. Images in medical applications is complex and changes from to switch between applications. The medical images display information on the characteristics of the structure, organs, and physiological characteristics. In order to have high-quality medical diagnostic imaging, image processing is necessary. The range of image processing and analysis of medical

S. Gao (✉)
School of Information Science and Technology, Northwest University, Xi'an 710127, China
e-mail: shiruigao@163.com

W. Du (ed.), *Informatics and Management Science V*, Lecture Notes in Electrical Engineering 208, DOI: 10.1007/978-1-4471-4796-1_88,

applications is to improve the quality of acquired images, and extract quantitative information from an efficient medical imaging data and accurate manner.

Matrix Laboratory (MATLAB) is a high-performance interactive software package developed by MathWorks Inc. (2009) scientific and engineering computing [2]. MATLAB allows matrix operations, algorithm implementation, simulation, drawing functions and data, signal and image processing, image processing toolbox. It allows several ways of medical imaging, quantitative analysis and visualization of single-photon emission computed tomography (SPECT), positron emission tomography (PET) or hybrid systems (SPECT/CT examination), computed tomography system (CT) scan is included in the SPECT system [3]. The Image Processing Toolbox is a comprehensive reference standard algorithms and graphical image processing, analysis, visualization, and algorithm development tools. It offers the possibility of noisy or degraded image restoration, and improves image clarity, feature extraction and analysis of shapes and textures, and registers the two images. Therefore, with all the features, we can use MATLAB to collect to perform any sophisticated analysis of the image. Most toolbox functions are written in MATLAB's open language and this provides users with the opportunity to check the algorithms, modify the source code and create a custom function.

## 88.2 Principles and Methods of Medical Imaging

Medicine is the human physiology and biochemistry about science section of the nature of radiopharmaceuticals and clinical data. According to the emphasis on pharmaceuticals (tracer) and the entire complex for each patient examination, radionuclide, and then sent to the patient intravenously or swallowing or even inhaling. Radioactivity follows the physiological pathways; it focuses on specific organs and tissues of the time of the short period [4]. Then, the patient is positioned in the medical device and the image caused by the human body can detect biological distribution of the radiation emitted by radioactive drugs.

There are two main approaches in medicine, the patient's imaging, imaging plane images, moving images or SPECT and PET. In the past decade, the hybrid system has been developed to integrate the SPECT or PET and CT technology lead in of SPECT/CT and PET/CT. This chapter will focus on the $\gamma$ camera planar imaging, SPECT and SPECT/CT in the implementation of MATLAB code.

The medical examination of the range is quite extensive. It includes, inter alia, the patients in the study. Basic image analysis methods, cardiac muscle, brain, kidney, thyroid, lung tumor imaging (such as neuroblastoma cell tumor) medical research, including regional property, boundary analysis, curvature analysis, or a straight line and circle detection image processing services image reconstruction technology, viewing the improvement of image quality and image preparation quantitative analysis of the results referring to the examination of the data for the following using MATLAB to clear and display algorithms, medical image processing and analysis of these studies MATLAB practical.

Our goal is a reliable image quality plays an important role images of organs in medical imaging is expected to provide an accurate diagnosis or treatment. The physical characteristics used to describe the image quality [5]: (1) contrast (2) spatial resolution (3) noise. The image contrast corresponding to different concentrations of the intensity difference in the patient's activities. High diagnostic accuracy of medical imaging must be high contrast.

## 88.3 The Analysis and Processing of Medical Image

The image processing is a set of technologies, image data analysis and processing algorithms and tools to improve the interpretation of some image information more useful. Image processing allows the extraction of useful parameters, and increase the likelihood of detection of small lesions more accurately. In medical image processing has three main objectives [6]: (1) reconstruction tomography (SPECT) imaging technology; (2) quality to improve the image contrast, uniformity and spatial resolution view; (3) in order to extract useful diagnostic qualitative and quantitative image ready message.

### *88.3.1 Digital Imaging*

In all modern medical imaging systems, image display an array of discrete picture elements (pixels) in two-dimensional (2D) and is known as a digital image. In the digital image each pixel has an intensity value and the address of the location (Fig. 88.1). In medical image pixel values to display the record counts in it. Benefit compared to the simulation of a digital image from the digital image data for further computer processing.

Digital image of the characteristic matrix size, pixel depth and resolution. The matrix size was determined by the number of columns (m) and the image of the number of lines (N) matrix ($M \times N$). The size of the matrix is selected by the operator. In general, getting better as a matrix resolution, the size increases. Medical imaging matrix, now ranging from $64 \times 641\,024 \times 1024$ pixels. Pixels or bit depth of each pixel in each pixel digital color level image. Each pixel can take 2 K different values, where k is the bit depth image. This means that an 8-bit image, each pixel can have from the different layers of color (gray level) of 1–28 (= 256) Medical Imaging is often expressed as 8–16 images.

The long-term image resolution refers to the number of pixels per unit length of the image. The spatial resolution in digital images depends on the pixel size. The pixel size calculation of view (FOV) divided by the entire matrix of pixels field.

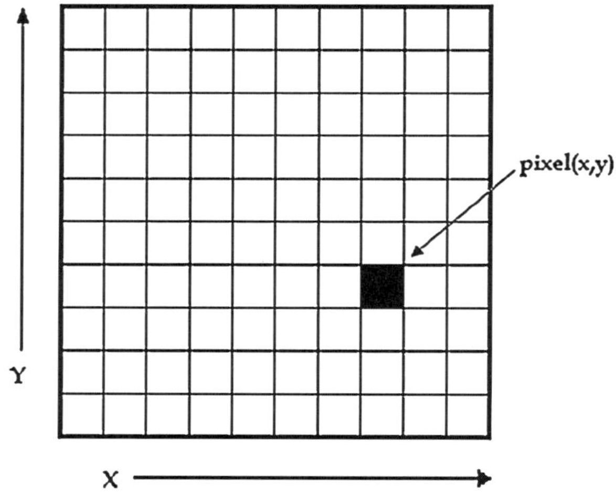


**Fig. 88.1** Digital image is a two-dimensional array of pixels

For a standard FOV, matrix size increases to reduce the pixel size to be able to see the details to improve.

## *88.3.2 MATLAB Type of Digital Imaging*

MATLAB provides a simple function, you can read a variety of image file format, and supports a number of color maps. Based on file type and color space, the returned matrix is two-dimensional matrix of intensity values (gray-scale images) or 3D matrix of RGB values. Medical Imaging grayscale or true color image (RGB or red, green, blue).

## *88.3.3 Image Processing Techniques Based on MATLAB*

Image processing technology, including all of the tools used to change or analysis of an image based on individual needs. This section describes the most extensive implementation of image processing in medical imaging technology. Examples, mostly from the medical kidney graphic images, radiographs is a simple object, display the image processing application of MATLAB tools.

## *88.3.4 Contrast Enhancement*

The problem is one of the first image processing and contrast enhancement. Acquired image usually does not propose the necessary objects contrary. The contrast improvement is absolutely necessary organ shape, boundary and internal functions can be better portrayed. In addition, institutions division, can be achieved in many cases, does not eliminate the background activity.

Contrast processing command is imadjust. Use this, if necessary, the image contrast can be enhanced or degraded. In addition, a very useful result of the object of interest may be color, especially in the gray-scale image inversion, can be outlined. The general features of contrast enhancement of the following:

J=imadjust (I, [low_in_high_in], [low_out_high_out], gamma);

Assume that J, the relationship between the new image, I, the first image and the shape of the curve described by the gamma described values, I and J. If gamma is omitted, it is considered to be 1.

### 88.3.5 The Organ Contours

In many medical imaging, organ boundary is not clear, due to low resolution or the existence of a high percentage of noise. To command to draw the organ contours imcontour be used in medical imaging. In addition, the variable n defined equidistant contours required. This variable count is closely related to strength. For the higher value of n, the line drawn between the smaller spaces and depicts the stripes of different intensity. And contour type, you can specify. For example, when the outline of a five contour drawn with a solid line is the desired result, the overall function is as following:

Figure, imshow(I)
J=imcontour(I,5,'—');
Figure,imshow(J)

Among them, the J and I represent the final and initial images and symbols ('-') represents the solid line drawing. For example, the initial image contour with N = 15 and N = 5, respectively, as follows (Fig. 88.2).

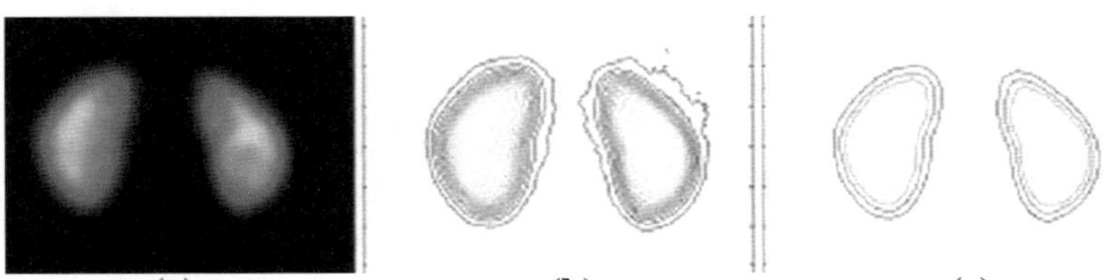

(a) (b) (c)

**Fig. 88.2** **a** Original image depicts the kidney. **b** N = 15 the contours of organs. **c** N = 5 organ contours

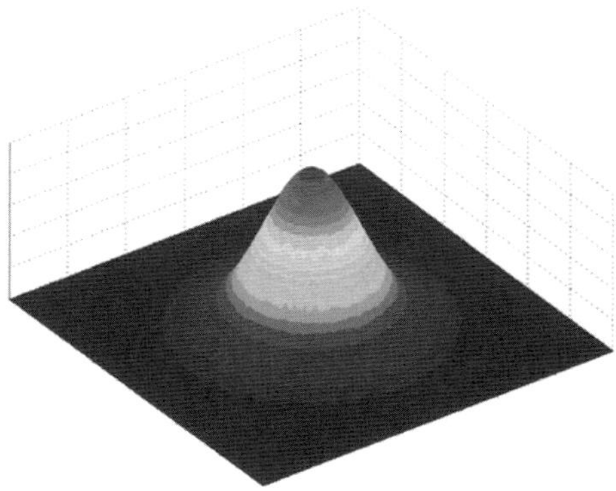

**Fig. 88.3** The 2D Gauss function

### *88.3.6 Gaussian Filter*

Gaussian filter is a linear low-pass filter. Or the curve side of the tapered part of the Gaussian filters mask forms the high point in the center and symmetric (Fig. 88.3). The application of the Gaussian filter to produce, each pixel in the image, the weighted average, such a central pixel more than the results of the edge of the pixel mask. According to the weight of the Gaussian function (Eq. 88.1) [7]:

$$f(x) = \frac{1}{\sigma\sqrt{2\pi}} e^{1(x-\mu)^2/(2\sigma)^2} \tag{88.1}$$

Among them, $\mu$ is for the average value and $\sigma$ for standard deviation sigma.

Smoothness depends on the standard deviation. Larger standard deviation, smooth image portrayed. The Gaussian filter is very effective at reducing impulsive and Gaussian noise. Gaussian noise due to random variations in intensity and distribution follows the Gaussian curve.

### *88.3.7 MATLAB Filter*

In Matlab, the use of image processing toolbox, we can design and implementation of the filter image data. For linear filter, MATLAB provide fspecial command to produce some intended a common 2D filter.

```
h=fspecial (filtername, parameters)
```

The following example describes the command packet, you can use the mean (average) radiographs of the convolution kernel of the application size of the filter (3 × 3, 9 × 9, 25 × 25 average filter) (Fig. 88.4).

```
h=fspecial('average', [3 3]);
b=imfilter(a,h);
figure, imshow(b);
```

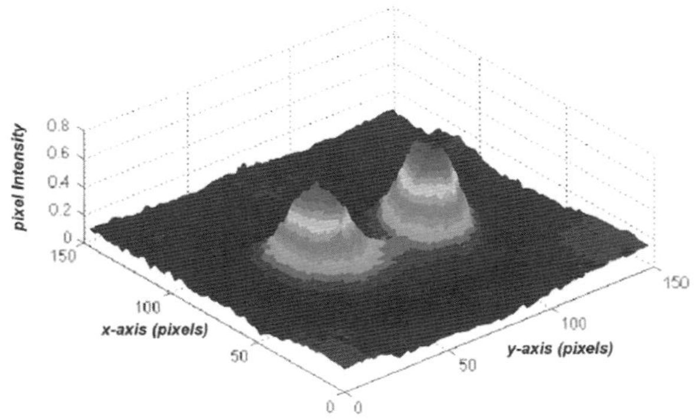


**Fig. 88.4** pixel intensity surface plot represents the identity in X and Y axis, and pixel represents the intensity of the pixel in the z-axis

```
i=fspecial('average', [9 9]);
c=imfilter(b,h);
figure, imshow(c);
j=fspecial('average', [25 25]);
d=imfilter(c,h);
figure, imshow(d);
```

## 88.4 DICOM Image Processing

MATLAB digital medical image processing, the development of a standard is the same in order to enable users to retrieve the image transfer digital images and the related use of a standardized way all the information from different imaging way. The geometry of three-dimensional image data in accordance with the header contains the information. DICOM files there. The DCM expansion. In medicine, the most common and supported formats for the three-dimensional data is stored using the DICOM volume split into pieces (such as heart or kidney), and save each as a simple DICOM image slice. Encoded digital slides can be distinguished in the file name or by specific DICOM tags. MATLAB support DICOM files and DICOM image processing is a very useful tool. DICOM images are grayscale images. The name of the hypothesis "kidney" to read, to read DICOM images of the image data in DICOM files of the command dicomread in the following functions.

```
I=dicomread('kidneys.dcm');
```

To read DICOM file meta-data, command the dicominfo. The latter returns a MATLAB structure, where each field contains a specific DICOM metadata. Such as the same DICOM images,

```
info=dicominfo('kidneys.dcm')
```

Loaded images can now modify and processed in any way desirable. Many times a word or letter, describe the piece, or a projected image appears in these can create a new image to delete, and do not believe. To modify or write image data or metadata in the DICOM format files, use dicomwrite functions. The following command to write the image I or K DICOM file kidneys_file.dcm of thyroid_file.dcm and DICOM files.

## 88.5 Conclusion

The application of medical diagnostic imaging, image processing and analysis to improve the quality of images acquired as well as to provide quantitative data useful for patient treatment and care. Advanced image processing and analysis techniques found widespread use in medicine. MATLAB and Image Processing Toolbox also enabled the quantitative analysis of medical imaging and visualization plane or acquisition angle projection image reconstruction tomography (SPECT, PET) slices and 3D volume surface rendering image.

## References

1. Li F, Ju L, YunHua D (2005) VB and Matlab mixed programming method. Softw Technol 5:110–112
2. Feisi Technology R&D Center (2003) Matlab 6.5 application programming interface. Electronic Industry Press, Beijing vol 64, pp 1132–1134
3. Yufang Y, Zhong-sheng Y, Yu-Shuang Z (2005) VB programming based on COM mixed with Matlab. Comput Eng Design 1:61–65
4. Liang W, Fan H, Dan W (2006) VB and Matlab simulation of mixing programming application. Comput Simul 12:104–106
5. Lishan M, Dapeng D, Liu G (2009) Visual basic 6.0 advanced programming skills. Electronic Industry Press, Beijing vol 213, PP 412–415
6. Wang H (2007) DICOM medical image file information extraction and the realization of the images show. Med Equipment Inf 9:1–3
7. Zhao L (2002) Matlab6. X image processing. Tsinghua University Press, Beijing vol 56, pp 555–557

# Chapter 89
# Automatic Identification Water Flooding Level of Oil Layer Based on Fluorescence Microscopic Image Processing Technology

**Huijian Wen, Xueying Li and Guangjuan Fan**

**Abstract** Based on fluorescence microscopic image data of hermetic coring well, this paper realizes fluorescence image quantification interpretation for different reservoir and water-flooded level, through studying the fluorescent color, relative intensity and change rule of the light-emitting area in different water-flooded levels, establishing quantitative standards of fluorescent color and fluorescence intensity. Taking advantage of images of colorimetric principle, cluster analysis was used for analyzing the fluorescence color and each color of the wavelength and the relative strength was quantified in this paper. By means of image processing, count the pixels with similar fluorescent color to identify the light-emitting area of fluorescence, then draw quantitative spectrum graph of fluorescent image, finally realize the automatic identification of water flooded degree.

**Keywords** Relative intensity · Fluorescent image · Water flooding level

H. Wen (✉) · X. Li · G. Fan
College of Earth Sciences, Northeast Petroleum University, Daqing, Heilongjiang 163318, china
e-mail: whjdqpi@163.com

X. Li
e-mail: lxy8403@sina.com

G. Fan
e-mail: fangj99@163.com

W. Du (ed.), *Informatics and Management Science V*, Lecture Notes in Electrical Engineering 208, DOI: 10.1007/978-1-4471-4796-1_89,

## 89.1 Introduction

Evaluation of water flooding layer is one of the most key links in reservoir description, and also one of the most explanation difficulties. For the moment, logging information to explain water flooding layer is mainly used in production, but logging information is indirect and multifactorial, which brings lots of difficulties to water flooding layer logging interpretation, especially when enter the later stage of a high water cut, the top potential objects begin to transfer to thinner, worse oil layer, thin layer and heterogenous interlayer like pelitic, calcic have a big effect on logging information, but effective thin and bad well logging interpretation model cannot be established at present, the coincidence rate of water flooding explanation is very low. In this situation, technology of geochemical analysis and fluorescence microscope which is gradually developing in recent years, provide a new type of technical means for evaluation of water flooding layer [1, 2]. Many domestic authors do much further research work for this. The achievements to be obtained lay the solid foundation for using fluorescent image observation and water flooding level research [3].

## 89.2 Geofluid with Fluorescence Separability

The research and theory analyses indicate that the integral fluorescent characteristic of the geofluid is strongly linked to its relative composition and thickness distribution [4, 5], different geofluid owns different fluorescent colour and fluorescent luminance: the fluorescent colour of oleaginous asphalt always appears yellow and green–green and yellow–yellow, and its luminance is high, colloidal asphalt appears orange and yellow, tan, brown, red and brown and so on, its fluorescence luminance is relatively low, the fluorescent colour of bitum asphalt is darker, often appears ton, and its fluorescent luminance is very dim, because of dissolving a slight amount of aromatic substance, water appears green in high purity. Therefore, different geofluid owns fluorescence separability. This provides a secure basis for using fluorescent images to do qualitative interpretation of water–oil beds and water flooding layer.

## 89.3 The Features Changing Rules of Fluorescence Image in Different Water Flooding Levels

### *89.3.1 The Relationship Between the Fluorescent Colour and Intensity of Fluid Components and Water Flooding Levels*

Expressed by fluorescent image feature like fluorescent colour and fluorescent intensity within oil layer in different water flooding levels, is that oleaginous

**Fig. 89.1** Ideal schemes about the relationship between the changing rule of fluorescent colour and fluorescence intensity in rock samples and three kinds of pole fluid component concentration dynamic change

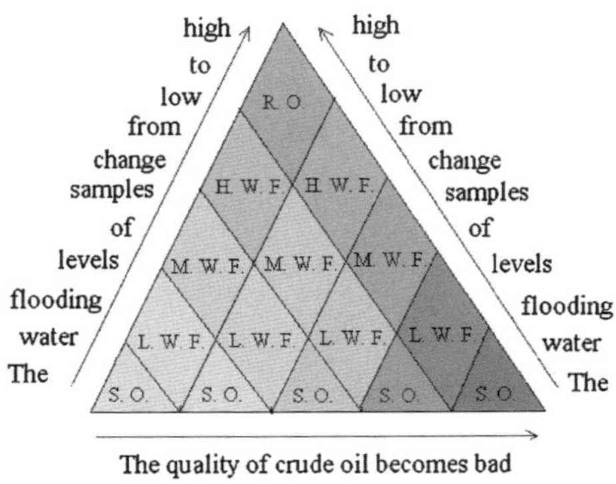


asphalt, colloidal asphalt when generated by the action of washing and dilution of the water, its dynamic changing rules can be described synthetically in Fig. 89.1 which appears an idealized model. In the Fig. 89.1, the three vertex coordinates of the triangle represent three fluid components of water, oleaginous asphalt and colloidal asphalt, the changes of the relative percent contents of oleaginous asphalt, colloidal asphalt and water are provided in three sides. The colour and luminance in the triangular plot shows that when the percentage composition of three kinds of pore fluid components changes, its changing rule of the fluorescent colour and fluorescent intensity of its sample. That is, when the water ratio of the sample changes from low to high, that is to say, the water flooding levels of the sample changes from weak to strong in the oleaginous asphalt direction, fluorescent colour gradually transit from yellow to light yellow–yellow and green–green, fluorescent intensity follow the rule which is from moderate to stronger–moderate bias weak–weak, when crude oil quality becomes weak, it changes in the direction of from oleaginous asphalt to colloidal asphalt, fluorescent colour transits in the way of yellow–brown and yellow–brown, at the same time, fluorescence intensity is becoming weaker gradually.

### *89.3.2 The Relationship Between the Fluorescent Relative Light-Emitting Area and Water Flooding Level*

Fluorescent light-emitting area is an important indicator to judge water flooded level. To confirm the concentration distribution range of the rock pores fluid components in different water flooding levels, to divide fluorescent light-emitting area in different water flooded levels, we suppose interstitial surface area is 1 in the fluorescent image, while compare the total percentage of pore area of the light-emitting area of oleaginous asphalt, colloidal asphalt, water to the relative light-emitting area of all kinds of pore fluid. Three main kinds of fluid components like oleaginous asphalt, colloidal asphalt and water, which are in 100 water displacing oil samples, their relative light-emitting area are made statistics and analysis in different water flooding levels, in order to confirm the percentage of all kinds of

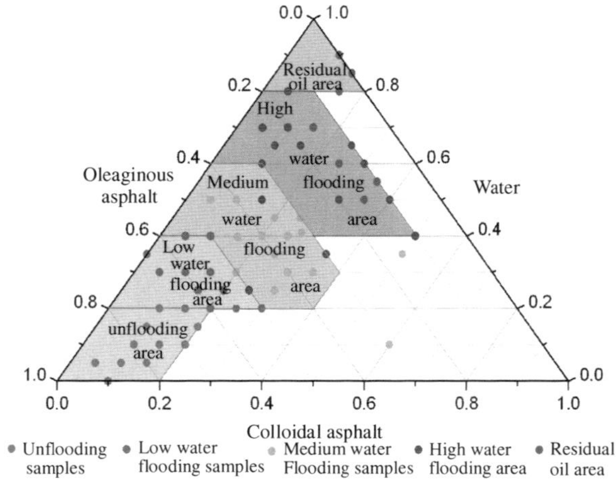


**Fig. 89.2** Distribution about fluorescence relative light-emitting area of fluorescence samples in different water flooding degree

fluid components in pore space. Vertex coordinates represent the content of oleaginous asphalt, colloidal asphalt and water in pore space, the distribution interval figure of fluorescent relative light-emitting area of all kinds of interstitial fluid in different water flooding levels is established at last (Fig. 89.2). The figure shows that when water flooding levels increase, apparent containing water area in the pore space is increasing gradually, this phenomenon fits with the fact, and this changing rule lays the foundation for the quantitative judgment of the water flooding level of the oil samples.

### *89.3.3 Fluorescent Colour and Quantitative Standards of the Fluorescence Intensity*

The essence of the generation of fluorescence is when electrons in different orbital is excited, they jump from a low energy orbital to a higher one, then they release the energy that they have absorbed in light quantum form. Ultimately, the launch of fluorescence material is a mutual switching process of orbital electron in different energy levels, this process can produce fluorescence with different wavelengths, different intensity according to the different energies they have absorbed. This physical process can describe in use of Einstein—Planck law:

$$\Delta E = hc/\lambda \tag{89.1}$$

$h$ Planck constant,
C velocity of light,
$\lambda$ wavelength of blip,
$\Delta E$ the energy between minimum single lines excited state $S_1$ and the ground state $S_0$

**Table 89.1** Wavelength coverage quantitative indicators of different fluorescence colour

| fluorescence colour | Blue–green | Yellow–green | Green–yellow | Light green–yellow | Yellow | Light yellow | Orange–yellow | Yellow–orange |
|---|---|---|---|---|---|---|---|---|
| Wavelength coverage (nm) | <500 | 500–540 | 540–560 | 560–570 | 570–585 | 585–600 | 600–605 | 605–610 |

**Table 89.2** Fluorescence relative intensity quantitative classification index

| Relative intensity | Stronger | Strong | Moderately strong | Moderately weak | Weak | Weaker |
|---|---|---|---|---|---|---|
| Level (%) | 70–100 | 60–70 | 50–60 | 40–50 | 30–40 | <30 |

From Eq. (89.1) we can include the size and scope of $\lambda$ express different fluorescent colours, the height of $\Delta E$ express different fluorescence intensity. In practical applications fluorescence is expressed by fluorescence relative intensity [6].

The relationship of the transformation energy of the colour, Relative intensity takes the radiation energy of green light which owns a wavelength of $\lambda = 555$ nm, or light intensity is 100 %, to regard it as a reference point, and compare with light intensity of other wavelength, to obtain relative intensity in different wavelengths. From the analysis above, colour of quantitative characterization are established by wavelength, therefore, to realize the qualitative interpretation of the fluorescence image, quantified standards of fluorescent colour and intensity are established. Wavelength quantitative standard of different fluorescent colour is shown in Table 89.1, while fluorescence relative intensity quantitative classification is shown in Table 89.2.

### *89.3.4 Auto-Explanation of Fluorescent Image*

Entering the figure which is caught by fluorescence microscope into the computer and keep it as a real color image, Pore fluid components in the distribution and concentration is very complex, so similar pore fluid owns similar colour, that is, there are many pixel colours close but components of three primary colours owns a little difference, therefore fluorescent image interpretation must be carried out before the image colour clustering analysis, given the types of pore fluid and the distribution of the pore space.

### *89.3.5 Colour Clustering Analysis*

The means of clustering is to choose green, yellow–green, light yellow, brown–yellow, brown as initial colour sample at first, then look the minimum distance as polymerization principles function gradually polymerize, types from many to little, until the right classification is found. To a certain colour value, choose the pixels similar to itself in order to form colour group which owns a certain wavelength range.

### 89.3.6 Acquire the Wavelength and Relative Intensity of Colour Groups After Clustering

According to every big kinds of pixel colours after clustering, reading the RGB colour component in the computer, it will be seen approximatively as a composite colour of the stimulus value, using the formula below to calculate X,Y,Z three component stimulus values of International Lighting Committee CIE-XYZ colour space and X,Y,Z chromaticity coordinates [7].

Since in the fluorescent images mainly oriented green, yellow–green, yellow, based on Y chromaticity coordinates components, searching corresponding relation contrast table of International Commission on III umination (CIE) different wavelengths and chromaticity coordinate x, y, z, looking for colour wavelength which is matching Y chromaticity component the most, and if there is no wavelength which can match Y chromaticity component, then choosing the linear interpolation to get the pixel which associates with wavelength $\lambda$ in the adjoining wavelength ranges(reference x component), to repeat the process for every types of colour group of each pixel, to get the wavelength distribution ranges of the colour group.

After finding the wavelength $\lambda$ corresponding with the pixel, to search the CIE-RGB colour space, look for the $\bar{r}(\lambda).\bar{g}(\lambda).\bar{b}(\lambda)$—Three component stimulated values of CIE-RGB colour space which respond to wavelength $\lambda$,, then relative intensity of each colour can be calculated by the formula below.

$$\begin{aligned} C(\lambda) &= \bar{r}(\lambda)(\overrightarrow{R}) + \bar{g}(\lambda)(\overrightarrow{G}) + \bar{b}(\lambda)(\overrightarrow{B}) \\ &= \bar{r}(\lambda) + 4.5907\bar{g}(\lambda) + 0.0601\bar{b}(\lambda) \end{aligned} \tag{89.2}$$

Normalize each colour group or the relative intensity which is in the determined wavelength range, to select advantage values as fluorescence relative intensity within the wavelength range, after acquiring the relative intensity of all colour groups, to smooth the spectrum Table, then stretch and release relatively spectrum peaks of each colour group, to make it have obvious spectrum characteristics. According to the principles above, using the quantitative parameters data like fluorescence wavelength, relative intensity, light-emitting area (the number of pixels with the same colour), drawing the quantitative spectrum Table of fluorescent image. From the spectrum Table we can see, for the oil layer that is medium water flooding, which is with high relative intensity oil and water emulsion peak in wavelength of 555–565 nm (bright yellow–green), and it reflects the relatively contents of the injected water in the pore space, in the scope of 575–585 nm (yellow, green–yellow), the oil peak with medium intensity can be seen, which reflects the content of the remaining oleaginous asphalt in the rock pore space, glial asphalt peak is relatively very low within 605–620 nm (orange, brown, red–brown), and it reflects the content of glial asphalt and asphaltene asphalt in the rock pore space. Therefore, on the basis of wavelength distribution characteristics of fluorescent image spectrum Table, relative intensity distribution characteristics and fluorescence area, fluorescence image quantitative interpretation and judgment of water flooding layer or water–oil layer can be done.

## 89.4 Conclusion

Fluorescent characteristics of different water flooding level in oil layer is a result of the interaction by the continuous dynamic changing of consistence, decided by three kinds of pole fluid like oleaginous asphalt, glial asphalt and water in the pore space, therefore oil layer water flooding levels can be quantitative divided by fluorescence intensity, fluorescence wavelength, apparent containing water area.

Fluorescent images directly register much information about oil layers in different water flooding level; it's a real reflection of the underground oil layer. It possesses directness, authenticity, visuality, therefore this information can be quantitative gotten from fluorescence microscopic scale, and it can increase the accuracy of the water flooding layer interpretation.

Fluorescent images consists of pixels of different colours, different luminance, while colour and luminance can reflect the physical property of fluorescence blip, image colorimetric theories can be adopted to quantify fluorescence physical parameters, to obtain the spectrum Table of fluorescence image, according to the characteristics of the spectrum and light-emitting area of each pore fluid, quantitative interpretation of the oil layer water flooded levels can be done automatically.

## References

1. Mingfa Y, Weidong J, Yanzhi L (2003) Apply geochemical well logging and discussion of P-K analysis technology comprehensive evaluation of water flooding. Well logging 14(3):20–25
2. Baskin R, Huang J (1995) Predicting gas.oil and water intervals in Niger Delta reservoirs using gas chromatography. AAPG 79(3):337–350
3. Chen, J-D, Wada, N (1986) A new technique for visualizing the distribution of oil, water, and quartz grains in a transparent, three-dimensional, porous medium. SPE13349: SPE Form Eval 1(2):205–208
4. Wenhui W, Zhan W, Liangzhu Z (2001) Based on the content of colour image characteristics extraction method. Comput Aided Design Graph J 13(6):566–599
5. Chengfa H (1993) Printing colour science and colorimetry. Printing Industry Publishing House, Beijing vol 442, pp 71–125
6. Jie Y, Lihua C (2002) Using fluorescence spectrum to determine and compar. Oil Explor Dev 29(6):67–69
7. Zhou B, Junyi S (2004) Peng Qinke. Based on the calculation of colour quantitative algorithrm of the colour couple. Comput Eng 30(14):24–26

# Chapter 90
# An Effective Image Segmentation Algorithm Method Based on Compound Morphology Filter and Modified Watershed

**Weifeng Wang, Huifeng Yan and Mingliang Zhou**

**Abstract** To describe an effective method for image segmentation algorithm, which is based on the compound morphology filter and modified watershed? First of all, we define the compound morphology filter with opening-closing operators and closing-opening operators, which overcomes the over-segmentation of the traditional water shed algorithm. Secondly, we design a new morphology water shed with inner and external marks, and also define the regional minima to inner marks from the low frequency components of the gradients and external marks between the region; the inner and external marks change along with the image information, thus which has realized to the image information adaptive segmentation. Experimental result show that the new algorithm can overcome under the low contrast is not high enough accuracy in the traditional morphology watershed segmentation algorithm, which could accurately obtain the image edge and contour.

**Keywords** Image segmentation · Morphology filter · Water shed algorithm

W. Wang (✉) · H. Yan · M. Zhou
College of Mobile Telecommunications, Chongqing University of Posts and Telecom, Chongqing 400065, China
e-mail: 21973433@qq.com

H. Yan
e-mail: 4423360@qq.com

M. Zhou
e-mail: zml-0913yy@163.com

W. Du (ed.), *Informatics and Management Science V*, Lecture Notes in Electrical Engineering 208, DOI: 10.1007/978-1-4471-4796-1_90,

## 90.1 Introduction

Digital image segmentation is an important tool for feature extraction and measurement. Image segmentation is one of the most widely used steps in the process of reducing images to get useful information. Currently used technology of medical image segmentation are: the threshold-based segmentation, the segmentation based on contour model, based on the segmentation and clustering based on the morphology and the theory of multi-scale segmentation technology [1, 2]. One of the morphological watershed algorithms based on the segmentation method for its calculation speed and accuracy of the edge position was widely used in medical image segmentation, but the method is very sensitive to noise, low-contrast images for the loss of important contours easily. In this paper, based on morphological filtering and improved composite watershed algorithm, to reduce the image segmentation process of the border location of the problem of inaccurate and at the same time, more quickly determine the regional boundaries, to improve the efficiency of image segmentation [3].

## 90.2 Morphological Filter

For sets A and B in $Z^2$ the erosion of A by B, denoted A $\oplus$ B, is defined as

$$\mathrm{A} \oplus \mathrm{B} = \{\, \mathrm{z} \big| (\hat{\mathrm{B}})_{\mathrm{z}} \bigcap A \neq \emptyset \} \tag{90.1}$$

This equation is based on obtaining the reflection of B about its origin and shifting this reflection by z. For sets A and B in $Z^2$ the erosion of A by B, denoted $A\Theta B$, is defined as

$$A\Theta B = \{z|(B)_z \subseteq A\} \tag{90.2}$$

This equation indicates that the erosion of A by B is the set of all points z such that B, translated by z, is contained in A. The opening of set A by structuring element B, denoted A○B, is defined as

$$A \circ B = (A\Theta B) \oplus B \tag{90.3}$$

The opening A by B is the erosion of A by B, followed by a dilation of the result by B. The losing of set A by structuring element B, denoted A·B, is defined as

$$A \cdot B = (A \oplus B)\Theta B \tag{90.4}$$

Which says that the closing of A by B simply the dilation of A by B, followed by the erosion of the result by B. That may exist under low-resolution pseudo-edge. B of A, k times the expansion is defined as:

$$[A \oplus B]^k = \underbrace{((\cdots(A \oplus B) \oplus B \cdots) \oplus B)}_{k} \quad (90.5)$$

B of A, defined as k times the corrosion:

$$[A\Theta B]^k = \underbrace{((\cdots(A\Theta B)\Theta B \cdots)\Theta B)}_{k} \quad (90.6)$$

K is that the number of expansion or corrosion, B often become a structural element, the actual use often choose 7 × 7 cross mask. When the use of the structural elements of the images and the expansion of k times after corrosion, the corrosion of those parts out and the expansion part is the image from the edge of the region with $\Omega$:

$$\Omega = [A \oplus B]^K - [A\Theta B]^k \quad (90.7)$$

The real burden will fall on the edge of the zone. In the marginal zone of the image corresponding to the region, there are many image edge detection methods. However, due to relatively low-contrast image of the edge of ambiguity, the use of Canny operator methods are more sensitive to noise, while a lot of false edge detection, edge fracture phenomenon occur frequently [4].

## 90.3 Improved Watershed Image Segmentation

The principal objective of segmentation algorithms is to find the watershed lines [5]. The basic is simple: Suppose that a hole is punched in each regional minimum and that the entire topography is flooded from below by letting water raise though the holes at a uniform rate. When the rising water in distinct catchment basins is about is about to merge, a dam is built to prevent the merging. The flooding will eventually reach a stage when only the tops of the dams are visible above the water line. These dam boundaries correspond to the divide lines of the watersheds. Therefore, they are the (continuous) boundaries extracted by a watershed segmentation algorithm.

Let $N_1, N_2 \ldots N_R$ be sets denoting the coordinates of the points in the regional minima of an image g(x,y). Let C $(N_i)$ be a set denoting the coordinates of the points in the catchment basin associated with regional minimum $N_i$. The notation min and max will be used to denote the minimum and maximum values of g(x, y). Finally, let T[n] represent the set of coordinates (s, t) for which g (s, t) < n. That is, T[n] = {(s, t) | g(s, t) < n}. The topography will be flooded in integer flood increments, from n = min + 1 to n = max + 1. Geometrically, T[n] is the set of coordinates of points in g(x, y) lying below the plane g(x, y) = n. Let $Cn(N_i)$ denote the set of coordinates of points in the catchment basin associated with minimum Ni that are flooded at stage n. $C_n(N_i) = C(N_i) \cap T[n]$. $Cn(N_i) = 1$ at location (x, y) if (x, y) ∈ T[n] AND (x, y) ∈ T[n]; otherwise $C_n(N_i) = 0$. The geometrical interpretation of this result is straightforward. We are simply using the AND operator to isolate at stage n of flooding the portion of binary image in T[n] that is associated with regional minimum Ni.

Based on the method of marking the key is to find a set of all regions associated with the image of the seed point, from the original gradient image to find the low-frequency components associated with the region as a marker of local minimum, the approach taken for noise reduction is the first direct image of the side map, and then the gray level histogram distribution according to size to determine the probability of local extremum points as seed points. At the same time, we definition of internal markers in this case: (1) a region that is surrounded by points of higher "altitude"; (2) such that the points in the region from a connected component; (3) in which all the points in the connected component have the gray-level value. After smoothing the image on the use of the watershed algorithm, and to limit these internal markers can only be allowed to local minimum, thus dividing line, as an external marker. External marking image is separated into different regions; each region contains a unique part of the internal marker and the background.

## 90.4 Experimental Results and Discussion

As can be seen from the experimental results, the use of canny operator methods is more sensitive to noise, while a lot of false edge detection (Figs. 90.1, 90.2, 90.3, 90.4).

In this paper, in the form of composite image filter, and then bound algorithm only tags that contain a specific region of the watershed to deal with, as opposed to traditional watershed algorithm, the effect of segmentation algorithm a clear asset.

**Fig. 90.1** Original image

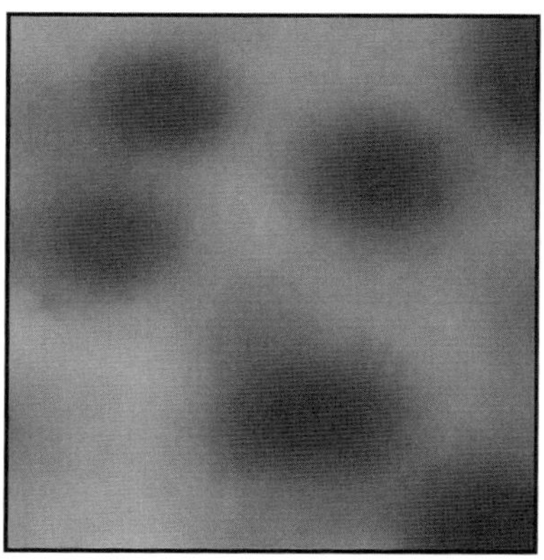

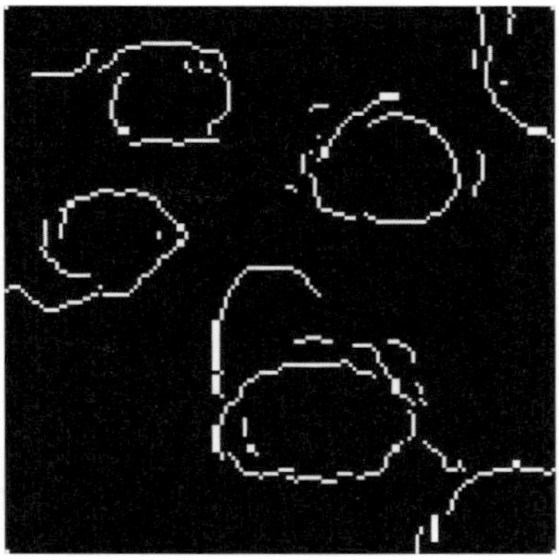

**Fig. 90.2** Canny edge detector

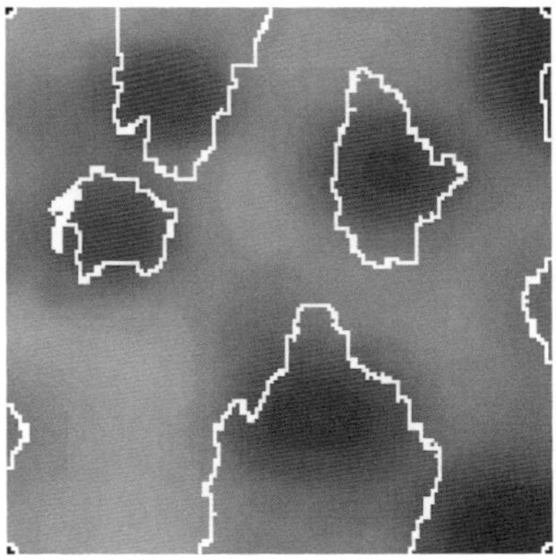

**Fig. 90.3** Traditional watershed segmentation

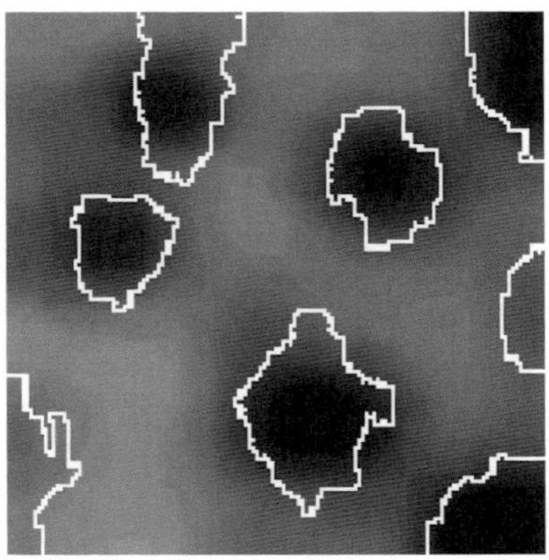

**Fig. 90.4** Algorithm results

## 90.5 Summary

In this paper, efficient algorithms to overcome noise other conditions resulting from the gradient of false information. The new algorithm can overcome under the low contrast is not high enough accuracy in the traditional morphology watershed segmentation algorithm. I will continue on the low-contrast image contour extraction algorithm.

## References

1. Aylward SR, Bullit E (2002) Initialization, noise, singularities, and scale in height ridge traversal for tubular object centerline extraction. IEEE Trans Med Imaging 21:61–75
2. Beucher S, Meyer F (1993) The morphological approach to segmentation: the watershed transformation. In: Dougherty ER (ed) Marcel Dekker, New York, vol 12, pp 433–481
3. Zahlten C, Jurgens H, Evertsz CJG, Leppek R, Peitgen H-O, Klose KJ (1995) Portal vein reconstruction based on topology. Eur J Radiol 19:96–100
4. Dumoulin CL, Hart HR (1986) Magnetic resonance angiography. Radiology 161:77–720
5. Passat N, Ronse C, Baruthio J, Armspach JP, Maillot C, Jahn C (2004) Atlas-based method for segmentation of cerebral vascular trees from phase-contrast magnetic resonance angiography. In SPIE Image Process 5370:420–431

# Chapter 91
# A New Graph-Based Image Segmentation Algorithm

**Qian Zhang, Fujian Feng, Lin Xin and Lin Wang**

**Abstract** Based on graph theory, we choose two-dimensional Gaussian as a dynamic adaptive index for weighting function, difference function Dex for inter-area and Din for one area were defined by structural similarity index (SSIM), the function determines different area to be merged or segmented is achieved. The algorithm was implemented on Mat lab successfully. Experimental results show that the algorithm the segmentation is better than others in effect and calculate time.

**Keywords** Image segmentation · Graph · Two-dimensional gaussian distribution · SSIM

## 91.1 Introduction

Image segmentation is not only important research aspects of computer image processing and pattern recognition, but also the difficulty of image understanding and analysis. It gains the people's interest and attention in theoretical research and practical applications, and lot of the classic image segmentation formed. Such as: Liu Jun [1], etc. using histogram threshold for image segmentation, Abraham Duarte [2] proposed region merging algorithm for image segmentation, the local spectral histogram image segmentation methods was employed in Liu X's image

Q. Zhang (✉) · F. Feng · L. Xin · L. Wang
Pattern Recognition and Intelligent Systems Laboratory of Guizhou, Guiyang, Guizhou, China
e-mail: zhqwl@126.com

Q. Zhang · L. Wang
Academic Affairs Office of Guizhou University for Nationalities, Guiyang, Guizhou, China

W. Du (ed.), *Informatics and Management Science V*, Lecture Notes in Electrical Engineering 208, DOI: 10.1007/978-1-4471-4796-1_91,

and Texture Segmentation Using Local Spectral Histograms [3]. In addition, there is image segmentation based on evolutionary algorithm, the image segmentation based on the multi-scale Markov random field, the image segmentation based on the edge. The image segmentation based on optimization function, the image segmentation based on watershed, the image segmentation based on cluster analysis, the image segmentation based on fuzzy set theory, the image segmentation based on wavelet transform, the image segmentation based on neural networks and so on. The image segmentation based on graph understands and analysis image from the global characteristics of image, and it is more accord with machine vision systems, so it was developed rapidly recent years.

Image segmentation base on graph was proposed first by Leahy in 1993, An Optimal Graph Theoretic Approach to Data Clustering: Theory and its Application to Image Segmentation [4] subsequently Shi and Malk established a graph-based standard segmentation model [5] then the standard model has been improved respectively: minimum segmentation model, nest segmentation model, optimal segmentation model and so on. Pedro's Efficient Image Segmentation based on graph [6] in 2004 discussed an efficient image segmentation method based on graph; the method is calculated easily, and achieved non-local nature of the graphics. But this method is not the best for its measure function defect, the weighting function can be improved and also assessment functions. Literature [7] improved it, but the effect is not satisfactory. We improved the measure the function, and proposed a new image segmentation algorithm based on graph. The method maintains the advantages of the original algorithm, and improves segmentation effect, accuracy and speed.

## 91.2 Graph Theory and Image Segmentation

Image segmentation is to divide an image into several areas according to some algorithm and the nature of a certain area, the concerned part is extracted from the image in the end, followed by further analysis and processing. Image segmentation also have other names, such as target profile technology, target detection, threshold techniques, target tracking technology, all these technologies is image segmentation techniques in fact.

Image segmentation based on graph theory is graph partitioning problem, an image can be represented as an weighted and undirected graph $G = <V,E>$, vertices: $vi \in v$, V is the set of elements, $(vi, vj) \in E$ is defined as a pair of vertices of the edge, while $w(vi, vj)$ is defined the corresponding weight, that the differences degree between $vi$ and $vj$, it is an non-negative measure. In image segmentation, $vi$ is the pixel in V, and $w(vi, vj)$ is defined as the difference degree between two pixels. V is divided into separate regions $ci$; therefore, $ci \in v$, $\sum_{i=1} ci = v, ci \cap cj = 0$ $ci$ is a connected region. The image segmentation principle: elements in the same area as similar as possible, elements in different areas as distinct as possible in different regions. That is, weight of same area is little; weight of different areas is large.

## 91.3 Image Segmentation

It is assumed that the color image is processed into the gray image.

### 91.3.1 Definitions

**Definition** 1 Edge weighting function: the absolute value of the gray-scale differences of pixel $p_i$ and $p_j$, then computer it and the distance of pixel $p_i$ and $p_j$.

$$w(p_i, p_j) = \mu(p_i, p_j)|I(p_i) - I(p_j)| + d(p_i, p_j) \tag{91.1}$$

where: $I(p_i)$ is defined as the pixel gray level, and $d(p_i, p_j)$ is defined as Euclidean distance between $p_i$ and $p_j$.

$$d(p_i, p_j) = \sqrt{(x_i - x_j)^2 + (y_i - y_j)^2} \tag{91.2}$$

where: $p_i$ is represented $(x_i, x_j)$, so as the $p_j$, when the $x$ and $y$ direction are not the same size, we calculate the feature vectors of the high one, and employ principal component analysis to make it with the same size as low one. That the $x$ is $\xi_1, \xi_2, \ldots, \xi_r$, while the $y$ is $\xi_1, \xi_2, \ldots, \xi_m$, we have to calculate the dimension $n$, where $n = \min(r, m)$.

$U(p_i, p_j)$ is weighting coefficients, and it is an adaptive two-dimensional Gaussian factor:

$$u(p_i, p_j) = \frac{1}{\sigma_i \sigma_j \sqrt{1 - r^2}\sqrt{2\pi}} \exp\left\{ -\frac{1}{2\left(1 - r^2\right)} \left[ \frac{(i - u_i)^2}{\sigma_i^2} - \frac{2r(i - u_i)(j - u_j)}{\sigma_i \sigma j} + \frac{(j - u_j)^2}{\sigma_j^2} \right] \right\} \tag{91.3}$$

where, $u_i$ and $u_j$ correspond to the expectations of each direction pixel gray-level; while $\sigma_i$ and $\sigma_j$ correspond to the standard deviation of each direction pixel gray scale;

$$r = \frac{\operatorname{cov}(i, j)}{\sqrt{x^2 + y^2}} \tag{91.4}$$

Adjustability was presented to references 6 and adaptive to references 6. In the references 7, $U$ is an empirical constant, so it does not have adaptability; consequently we used the adaptive function of $p_i$ and $p_j$ here.

**Definition** 2 Structural similarity index (SSIM): for two divided regions of similarity are defined as the SSIM, the difference (DEX) can be defined as a 1-SSIM. If there is no such edge, we defined it as $\infty$.

$$\mathrm{SSIM} = \frac{(2u_i u_j + c_1)(2\sigma_i \sigma_j + c_2)}{(u_i^2 + u_j^2 + c_1)(\sigma_i^2 + \sigma_j^2 + c_2)} \tag{91.5}$$

$$\mathrm{Dex}(c_1, c_2) = \text{1-SSIM} \tag{91.6}$$

where $u_i$ and $u_j$ correspond to the expectations weights of region $c_1$ and $c_2$; While $\sigma_i$ and $\sigma_j$ correspond to weights standard deviation of region $c_1$ and $c_2$ region.

**Definition** 3 $\mathrm{C} \subseteq \mathrm{v}$ an internal difference defines as minimum weight spanning tree mean.

$$\mathrm{Din(C)} = \frac{1}{N} \sum_{e \in MST(C,E)} w(e) \tag{91.7}$$

where, $MST(C, E)$ is defined as the shortest path set of edges [8], we invited the Dijkstra algorithm here, $N$ is the number of edges.

**Definition** 4 Detection function: used to determine the two regions are merged or split.

$$\mathrm{Edge}(c_1, c_2) = \begin{cases} \mathrm{T} & \mathrm{Dex}(c_1, c_2) > \min[\mathrm{Din}(c_1), ]Din(c_2)] + \pi(c_1,c_2) \\ \mathrm{F} & \text{otherwise} \end{cases} \tag{91.8}$$

where, $\pi(c_1, c_2)$ is a difference great extent systems of control area, which must be less than the smallest between the inner differ.

$$\pi(c_1,c_2) = \left\{ \begin{array}{ll} \eta(c_1) & \mathrm{Din}(c_1) < \mathrm{Din}(c_2) \\ \eta(c_2) & \text{otherwise} \end{array} \right\} \tag{91.9}$$

$$\eta(c) = kw_{even} / |c| \tag{91.10}$$

where: $|c|$ is the number of elements in the region $c$, $w_{even}$ is the average weight in $c.k$ is the observation scale, the larger $k$ value can be beneficial to split a large area [7].

$$\mathrm{w}_{even} = \frac{1}{|c|} \sum_{e \in Ec} w(e) \tag{91.11}$$

### 91.3.2 Segmentation Algorithm

Image segmentation based on intra-regional differences as small as possible, the differences between different regions as large as possible, we determined two regions are split or merged by the Algorithm determine function.

Description of the algorithm:

1. Any pixel $p_i$ corresponds to a vertex $vi, \in V$. We can get set $E$ of 8-adjacent by the definition of graph theory and thus given an image of the graph that $G = (V, E)$, we assume that graph G with n vertices and m edges;
2. Descend weight edges, constitute a new sequence $O(o_1, o_2, o_3, \ldots o_m)$;
3. Take a partition $Seg^0$ to start, each vertex of $Seg^0$ form a new region;
4. Make $q = 1, 2, 3, \ldots, m$. And recycle to the next step work;
5. Assume that $v_i$, $v_j$ is the orderly sequence of $O$ in the first $q$ vertices connected to each other, expressed as $oq = (v_i, v_j)$ $c_i^{q-1}$ is one of the area $seg^{q-1}$ which include $v_i$, $c_j^{q-1}$ is one of the area $seg^{q-1}$ which include $v_j$. If $c_i^{q-1} \neq c_j^{q-1}$ and $Edge(c_i, c_j) = F$, that is $w(oq) < \min[Din(c_i^{q-1}) + \eta(c_i^{q-1}), Din(c_j^{q-1}) + \eta(c_j^{q-1})]$, $c_i^{q-1}$ and $c_j^{q-1}$ of $seg^{q-1}$ were merged into $seg^q$, otherwise $seg^q = eg^{q-1}$.
6. Return $seg = seg^m$;
7. According to $seg = (c_1, c_2, \ldots c_r)$, we can employee segmentation to the original image, and achieve results at end.

## 91.4 Test Data and Results

The Test is conducted in the following circumstances.

Lenovo ordinary PC: (CPU Intel E3300 (dual core), memory 2G) 200 images were processed by the other methods and ours so that we can compare their effect (Fig. 91.1).

The effect of our study is better than that in references 7 and 6, The reasons:

- The first one: the edge weight function coefficients $\mu(pi, pj)$ was adapted, the weight function in reference 7 as a constant, neglect the system's adaptability and optimum adjustment; so we employed adaptive function $\mu(pi, pj)$ to improvement of effect;

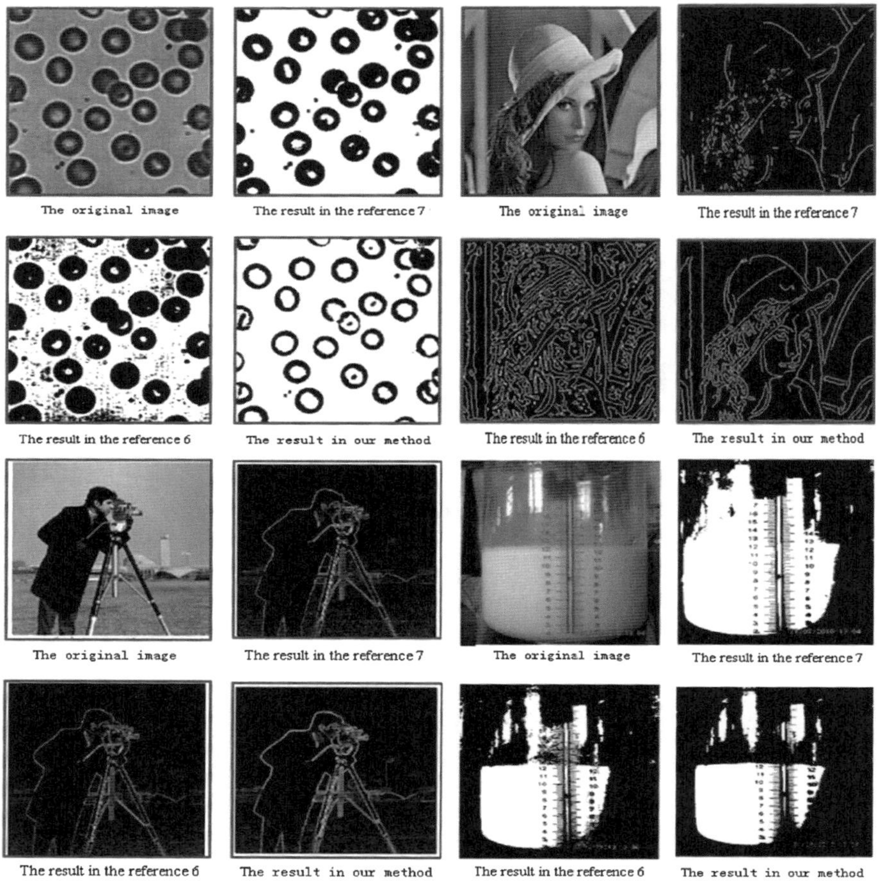


**Fig. 91.1** Results from different image processes

- Minimum spanning tree algorithm was used in the references 6 and 7 for the definition of internal gap, but the shortest path map was in the test, so that it is more in line with actual needs;
- For the threshold, similar function was used in this paper which is similar to similarity function, taking the considering the differences different categories into account.

The time complexity of these algorithms (Table 91.1).

**Table 91.1** The average split time

| Reference | Reference 6 | Reference 7 | Our paper |
|---|---|---|---|
| Average split time for 200 images | 2.3 s | 2.0 s | 1.9 s |

## 91.5 Conclusion

We presented a new image segmentation algorithm which based on graph and the other algorithms, we take the adaptive function as the factor and the internal gap the minimum spanning tree was employed, the method for threshold was improved than the other references. As a result, the segmentation effect is better than the other, time complexity is also greatly improved.

## References


1. Liu J, Wu J (2010) An improved image segmentation based on gradient-threshold histogram method. Comput Digital Eng 38(4):131–133
2. Duarte A, Sanchez A, Fernandez F, Antonio S (2006) Monte mayor Improving image segmentation quality through effective region merging using a hierarchical social met heuristic. Pattern Recogn Lett 55(26):1239–1251
3. Liu X, Wang D (2006) Image and texture segmentation using local spectral histograms. IEEE Trans Image Process 15(10):3066–3077
4. Leahy R (1993) An optimal graph theoretic approach to data clustering: theory and its application to image segmentation. IEEE Trans Pattern Anal Mach Intell 15(11):1101–1113
5. Shi J, Malk J (2000) Normalized cuts and image segmentation. IEEE Trans Pattern Anal Mach Intell 22(8):888–905
6. Felzenszwalb PF, Huttenlocher DP (2004) Efficient graph-based image segmentation. Int J Comput Vision 59(2):167–181
7. Zhang T (2011) An improved image segmentation method based on graph. J West China Univ Nat Sci 30(1):61–64
8. Liu DB, Hou MS, Qu ZX, Wu H (2011) An efficient dynamically updated algorithm for shortest path tree. Comput Sci 38(7):96–99

# Chapter 92
# An Approach for Image Retrieval Based on Support Vector Machines

**Guoyong Wang, Wen Cui and Chen Sun**

**Abstract** Various approach including artificial neural networks have been used to classify a large image database efficiently and shown to be highly successful in this application area. This paper presents a new, scaling and rotation invariant encoding scheme for shapes. Support vector machines (SVMs) are used for the classifications of shapes encoded by the new method. This paper examines the performance of the proposed method by comparing it with that of multilayer perception, one of the artificial neural network (ANNs) techniques, based on real real-world image data. The experiment shows that the results of one-class SVMs outperform those of ANNs.

**Keywords** Image retrieval • Support vector machines • Artificial neural network

## 92.1 Introduction

With the popular of the Internet, people can acquire various types of information any time from anywhere. In spite of a search engine is quite competent in locating textual data, its development is still in the infancy stage in finding image-based

G. Wang (✉) · W. Cui
Computer Information Engineering Department, LuoYang Institute of Science and Technology, Luoyang, Henan, China
e-mail: wgy@lit.edu.cn

W. Cui
e-mail: cuiwen@sina.com

C. Sun
School of Computer Science and Technology, Huazhong University of Science and Technology, Wuhan, China
e-mail: katyusha2008@126.com

W. Du (ed.), *Informatics and Management Science V*, Lecture Notes in Electrical Engineering 208, DOI: 10.1007/978-1-4471-4796-1_92,

data. Consequently, future search engines must be able to accept entering a desired picture (e.g., a shoe image), and return similar pictures in that category when asked [1]. This new demand has stirred increasing interest in shape-based image retrieval.

Recently the support vector machine (SVM) has been introduced as a new technique for solving a variety of learning, classification and prediction problems. We therefore employ the SVM to solve classification problems based on previous sample [2]. Experimental testing has shown that the SVM performance is better than that of the ANN in classification. Accordingly, the second objective of this study is to explore the feasibility of applying SVM and ANN for image classification. In essence, these two data mining techniques are composed of nothing more than function approximation and optimization algorithms [3]. Moreover, we examine the robustness of SVM against different values of parameter by altering two important parameters to understand the effects on the image classification quality.

## 92.2 Related Works

A useful technique for constructing nonlinear features is kernel methods, where each feature is a function of the current input and one of the example input (such as their distance). This idea has recently received much attention because of the introduction of support vector machines (SVMs) and the renewed interest in Gaussian processes. SVMs, introduced by Vapnik and his collaborators, were originally formulated for binary-classification problems. The resulting method is similar to penalized maximum-likelihood estimators using logistic models (also known as conditional maximum entropy classification models). Later, SVMs were extended to regression problems, and the resulting formulations are similar to ridge regression.

ANN is a biologically inspired form of distributed computation. It is composed of simple processing units, or nodes, and connections between them. The connection between any two units has some weight, which is used to determine how much one unit will affect the other. A subset of the units acts as the input nodes and another subset acts as the output nodes, which perform summation and threshold. The ANN has been successfully applied in different settings, including network reliability, sports winning prediction, medical, marketing, retail, banking and finance [4].

SVM is originated as an implementation structural risk minimization (SRM) principle. The basic idea in SVM is to transform the data into a higher dimensional space and find the optimal hyper plane in the space that maximizes the margin between classes. The simplest SVM deals with a two-class problem—in which the data is separated by a hyper plane defined by a number of support vectors [5].

## 92.3 Methodology

SVMs were originally designed for binary classifications. Many real-world problems, however, have more than two classes. How to effectively extend it for multi-class classification is still an on-going research issue [6]. Most researchers view multi-class SVMs as an extension of the binary SVM classification problem. However, this method does not have an established bound on the generalization error. Two commonly used approaches are one-against-all and one-against-one methods. The one-against-all method separates each class from all others and constructs a combined classifier while the one-against-one method separates all classes' pair wise and constructs a combined classifier using voting schemes.

The one-against-all method is probably the earliest used implementation for SVM multi-class classification. It constructs $k$ SVM models where $k$ is the number of classes. The $p$th SVM is trained with all of examples in the $p$th class with positive labels, and all other examples with negative labels. The decision function chooses the class of a sample that corresponds to the maximum value of $k$ binary decision functions specified by the furthest positive hyper plane. This approach is computationally expensive because we need to solve $k$ quadratic programming optimization problems with sample size $l$.

To detect corner points, we employ the curvature variance. A chain code is similar to the first derivative of a curve at a point. A curvature is the second derivative of a curve at a point. Thus, a curvature is equal to the difference between the first derivative of the leaving and entering portion of the chain codes. To effectively identify corner points, we define the total curvature variance $TC_i$ at a point Pi with its chain code $C_i$ as:

$$TC_i = \sum_{j=-1}^{j=1} Mod(C_{i+j} - C_{i+j+1}) \tag{92.1}$$

The function $Mod(C_{i+j} - C_{i+j+1})$ is defined as $\Delta C_{nm} = C_n - C_m$

If $ABS(\Delta C_{nm}) > 4$ then If $\Delta C_{nm} > 4$ then $Mod(C_n - C_m) = \Delta C_{nm} - 8$ else $Mod(C_n - C_m) = \Delta C_{nm} + 8$ else $Mod(C_n - C_m) = \Delta C_{nm}$

After we calculate $TC_i$ for every point on a curve, we compare it with the following criteria from:

1. The total curvature variance $TC_i > 1$ (lie beyond 45°),
2. $ABS(TC_i) = 1$ (lie in 45) and $(C_i \neq -C_{i-1})$ and $((C_{i-1} = C_{i-2} and C_{i-2} = C_{i-3}) or (C_i = C_{i+1} and C_{i+1} = C_{i+2}))$,
3. $(TC_i) = 0$ (look like a straight line) and $(C_i \neq -C_{i-1})$ and $((C_{i-1} = C_{i-2} = C_{i-3}\ and C_i = C_{i+2}) or (C_i = C_{i+1} = C_{i+2} and C_{i-1} = C_{i-3}))$,
4. $(TC_i) = 0$ (look like a straight line) and $(C_{i-1} \neq -C_{i+1})$ and $((C_{i-1} = C_{i-2} = C_{i-3} = C_{i-4} and C_{i+1} = C_{i+2}) or (C_{i+1} = C_{i+2} = C_{i+3} = C_{i+4} and C_{i-1} = C_{i-2}))$.

A point $P_i$ that satisfies one of the criteria is considered as a corner point.

We can obtain similarity measures for SVM and ANN. The first similarity measure is the total normalized area (TNA)—areas covered under the spectrum. The second measure is the perimeter of the spectrum. This is also the total normalized moment variance (TNMV). The computations are shown as follows: where $d$ denotes the number of dominant points, $M$ denotes the moment at point $i$, $\Delta M$ denotes difference of the moment between two consecutive dominant points, $NP$ denotes the normalized perimeter length, and $\Delta D$ denotes the normalized perimeter difference between two dominant points.

$$TNA = \sum_{i=0}^{i=d} \frac{(M_i + M_{i+1}) * \Delta D}{2}, \quad \Delta D = NP_{i+1} - NP_i \tag{92.2}$$

$$TNMV = \sum_{i=0}^{i=d} \sqrt{(\Delta M)^2 + (\Delta D)^2}, \quad \Delta M = M_{i+1} - M_i \tag{92.3}$$

There are two ways to encode the multi-class classification: (1) single output unit, assign outputs of, say, 0.2, 0.4, 0.6 and 0.8 to encode these four possible values for four categories and (2) several distinct output units, each representing one of the possible categories. The output unit corresponding to maximum of these probabilities outputs 1, while those remaining output 0. This is often called a 1-of-n output encoding. The advantage of single-output encoding scheme is that the class encoded by 0.2 is neighbored to the class encoded by 0.4, and 0.2 and 0.8 are "less" neighbored. However, if there is no a priori knowledge about such a neighborhood relation, it should not be introduced by the classifier. On the other hand, there are two motivations for choosing the 1-of-n output encoding over the single unit option. First, it provides more degrees of freedom to the network for representing the target function. Second, in the 1-of-n encoding the difference between the highest-valued output and the second highest can be used as a measure of the confidence in the network prediction [7].

## 92.4 Results and Discussion

### *92.4.1 Dataset*

The data used here was originated from Sebastian's study [8]. Sebastian's study contains 40 categories of shapes. We choose the first 353 image in 16 categories. Our test procedure is outlined in Fig. 92.1. First, an image is converted through chain code extraction, corner point detection, dominant point detection, spectrum of an image and 14 similarity measures. Second, for each category, we reserve 25 % of the data for testing and the rest of data are for training and validation. By feeding training and validation set to SVM and ANN, we can obtain the learning

**Fig. 92.1** The experimental procedure of SVM and ANN

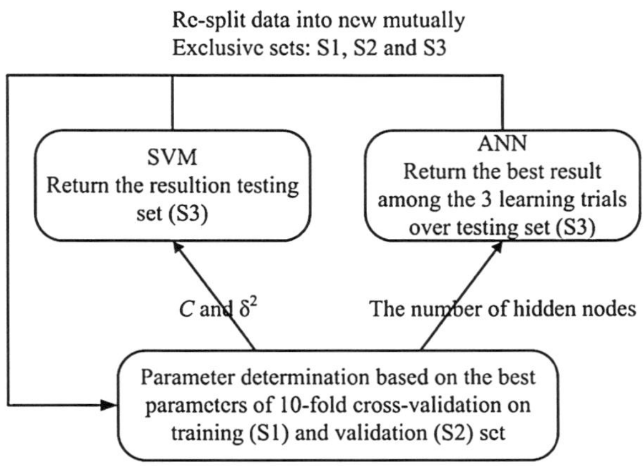


models of SVM and ANN. When the training process is finished, unseen data instances are fed to SVM and ANN.

A double-loop cross-validation (CV) procedure is used to estimate the performance of SVM and ANN. We determine the best parameter values for both SVM and ANN by using 10-fold CV technique on training and validation sets in the inner-loop procedure. Since the final goal is to obtain the performance on new data, we obtained the experimental results by using 5-fold CV technique on testing set in the outer-loop procedure. Because the results of ANN are sensitive to the initial weight configuration, we perform three learning trails with different initial weight configurations to obtain the best result (see Fig. 92.2).

## 92.4.2 SVM Results

By varying the number of categories from 8 to 40, we compared the predictive performance of SVM and ANN. To further test the effectiveness of our coding

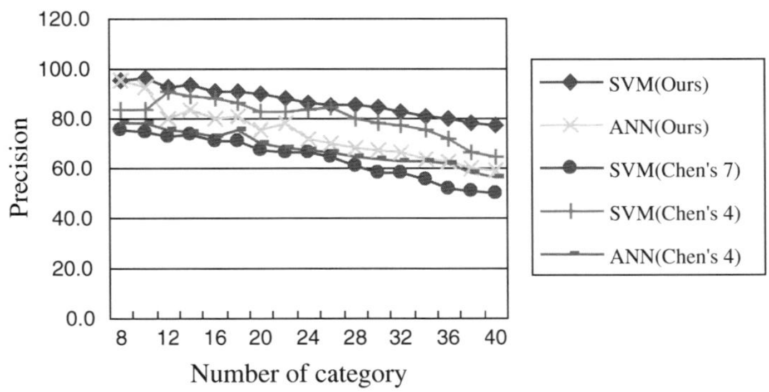


**Fig. 92.2** SVM and ANN performance

scheme, we compared the predictive performance of our method with Chen's method [4]. Basically, Chen's curve moment invariants are quite similar to Hu's [8] area moment invariants. Instead of the whole shape area, Chen's seven moment invariants only require the computation along the shape boundary. The experiments showed that the last three moment invariants were not stable and would cause the over-fitting problem. Accordingly, we perform the testing by using only the first four moment invariants.

The typical kernel functions are the polynomial kernel $k(x,y) = (x \times y + 1)^d$ and the Gaussian kernel $k(x,y) = \exp\left(-(x-y)^2 \middle/ \delta^2\right)$, where $d$ is the degree of the polynomial kernel and $\delta^2$ is the bandwidth of the Gaussian kernel. In our experiment, we chose the Gaussian kernel as our kernel function because it tends to achieve better performance. The parameters that must be determined are the kernel bandwidth $\delta^2$ and the margin $C$. In determining these two parameters, 10-fold CV technique was used to choose parameters that yield the best result. Subsequently, this set of parameters was applied to the test dataset. The parameters tried in the 10-fold CV process were $\delta^2 \in \{2, 1, 0.5, 0.1, 0.01, 0.001, 0.0001\}$ and $C \in \{1000, 750, 500, 100, 50, 2\}$. A SVM implementation called LIBSVM was used in this work. We used the LIBSVM because it uses state-of-the-art optimization method SMO for the solution of multi-class SVM problem. The LIBSVM uses the one-against-one method because its training time is shorter than the one-against-all method and its performance is also comparable to one-against-all method [9]. The SVM and ANN results were shown in Table 92.1.

### 92.4.3 Results of Image Classification

Figure 92.2 showed that SVM yielded higher predication accuracy than ANN because its prediction rates are always higher than that for the ANN model, irrespective of the coding schemes. Moreover, we also observed that our coding method is more effective than Chen's method because the corresponding algorithm can achieve higher performance.

To further test the statistical difference, we conducted a series of rank sum paired test between SVM and ANN, and our and Chen's coding method. The results are shown in Table 92.1 reach the same conclusion that SVM statistically

**Table 92.1** Wilcoxon rank sum paired test between our and Chen's first four moment invariants

| | Ours–chen's | |
|---|---|---|
| | SVM | ANN |
| Z-statistic | 3.62 | 3.62 |
| $p$-value | <0.001 | <0.001 |

**Table 92.2** Sensitivity of SVMs to parameters

| C \ δ | 5 | | 2 | | 1 | | 0.5 | | 0.1 | |
|---|---|---|---|---|---|---|---|---|---|---|
| | Train | Test | Train | Test | Train | Test | Train | Test | Train | Test |
| 1 | 70.46 | 78.64 | 67.56 | 75.61 | 63.61 | 69.32 | 55.01 | 60.91 | 60.12 | 63.29 |
| 10 | 74.02 | 81.06 | 75.09 | 81.91 | 76.32 | 83.51 | 74.16 | 81.63 | 65.12 | 68.78 |
| 50 | 74.31 | 80.12 | 75.41 | 81.11 | 75.83 | 82.57 | 77.02 | 83.56 | 74.32 | 80.09 |
| 100 | 74.21 | 80.01 | 74.92 | 81.03 | 74.42 | 81.06 | 76.98 | 82.61 | 74.42 | 81.58 |
| 1,000 | 73.35 | 80.01 | 74.02 | 80.02 | 74.81 | 80.12 | 76.72 | 81.09 | 76.92 | 82.05 |
| Avg | 73.27 | 79.97 | 73.41 | 79.94 | 73.01 | 79.32 | 72.01 | 77.96 | 70.18 | 75.16 |

performed better than ANN for the image retrieval (both $p<0.001$) and our coding scheme is more effective than Chen's method (both $p<0.001$).

### 92.4.4 Sensitivity of SVMs to Parameters

In this section, we would like to understand the effects of changing two important parameters (kernel parameters $\delta^2$ and margin $C$) on the quality of image retrieval for 40 categories. We ran 25 experiments for this purpose (see Table 92.2). We have changed $\delta^2$ from 5 to 2, 1, 0.5 and 0.1, while changing $C$ from 1 to 10, 50, 100 and 1,000, respectively. The quality of the image retrieval is measured by the prediction accuracy on both training and testing dataset.

We observed that when $\delta^2$ is at 0.1 and $C$ is at 1, the SVM's results achieved the worst performance on both training set (29.39 %) and testing set (33.16 %). The reason lies in that a small value for $C$ will under-fit the training set. While most of parameter combinations yield quite stable results. This suggests that SVM is quite robust against parameter selections, especially for the kernel parameter.

## 92.5 Conclusions

The image retrieval issue has become more important, inducing researchers to continue developing effective measures to retrieve images. In particular, they must consider the image rotation and scaling conditions because such conditions are quite common. To address this issue, we proposed a set of similarity measures that are invariant with respect to image rotation and scaling. Our preliminary experiments with the Sebastian's data showed that these measures are invariant and also quite effective in that they help SVM to achieve high classification rates. Moreover, our coding method is comparable to previous coding scheme in terms of SVM and ANN performance. Generally [10], the SVM classifier yielded higher predication accuracy than ANN, irrespective of the coding schemes. However, in

ANN the existence of several local minima constitutes a drawback. This does not mean that SVM can achieve results that are over-fitting free. The determination of these parameters is not an easy task and improper parameter selection might cause the over-fitting problem. Because the choice of these parameters is problem-dependent, it is difficult to have an ANN to produce optimal solutions at all times. We also found that the ANN trained with gradient-based optimization depends heavily on the initial weights [11]. Moreover, as we increased the number of hidden nodes in the network structure, we found that the computation time increases exponentially because the complexity of the network structure increases with it.

**Acknowledgments** This research was supported by the Natural Science Foundation of Luoyang Institute of Science and Technology (Grant No. 2008QZ28).

## References

1. Burges CJC (1998) A tutorial on support vector machines for pattern recognition. Data Min Knowl Disc 2:955–974
2. Cai YD, Lin XJ, Xu XB, Chou KC (2002) Prediction of protein structural classes by support vector machines. Comput Chem 26:293–296
3. Chen WH, Hsu SH, Shen HP (2005) Application of SVM and ANN for intrusion detection. Comput Oper Res 32:2617–2634
4. Cristianini N, Shawe-Taylor J (2000) An introduction to support vector machines and other kernel-based learning methods, vol 33. Cambridge University Press, Cambridge, pp 145–188
5. Freeman H (1974) Computer processing of line drawing images. ACM Comput Surv 6(1):57–97
6. Latecki LJ, Gross A, Melter R (2002) Shape representation and similarity for image databases. Frontiers Artif Intell Appl 104(35):1–2
7. Inesta JM, Buendi M, Sarti MA (1998) Reliable polygonal approximations for imaged real objects through dominant point detection. Pattern Recogn 31:685–697
8. Santini S, Jain R (1999) Similarity measures. IEEE Trans Pattern Anal Mach Intell 21(9):871–882
9. Nishida H (2002) Structural feature indexing for retrieval of partially visible shapes. Pattern Recogn 35:55–67
10. Zhu RB, Wang JQ (2011) Power-efficient spatial reusable channel assignment scheme in WLAN mesh networks. Mob Netw Appl 78:233–256
11. Zhu RB, Wang JQ, Ma MD (2008) Intelligent MAC model for traffic scheduling in IEEE 802.11e wireless LANs. Appl Math Comput 205(1):109–122 (Elsevier press)

# Chapter 93
# High Accuracy Handwritten Chinese Character Recognition Based on Support Vector Machine and Independent Component Analysis

**Zhiguo He, Yuquan Zhong and Yudong Cao**

**Abstract** This paper proposed a new method for handwritten Chinese character recognition based on a combination of independent component analysis (ICA) and support vector machine (SVM). First, we extracted independent basis images of handwritten Chinese character image and the projection vector by using fast ICA algorithm, and obtained the feature vector. Then, we used two stage classification methods based on SVM for classification. The scheme took full advantage of good extraction local features capability of ICA and strong classification ability of SVM, thus increasing the system's recognition rate. The experiments show that the feature extraction method based on ICA is superior to that of gradient-based, and the two stage classifiers based on SVM is better than that of modified quadratic discriminant function. On HCL2000, a handwritten Chinese character database, the recognition accuracy of 99.87 % has been achieved.

**Keywords** Handwritten chinese character recognition · Independent component analysis · Support vector machine · Feature extraction

Z. He (✉) · Y. Zhong · Y. Cao
School of Computer Science, Panzhihua University, Panzhihua, China
e-mail: zhiguohe@126.com

Y. Zhong
e-mail: zhongyuquan@126.com

Y. Cao
e-mail: caoyudong@126.com

W. Du (ed.), *Informatics and Management Science V*, Lecture Notes in Electrical Engineering 208, DOI: 10.1007/978-1-4471-4796-1_93,

## 93.1 Introduction

Handwritten Chinese character recognition (HCCR) is an important research topic in pattern recognition, which is widely used in the automatic input of Chinese electronic data processing, in the Chinese text compression, in office automation and computer-aided teaching, etc. It can bring about huge economic and social benefits, but it is also one of the more difficult issues in the field of pattern recognition. Because Chinese characters set [1–3] is very large; the structure of Chinese characters is very complex; Many Chinese characters have high degree of similarity; the writing styles for the same character are many kinds and have large shape variations. These four reasons make HCCR very difficulty. At present, HCCR system has not yet reached satisfactory results, especially for Chinese characters with cursive script. HCCR is divided into four steps: pre-processing, feature extraction, classification and post-processing, among which feature extraction method and classifier is an import factor for recognition performance. To a large extent, the accuracy of an overall recognition system depends on the discriminative capability of features and generalization performance of a designed classifier. Determining how to extract stable and good separable feature for Chinese character is an important research direction. The bottle-neck of feature extraction is the instability of the feature between the different samples of the same Chinese character, so the key for HCCR is to accurately describe the details of the differences for the same Chinese character caused by different writing styles. At present, in HCCR, the method for feature extraction can be mainly divided into two categories: one is based on structural features, which is rarely used because they are difficult to extract and very sensitive to noise; the other is based on statistical features which is widely used in HCCR. Widely used statistical feature extraction methods include gradient features [4] and features based on independent component analysis (ICA) [5]. However, the gradient feature exist deficiencies such as large amount of calculation and high dimension features. Although principal component analysis (PCA) was used to dimensionality reduction, but it is a method based on second-order statistical characteristics and its purpose is to remove the correlation between the components of the image. A large number of studies have shown that the most important information of image is existed in the high-order statistics of image pixels, but the dimensionality reduction method based on PCA did not use the high-order statistical characteristics of images [6]. While ICA is an analysis method based on higher-order statistical characteristics of signal, it is more fully taken into account the statistical independence of the probability density function of the signal. Principal component obtained by PCA is only de-correlation (orthogonal to each other), while ICA not only achieves de-correlation, but also the higher-order statistics obtained are mutually independent. In PCA, the signal to be processed is generally assumed the Gaussian distribution; while in ICA, the signal is assumed non-Gaussian signal which is more in line with the realistic problems. The goal of ICA is separating out the independent component by using linear transform, to remove or minimize the degree of statistical dependence in the image.

At the same time, in HCCR system, the currently widely used classifiers are support vector machine (SVM) classifier [7] and modified quadratic discriminant function (MQDF) [8]. SVM is a new machine learning methods developed in recent years, which uses the principles of structural risk minimization in statistical learning theory, showed strong superiority in dealing with the classification of high-dimensional space and showed strong nonlinear classification ability. While MQDF requires the distribution for each class to be a Gaussian distribution, which often does not match with the reality. Thus, in this paper, SVM is used for classification for Chinese characters. Based on the above analysis, this paper proposes a new method for HCCR. First, we use ICA method for feature extraction for Chinese character image, and then use SVM for classification. Our scheme has reached a higher recognition rate compared to gradient-based feature extraction method and MQDF classifier.

## 93.2 ICA and Feature Extraction for Handwritten Chinese Character

### *93.2.1 ICA*

The ICA model can be described as:

$$X = AS \tag{93.1}$$

The model describes how the observed data X is obtained by mixing the source S. The source variable S is a hidden variable which can not be directly observed, and the mixing matrix A is also unknown. All the data can be observed is only the random variable X, so it is necessary to estimate the mixing matrix A and the source S. ICA is based on a simple assumption: the source variable S is statistically independent and non-Gaussian distribution. In this basic model, the distribution is unknown.

Extraction image features by ICA is to find a separation matrix W by using linear transform of the observed image, to make the component decomposed by the linear transform as mutually independent as possible and approximation of S. Y is an estimation of S, that is, Y is the extracted feature vector of the image:

$$Y = WX \tag{93.2}$$

where

$$W = A^{-1} \tag{93.3}$$

Through the establishment of the linear model of an image, we can be applied ICA technology to separation the independent component of the observed image, to extract image features, making the separated independent component Y as statistically mutually independent as possible. Y is the estimated mutually independent coefficient, and A is the basis image obtained.

Solving the separation matrix W, we used the FastICA algorithm [9]. The algorithm has the advantage of fast convergence, not requiring a known probability distribution in advance and independent component can be solved one by one, etc., which can reduce the computational cost. FastICA algorithm is to find a direction, namely the unit vector W, through the system's learning to make the projection of $W^T$ X has the largest non-gaussianity. The method of solving the independent variable S is to find the vector W, to make W maximum the non-gaussianity of $W^T$ X. There are a lot of methods to measure the non-gaussianity. We used negentropy, that is:

$$J(W) = [E\{G(W^T X)] - E\{G(v)\}^2 \quad (93.4)$$

where

X is the observation vector,

W is the weight vector,

V is a Gaussian variable with zero mean and unit variance,

G is a nonlinear function.

We took the following functions:

$$G(x) = -\exp\left(-\frac{x^2}{2}\right) \quad (93.5)$$

## 93.2.2 *Feature Extraction for Handwritten Chinese Characters Images*

It is necessary to do some preprocessing before Chinese character samples is separated by ICA. Chinese characters are to first normalized, then to whiten and to zero mean value of the input signal. After the processing, the FastICA algorithm was used to solve the separation matrix W. After solving the separation matrix W, we can obtain the basis image Y by using Eq. (93.2). According to the definition of the ICA model, there are: an image X can be obtained by linear combination of the basis image Y, namely:

$$x = \sum_{i=1}^{n} a_i s_i \quad (93.6)$$

where $a_i$ is the feature vector of the Chinese character image $x$.

Similarly, for any unclassified Chinese characters image, you can use the same method by projection Chinese characters image to the space of basis image Y, then obtaining all projection coefficients and its feature vector.

## 93.3 Two Stage Classification Based on Support Vector Machine

SVM is a new kind of learning machine based on statistical learning theory. Statistical learning theory is a theory of studying statistical learning for small samples, its main idea is: for linearly inseparable data, first, the input space is mapped into a high-dimensional space by a nonlinear map, to make the data becomes linearly separable or nearly linearly separable in this space, then in this high-dimensional space to obtain the optimal linear classification surface. All operations in the feature space are carried out by the inner product kernel function of the input space. If an inner product kernel function is selected, then it defines a feature space.

SVM is a learning machine following the principle of structural risk minimization. It was first proposed for two-class classification problems, its goal is obtain a linear classification hyper-plane which not only makes two classes separated, but make the separation intervals maximum for the two classes. Assume that the training sample set is: $(x_i, y_i),\quad \mathrm{i} = 1, 2, \ldots, \mathrm{n}$, $\mathrm{x} \in \mathrm{R^d}$, $\mathrm{y} \in \{-1, +1\}$ is category symbol. By solving a condition optimization problem, we can get the optimal classification function:

$$\mathrm{f(x)} = \mathrm{sgn}\left\{\sum_{\mathrm{i}=1}^{\mathrm{n}} \mathrm{a_i y_i k(x_i, x)} + \mathrm{b}\right\} \tag{93.7}$$

where

$x_i$ is the support vector;
$x$ is the vector to be classified.

When using SVM for handwritten Chinese character recognition with a large number of categories, it is an important issue to reduce the computational cost and storage cost either in training or recognition stage. If we selected too many samples, the training was too time-consuming, and the number of support vectors is also increased significantly. Based on the above analysis, we adopted two stage classification strategy, the classifier based on Euclidean distance as a pre-classification, the training sample for training SVM using only the candidates obtained by the distance classifier, which made the computational cost and storage cost reduced greatly, and reduced the number of support vectors accordingly. To Train the support vector machine, we used the algorithm proposed by Dong [7].

## 93.4 Experiments and Results

In this study, we use HCL2000 database, an offline handwritten Chinese character standard database, which is funded by National 863 Program of China, created by Pattern Recognition Laboratory of Beijing University of Posts and Telecommunications. It has become the most influential database for handwritten Chinese character recognition, contains 3,755 frequently used simplified Chinese characters written by 1,000 different persons. It has the characteristics of large sample size, mutual inquiries between sample database of Chinese characters and information database of penman. All the sample of HCL2000 is normalized binary samples, with size 64 (height) by 64 (width). Part of the samples was shown in Fig. 93.1. We chose 700 samples located in xx001–xx500 of HCL2000 as training samples, chose 300 samples located in hh001–hh300 of HCL2000 as test samples. When recognizing Chinese characters, we first used the rough classification (Euclidean distance classifier), and then chose the first 35 candidates obtained by the rough classification as the fine classification (SVM classifier). The candidate characters after fine classification were the final result.

When extraction independent component by using FastICA algorithm, the number of independent components can not be choose too much nor too little. If we chose too many, it may contain a large quantity of noise signals; if too little, it may lose too much feature information of the image. The experimental results show that when the number of principal components is 89, the recognition rate has been reached the highest, shown in Fig. 93.2.

Feature extraction based on ICA was compared with the widely used gradient feature in HCCR. The method for extraction gradient feature please referred to literature [4] and its dimension is 256. In order to facilitate comparison, both

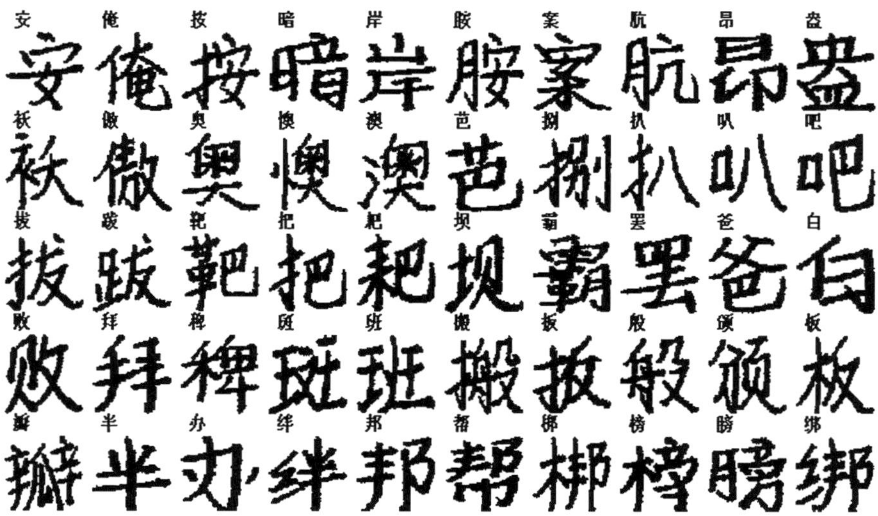

**Fig. 93.1** Some sample images of HCL2000

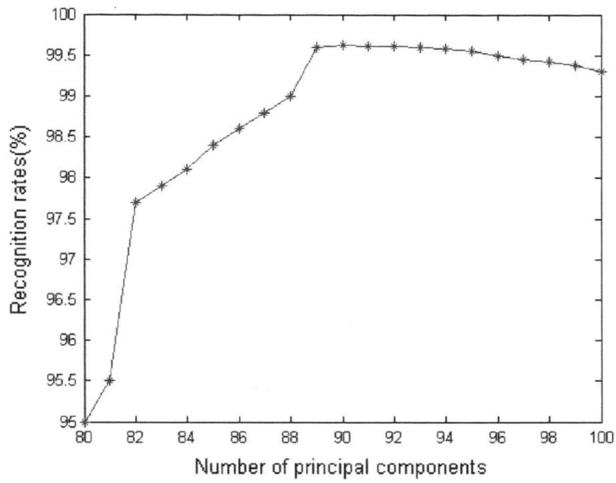


**Fig. 93.2** The relationship between the number of principal components and its recognition rate

methods used the same MQDF classifier and the results shown in Table 93.1. The experiment was performed on Intel Pentium 4 3.2G with MATLBA 7.0 system equipped with 768 megabytes RAM. From Table 93.1, we known that the recognition rate based on ICA method is higher than that of gradient feature, but the time for feature extraction is quite different. The time for extraction gradient features is about 5 s; while extraction feature based on ICA needs a long time and the time is concerned with the number of principal components and is approximately 2 min when the number of principal components extracted is 89. Recognition time is in a few seconds for each Chinese character with MQDF classifier.

In order to reduce the computational cost and speed up the classification, we use the Euclidean distance classifier as a pre-classifier, and training support vector machine used only the candidate character set obtained by the distance classifier as training samples, which can make the computational cost and storage cost greatly reduced, also make the number of support vectors reduced accordingly. The experiments show that when 35 candidates were chosen, the system has reached very high cumulative recognition rate. For SVM classifier, the one-against-others method was used to construct 35 classifiers. We used polynomial and radial basis function as kernel function, respectively. The parameter selection for kernel function is a rule of thumb. We found by experiments that its recognition rate is

**Table 93.1** Recognition performance of HCCR affected by different feature extraction method

| The method for feature extraction | The time for feature extraction (The number of principal component) | Recognition rate (%) |
|---|---|---|
| Feature based on ICA | 90 s (85) | 97.57 |
| | 127 s (89) | 99.45 |
| | 158 s (94) | 98.72 |
| Feature based on gradient | 5 s | 97.27 |

**Table 93.2** Recognition rate with different classifiers

| Classifiers | Recognition rate (%) |
|---|---|
| MDQF | 99.45 |
| SVM based on two stage classification | 99.87 |

very high by using RBF kernel function. When is 3, it reached the highest recognition rate. The RBF kernel function used is as follows:

$$K(x, x') = \exp\left(-\frac{||x - x'||^2}{2\sigma^2}\right) \tag{93.8}$$

The classification method used in this study was compared with the currently widely used MDQF classifier. When using the same method for feature extraction (based on ICA and the selected principal components is 89), the results was in Table 93.2. Table 93.2 shows that the two stage classifiers used in this study is significantly better than MDQF classifier.

The experimental results show that feature extraction by ICA is superior to the gradient feature. This is because the information between the Chinese character images has certain relevance, is not mutually independent, and this lead to a correlation between the features of different categories, which made the classification accuracy not high. But component obtained by ICA is mutually independent and removes the correlation between the features to a certain extent, thus it may improve the classification accuracy. Meanwhile, the two stage classifiers based on SVM is superior to the widely used MQDF classifiers, because SVM has strong classification capability.

## 93.5 Conclusion

ICA based on the higher-order statistical correlation between data, extracted internal features of the image, made full use of the statistical characteristic of the input data. ICA as an extension of PCA, it focuses on the higher-order statistical characteristics between data, each of the transformed components is not only unrelated, but also as statistically independent as possible. Therefore, ICA can be more fully reveal the essential features of the input data. But for Chinese character images, a lot of important information is contained in the high-order statistics between the pixels of the image. So the feature extracted by ICA for Chinese character is significantly better than the currently widely used gradient feature. At the same time, the two stage classifier, adopted in this paper, can not only reduce the computational cost for classification with large number of categories by using SVM classifier, but also improve the recognition rate of the entire system. Experiments show that our scheme is superior to the widely used MQDF classifier. But using ICA for feature extraction exist the following deficiencies: the computational cost is very high; iteration of the algorithm depends on the selection of the initial value; the time for feature extraction is very long. All these needs further study.

**Acknowledgments** The work was partially supported by Key Projects of Education Department of Sichuan Province under Grant No. 10ZA186.

## References

1. He ZG, Cao YD (2008) Survey of offline handwritten chinese character recognition. Comput Eng 34(15):201–204
2. Liu CL, Fujisawa H (2008) Classification and learning methods for character recognition: advances and remaining problems. Stud Comput Intell 90:139–161
3. Zhu CH, Shi CY, Wang JP et al. (2011) Study of offline handwritten chinese character recognition based on dynamic pruned FSVMs. In: International conference on electrical and control engineering, vol 123. pp 395–398
4. Liu CL (2007) Normalization-cooperated gradient feature extraction for handwritten character recognition. IEEE Trans Pattern Anal Mach Intell 29(8):1465–1469
5. Rui T, Shen C, Ding J et al (2005) Handwritten digit character recognition by model reconstruction based on independent component analysis. J Comput Aided Des Comput Graph 17(3):455–460
6. Bartlett MS (1998) Face image analysis by unsupervised learning and redundancy reduction, vol 12(31). Dissertation, University of California, San Diego, pp 93–99
7. Dong JX, Krzy_zak A, Suen CY (2005) An improved handwritten chinese character recognition system using support vector machine. Pattern Recogn Lett 26:1849–1856
8. Dai R, Liu CL, Xiao B (2007) Chinese character recognition: history, status and prospects. Frontiers Comput Sci China 1(2):126–136
9. Hyvärinen A, Oja E (2000) Independent component analysis: algorithms and applications. Neural Netw 13(4):411–430

# Chapter 94
# Image Compression Algorithm Based on Lifting Wavelet

**Yong Li, Dejian Kong and Hengji Du**

**Abstract** Wavelet transform has good characteristics of time and frequency, and it is very suitable for the analysis of mutant signal. So as a good method of image analysis, wavelet transform has been widely used in extraction of edge and compression of image. But large computation of wavelet transform, limits its application in high-speed and real-time signal processing field. Compared with traditional wavelet transform, lifting algorithm does not depend on the Fourier transform. Lifting Wavelet transform is very suitable for implementation of hardware and will reduce the computing complexity. This paper designs a method of image processing based on 5/3 lifting wavelet transform and realizes the simulation on FPGA. The lifting algorithm implemented in hardware has lower computational complexity and save the memory space. The experimental results show that 5/3 lifting wavelet transform implemented on FPGA can be effective on digital image compression.

**Keywords** Lifting wavelet transform · FPGA · Image processing · System generator

## 94.1 Introduction

Wavelet transform is a new mathematical tool after the emergence of Fourier transform; it has been widely studied and used in signal processing field. By changing the shape of time window and frequency window, wavelet transform

Y. Li (✉) · D. Kong · H. Du
College of Computer Science and Engineering, Qujing Normal University, Qujing 655011, China
e-mail: qjsfxyly@163.com

W. Du (ed.), *Informatics and Management Science V*, Lecture Notes in Electrical Engineering 208, DOI: 10.1007/978-1-4471-4796-1_94,

makes a better solution between the time resolution and frequency resolution. Due to overcome the shortcomings of Fourier-transform only with a single resolution in time frequency domain, wavelet transform has good local characteristics both in time and frequency domain [1]. High order wavelet does good effects in reducing the image encoding time, improving the compression ratio and reducing the distortion degree. Therefore, the wavelet transform makes a very superior performance in image processing. As the need for a large number of calculations, implementation of wavelet transform in hardware is a key problem in the practical engineering application.

Daubechies and Sweldens proposed lifting scheme wavelet transform [2]. Lifting algorithm does not depend on the Fourier transform and can realize signal analysis of the frequency domain in spatial domain. Lifting wavelet transforms high-pass and low-pass filter into a series of prediction and update steps. Thereby, it simplifies the implementation of wavelet decomposition and reconstruction structure, reducing the computational complexity. This paper achieves 5/3 lifting wavelet algorithm based on pipeline operation of addition and shift operation on FPGA, then uses the algorithm to deal with the mathematical image signal. 5/3 lifting wavelet transform is implemented in System Generator (high-performance DSP system development tools), which will shorten the cycle of design.

## 94.2 The Lifting Scheme of Wavelet Transform

Wavelet transform does a displacement $b$ to the basic wavelet function $\psi(t)$, and then does the inner product with $f(t)$-the signal to be analyzed in different scales $a$:

$$w_f(a,b) = \frac{1}{\sqrt{a}} \int_{-\infty}^{+\infty} f(t)\,\psi^*\left(\frac{t-b}{a}\right) dt \; a > 0 \tag{94.1}$$

Traditional wavelet analysis is based on the Fourier analysis; it is limited by the Fourier analysis. Framework of Traditional multi-resolution wavelet analysis was constructed as the first generation wavelet. Sweldens proposed a new wavelet construction method without relying on Fourier Transform, which is lifting wavelet, known as second generation wavelets. The wavelet lifting wavelet expands the research of wavelet analysis greatly and can be used as needed to construct the wavelet lifting scheme. Lifting wavelet algorithm includes 3 main steps: split, predict and update.

Split: In this step, the input signal $x(n)$ is divided into two subsets, namely the even sequence signal $x_e = x(2n)$ and the odd order sequence signal $x_o = x(2n+1)$.

$$x_o(n) = x(2n+1) \tag{94.2}$$

$$x_e(n) = x(n+1) \tag{94.3}$$

Predict: In this step, even sequence of signals predicts the odd signal by multiplying a forecast parameter P. The difference of the original odd signal and the predictive value is the high frequency coefficients $d(n)$.

$$d(n) = x_o(n) - p(x_e(n)) \tag{94.4}$$

Update: An update operator is structured to update the even sequence. The result is low frequency coefficients $c(n)$. The step is also regarded as a low-pass filter and the processed signal has less high frequency component than the previous scale.

$$c(n) = x_e(n) + U(d(n)) \tag{94.5}$$

Reconstruction of lifting wavelet transform also has three steps: anti-updates, anti-forecast and merge.

## 94.3 5/3 Lifting Wavelet Transform

Lifting scheme of 5/3 wavelet and 9/7 wavelet is often used in the field of signal processing [3]. There are many characteristics about 5/3 lifting wavelet. First, 5/3 lifting wavelet has fast computing speed, it can achieve calculation of transform and inverse transform by integer addition and shift, while 9/7 lifting wavelet needs floating-point addition and multiplication; second, 5/3 lifting wavelet needs lower memory space, and can save the calculation results with 16-bit integer, while 9/7 use 32-bit floating-point. Formula of 5/3 lifting wavelet transform is as follow:

$$d_n^{(0)} = x_{2n+1} \tag{94.6}$$

$$s_n^{(0)} = x_{2n} \tag{94.7}$$

$$d_n^{(1)} = d_n^{(0)} - \left| \frac{s_n^{(0)} + s_{n+1}^{(0)} + 2}{4} \right| \tag{94.8}$$

$$s_n^{(1)} = s_n^{(0)} - \left| \frac{d_n^{(1)} + d_{n-1}^{(1)}}{2} \right| \tag{94.9}$$

In the above formula, denotes the integer operation. Lifting scheme can be modified to get a wavelet transform of integer to integer. From the above equation, the lifting scheme can achieve calculation of in-place that is the coefficients calculated in the wavelet transform can be a direct replacement for the original data without the need for additional data storage space. This feature makes the lifting scheme easy to realize on FPGA [4, 5].

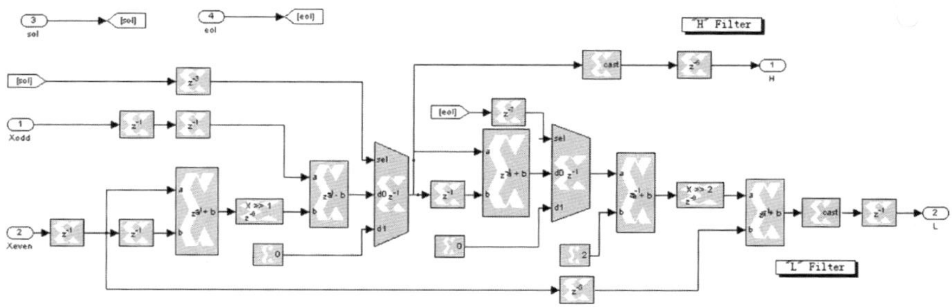


**Fig. 94.1** Design of 5/3 lifting wavelet transform module

## 94.4 Design of Lifting Wavelet Transform

This article designs 5/3 lifting wavelet algorithm in the System Generator platform of Xilinx company. System Generator is an advanced FPGA development tools used to design high-performance of DSP systems. It can transform design of abstract mathematical algorithms of DSP system into a reliable integrated hardware system. The design model implemented on the System Generator can automatically generate FPGA hardware description language (VHDL/Verilog). Design of DSP system on FPGA can be done in System Generator without the use of RTL (level hardware language), thereby this will reduce the cycle of design. In order to verify DSP design model established in System Generator. It needs to use ModelSim for simulation.

According to the formula of the 5/3 lifting wavelet transform, design model of 5/3 lifting wavelet algorithm is shown in Fig. 94.1. The input signal is odd sequences and even sequences of the digital image signal, the output signal is the high and low coefficient of lifting wavelet transform. The module mainly used in the design model includes the delay, the adder and the shifter. The System, through the shifter, shifts to right with 1 bit and 2 bit to achieve the operation of divided by 2 and 4, this avoids the use of multipliers, improves operational efficiency. This article uses the "Delay" delay module to achieve control of the timing, and achieve a deal with the boundary data, through Multiplexer and filling the border with zero.

## 94.5 Simulation of Lifting Wavelet Algorithm

This section will convert 5/3 wavelet model designed in System Generator into VerilogHDL hardware description language, and do functional simulation through Modelsim tool. Simulation of 5/3 lifting wavelet results in Fig. 94.2. The original image signal is spited into odd sequence and even sequence respectively, as two input signals of 5/3 lifting wavelet transform. Figure 94.3 is the original image, Fig. 94.4 is the image represented by high frequency and low frequency wavelet

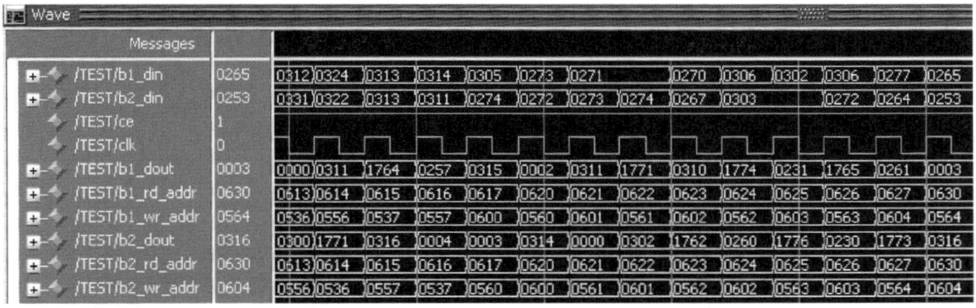


**Fig. 94.2** Simulation of 5/3 lifting wavelet

**Fig. 94.3** Original image

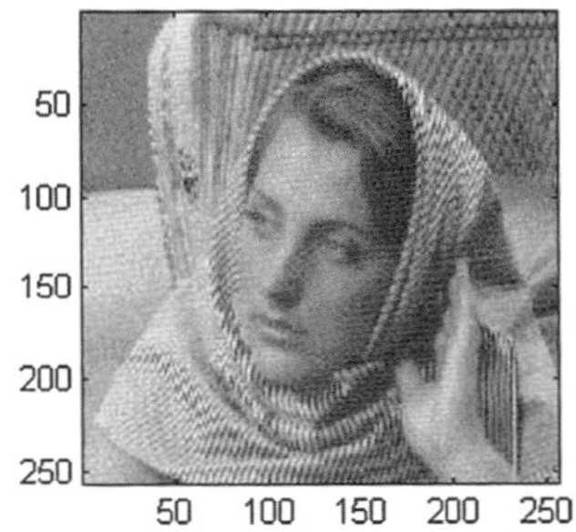


**Fig. 94.4** High and low frequency coefficient

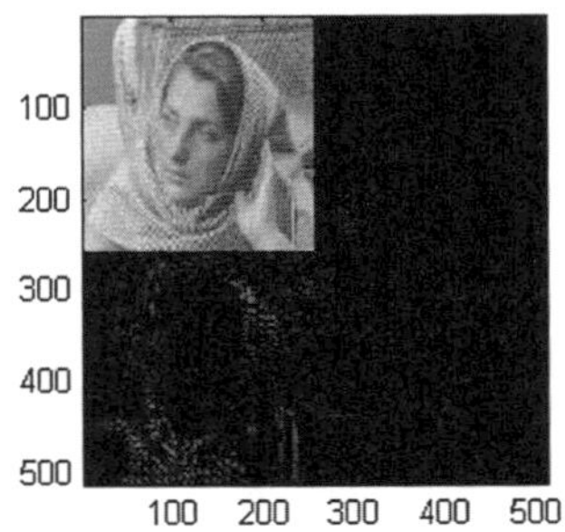


coefficient after lifting wavelet transform. A series of sub-image with different resolution can be obtained after decomposition of lifting wavelet. The frequency of sub-image with different resolution is not the same. The values of most point in image with High resolution (High frequency) are close to 0, so the image pixels with High resolution are black. The low frequency part is the most important part to performance of an image, so removing the high frequency component in the image while retaining only the low frequency part can realize compression of image. This First compressed image is showed in Fig. 94.5. The first compressed image extracts of the low-frequency information in the original image after first decomposition of wavelet, the compression ratio is 1/2; the second compressed image is showed in Fig. 94.6. It extracts of the low-frequency information after second decomposition of wavelet, the compression ratio is 1/4.

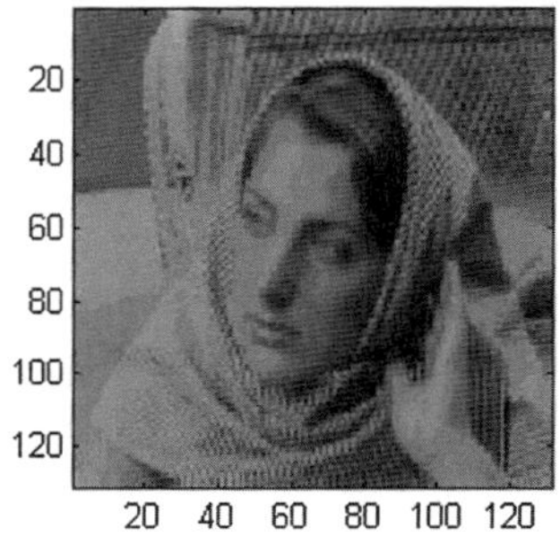


**Fig. 94.5** First compressed image

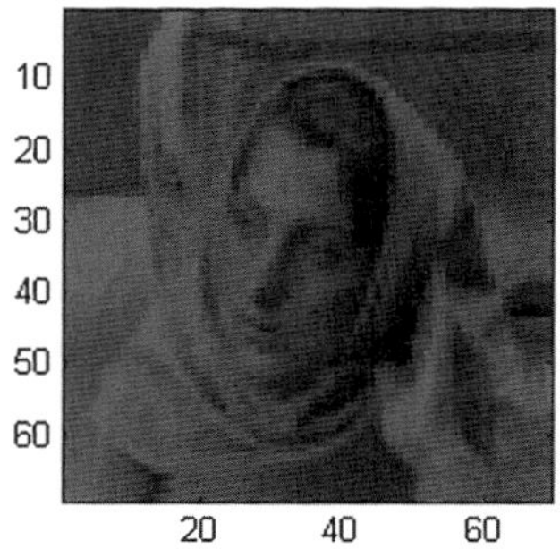


**Fig. 94.6** Second compressed image

**Table 94.1** Performance analyzes of 5/3 lifting wavelet

| Name of wavelet | CR | PSNR |
|---|---|---|
| 5/3 lifting wavelet (Matlab) | 3.9537 | 38.1670 |
| 5/3 lifting wavelet (FPGA) | 3.9489 | 37.2108 |

This paper analyzes the performance of 5/3 wavelet in the peak signal to noise ratio (PSNR) and compression ratio (CR), as shown in Table 94.1. However, Table 94.1 shows the performance of 5/3 lifting wavelet implemented in Matlab and 5/3 wavelet implemented on FPGA is almost the same both in the PSNR and CR. The difference is caused by shifting and integer operation in the implementation process of algorithm.

## 94.6 Conclusion

The paper makes a through research on lifting wavelet transform in the aspect of image processing. 5/3 lifting wavelet algorithm is designed in the System Generator platform. At last, the paper does functional simulation through Modelsim tool. The results show that 5/3 lifting wavelet transform implemented on FPGA can be effective on digital image compression.

## References

1. Daubechies I, Sweldens W (1998) Factoring wavelet transforms into lifting steps. Fourier Anal Appl 4(3):245–267
2. Wang L-Y, Ji Y (2011) FPGA solution of 5/3 lifting DWT based on DSP builder. Comput Digital Eng 259(39):175–179
3. Kehtarnavaz N, Kim N (2005) DWT-based scene adaptive color quantization. Real-Time Imaging 11(5):443–453
4. Zhou W (2006) MATLAB wavelet analysis advanced technology. China, Xi'an 1:83–85
5. Wang B, Guo Z (2010) Simulation and FPGA implementation for 5/3 lifting wavelet transform. Comput Knowledge Technol 2(6):397–399

# Chapter 95
# Super-Resolution Image Reconstruction Based on Improved POCS Algorithm

**Juan Li, Jin Wu, Guang Hu and Shen Yang**

**Abstract** Projections onto convex sets (POCS) algorithm is a widely used super-resolution image reconstruction method. Aiming at the edge ringing effect of traditional POCS algorithm, this paper analyzes the basic reason causing the effect, and adopts an improved POCS algorithm to reduce it. In the improved algorithm, the Point Spread Function (PSF) centered at any edge pixel is weighted, making the far the position of the PSF coefficient is from the edge, the smaller the corresponding PSF coefficient is, and the coefficients remain unchanged along the edge direction. This paper uses wavelet transform modulus maxima method to detect image edges. Considering the edge ringing effect is not only relevant to the edge pixels, but also relevant to their neighboring pixels, we dilate the edge-detected image with a structuring element so as to obtain thicker image edges. Experimental results show that our method greatly reduces the edge ringing effect at little cost in terms of image sharpness, so we can get a better reconstruction image.

**Keywords** Super-resolution image reconstruction · Projections onto convex sets · Wavelet transform modulus maxima · Edge ringing

J. Li (✉) · J. Wu · S. Yang
College of Information Science and Technology, Wuhan University of Science and Technology, Qingshan District, Wuhan, 430081 Hubei, China
e-mail: lilian0122@163.com

G. Hu
Electrical Engineering Department, Huazhong University of Science and Technology, Wuhan, China

W. Du (ed.), *Informatics and Management Science V*, Lecture Notes in Electrical Engineering 208, DOI: 10.1007/978-1-4471-4796-1_95,

## 95.1 Introduction

Super-resolution image reconstruction technology utilizes the different but similar information in the continuous multi-frame low-resolution images combined with prior knowledge to reconstruct the high-resolution image by means of signal processing. Super-resolution image reconstruction has important application in the fields of video, remote sensing, medicine, safety monitoring, and so on. Super-resolution image reconstruction algorithms can be divided into two classes: Frequency domain algorithms and spatial algorithms. Spatial algorithms use the general observation model, and have good adaptability and reconstruction result. Spatial algorithms are the main research direction of super-resolution reconstruction technology, in which, Projections onto convex sets (POCS) algorithm is one of the most spatial promising algorithms [1, 2].

This paper is organized as follows: Sect. 95.2 introduces the basic theory of POCS algorithm. Section 95.3 analyzes the edge ringing effect of traditional POCS Algorithm, and describes its improvement algorithm. Experimental results are given in Sect. 95.4, and conclusions are drawn in Sect. 95.5.

## 95.2 POCS Algorithm

In the super-resolution image reconstruction, we should first build an image acquisition model relating the ideal high-resolution image and the corresponding low-resolution observation sequence. The general image acquisition model [3] can be expressed as

$$g_i(m,n) = \sum_{(k,l)} f(k,l)h_i(m,n;k,l) + v_i(m,n) \tag{95.1}$$

where $g_i(m,n)$ denotes the ith frame of a sequence of low-resolution observation images, $f(k,l)$ denotes the ideal high-resolution image, $h_i(m,n;k,l)$ denotes the Point Spread Function (PSF) for the ith frame, which includes the degradation caused by warping, blurring and decimating, $v_i(m,n)$ denotes the additive noise for the ith frame.

We define the following closed, convex constraints [4] (one for each observed low-resolution image pixel at each frame)

$$C_{m,n;i} = \left\{f(k,l) : \left|r_i^{(f)}(m,n)\right| \leq \delta_0\right\} \tag{95.2}$$

where

$$r_i^{(f)}(m,n) = g_i(m,n) - \sum_{(k,l)} f(k,l)h_i(m,n;k,l) \tag{95.3}$$

and $\delta_0$ represents the confidence that we have in the observation and is set equal to $c\sigma_v$, where $\sigma_v$ is the standard deviation of the noise and $c \geq 0$ is determined by an appropriate statistical confidence bound.

The projection of an arbitrary $x(k,l)$ onto $C_{m,n;i}$ is defined as

$$P_{m,n;i}[x(k,l)] = \begin{cases} x(k,l) & -\delta_0 \leq r_i^{(x)}(m,n) \leq \delta_0 \\ x(k,l) + \frac{r_i^{(x)}(m,n) - \delta_0}{\sum\sum h_i^2} h_i(m,n;k,l) & r_i^{(x)}(m,n) > \delta_0 \\ x(k,l) + \frac{r_i^{(x)}(m,n) + \delta_0}{\sum\sum h_i^2} h_i(m,n;k,l) & r_i^{(x)}(m,n) < -\delta_0 \end{cases} \tag{95.4}$$

## 95.3 Reduction of Edge Ringing Effect

Around the boundary between the dark region and the bright region of the reconstructed high-resolution image using POCS algorithm, the colors of the pixels on the side of the dark region become darker, and the colors of the pixels on the side of the bright region become brighter, namely producing the edge ringing effect 5, the reason for which is analyzed as follows [5]. Each edge pixel at the dark region of the observation image is projected onto the currently estimated high-resolution image, the projected position is also on the edge. The PSF is overlaid on the projected position, there will be bright colored pixels within the PSF window, so the simulated degradation value will be larger than the real observation value. The value of the pixel on the projected position will be modified to a smaller value, making the color of the edge pixel at the dark region become darker. On the contrary, the values of the edge pixels at the bright region will be modified to a larger value, making the colors of the edge pixels at the bright region become brighter. According to the above analysis, we know that the edge ringing effect is not only relevant to the edge pixels, but also relevant to their neighboring pixels [6].

Therefore, we should conduct edge detection on the currently estimated high-resolution image before image modification. For each edge pixel and its neighboring pixels, we use the PSF different from that used by non edge pixels during image modification, making the coefficients of the modified PSF remain unchanged along the edge direction and decrease in the direction orthogonal to the edge [7].

This paper uses wavelet transform modulus maxima method to detect image edges. The method can detect rich edge details and reduce the effects of noise, meanwhile, the edge location is precise.

We dilate the edge-detected image with a structuring element so as to obtain thicker image edges. In this way, the edge pixels and their neighboring pixels can be detected more accurately. This paper selects a square structuring element.

Figure 95.1 shows a 5-by-5 neighborhood, suppose its center pixel is the detected edge pixel. In a small neighborhood, the detected edge approximates a line. The equation of the edge line can be written as

$$G_x \cdot x + G_y \cdot y = 0 \tag{95.5}$$

where $G_x$ is the derivative of the center pixel along the x direction, $G_y$ is the derivative of the center pixel along the y direction.

The PSF centered at any edge pixel as well as its neighboring pixels is weighted by a weighting function, making the far the position of the PSF coefficient is from the edge, the smaller the corresponding PSF coefficient is, and the coefficients remain unchanged along the edge direction.

Suppose the coordinate of the PSF coefficient to be $(x, y)$ and $d$ to be the distance between $(x, y)$ and the edge line. $d$ is defined as

$$d = \frac{|G_x \cdot x + G_y \cdot y|}{(G_x^2 + G_y^2)^{1/2}} \quad (x, y) \in S_h \tag{95.6}$$

where $S_h$ is the support region of the PSF.

For the above purpose, the weighting function is selected as a decaying exponential function

$$c(x, y) = e^{-\lambda d} \quad (x, y) \in S_h \tag{95.7}$$

where $\lambda$ is a constant, which adjusts the descend rate of the function value.

The modified PSF is

$$h_1(x, y) = h(x, y) \cdot c(x, y) \quad (x, y) \in S_h \tag{95.8}$$

To be specific, for the horizontal edge, the vertical edge, the positive diagonal edge and the negative diagonal edge, we can select the following PSF respectively:

**Fig. 95.1** Diagram of edge detection

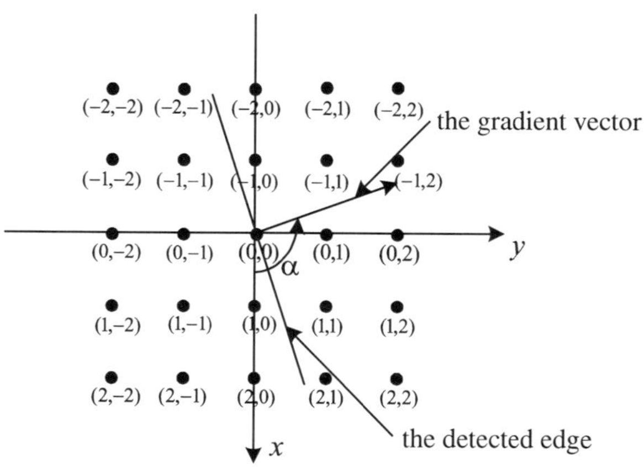

**Fig. 95.2** High-resolution image

$$h_1(x,y) = \begin{cases} h(x,y)\cdot e^{-\lambda|x|} & \text{horizontal edge} \\ h(x,y)\cdot e^{-\lambda|y|} & \text{vertical edge} \\ h(x,y)\cdot e^{-\lambda\frac{|x+y|}{\sqrt{2}}} & \text{positive diagonal edge} \\ h(x,y)\cdot e^{-\lambda\frac{|-x+y|}{\sqrt{2}}} & \text{negative diagonal edge} \end{cases} \quad (95.9)$$

Finally, $h_1(x,y)$ should be normalized.

## 95.4 Experimental Results

We carry out super-resolution image reconstruction experiments on many groups of low-resolution image sequences so as to verify the performance of our method, this paper selects two experiments of which.

In the first experiment, a high-resolution image of size 256 × 256 (as shown in Fig. 95.2) is first blurred with a 2D Gaussian filter of size 9 with standard deviation 1.5, and then down-sampled by a factor of 2 whose sampling mode is shown in Fig. 95.3, finally added with a Gaussian white noise with mean 0 and deviation 0.1. The four low-resolution frames of size 128 × 128 are shown in Fig. 95.4.

Figure 95.5 is the reference frame using bilinear interpolation method, Fig. 95.6a is the reconstructed high-resolution image using traditional POCS algorithm, Fig. 95.6b is the local image zooming of Figs. 95.6a, 95.7a is the

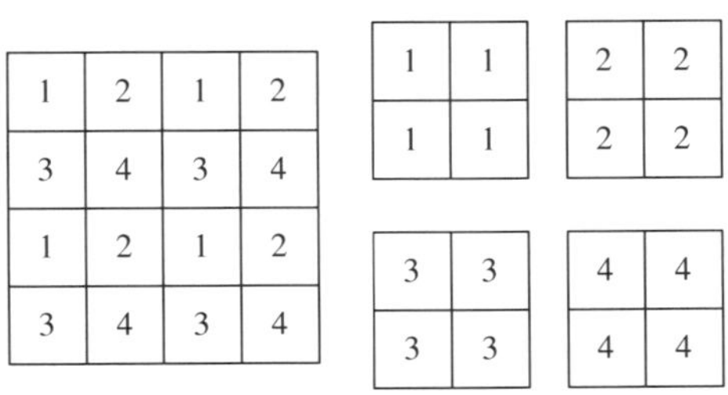


**Fig. 95.3** Sampling mode

**Fig. 95.4** Low-resolution frames

**Fig. 95.5** Bilinear interpolation

reconstructed high-resolution image using our method, Fig. 95.7b is the local image zooming of Fig. 95.7a.

In the second experiment, a high-resolution image of size 500 × 500 (as shown in Fig. 95.8) is first blurred with a 2D Gaussian filter of size 9 with standard deviation 1.2, and then down-sampled by a factor of 2, finally added with a Gaussian white noise with mean 0 and deviation 0.1. The four low-resolution frames of size 250 × 250 are shown in Fig. 95.9.

Figure 95.10 is the reference frame using bilinear interpolation method, Fig. 95.11a is the reconstructed high-resolution image using traditional POCS algorithm, Fig. 95.11b is the local image zooming of Figs. 95.11a, 95.12a is the

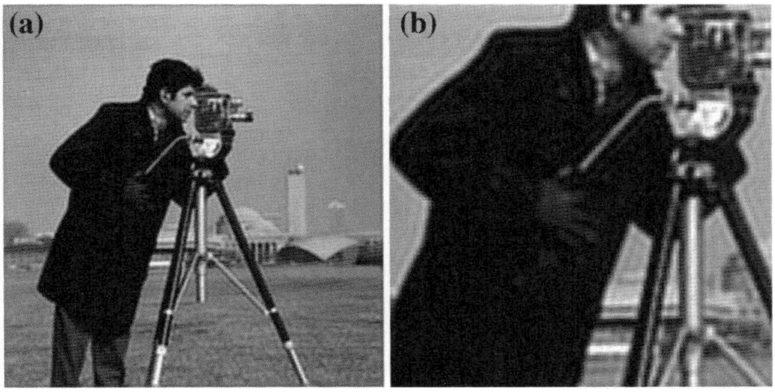


**Fig. 95.6** **a** Traditional POCS algorithm. **b** Local image zooming of (**a**)

**Fig. 95.7** **a** Our method. **b** Local image zooming of (**a**)

**Fig. 95.8** High-resolution image

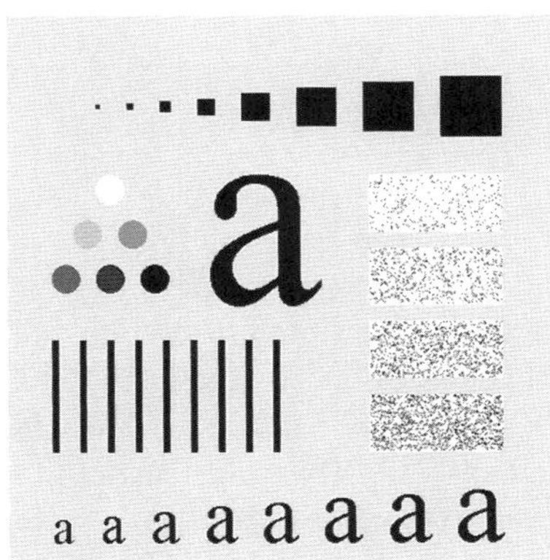

**Fig. 95.9** Low-resolution frames

**Fig. 95.10** Bilinear interpolation

reconstructed high-resolution image using our method, Fig. 95.12b is the local image zooming of Fig. 95.12a.

We can see that the result from traditional POCS algorithm has many ringing effects. However, edge ringing is greatly reduced using our method at little cost in terms of image sharpness.

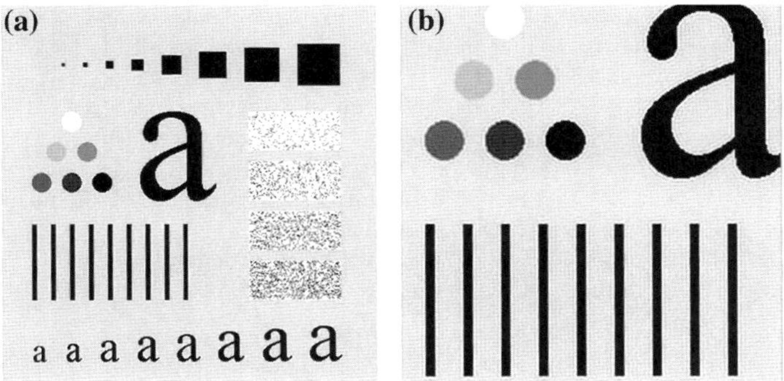


**Fig. 95.11** **a** Traditional POCS algorithm. **b** Local image zooming of Fig. 95.11a

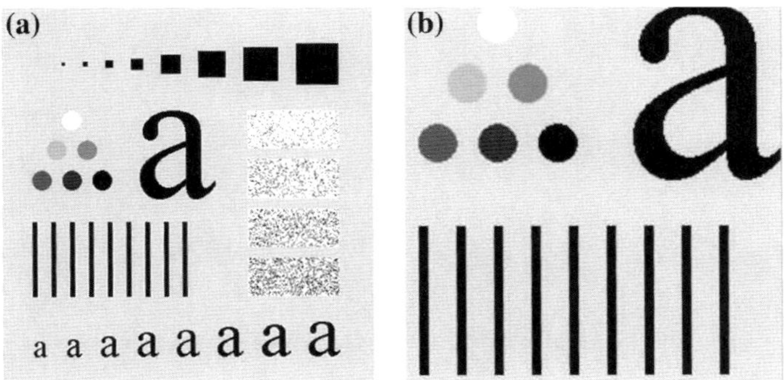


**Fig. 95.12** **a** Our method. **b** Local image zooming of Fig. 95.12a

## 95.5 Conclusion

This paper researches the super-resolution image reconstruction technology based on POCS algorithm. Traditional POCS algorithm can produce edge ringing effect. This paper analyzes the basic reason causing the effect, and adopts an improved POCS algorithm to reduce it. Contrastive experiments between the improved and traditional POCS algorithm are conducted, which verifies the validity of the improved POCS algorithm.

## References

1. Youla DC, Webb H (1982) Image restoration by the method of convex projections: part 1-theory. IEEE Trans Med Imag Mi- 1(2):81–94
2. Sezan MI, Stark H (1982) Image restoration by the method of convex projections: part 2-applications and numerical results. IEEE Trans Med Imag Mi 1(2):95–101
3. Elad M, Feuer A (1997) Restoration of a single super resolution image from several blurred, noisy, and under sampled measured images. IEEE Trans Image Process 6(12):1646–1658

4. Tekalp M, Ozkan MK, Sezan MI (1992) High-resolution image reconstruction from lower-resolution image sequences and space-varying image restoration. IEEE Int Conf Acoustics Speech Signal Processing 3:169–172
5. Patti AJ, Altunbasak Y (2001) Artifact reduction for set theoretic super resolution image reconstruction with edge adaptive constraints and higher-order interpolants. IEEE Trans Image Process 10(1):179–186
6. Li H-F, Du M-H (2003) Super resolution image restoration based on improved POCS algorithm. J South China Univ Technol 31(10):24–27
7. Jing Y, Kai-Na S, Chuang-Bai X (2007) Edge artifact reduction for super-resolution image reconstruction. Acta Automatica Sinica 33(6):577–582

# Chapter 96
# Research on Images Identification Technology Based on Neural Network

**Yiqiu Xu, Zhanbo Liu and Wei Zhang**

**Abstract** Image identification, with a mass of information computations, needs high speed and precision. The real-time and robustness of neural network accord the demands of the images identification. Aiming at the question that BP network easily to get bogged down in the partial dinky weakness, this paper proposed an improved neural network method, which can avoid the partial dinky and achieve the global minimum by adding the momentum factor in weight increment.

**Keywords** Image identification · Neural network · BP algorithm

## 96.1 Introduction

With the rapid development of communication and information processing technology, digital image technology is widely used in all kinds of fields of industrial production and agricultural production, and is gradually affecting each aspects of our life [1]. In the field of digital image processing, digital image storing and transmission technology has already been very ripe and has been used extensively [2]. However, compared with digital image storing and transmission technology, the technology of image analysis and understanding is far behind neither in theory, nor in application. Image recognition as the key problem of image analysis and understanding area has being the studying focus and difficult point in recent years, which has been becoming the main contest of the classification and description of the

Y. Xu (✉) · Z. Liu · W. Zhang
Educational Technology and Information Center, Mudanjiang Medical University, Mudanjiang, China
e-mail: 14823470@QQ.com

W. Du (ed.), *Informatics and Management Science V*, Lecture Notes in Electrical Engineering 208, DOI: 10.1007/978-1-4471-4796-1_96,

image [3]. The image recognition consists of three steps: data recognition, data processing, and data classification [4, 5]. The four ways of image recognition are statistical matrix recognition, structural matrix recognition, hazy image recognition and intelligent matrix recognition [6, 7]. The traditional image recognition technique is mostly based on the foundation of large-scale calculating, such as statistical matrix recognition and structural matrix recognition [8]. However, there are too much contradiction between the amount of calculation and the accuracy of recognition. Artificial neural network provided a new solution for this problem [9]. As a generalized mode of intelligent matrix recognition, the artificial neural network had the feature of parallelism, distributional memory, good fault tolerance, auto-adaptation, associative memory, and good non-linear processing, has achieved many achievements that could not be reached by many other conventional ways in the domain of matrix recognition. And the image recognition technology of the neural network is a newly developed image recognition technology together with the modern computer technology, the image processing, the artificial intelligence, and the matrix recognition theory. It is a new way of image recognition which combines the neural network algorithm and the traditional matrix recognition. In this paper, the use of neural networks that have access to the digital image to identify and will be the introduction of BP neural network image recognition in the field. This paper firstly introduces the theory of image recognition and neural network, and then analyzes the structure of BP neural network and the BP algorithm, on the basis of which proposed the application of BP neural network in image recognition. Finally, in order to solve the question that BP neural network is easily to get bogged down in the partial dinky weakness, this paper proposes an improved neural network method which can avoid the partial dinky and achieve the global minimum.

## 96.2 Image Recognition and Artificial Neural Network

### *96.2.1 Image Recognition*

Image recognition refers to recognize the targets from the images by analyzing digital image. An image recognition system or algorithm generally includes three components, namely: Image acquisition and preprocessing, image feature extraction and feature classification. The structure of the image recognition system is shown in Fig. 96.1.

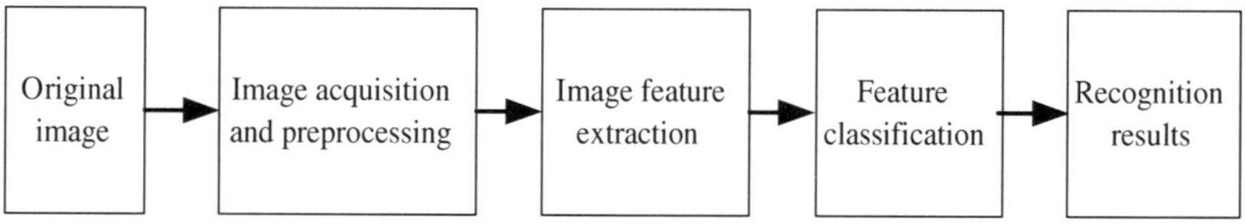


**Fig. 96.1** The structure of the image recognition system

- Image acquisition and preprocessing. Image acquisition refers to convert the image to electrical signals through electronic equipment; image preprocessing refers to eliminate the noise and distortion of the image, weaken the irrelevant features and strengthen the essential features of the image. If the image contains multiple targets, it can be divided.
- Feature extraction. It refers to extract the feature from the image after preprocessing. The image recognition result is determined the result of the feature extraction.
- Feature classification. The function of this part is mapping image feature space mapping to the type of space to get the recognition results.

## *96.2.2 Artificial Neural Network*

### 96.2.2.1 The Characteristic of Artificial Neural Network

An artificial neural network (ANN) is a mathematical model or computational model that is inspired by the structure and/or functional aspects of biological neural networks. A neural network consists of an interconnected group of artificial neurons, and it processes information using a connectionist approach to computation. In most cases an ANN is an adaptive system that changes its structure based on external or internal information that flows through the network during the learning phase. Modern neural networks are non-linear statistical data modeling tools. They are usually used to model complex relationships between inputs and outputs or to find patterns in data. An artificial neuron consists of three basic elements, namely: the connection weights, summation unit and a nonlinear activation function. The structure of artificial neuron model is shown in Fig. 96.2.

1. Connected weight. It stands for the connection strength. Positive value indicates excitation; negative value indicates dampening.
2. Summation unit. It can calculate the weighted sum of all the input information.
3. Excitation function, as known as transfer function. It can limit the output of neurons within a certain range.

The relationship between input and output of artificial neuronal is described as follows:

$$u_K = \sum_{i=1}^{n} w_{ik} x_i \tag{96.1}$$

$$y_K = f(u_K + b_K) \tag{96.2}$$

where, $x_i\ (i = 1, 2, \ldots, n)$ stands for input signal; $w_{ik}\ (i = 1, 2, \ldots, n)$ stands for the weight of neurons $k$; $u_K$ stands for the output of linear combiner of input signal; $b_K$ stands for threshold of neurons; $f(\cdot)$ is excitation function, $y_K$ is output signal of neurons.

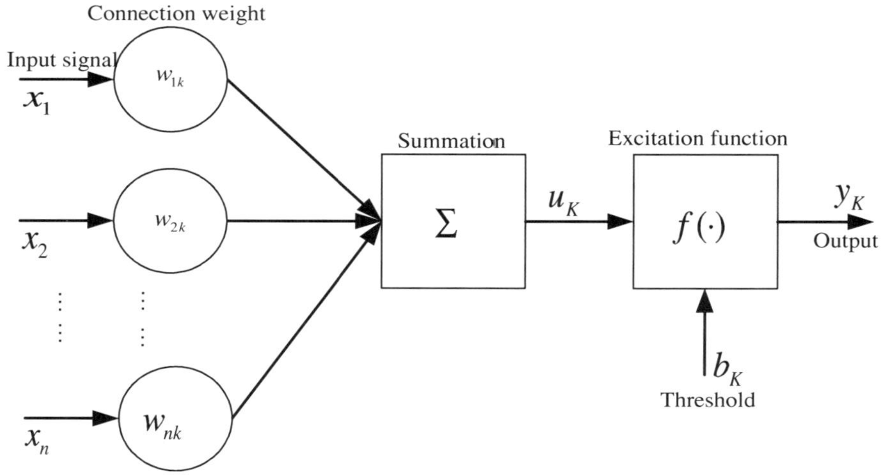


**Fig. 96.2** The structure of artificial neuron model

#### 96.2.2.2 The Structure of Artificial Neural Network

The structure of neural network mainly refers to connection mode of neuron node. Generally, it includes four structures, namely: feedforward network, feedback network, combined network and hybrid network.

1. Feedforward network. As shown in Fig. 96.3, it is a 3-layers feedforward network. It consists of input layer, hidden layer and output layer. The connection exists only in the adjacent neurons and there is no feedback in each neuron.
2. Feedback network. The characteristic of feedback network is that the input for the neuron includes external input and feedback input and each neuron has counting function. (Fig. 96.4).
3. Combined network. The characteristic of combined network is that all the neurons can be used as input and also used output, and any neurons may be connected to the other neurons. The structure of combined network is as follow (Fig. 96.5).

Hybrid network can be regarded as a special feedforward network. The neurons in the same layer can connect each other. The structure of hybrid network is as follow (Fig. 96.6).

## 96.3 BP Neural Network

BP neural network is multi-layer feedforward network which is based on BP algorithm. BP algorithm is designed to solve weight optimization of multi-layer feedforward neural network. Its essence is to calculate the minimum error

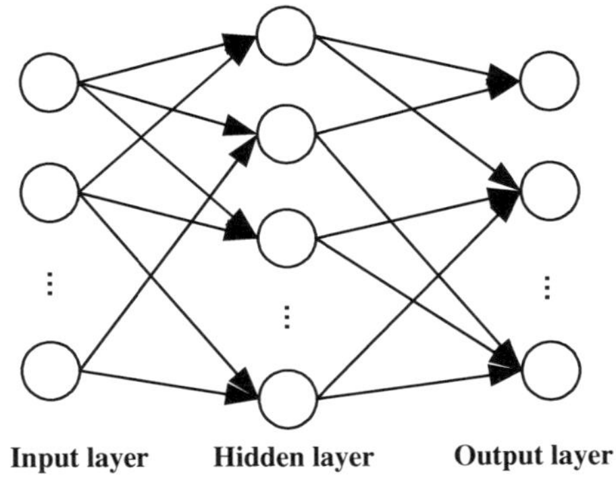


**Fig. 96.3** Feedforward network

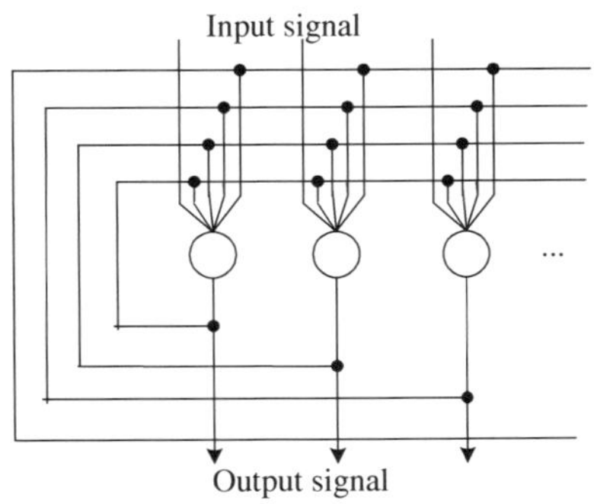


**Fig. 96.4** Feedback network

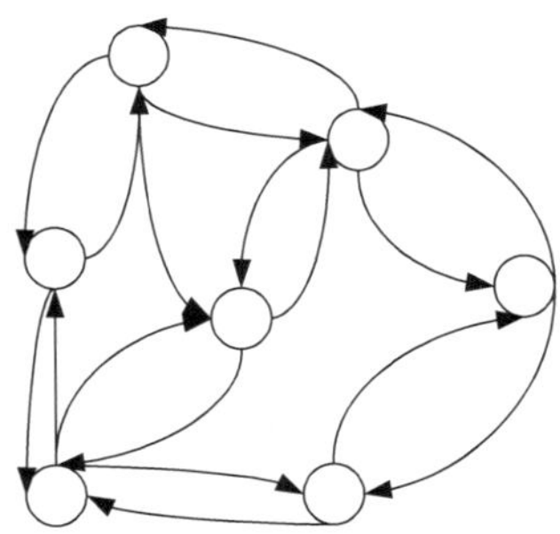

**Fig. 96.5** The structure of combined network

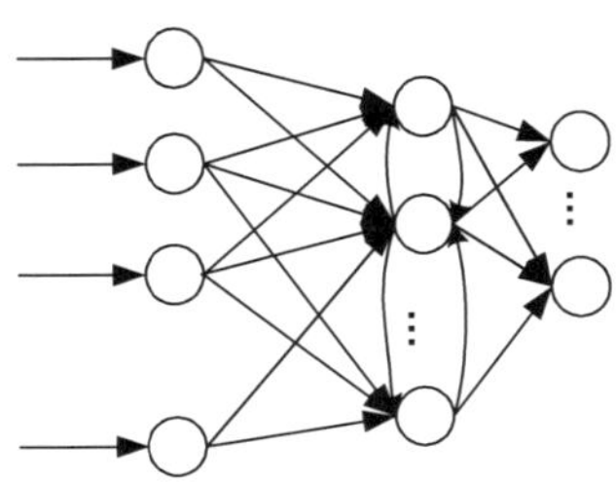

**Fig. 96.6** The structure of hybrid network

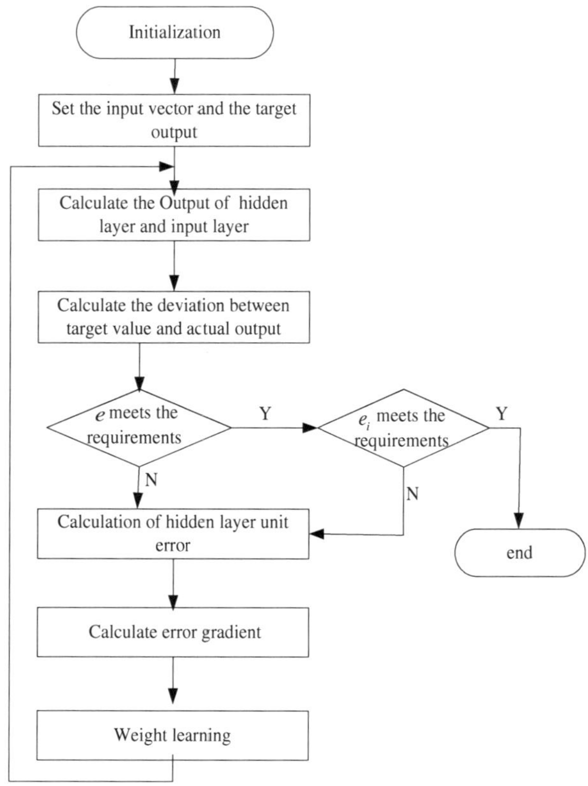


**Fig. 96.7** Process of BP algorithm

function. This algorithm used steepest descent method of non-linear programming, and it modified the weight coefficient according to the negative gradient direction of error function. Process of BP algorithm is shown as follows (Fig. 96.7).

BP neural network is the most widely used, but there are still some deficiencies:

- The learning speed of BP algorithm is show. The essence of BP algorithm is gradient descent method. However, the objective function is optimized is very complicated, so the BP algorithm is inefficiency.
- The BP algorithm is a local search optimization method, so it is likely to fall into the local extreme.
- It is difficult to resolve the contradiction between the case scale and network scale of application problems.
- It is difficult to select appropriate network structure leading to large calculation and low efficiency.
- It is difficult to solve the contradiction between network prediction ability and training contradiction.

## 96.4 Improved BP Network Algorithm

Aiming at the question that BP network easily to get bogged down in the partial dinky weakness, this paper proposed an improved neural network method. It use momentum method and learning rate adaptive adjustment strategy to reduce the network sensitivity to the error surface and avoid to get the partial dinky. The BP algorithm is a simple steepest descent static optimization algorithm and the convergence is slow. Improvement principle of momentum weight adjustment is described as follow.

$\Delta w(k+1) = \eta(1-\alpha)D(k) + \alpha D(k-1)$, where $D(k) = \frac{-\partial E}{\partial w(k)}$ is the negative gradient of $k$ time, $D(k-1)$ is the negative gradient of $k-1$ time, $\eta$ is earning rate, $\eta > 0$ and $\alpha$ is momentum factor, $0 \leq \alpha < 1$. It reduced the oscillating trend of learning process and improves the convergence.

We use learning rate adaptive adjustment strategy to resolve the problem of slow convergence of BP algorithm. Learning rate adaptive adjustment strategy described as follow:

$$\eta(k+1) = \begin{cases} 1.08\eta(k) & SSE(k+1) < SSE(k) \\ 0.8(k) & SSE(k+1) > 1.06SSE(k) \\ \eta(k) & other \end{cases} \tag{96.3}$$

When the new error is over the old error at certain level, learning rate will the reduced. When the error is less than the old error, learning rate will increase. This method ensures that the BP neural network is always with maximum acceptable learning rate for training to make learning rate adaptive adjustment in the iterative process. The objective function approaches the minimum along different directions according to the reasonable learning rate.

## 96.5 Conclusions

The technology of neural network image recognition is a kind of new style image recognition technology, which is developed due to the theory of present computer technology, image processing, artificial intelligence and pattern recognition. The real-time and robustness of neural network accord the demands of the images identification. Aiming at the question that BP network easily to get bogged down in the partial dinky weakness, this paper proposed an improved neural network method, which can avoid the partial dinky and achieve the global minimum.

## References


1. Fei Y, Kunming W, Xin M, Shuangdong Z (2003) Application of BP neural network classifier for road traffic sign recognition. Comput Eng 10:73–77
2. Yao Z, Fei M, Li K (2007) Recognition of blue–green algae in lakes using distributive genetic algorithm-based neural networks. Neuro Comput 4:641–647
3. Peng S, Wang J (2005) Methods of image recognition based on neural networks. Electron Sci Technol 01:280–288
4. Hao-dong Z, Yong Z (2010) Image recognition model combining rough set with neural network. Comput Eng Appl 13:38–46
5. Cai-xia G, Jian-chang Y, Shou-feng J (2006) Recognition algorithm of the foreign fibers in cotton by BP neural networks. J Xi'an Univ Eng Sci Technol 20(5):542–544
6. Dan-qing W, Zhi-jun L (2005) Pattern recogition based on hopfield neural network. J Wuhan Yejin Univ Sci Technol 04:231–236
7. Mingdong X, Li G, Guoxuan Z, Shijian L (2005) A new method of image recognition. Comput Eng 9:182–184
8. Ding-qiang Y, Shu-ping X, Jia-fu J (2008) Texture image recognition based on differentia evolution probabilistic neural network. Comput Eng Appl 11:289–295
9. Asvestas P, Matsopoulos FK, Nikita KS (2007) A power differentiation method of fractal dimension estimation for 2_D signals. J Vis Commun Image Represent 9:392–400

# Chapter 97
# Dynamic Difficulty Adjustment by Facial Expression

**Nan Xiang, Lili Yang and Mingmin Zhang**

**Abstract** To enhance players' game experience is one of the most important targets in game exploring. In this paper we integrated a dynamic game difficulty adjustment (DDA) method into Tetris that using Active Shape Model (ASM) and HMM to recognize emotion states of players' from camera and utilizing Kalman filter to dynamically detect the experience of players. According to the users' experience then adjusted the speed of game. Experiments shown that our DDA method could give player better game experience.

**Keywords** Affective computing · Dynamic difficulty adjustment · Facial expression · Game intelligence · Kalman filter

## 97.1 Introduction

In the past decade, researchers and game explorers lay their efforts mainly on the visual realism of game performance to enhance the players' game experience. As a result, game performance on 3D model rendering, character's animation, NPC's intelligence, fluid simulation, social interaction environments etc. have made remarkable achievements. To increase players' experience, a computer game must be balanced well. For instance, a game must provide meaningful choices, the role of chance should not be so great that player skills become irrelevant, and players must perceive the game to be fair [1], otherwise the player should lose their interesting. This is so called Dynamic Difficulty Adjustment (DDA) and Dynamic

---

N. Xiang · L. Yang (✉) · M. Zhang
Library of Chongqing University of Technology, Chongqing 400054, China
e-mail: lwuying1218@163.com

W. Du (ed.), *Informatics and Management Science V*, Lecture Notes in Electrical Engineering 208, DOI: 10.1007/978-1-4471-4796-1_97,

Game Balancing (DGB). In addition, players' emotion plays an important role in game experience. The gaming community has recently recognized the importance of emotion in the development of more engaging games, and the area of affective gaming is receiving increasing attention [2]. Many psychophysiological studies have been made to investigate different traits of game-play experience and several games have been developed in laboratories exploring the possibility of adapting the game-play to the player's state [3]. This provides us a practicable method to avoid undesired player emotions such as frustration by dynamically adjusting game parameters especially game difficulty level.

In common video games, the level of game difficulty cannot be changed within each stage of game. Although there is a training model for player to learn the game mechanics, however once the player set the game difficulty, it is easy for him to feel frustrated as the mission is too difficult to accomplished or too tedious without any challenges. Most large games adjust the game parameters in their new versions which often take a long time to be released. Currently, players' in-game performances such as the rate of successful shots or hits, number of life points, time left to complete a task are used for dynamic game balancing [3–8]. However, as players' hobbies, performance and responses are various with their emotional states which also dynamically changes with the game progress, game adaptation should take the emotional response of the player into account more than in-game performance. Physiological signals such as Blood Volume Pulse (BVP), Heart Rate (HR) and Electrodermal Activity (EDA) are used to detect the arousal of emotion. Unfortunately, all these method need to contact the equipment to players, as a result, players will probably not accept this kind of detection in game.

Take these into account, we provide an emotion based dynamic game adjusting prototype Emotetris which utilizes face expression captured from camera to detect the players' emotional states and then adjust game difficulty level dynamically according to these emotional states. Our method is grounded on the user-centric design by adjusting the difficulty level of the game to maximize the user's experience which is deviated from the traditional game-centered system. Traditional systems based on the gradual and consistent augmentation of the difficulty level treat all users equally and ignore the broad diversity of their gaming skills, and preferences [9].

## 97.2 Background

To enhance players' game experience is one of the most important targets in game exploring. In Ermi's study, immersion is composed by sensory immersion, challenge-based immersion and imaginative immersion [10]. Wijnand considered flow, an optimal state of enjoyment where people are completely absorbed in the activity. In the gaming domain, immersion is mostly used to refer to the degree of involvement or engagement one experiences with a game [11]. Lennart explored well-accepted common meanings of certain user experiences such as flow and immersion instead of a more through understanding of loosely defined subjective experiences [12]. Kiel pointed two kinds of frustration during games, the at-game-frustration and in-game-frustration. The first is due to lack of skill during game playing and the second in caused by difficult game levels [13]. The players' experiences study is not limited in traditional PC games but also extend to other platforms such as mobile games [14, 15].

The relations between physiological signals and players' emotional activities have been studied in many works [9, 16–20]. These researches showed that high EDA values correlate with high arousal and that a high level of arousal can be indicative of a high level of challenge, frustration, and/or excitement. Players are engaged in low speed of heart rate and frustrated in high speed of heart rate. Although these signals could be used to detect the arousal of emotional activates, yet it is hard to distinguish some distinct experiences which have the same physiological reactions, for example, being frustrated and being immersed.

The other methods are to utilize game AI to dynamically adjust the difficulty level. Gaussian Mixture Module and multivariate pattern mining were used to model the player's reaction pattern [21, 22]. NPCs behaviors are controlled by reinforce learning algorithm [6, 7]. They did not, however, change the game environment or adjust the difficulty of the game level during play. By conducting a cheap, abstract simulation of the player's progression through state space, Hunicke [4] used Hamlet system to predict when the player is repeatedly entering an undesirable loop, and help them get out of it. Joost [23] proposed an adaptation approach that uses expert knowledge for the adaptation. They used a game adaption model and organized agents to choose the most optimal task for the trainee, given the user model, the game flow and the capabilities of the agents. Hom [24] used AI techniques to design balanced board games like checkers and Go by modifying the rules of the game, not just the rule parameters.

## 97.3 Program

In this section we first introduce how to generate the facial feature points and then construct HMMs to distinguish emotions.

### *97.3.1 Facial Feature Detection and Emotion Recognition*

System adopts the Active Shape Model (ASM) animation algorithm to facial feature point detection. Model total includes 77 feature points as shown in Fig. 97.1. In order to establish facial feature point, the system divides face into upper, middle and lower areas, thus other points of face can be determined through relevant constraints on linear interpolation. To estimate the emotional states of players' captured from camera we adopted HMM. We exploited the facial feature

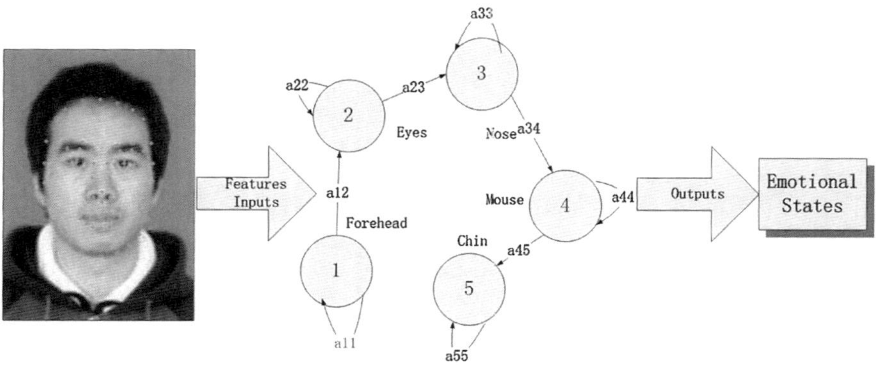


**Fig. 97.1** Facial feature detection and emotion recognition

points as the observations of HMMs and then the players' emotions could be generated from facial images using HMMs by assigning each of these regions to a HMM. For more details about the emotion recognition process please find our previous research [25].

## *97.3.2 Facial Feature Detection and Emotion Recognition*

After generating the facial expressions of players', we would calculate his or her game experience with their game performances. The numbers of lines that had been eliminated and had not been eliminated represented as $le_n$ and $lr_n$, by player over time, were utilized to represent their in-game performance. Ground on the theory of Kalman filter, the expressions of player were treated as an observational variable to adapt the prediction of experience.

Assuming $S_n = \{bore, relas, engaged, frustrated\}$ represents the player's state when the nth diamond was dropping, $P_n = (le_n, lr_n)$ is a vector denoted the in-game performance state, then

$$S'_n = AP_{n-1} + U \tag{97.1}$$

where $S'_n$ is the predicted player's experience state, and $A$ is translation matrix, $U$ is covariance.

$$S_n = S'_n + (BE_n - AP_{n-1})kg \tag{97.2}$$

$B$ is control matrix, $E_n$ represents the expression of player's face, $kg$ denotes the Kalman gain.

$$kg = BQ'_nB'/(BQ'_nB' + V) \tag{97.3}$$

$$Q'_n = AQ_{n-1}A' + U \tag{97.4}$$

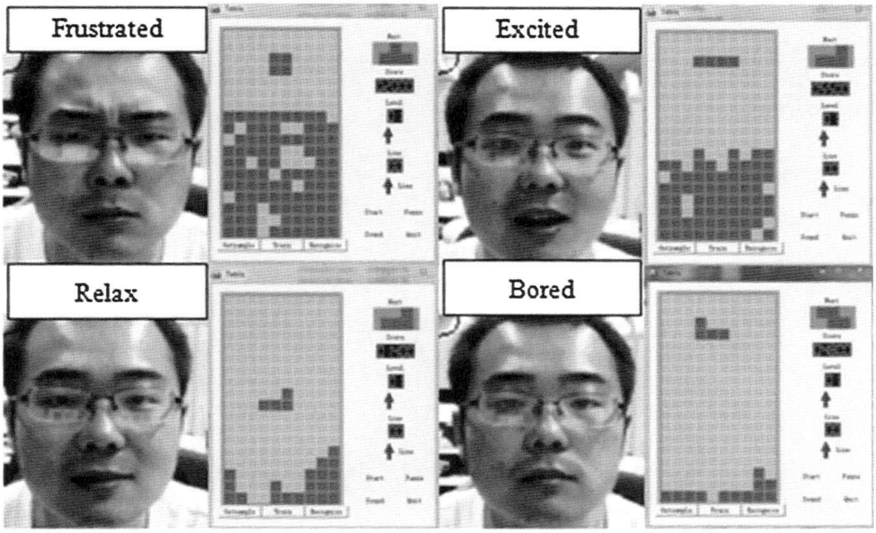


**Fig. 97.2** The sample facial expressions and game results in four kinds of game mood

$Q_n$ represents the optimal covariance. All these parameters are set according to experiments.

## 97.4 Experimental Results

In this section we would show the adjustment strategy in our DDA system and give the user study result.

### *97.4.1 Recognition and Adjustment*

The computer that we run this system has Dual-Core E5300 CPU and 4G memory. Every new player's expression would be sampled and trained before starting game.

We integrated difficulty adjustment into Tetris to evaluate the performance of algorithm. The speed of dropping diamond is the parameters to be adjusted for it can affect players directly. Figure 97.2 showed the player's standard facial expression when he was frustrated, flow and excited, flow and relax and bored according to the game results. The adjustment strategy can be found in Table 97.1, when player is frustrated, the game should slow down. Flow means a good game experiences and the speed should be kept on. When players are bored they may need change the speed or a rest. Good in-game performances do not means speed up and bad in-game performances do not mean slow down, they are related to the player's game mood. For example, players may consider eliminating many lines in one time.

**Table 97.1** Adjustment strategy

| | Frustrated | Flow and excited | Flow and relax | Bored |
|---|---|---|---|---|
| Good performances | Slow down | Keep on | Speed up/keep on | Speed up |
| Bad performances | Slow down | Keep on | Keep on | Need a rest |

### *97.4.2 User Study*

To estimate the players' experience, 20 volunteers who had no research background on dynamic difficulty adjustment were participated in our experiment. They first played the game with difficulty adjustment only according to their in-game performances and then play in the expression based adjustment mode. After trying two kinds of game mode, they were asked two questions: does the game make adjustment in time and which mode is better? For the first question, 16 players thought the game could make in time adjustment when they was frustrated or bored. For the second question, 14 players considered the expression based game adjustment is better than in-game performances based adjustment in bringing them better game experiences.

## 97.5 Conclusion

In this paper, we provided an effective method in improving players' game experiences. Our method combined the in-game performances and facial expressions of players to dynamically adjust the game difficulty. Experiments shown that, only employing in-game performances cannot make a just decision in changing game parameters. Dynamic difficulty adjustment can attract players' attention when they were bored and release the pressure when they were frustrated. However, how to change the game parameters in a certain situation is due to the custom of different players. Our next work was to propose special solution for different players in different game styles.

**Acknowledgments** This research is supported by "Key technology and software development of digital mining", with grant no. 2009AA062704.

## References

1. Tijs T, Brokken D, IJsselsteijn W (2008) Dynamic game balancing by recognizing affect. Fun and Games 3:88–93
2. Hudlicka E (2009) Affective game engines: motivation and requirements, ICFDG 2009. ACM 22:299–306

3. Rani P, Sarkar N, Liu C (2005) Maintaining optimal challenge in computer games through real-time physiological feedback. In: Proceeding of the 1st international conference on augmented cognition 40:184–192
4. Hunicke R, Chapman V (2004) AI for dynamic difficulty adjustment in games. In: Proceedings of the challenges in game AI workshop, 19th national conference on artificial intelligence (AAAI '04) 8(3):91–96
5. Hunicke R (2005) The case for dynamic difficulty adjustment in games. In: Proceedings of the 2005 ACM SIGCHI international conference on advances in computer entertainment technology, ACM 33(4):429–433
6. Spronck P, Sprinkhuizen-Kuyper I, Postma E (2004) Difficulty scaling of game AI. In: Proceedings of the 5th international conference on intelligent games and simulation 2004 (GAME-ON 2004) 3:33–37
7. Andrade G, Ramalho G, Gomes AS, Corruble V (2006) Dynamic game balancing: an evaluation of user satisfaction. In: Proceeding of the 2006 AI and interactive digital entertainment conference, AIIDE'06, vol 12(3). The AAAI Press, pp 3–8
8. Yannakakis GN, Hallam J (2004) Evolving opponents for interesting interactive computer games. From animals to animats 8:499–508
9. Yun C, Shastri D, Pavlidis I, Deng Z (2009) O'game, can you feel my frustration?: improving user's gaming experience via stresscam. ACM 33(122):2195–2204
10. Ermi L, Mayra F (2005) Fundamental components of the game play experience: analyzing immersion. In: Worlds in play: international perspectives on digital games research 2(3):15–27
11. IJsselsteijn W, de Kort Y, Poels K, Jurgelionis A, Bellotti F (2007) Characterizing and measuring user experiences in digital games. In: Proceedings of the international conference on advances in computer entertainment technology 112(33):399–403
12. Nacke LE, Lindley CA (2010) Affective ludology, flow and immersion in a first-person shooter: measurement of player experience. Arxiv preprint arXiv 12(31):248–255
13. Gilleade KM, Dix A (2005) Using frustration in the design of adaptive videogames. In: Proceedings of the 2004 ACM SIGCHI international conference on advances in computer entertainment technology 22:228–232
14. Baillie L, Morton L, Moffat DC, Uzor S (2011) Capturing the response of players to a location-based game. Personal Ubiquitous Comput 15(1):13–24
15. Ermi L, Mayra F (2005) Challenges for pervasive mobile game design: examining players' emotional responses. In: Proceedings of the 2005 ACM SIGCHI international conference on advances in computer entertainment technology 21:371–372
16. Drachen A, Nacke LE, Yannakakis G, Pedersen AL (2010) Correlation between heart rate, electro dermal activity and player experience in first-person shooter games. In: Proceedings of the 5th ACM SIGGRAPH symposium on video games 7(10):49–54
17. Nacke L, Lindley CA (2008) Flow and immersion in first-person shooters: measuring the player's gameplay experience. In: Proceedings of the 2008 conference on future play 3(21):81–88
18. Tijs T, Brokken D, IJsselsteijn W (2009) Creating an emotionally adaptive game. Entertainment Computing-ICEC 20(08):122–133
19. Yannakakis GN, Hallam J, Lund HH (2008) Entertainment capture through heart rate activity in physical interactive playgrounds. User Model User-Adapted Interact 18(1–2):207–243
20. Ravaja N, Saari T, Salminen M, Laarni J, Kallinen K (2006) Phasic emotional reactions to video game events: a psychophysiological investigation. Med Psych 8(4):343–349
21. Lee S, Jung K (2006) Dynamic game level design using gaussian mixture model, book dynamic game level design using gaussian mixture model, series dynamic game level design using gaussian mixture model, vol 32(2). Springer, pp 955–959
22. Chiu KSY, Chan KCC (2008) Using data mining for dynamic level design in games, book using data mining for dynamic level design in games, series using data mining for dynamic level design in games, vol 33. Springer, pp 628–637

23. Westra J, van Hasselt H, Dignum F, Dignum V (2009) Adaptive serious games using agent organizations. Agents for Games and Simul 1(8):206–220
24. Hom V, Marks J (2007) Automatic design of balanced board games. In: Proceeding of the 3rd conference on artificial intelligence and interactive digital entertainment 3(4):25–30
25. Pan Z, Li H, Zhang M, Ye Y, Cheng X, Tang A, Yang R (2009) Photo realistic 3D cartoon face modeling based on active shape model. Trans Edutainment II 12(32):299–311

# Chapter 98
# Research on Detection Technology of Image Spam

**Lixin Tao, Hean Liu, Tao Zhang, Ning Zhong and Shuguang Wu**

**Abstract** This document calculates and discusses circular groove angle (CGA) affects on journal bearing performance. CFD method has been used for calculate N–S equations and cavitation effect has been considered. The results show that as CGA increases, the maximum pressure is decreased; the capacity and friction moment are also decreased. The maximum pressure position is increased from 1 to 2 when CGA is 180°. This paper will provide a theoretical basis for journal bearing groove optimal design.

**Keywords** Image type spam · Characteristics analysis · Classification algorithm · Spam image

## 98.1 Introduction

With the rapid development of network technology, the number of the internet users in China is also growing increasingly, which led to the sharply expanded spam, and had a great impact people's normal life and production. In order to

L. Tao (✉) · T. Zhang
Institute of Computer and Communication, Hunan University of Technology, Zhuzhou 412007 Hunan, China

H. Liu
Network Information Center, Hunan City University, Yiyang 413000, China
e-mail: 33894549@qq.com

N. Zhong · S. Wu
Modern Education Technology Center, Hunan University of Technology, Zhuzhou 412007 Hunan, China

W. Du (ed.), *Informatics and Management Science V*, Lecture Notes in Electrical Engineering 208, DOI: 10.1007/978-1-4471-4796-1_98,

avoid the detection of general spam filtering system, the manufacturers are constantly changing the pattern in different way, while the defenders also are constantly using different new technologies to defend against it [1]. Image type spam filter detection technology is a typical representative of this new spam. Spammers often use some coding, replacement means in the case of does not affecting the reading of the information, while at the same time not to expose their own hidden to escape the detection of the interference filter, such as distorting the image, adding the interfering elements, etc. As to the number of scale growth and various forms of content of image spam, we will continue to exalt innovation to improve spam detection, advanced with the times [2]. So, by studying the functions of the difference between the text area of spam image and the spam image we can find an algorithm research which is more suitable for the spam image with strong interference to extract the corner information [3]. The performance and identification of this algorithm have a higher degree with a wide range of practical.

## 98.2 The Detecting Difficulties on Image Type Spam

There is a big difference between the image type spam and the text type spam. With the rapid development of Internet, more and more people use e-mail, micro blogging, blog and other channels to communicate and exchange information. Through analyzing the image features of the spam, the main difficulties of identifying image type spam can be grouped into the following four areas.

### *98.2.1 The Image Type Spam is Lack of Uniform Definition*

Due to the rapid development of network technology; it's open and global shortcuts, making the dissemination of the information network is also a form of diversification [4]. All along, the concept of spam is relatively vague, inconclusive. The Spammer is often spread spam via e-mail. From the perspective of the Internet, the information which is not useful for the network user can be called spam. Telecommunications industry standard "to guard against Internet junk e-mail technical requirements" clearly define the spam as [5]: (1) The recipients receive the advertising, electronic publications and a variety of promotional e-mail without prior request or agreement; (2) Hiding the identity of the sender and the e-mail address; source of information, sender, and routing information; (3) E-mail with forged source of information, sender, and routing information, however, there is no unified and clear definition of which images can be called the spam image.

### 98.2.2 The Lack of Corpus of Image Type Spam

Standard and recognized semantic features database could provide a reliable source of data for the study, and a unified set of tests for each classification algorithm, and then reach a different evaluation. While the image type spam corpus is very small, for which we regret.

### 98.2.3 Human Disturbance

Spam as the main network information wantonly spread in the search engines, major blog forum, BBS forums, which we can see it everywhere. When we open a Web page, major image spam will be pop up; in order to stop it, the spammer often use jamming technology to add human disturbance. The spam added in human disturbance does not affect the readability of the image information.

### 98.2.4 The Inefficiency of Processing Images

For the phenomenon which spam is widely increase with age has been a serious interference in human's everyday life, to this end the anti-spam filtering technology will present higher real-time requirements. While the present image processing techniques' processing time is too long, and the image size changes from linear change—even some type of image spam recognition algorithm in dealing with each message will spend 2–3 h, such problem will directly impact on the practicality of the algorithm.

## 98.3 Classification Algorithm of the Image Type Spam

In the classification algorithm of the evaluation of spam, the correct rate, recall rate, and precision rate are more commonly used. The evaluation criteria of spam are suitable for e-mail image classification algorithm [2]. Setting an e-mail image classification results as follows in Table 98.1.

The above evaluation can be calculated by the (98.1)–(98.5):

**Table 98.1** Results judgment of image type spam

| | Spam images | Normal images |
|---|---|---|
| System is judged to be spam image | TP | FP |
| System is judged to be normal image | FN | TN |

$$R = \frac{TP}{TP + FN} \tag{98.1}$$

$$P = \frac{TP}{TP + FP} \tag{98.2}$$

$$Accur = \frac{TP + TN}{TP + FP + FN + TN} \tag{98.3}$$

$$Er = 1 - Accur \tag{98.4}$$

$$F = \frac{2PR}{P + R} \tag{98.5}$$

Some scholars have been done a little contrast test used a different classification algorithm. The Seven Krasner collected 3,600 different spam images and regular image type spam for Corpus. He selected the feature set in which the file and image properties are different, to support the comparison between the vector machine algorithm and the C4.5 decision tree algorithms, are found that the formers lightly better performance than the latter.

As to different feature sets for the algorithm, its advantages and shortcomings, we can take various algorithms for comparison to determine classification in accordance with performance indicators to evaluate, and then to select the best kind of algorithm.

## 98.4 Algorithm Description

### *98.4.1 The Judgment of Corner Possibility*

There are three categories which should not be corner among all of the pixels in the grayscale edge image: (1) If the point pixel value is 0, the point will be background color or interference color; (2) If the pixel value is 1, its eight surrounding points' pixels values are 1; (3) On the B ring of the circular template there exist only 3 point whose pixel values is the sequence 0. When the B ring appears the sequence 1,0,0,0,1, the pixel is not a corner.

### *98.4.2 The Distribution of Pixels in the Circular Template*

When all the pixels in the circular template and the center of the circle pixel (i, j) belong to the same text, they do not adjacent to the edge of the text, as long as we add the edge pixel (i, j) which belongs to edge points in a connected region to the circular template.

### 98.4.3 The Calculation of the Approximate Angle

At first, we should calculate approximate angle value of the *n* effective “0” domain in each ring. Secondly, we should calculate the approximate angle values of the effective “0” domain. Thirdly, we should calculate the maximum angle values for all valid “0” domain, and then calculate the approximate angle value, adding the value recorded into the candidate corners point matrix. Finally, we should returned to the first step, and calculate the approximate angle value.

### 98.4.4 Inhibition of Non-maxima

As to the mid-point (i, j) of the candidate corner point matrix, when the angle value is greater than 0, the steps to determine whether the point is a corner or not are as follows:

1. Selecting the circular template and the same calculating angle values as a local area, and then taking the same second step to calculate the pixel maps.
2. Selecting the corresponding values and points of all the point angle value in dot matrix whose point angle values are “0”, and then calculate the average angle value in the circular area.
3. As to calculation on each points in candidate circular area and the average difference, if the absolute value between the other angle values in the circular area is less than the absolute value between the angle values in point (i, j) and the difference of the mean, the point (i, j) can be considered as the corner, otherwise not.

## 98.5 Experimental Results and Performance Analysis

All the test images selected in this paper are used true spam set of images, as the performance extraction when verify the corner detection algorithm in the image type spam at text corner, by using the corner detection and SUSAN algorithm to extract its image in the corner information.

From the image with slanted text, the SUSAN algorithm extracts 502 angles and gets 537 corners. While from the images with noise points and short-term interference, the SUSAN algorithm extracts 439 angles and gets 419 corners.

Corner detection results show that the corner location accuracy and the number of its algorithm close to the SUSAN algorithm [5]. After experimental analysis shows that SUSAN algorithm has its shortcutting because for the point of view, it missed the more “X” shape, and also considered some of the noise and short-term as corners. Thus the algorithm performance in this passage is superior to the SUSAN algorithm and is steadier. It was not only able to effectively positioning text corner, but also to calculate the approximate angle value in the corner angle. As to a variety of detection algorithms, the size of the angle always has been neglected, but plays an important role in processing a large number of text type spams of this perspective.

## 98.6 Conclusion

The traditional spam's content is text, and the form of it is old-fashioned, while the image type spam is much more fashionable which can hide the text in the figure. Thus need the constant improvement of detection technology. This paper focused on the difficulties existing in the detection of image type spam to give a better recognition algorithm on corner detection in image type spam. The experiments show that this kind of algorithms with a higher identification of recognition and a higher performance of the filter are really important to the recognition of the image detection filtering technology, when the extracted text corner information include a large mount of text type spam. With the rapid development of Internet, the form of the spam is open, efficient and diversity, so the requirements for the spam detection technology will be higher. In order to prevent the distressing brought by the spam, we still need to keep pace with time and develop constantly.

## References

1. Dredzem M, Gevaryahu R, Bachrach AE (2009) Learning fast classifiers for image spam. In: Proceeding of the 4th conference on E-mail and anti-span, vol 31(3), pp 56–70
2. Wu CT (2010) Embedded-text detection and its application to antrspan filtering, vol 56(32). University of California, California, pp 63–65
3. Yaling P, Zhguo Z (2008) Study of CFD aid design of ship water journal bearing of shaft. J Lubr Eng 133:72–76
4. Deligant M, Podevin P, Descombes G (2011) CFD model for turbocharger journal bearing performances. Appl Therm Eng 31:811–819
5. Gertzos KP, Nikolakopoulos PG, Papadopoulos CA (2008) CFD analysis of journal bearing hydrodynamic lubrication by Bingham lubricant. J Tribol Int 41:1190–1204

# Chapter 99
# Face Recognition Dimensionality Reduction Based on LLE and ISOMAP

**Tao Zhang, Shu Li, Shuguang Wu and Lixin Tao**

**Abstract** Local linear embedding (LLE) and isometric feature mapping (ISOMAP) are two basic patterns of nonlinear dimensionality reduction. Their respective strengths and weaknesses in face recognition deserve deep-going comparative study. Therefore, this paper is to test the two patterns' performance efficiency in different parameters, analyze and summarize the two dimensionality reduction pattern's characteristics and scope of application, apply LLE and main constituent analysis into face recognition and summarize probability of detection of face recognition.

**Keywords** Nonlinear dimensionality reduction · Local linear embedding · Isometric mapping · Face recognition

## 99.1 LLE Principle

A given data set $X = \{x_1, x_2, \ldots, x_N\}$ includes D dimensional real-valued vectors of N. And the data is sampled at a potential smooth manifold. It requires sufficient sample data points and each sampled data point and its adjacent points fall on a

T. Zhang · L. Tao
Institute of Computer and Communication, Hunan University of Technology, Zhuzhou, Hunan 412007, China

S. Li (✉)
Hunan City University, Yiyang 413000, China
e-mail: 56705185@qq.com

S. Wu
Modern Education Technology Center, Hunan University of Technology, Zhuzhou, Hunan 412007, China

W. Du (ed.), *Informatics and Management Science V*, Lecture Notes in Electrical Engineering 208, DOI: 10.1007/978-1-4471-4796-1_99,

local linear block or around the block of the potential manifold. Reconstruction of the adjacent point set of each data point will be a set of linear coefficients which can describe the local linear geometry nature of the manifold [1]. The easiest way to determine the number (K) of adjacent point of each data point is to use Euclidean distance. The total reconstruction error is determined by the following cost function $\varepsilon(W) = \sum_{i=1}^{N} ||x_i - \sum_{j=1}^{K} W_{ij}x_j||^2$ Weight matrix (W) is a $N \times N$ dimensional symmetric matrix. And the weight $W_{ij}$ represents the contribution of number j data point with reconstruction of number $i$ data point [2]. To calculate the weights needs to minimize the cost function meeting the following two constraints:

1. Each data point can only be reconstructed by its adjacent points. If number j data point is not the adjacent point of number $i$ data point, $W_{ij} = 0$;
2. All weight matrix columns of elements add up to 1. The optimum weights meeting these two constraints can be obtained by solving a least squares problem [3].

According to constraint 1, the reconstructed cost function can be written as: $\varepsilon(W) = \sum_{i=1}^{N} ||x_i - \sum_{j=1}^{K} W_{ij}x_j||^2$. Then, LLE builds neighborhood to reserve the mapping of high-dimensional observation vector $x_i$ into low-dimensional vector $y_i$ on a manifold. In the mapping process, firstly choose the D dimensional coordinate $y_i$, then remain $W_{ij}$ unchanged, and by optimizing $y_i$ achieve the target minimum of the following embedding cost function $\phi(Y) = \sum_{i=1}^{N} ||y_i - \sum_{j=1}^{N} W_{ij}y_j||^2$

## 99.2 ISOMAP Principle

The key step of the ISOMAP is how to calculate the geodesic distance between sample points. In ISOMPA, the approximation calculation of geodesic distance is as follows: the geodesic distance between sample point $x_i$ and its point in neighborhood is replaced by Euclidean distance between them. While that of sample point $x_i$ and its point beyond neighborhood is replaced by shortest path between them on the manifold [4]. For example:

Select neighborhood and construct the neighborhood graph G. Then calculate the Euclidean distance between each sample point $x_i$ and the rest of the sample points. When $x_j$ is one of the points of K nearest to $x_i$, they are considered as adjacent, that is, graph G has edge $x_ix_j$ (Such a neighborhood is called $K-$ neighborhood). Or when the Euclidean distance $d(x_i, x_j)$ between $x_i$ and $x_j$ is less than fixed value $\varepsilon$, the graph G is considered to have edge $x_ix_j$ (Such a neighborhood is called $\varepsilon-$ neighborhood). Suppose the weight of edge $x_ix_j$ is $d(x_i, x_j)$.

Calculate the shortest path. If the graph G ha s edge $x_ix_j$, suppose the shortest path is $d_G(x_i, x_j) = d(x_i, x_j)$,in contrast, $d_G(x_i, x_j) = \infty$.According to $l = 1, \ldots, N$,

$d_G(x_i, x_j) = \min\{d_G(x_i, x_j), d_G(x_i, x_l) + d_G(x_i, x_j)\}$, the shortest path of distance matrix will be $D_G = [d_G^2(x_i, y_j)]_{i,j=1}^N$ which consists of the square of the shortest path between the sample points of the graph G [5].

Calculate D dimensional embedding. Apply MDS to the distance matrix $D_G$. When

$$B = -(I - 1_N^T 1_N/N) D_G (I - 1_N 1_N^T/N)/2 \tag{99.1}$$

the largest Eigen value $\lambda_1, \ldots, \lambda_d$ of B and corresponding eigenvector $u_1, \ldots, u_d$ make up the matrix $U = [u_1, \ldots, u_d]$ so that $Y = diag(\lambda_1^{1/2}, \ldots \lambda_d^{1/2}) U^T$ is the result of D dimensional embedding.

## 99.3 Efficiency and Contrast of LLE and ISOMAP

MATLAB software has high efficiency in dealing with matrix and programming, so the MATLAB7.1 software is used to calculate program.

### *99.3.1 Effectiveness and Efficiency of LLE Dimensionality Reduction*

#### 99.3.1.1 Swiss Roll Data Sets

The number of data points is 800, that is, X = 800; target dimension after dimensionality reduction are two-dimensional, i.e. d = 2; construct the neighborhood graph with close neighbors K, K = 8, 12.

It can be seen from Table 99.1 when the selected number of neighbor points is different, the data point relations of low-dimensional space after dimensionality reduction is different, and the more the neighbor points are, and the longer the computing time is.

The number of data points is 800, 1,600 and 3,200, that is, X = 800, 1600, 3200; target dimension after dimensionality reduction are two-dimensional, i.e. d = 2; construct the neighborhood graph with close neighbors K, K = 8.

**Table 99.1** Dimensionality reduction timetable of swiss roll data sets (different neighbors)

| Dimensionality reduction time (s) | | | | | | | | Average time |
|---|---|---|---|---|---|---|---|---|
| 1 | 2 | 3 | 4 | 5 | 6 | 7 | 8 | |
| 0.86 | 0.907 | 0.843 | 0.813 | 0.844 | 0.859 | 0.843 | 0.843 | 0.8547 |
| 1.641 | 1.563 | 1.61 | 1.593 | 1.531 | 1.594 | 1.578 | 1.61 | 1.5954 |

Table 99.2 Dimensionality reduction timetable of swiss roll data sets (different sizes of data sets)

| Dimensionality reduction time (s) | | | | | | | | Average time |
|---|---|---|---|---|---|---|---|---|
| 1 | 2 | 3 | 4 | 5 | 6 | 7 | 8 | |
| 0.86 | 0.907 | 0.843 | 0.813 | 0.844 | 0.859 | 0.843 | 0.843 | 0.8547 |
| 3.109 | 3.109 | 3.141 | 3.031 | 3.063 | 3.078 | 3.078 | 3.109 | 3.078 |
| 12.062 | 11.359 | 12.406 | 11.531 | 11.531 | 11.343 | 12 | 11.969 | 11.7185 |

In Table 99.2, the time of dealing with 1,600 data points is 3.601 times the time to 800 data points, and the time of 3,200 data points is 3.807 times the time to deal with 800 data points.

#### 99.3.1.2 Twin Peaks Data Sets

The number of data points is 800, that is, X = 800; target dimension after dimensionality reduction are two-dimensional, i.e. d = 2; construct the neighborhood graph with close neighbors K, K = 8, 12.

The number of data points is 800, 1,600 and 3,200, that is, X = 800, 1600, 3200; target dimension after dimensionality reduction are two-dimensional, i.e. d = 2; construct the neighborhood graph with close neighbors K, K = 8.

By comparing Table 99.3 with Table 99.1, Table 99.4 with Table 99.2, it can be seen that based on the same conditions, LLE processing efficiency of different data sets is similar.

Table 99.3 Dimensionality reduction timetable of twin peaks data sets (different neighbors)

| Dimensionality reduction time (s) | | | | | | | | Average time |
|---|---|---|---|---|---|---|---|---|
| 1 | 2 | 3 | 4 | 5 | 6 | 7 | 8 | |
| 0.875 | 0.906 | 0.906 | 0.922 | 0.891 | 0.906 | 0.89 | 0.859 | 0.8952 |
| 1.672 | 1.719 | 1.625 | 1.687 | 1.704 | 1.75 | 1.719 | 1.593 | 1.6766 |

Table 99.4 Dimensionality reduction timetable of twin peaks data sets (different sizes of data sets)

| Dimensionality reduction time (s) | | | | | | | | Average time |
|---|---|---|---|---|---|---|---|---|
| 1 | 2 | 3 | 4 | 5 | 6 | 7 | 8 | |
| 0.875 | 0.906 | 0.906 | 0.922 | 0.891 | 0.906 | 0.89 | 0.859 | 0.8952 |
| 3.094 | 2.984 | 3.125 | 3.11 | 3.187 | 3.204 | 3.156 | 3.063 | 3.1033 |
| 11.516 | 11.469 | 11.547 | 11.516 | 11.547 | 11.422 | 11.578 | 11.531 | 11.5205 |

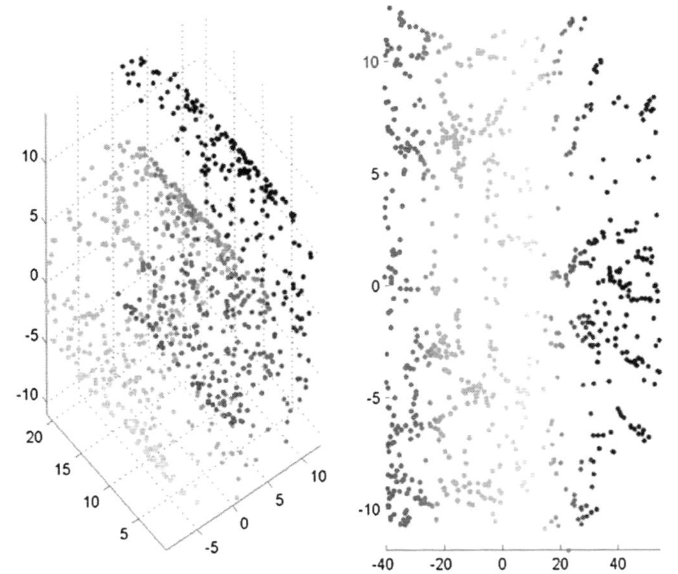


**(a)** Original data sets (800)

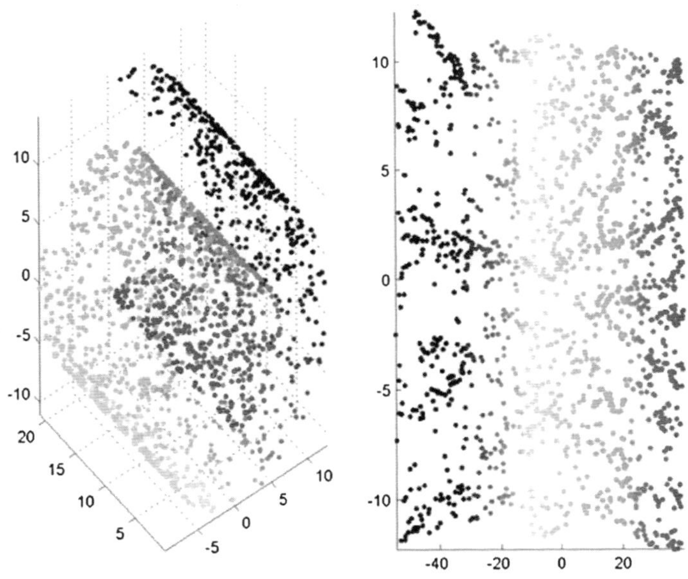


**(b)** K=8 **(c)** original data sets (1600) **(d)** K=8

**Fig. 99.1** Dimensionality reduction effectiveness figure of swiss roll data sets

### *99.3.2 Effectiveness and Efficiency of ISOMAP Dimensionality Reduction*

#### 99.3.2.1 Swiss Roll Data Sets

After the efficiency contrast of LLE, we find there is no need to repeatedly run. The time is almost the same, no more than one second. So ISOMAP efficiency experiments will take the approach that each condition is only run once.

In Fig. 99.1a, the computing time of the data sets is 55.819 s and Fig. 99.1c, 455.437 s. It is clear that the computing time of ISOMAP is a little long but the efficiency is not high.

## 99.4 Conclusion

Swiss roll is used as the test data, and the number of samples and close neighbors are respectively 800 and 12. The use of LLE and ISOMAP reduces above data sets from three-dimensional to two-dimensional and obtain the processing time 0.8952 s and 55.819 s respectively. When the number of samples adds to 1,600, the obtained processing time is 3.1033 s and 455.437 s respectively. It concludes that the processing time of LLE is better than that of ISOMAP.

## References


1. Zejie W (2008) Comparative analysis of two types of nonlinear reduced-dimensional manifold learning algorithm. Academic paper of Shanghai University of Engineering Science 22(1):54–59
2. Ziqiang W, Xu Q, Min K (2008) A summary of manifold learning algorithm. Comput Eng Appl 44(35):9–12
3. Qing L (2008) Semi-supervised handwriting recognition. Chinese Science and Technology University, 49:372–378
4. Varini C, Degenhard A, Nattkemper TW (2006) ISOLLE: LLE with geodesic distance. Neurocomputing 17:1768–1771
5. Roweis ST, Saul LK (2000) Nonlinear dimensionality reduction by locally linear embedding. Science 8:2323–2326

# Chapter 100
# Blind Watermarking Algorithm Based on Singular Block Value

**Ning Zhong, Zhike Kuang, Shuguang Wu and Lixin Tao**

**Abstract** This paper proposes a brand-new blind watermarking algorithm which is based on singular block value. Through combination of image decomposition technique and digital watermarking, it makes watermark technology extraction accomplished so as to make the embeddings and extraction of digital image watermarking much faster, and then reduces the decomposition process of image singular value, test and verify the robustness of this algorithm.

**Keywords** Block-SVD · Blind watermarking · Copyright protection · Robustness

## 100.1 Introduction

With rapid development of computer network technology and computer multimedia technology in recent years, E-commerce industries are rapidly emerging. Newly-generated service content and mode of operation bring more development opportunities and ways for various aspects, like business operation, scientific research and the people's entertainment life [1]. However, technology is a double-edged sword.

N. Zhong · S. Wu
Modern Education Technology Center, Hunan University of Technology, Hunan Zhuzhou 412007, China

Z. Kuang (✉)
Network Information Center, Hunan City University, Yiyang 413000, China
e-mail: 3270813@qq.com

L. Tao
Institute of Computer and Communication, Hunan University of Technology, Zhuzhou 412007 Hunan, China

W. Du (ed.), *Informatics and Management Science V*, Lecture Notes in Electrical Engineering 208, DOI: 10.1007/978-1-4471-4796-1_100,

Because the emergence of new technologies makes the pirates use more low-cost and more efficient technology to illegally copy and disseminate multimedia digital product information without the owner's authorization [1]. Network transmission of multimedia information security issues and copyright protection for digital products has become more serious. In order to solve the problems of the security of transmission of digital multimedia information and digital copyright protection, there have been a lot of protection technologies. Digital watermarking technology, as a new, more popular information security technology to digital multimedia copyright protection in the 1990s, is gradually attracting more and more people's attention and concern [2].

Digital watermarking technology, which combines a variety of techniques like date communication technology, signal processing techniques, computer technology and so on, is becoming a hot issue for the experts and scholars at home and abroad in recent years. In foreign countries, the digital watermark technology started earlier than the China. Related technologies' theory and algorithm research have attracted the scholars' attention all over the world for a long time ago. In China, the watermark research started late, but with great importance attached by the government, we have also made some achievement. Under such circumstance, we propose a new watermarking algorithm based on block singular value. After the using the singular value decomposition method combined with the image sub-block technology, watermark embedding and extraction have been done. The result shows that it can protect image copyright in a certain extent.

## 100.2 Digital Watermarking Technology's Principle and Model

The main idea of digital watermarking technology is that through a predetermined algorithm, using redundant data bits and randomness prevailing in the general digital works to embed some special information into the digital works without affecting the function and value of digital information content and without perceiving or noting by the human perceptual system, the embedded digital watermark information extracts through the appropriate design of the extraction algorithm [3]. Differing from the traditional encryption technology, digital watermarking technology cannot fundamentally prevent piracy activities. However, by the judgment of the digital watermark information, we can determine whether the digital products are protected or not, monitor protection data dissemination, distinguish the true digital products from the false and trace the information leakage source. To further enhance the security and confidentiality of digital watermarking, most of the watermarking algorithm usually makes use of the cryptography encryption system in practical applications. Through the using of a key or key joint in the watermark embedding and extraction processor, we can ensure the security of the watermark information [4]. Figures 100.1 and 100.2

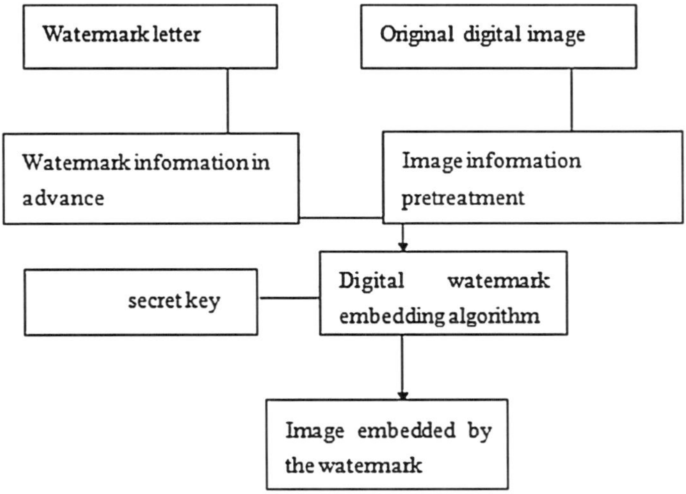


**Fig. 100.1** Digital watermark embedding model

**Fig. 100.2** Digital watermark extracting model

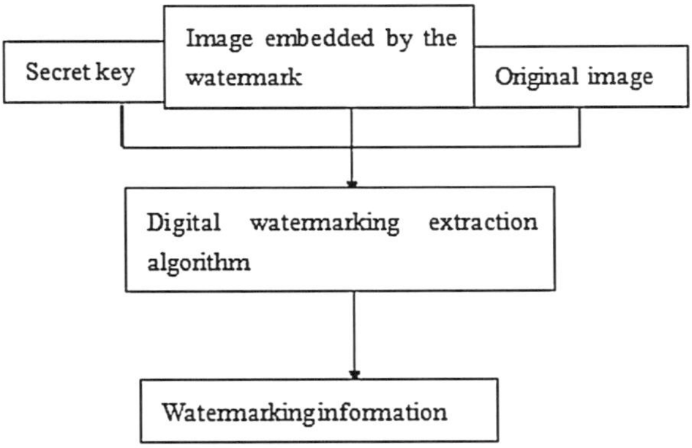


gives the basic model of the digital watermark embedding process and extraction process. The model consists of two parts, one is the watermark embedding process, the other is the watermark extraction process. Generally speaking, depending on the difference of algorithm design, water extraction output results can be carrier of the embedded watermark or also can be evaluation results to judge the watermark exists or carrier whether to be tampered. At the same time, we also can consider whether it needs the original image as an auxiliary in the extraction of the watermark based on the difference of the watermarking algorithm.

## 100.3 Singular Value Watermarking Algorithm Based on Block-SVD

### 100.3.1 Singular Value Watermarking Algorithm Based on SVD

From the point of view of image processing, a digital image can be seen as a matrix combined by many non-negative scalars, the singular value decomposition of this matrix has the following features: Its singular value does not dramatically change when the image is slightly disturbed. Its relatively small singular value changes, the image would not change a lot. Because of the singular value's good mathematical characteristics and good stability, singular value decomposition is not only used in signal decomposition and reconstruction, noise reduction, data fusion, target recognition, tracking, and neural network; but also used in article classification, image compression and digital watermarks.

Singular value watermarking algorithm based on SVD is a transform domain method. The idea is that the watermark is overlaid to the singular value of the original image, the error between the watermark image and the original image is easily estimated, the watermark energy and capacity control can be easy to solve, and it has strong robustness.

Matrix singular value decomposition is the orthogonal transform, and can be matrix diagonal. From the perspective of linear algebra, a grayscale image can be seen as a non-negative matrix [5]. If an image is indicated as A, the definition is $A \in R^{m \times n}$ which said that $R$ is the real number field, then the singular value decomposition matrix A is defined as follows: $A = USV^T$, U and R are orthogonal matrix, $U = [u_1, u_2, \ldots, u_n] \in R^{m \times m}$, $V = [v_{1,} v_2, \ldots v_n] \in R^{n \times n}$, $S \in R^{m \times m}$, it is a diagonal matrix, its diagonal elements meet $\sigma_1 \geq \sigma_2 \geq \cdots \geq \sigma_r > \sigma_{r+1} = \cdots$ $\sigma_m = 0$, in this equation, r is A's rank Which is equal to non-0 the number of singular value, so it has

$$\sum = \begin{bmatrix} \sigma_1 & 0 & 0 \\ 0 & \cdots & 0 \\ 0 & 0 & \sigma_m \end{bmatrix}$$

$\sigma_i$ is uniquely determined by the decomposition, called the A's singular value. It is the square root of $AA^T$, decomposition, $USV^T$, is called as the A's singular value decomposition.

The main theoretical basis for the application of singular value decomposition in digital image processing: image singular value stability is very good, that is, when the image is applied to small disturbances, the singular values do not vary widely, the performance of the singular value is the intrinsic characteristics of the image rather than the visual characteristics, and reflects the relationship between the image matrix elements.

## 100.4 Singular Value Watermarking Algorithm Based on Block-SVD

### 100.4.1 Watermark Embedding Process

The process of singular value watermarking algorithm based on block-SVD is follow:

Adjusting the original size for the preparation of blocking

Set the original image as A, matrix A's row and column is M and N respectively. If M and N is a multiple of 8, there is no need to resize. If not, it needs to resize to be a multiple of 8 and reset the row and column as $M'$ and $N'$, $p = M'/8$, $q = N'/8$. Adjusted method as follow: $M' = M + [8 - (M \bmod 8)]$, $N' = N + [8-(N \bmod 8)]$. Resizing the image to increase the image area to make it gray-scale value of 0.

Decomposing the singular value on each piece after blocking.

A is divided into non-overlapping $8 \times 8$ small pieces, noting every piece as $B_{ij} = f_{ij}(x, y)$, parameters are as follows $i = 1, 2 \ldots p$, $j = 1, 2 \ldots q$, $1 \le x, y \le 8$, factorization as:

$$B_{ij} = U_{ij} S_{ij} V_{ij}^T \tag{100.1}$$

Set $\sigma_{ij}$ as the maximum singular value in $S_{ij}$, extracting the largest singular value of each block to form a new matrix C, that is:

$$C = \begin{bmatrix} \sigma_{11} & \sigma_{12} & \cdots & \sigma_{1q} \\ \sigma_{21} & \sigma_{22} & \cdots & \sigma_{2q} \\ \vdots & \vdots & \ddots & \vdots \\ \sigma_{P1} & \sigma_{p2} & \cdots & \sigma_{pq} \end{bmatrix} \tag{100.2}$$

We can decompose the singular value to obtain the factorization as $C = USV^T$.

### 100.4.2 Extracting Process of Watermark

In the watermark detection process, if give matrix $U_1, S, V_1$ and the possible damaged watermark image $A^*$, the distorted watermark $W^*$ can be extracted by a simple inverse process. Watermark extraction steps are as follows:

1. The adjustment of watermark image size, as adjustment of the embedding process, makes the rows and columns the multiples of 8.
2. Block processing, as the process of embedding, divides A into non-overlapping $8 \times 8$ small pieces, then decomposing each sub-block to extract the largest singular value of each block to a new matrix.
3. Set extracting watermark as $W^*$, the formation process as follows:

$$C^* \Rightarrow U^* S_1^* V^{*T} \tag{100.3}$$

$$D \Leftarrow U_1 S_1^* V_1^T \tag{100.4}$$

$$W^* \Leftarrow (D^* - S)/a \tag{100.5}$$

### *100.4.3 Analysis of the Experiment*

From the above statistics and image analysis, it proves that Block-SVD watermarking algorithm has some resistance to a variety of attracts. When the watermarked images was rotated, the performance of sharpening attract was brilliant, the extracted watermark was distinct, NC value was high. Under the circumstance of the cutting and immediately noise attract, the definition is slightly low and distortion, but the whole similarity is high. It also proves that the Block-SVD watermarking Algorithm can resist kinds of attract and the robustness is very high. However, in the practical process of watermarking algorithm, we also need to pay attention to the time of watermark embedding (the demand of the extracting or embedding time is not so serious). If the embedded time is too long, the method is unpractical. Nevertheless, the embedded watermark experiment proves that the time in the Block-SVD watermarking algorithm is very short.

## 100.5 Conclusion

With the development of digital technology, information hiding technology is valued. Digital watermarking, as an important means for copyright protection, has been widely research and application. Using the singular value decomposition method and combining with the image block technology to complete the watermark embedding and extracting. Through the image block, reduce the singular value decomposition of image of the process, accelerate the digital image watermarking embedding and extracting, finally with VC++ realized the process, and show that the proposed algorithm is robust. The experiment shows that the newly-produced Block-SVD has the features of robustness, is suitable for image copyright protection and is practical.

## References

1. Wang B, Chen Q, Deng F (2008) Digital watermarking technology, vol 34. Xian University of Electronic Science and Technology Press, Xian, pp 13–25
2. Mei T, Siwei L (2010) Research the compression methods of the singular value watermarking algorithm based on SVD. J Beijing Jiaotong Univ 27(2):29–32

3. Yi K (2010) Digital image encryption and research of digital watermarking technology, vol 22. Zhejiang University Press, Hangzhou, pp 20–25
4. Sun S, Lu Z (2009) Digital watermark processing technology. Chin J Electron 5(12):22–27
5. Yang C (2000) Picture and voice compression techniques, vol 3. Zhejiang University Press, Hangzhou, pp 264–301

# Chapter 101
# A Novel Image Fusion Approach Combined Singular Value Decomposition with Averaging Operation

**Jing Luo and Fenghua Liu**

**Abstract** This paper presents a new image fusion algorithm, combined with business singular value decomposition (QSVD) with a simple average operation. Multi-focused image is the first averaging into a new image. The most error-contributing components in each error image are the most contribution to the portion of the image using QSVD Multi-focused to reduce mistakes. Each reduce error image, put forward a new kind of calculation singular vectors fusion image. Finally get to decide to fill each image fusion image through the calculation standard deviation. The experimental results, such as mutual information (MI), information entropy (IE), maintain edge information (Qabf) to the signal- noise-ratio (SNR) and root mean square error (RMSE) is used to assess algorithm. The experimental results show that the algorithm is a kind of high efficient development fusion algorithm.

**Keywords** Image fusion · Singular value decomposition · Averaging operation

## 101.1 Introduction

Collecting data from complex hardware of the visual scene increased ability to extract information from the exogenous environment. In order to more effective use of the hardware, the image fusion technology also needs to become more complicated. Based on image fusion feature, delegate multi-classifier combination,

J. Luo (✉) · F. Liu
Computer and Software Academe, Zhongyuan Institute of Technology, Zhengzhou, Henan 450007, China
e-mail: luojing@cssci.info

W. Du (ed.), *Informatics and Management Science V*, Lecture Notes in Electrical Engineering 208, DOI: 10.1007/978-1-4471-4796-1_101,

has been the focus of people's attention, increase pattern recognition. Characteristics of fusion algorithm advantages are obvious, because of the different characteristic vectors the different characteristics of the same pattern extraction always reflect mode. Through the optimization of these different features, and combined with, not only is effective information fusion of discrimination, but the redundant information maintenance and wiped out, to a certain degree. In this paper, we use QSVD, this is a kind of effective method of feature extraction of image fusion is a generalized QSVD singular value decomposition (holy). The singular value decomposition, widely used in mathematics [1, 2], and on the basis of numerical calculation is a very useful and universal tool. Many multivariate analysis technique is often used in signal processing is based on singular value decomposition of a certain matrix. For example, in the principal component analysis (PCA) got the singular value decomposition of a data matrix or covariance matrix. In a typical related analysis singular value decomposition of the calculation of the products, formed from the orthogonal projection two sets of data matrix. Often, however, a singular value decomposition of matrix we are to obtain is defined as a product or a index two or some matrix. A rectangular matrix decomposition of QSVD, consider limiting and good at the columns of the matrix. The yield QSVD weighted generalized least square method to estimate of a given matrix rank matrix with low, so, choose appropriate QSVD constraints, realize the linear methods (such as analysis, canonical correlation analysis, linear discriminate analysis, correspondence analysis).

### *101.1.1 QSVD*

The QSVD is a generalization of the singular value decomposition. Let $A \in R^{n\times p}$ and $B \in R^{m\times p}$ be given with $n \geq p$. Define $Y = [A^T|B^T]^T$ and assume that $rank(Y) = p$. There exist orthogonal matrices $U \in R^{n\times n}$ and $V \in R^{m\times m}$ and nonsingular matrix $X \in R^{p\times p}$ such that $A = USX^{-1}$, $B = VCX^{-1}$, where $S \in R^{n\times p}$ and $C \in R^{m\times p}$ have nonzero elements $\{s_1, s_2, \ldots, s_p\}$ and $\{c_1, c_2, \ldots, c_{\min\{m,p\}}\}$, respectively, only on their main diagonal with $0 \leq s_1 \leq \cdots \leq s_p \leq 1$ and $1 \geq c_1 \geq \cdots \geq c_{\min\{m,p\}} \geq 0$. Furthermore, it holds that $s_j^2 + c_j^2 = 1$ for $J = I, \ldots, \min\{m, p\}$ and $s_j = 1$for $j = m+1, \ldots, p$. If m < p, define $c_{m+1} = \ldots = c_p = 0$. Then $s_j^2 + c_j^2 = 1$ for all $J = I, \ldots, p$ the values $\sigma_j = s_j/c_j$ are called quotient singular values of the pencil $\{A, B\}$. If $c_j = 0$, then $\sigma_j = \infty$. The jth column $x_j$ of $X$ is the right generalized singular vector associated with $\sigma_j$. It follows that $c_j^2 A^T A x_j = s_j^2 B^T B x_j$. Hence, the finite generalized singular values are the square roots of the finite generalized eigenvalues of the pencil $\{A^T A, B^T B\}$ [3–5].

### *101.1.2 Image Fusion Using QSVD*

Let $F_1 \in R^{m\times p}$ and $F_2 \in R^{m\times p}$ be registered images to be fused. Denote their average image as A, i.e.

$$A = 0.5(F_1 + F_2) \tag{101.1}$$

Perform QSVDs on {A, $F_1$} and {A, $F_2$} respectively, namely:

$$A = U_1 C_1 W_1^T \quad and \quad F_1 = V_1 S_1 W_1^T \tag{101.2}$$

$$A = U_2 C_2 W_2^T \quad and \quad F_2 = V_2 S_2 W_2^T \tag{101.3}$$

Where $U_1^T = [u_1, u_2, u_3, \ldots u_m], u_1 \in R^{m\times 1}$, $W_1 = [w_1, w_2, w_3, \ldots w_p], w_1 \in R^{p\times p}$, $U_1$, $V_1 \in R^{m\times m}$, $C_1, S_1 \in R^{m\times P}$, $W_1 \in R^{p\times p}$, $U_1 U_1^T = V_1 V_1^T = I_m$, and $C_l^T C_l + S_1^T S_1 = I_p$ as explained in section II. Note that all matrices in Eq. (101.3) are defined in the same manner as in Eq. (101.2), but using image $F_2$ instead of $F_1$, let's define $E_1 = A - F_1$ and perform the SVD on $E_t$.

$$E_1 = A - F_1 = U_1 C_1 W_1^T - V_1 S_1 W_1^T = U_{e_1} V_{e_1}^T \tag{101.4}$$

Where $U_{e_1}^T = [u_1^e, u_2^e, u_3^e, \ldots u_m^e]$, $V_{e_1}^T = [v \quad v \quad v \quad \ldots \quad v]$, $u_i^e \in R^{m\times 1}$, $v \in R^{p\times 1}$ and $\sum_{e_1}$ has nonzero elements $\{\sigma_1^e, \sigma_2^e, \sigma_3^e \ldots \sigma_p^e\}$ only on its main diagonal. To reduce errors, let's replace the most contributing error part with the most contributing image part in the average image [6, 7]. This results in new matrices, $\tilde{U}_{e_1}$, $\tilde{\Sigma}_{e_1}$ and $\tilde{V}_{e_1}$. So we can define a new error matrix $\tilde{E}_1$:

$$\tilde{E}_1 \underline{\underline{def}} \tilde{U}_{e_1} \sim \tilde{V}_{e_1}^T \tag{101.5}$$

Where $\tilde{U}_{e_1}^T - [u_1, u_2^e, u_3^e \ldots u_m^e]$, $\tilde{V}_{e_1} = \left[\frac{w_1}{\|w_1\|}, v_2^e, v_3^e \ldots v_p^e\right]$ and $\tilde{\Sigma}_{e_1}$ has revised nonzero elements of $\{c_{11}/s_{11}, \sigma_2^e, \sigma_3^e \ldots \sigma_p^e\}$. Notice that the singular vectors and the first singular value in error matrix of Eq. (101.4) are replaced by using the elements of the average image obtained with QSVD in Eq. (101.2). Then, the $\tilde{U}_1$ term corresponding to reduced error matrix of (101.5) can be obtained using the Eq. (101.4) as the following Eqs. (101.6) and (101.7), where † implies the pseudo inverse:

$$\tilde{E}_1 \underline{\underline{def}} \tilde{U}_{e_1} \tilde{\Sigma}_{e_1} \tilde{V}_{e_1}^T \underline{\underline{def}} \tilde{U}_1 C_1 W_1^T - V_1 S_1 W_1^T \tag{101.6}$$

$$\Rightarrow \tilde{U}_1 = (\tilde{U}_{e_1} \tilde{\Sigma}_{e_1} \tilde{V}_{e_1}^T + V_1 S_1 W_1^T)(C_1 W_1^T)^{\dagger} \tag{101.7}$$

Thus, we can finally get a new image $\tilde{A}_1$ that includes the most important parts of average image A, while reducing the error as equation, i.e.

**Fig. 101.1** Pseudo code for the image fusion algorithm

```
A = (F1 +F2)/2;
fork=l: 1:2, [Uk,Vk, Wk., Ck,Sk)=gsvd(A,Fk,0);
    Ek =A - Fk;
 ifmax(Ek(:)) > e [Uek  Uek Uek ]= svd(Ek,0);
 Obtain Ũ_e1, Σ̃_e1, andṼ_e1 . Obtain Uk using equation (1.7)
     Ã_k = Ũ_k Σ̃_k W_k^T using equation (1.8)
        else Ã_k = A
       end
end
Choose the final infused image using equation (1.9)
```

$$\tilde{A}_1 = \tilde{U}_1 C_1 W_1^T \tag{101.8}$$

In the same way, let $E_2 = A - F_2$, then we can obtain a new image $\tilde{A}_2$ with registered image $F_2$ and average image$A$ as done with for $F_1$. Because the standard deviation of an image denotes the distributing status of the gray values relative to the mean gray scale, if the standard deviation is bigger, the gray value distributes more dispersedly, and the image is clearer. Hence, we take the standard deviation as a fusion rule to choose $\tilde{A}_1$ or $\tilde{A}_2$ as the final fused image as

$$F = \begin{cases} \tilde{A}_1, \ if \ \sigma_{\tilde{A}_1} \geq \sigma_{\tilde{A}_2} \\ \tilde{A}_2, \ if \ \sigma_{\tilde{A}_1} < \sigma_{\tilde{A}_2} \end{cases} \tag{101.9}$$

The new image fusion algorithm in this paper can be explained by MATLAB-like pseudo code as shown in Fig. 101.1.

## 101.2 Experimental Results

To test the algorithm proposed in this paper, several groups of images have been fused and evaluated by the objective tests explained in section IV. In order to compare the fusion quality, we fuse the test images using wavelet fusion [8] which has been the prevalent algorithm in recent years, and average fusion which is the simplest fusion method in Eq. (101.1). One of them with the size of 512 × 512 is shown in Fig. 101.2 and the objective test results are shown in Table 101.1, where RMSE and SNR are not available because a clean or focused reference image of Pepsi is not available. The algorithm presented in this paper is denoted by QSVD in the figures and tables. A test image is shown in Fig. 101.3, where (a) is the clean reference image and (b), (c) are two blurred images. Image (b) is blurred at the upper right comer and image (c) is blurred at the lower left comer.

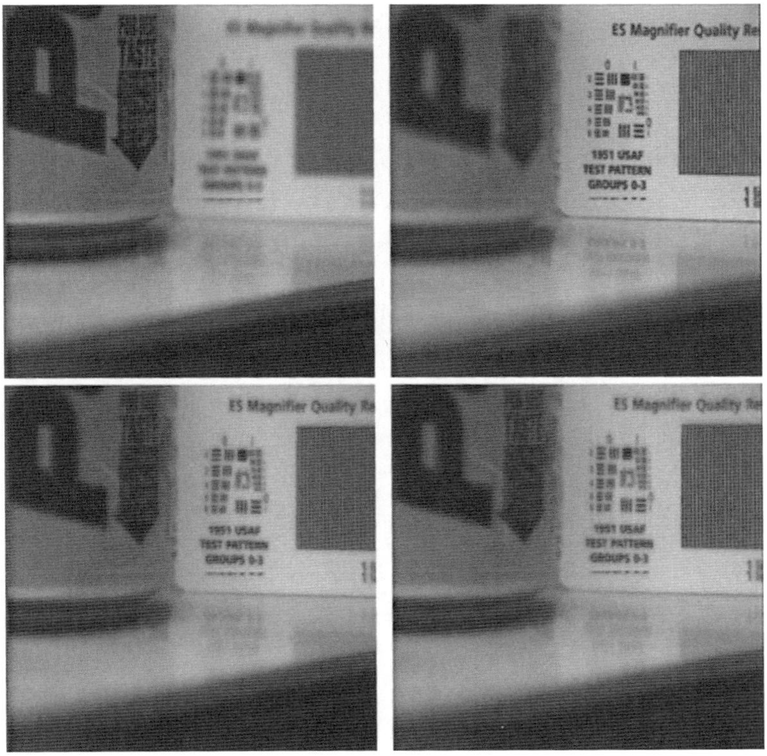


**Fig. 101.2** Image fusion comparison for pepsi image. **a** Focused on near part, **b** focused on far part, **c** wavelet fusion, **d** QSVD fusion

**Table 101.1** Comparison results of image fusions for Pepsi image

| Algorithms | MI | IE | $Q_{abf}$ |
|---|---|---|---|
| Average | 1.5161 | 7.0805 | 0.6295 |
| Wavelet | 1.5175 | 7.0882 | 0.6304 |
| QSVD | 1.5320 | 7.1027 | 0.6784 |

## 101.3 Conclusion

This paper put forward a new fusion algorithm is proposed on the basis of QSVD. Then, give the image fusion estimation, the results in the singular close matching the reference image. The experimental results show that the wavelet analysis and performance improvement average fusion when using an objective test process.

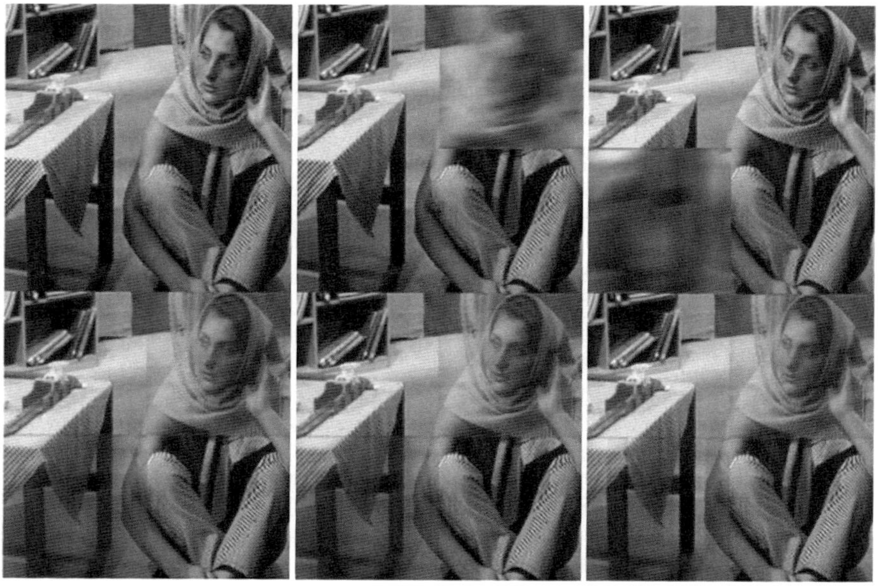

**Fig. 101.3** Image Fusion comparisons for Barbara image. **a** Clean image, **b** upper right blurred, **c** lower left blurred, **d** average fusion, **e** wavelet fusion, **f** QSVD fusions

# References


1. Hom RA, Johnson CR (1987) Matrix analysis, vol 35(2). Cambridge University Press, Cambridge p 77
2. Golub GH, Van Loan CF (1996) Matrix computation, vol 67(5) 3rd ed. John Hopkins University Press, Baltimore pp 110–118
3. Paige CC (1986) Computing the generalized singular value decomposition. Siam J Sc Stat Comp 7(4):126–146
4. Paige CC, Saunders MA (1981) Towards a generalized singular value decomposition. Siam J Num Anall 8(3):198–405
5. Betcke T (2008) The Generalized singular value decomposition and the method of particular solutions. Siam J Sci Comput 30(3):1278–1295
6. Hossny N, Nahavandi S, Creighton D (2008) Comments on "Information measure for performance of image fusion". Electron Lett 44(8):1066–1067
7. Xydeas CS, Petrovic V (2000) Objective image fusion performance. Electron Lett 36(3):308–309
8. Sun YQ, Li LP (2010) Research on wavelet base selection in infrared image fusion. J Comput Info Syst 6(1):2823–2831

# Chapter 102
# DCT-Based Blind Watermarking of 3D Models

**Xin Zhou, Xun Wang and Dingjun Huang**

**Abstract** To propose robust blind watermarking methods of 3D models based on DCT. First a 3D mesh models will be mapped to a 2D parametric mesh with a kind of planar parameterization method, geometric signals are then transformed into 2D signals. Then a DCT-based watermark scheme is proposed to embed the watermark into some DCT coefficients. The watermark can be detected without the original 3D models. Experimental results show that the embedded watermark is robust against various geometry signal processing.

**Keywords** Digital watermarking · 3D mesh model · Planar parameterization · Discrete cosine transform

## 102.1 Introduction

Digital media data have been widely distributed through Internet in recent years. It creates an urgent need for appropriate copyright protection schemes for digitized media. Digital watermarking provides a novel way to achieve effective copyright protection [1, 2]. Many publications on watermarking have been concentrated focuses on media types like still images, and video and audio streams [3–6]. As a

X. Zhou (✉)
Hangzhou Vocational and Technical College, Hangzhou, China
e-mail: zhouxin@hzvtc.edu.cn

X. Wang · D. Huang
Zhejiang Gongshang University, Hangzhou, China
e-mail: wx@mail.zjgsu.edu.cn

D. Huang
e-mail: huangdingjun@163.com

W. Du (ed.), *Informatics and Management Science V*, Lecture Notes in Electrical Engineering 208, DOI: 10.1007/978-1-4471-4796-1_102,

new media data type, 3D Models have been widely applied in industry in recent years [7–11]. Techniques for watermarking 3D models have received more attention especially for the purpose of copyright protection [12, 13]. A watermarking technique that requires the original 3D models to detect the watermark is called non-blind watermarking [14–17]. On the other hand, a blind scheme does not require the original 3D models to detect the watermark [18–21]. The blind watermarking technique is an important means for copyright protection [22–25].

In this paper, we propose a robust blind watermarking method of 3D models based on DCT. First 3D mesh models will be mapped to a 2D parametric mesh using a kind of planar parameterization method, geometric signals are then transformed into 2D signals. Then a DCT-based watermark scheme is proposed to embed the watermark into some selected DCT coefficients. The watermark can be detected without the original 3D models. Experimental results show that the embedded watermark is robust against various geometry signal processing.

## 102.2 The Proposed 3D Mesh Model Watermarking Scheme

Figure 102.1 shows the overview of our watermarking scheme.

As illustrated in Fig. 102.1, our 3D mesh model watermarking scheme consists of the following components:

### *102.2.1 Primary Component Analysis*

In order to detect watermark after similarty transform attack. The 3D mesh model must be adjusted automatically to a unique posture [26, 27]. Primary Component Analysis (PCA) method mentioned in [28] is used to ensure the 3D meshes have the unique posture before embedding and blind detection.

### *102.2.2 Planar Parameterization*

A 2D parametric mesh D isomorphic to M is constructed using a planar parameterization algorithm and the original geometric signal $F_M$ is transformed into 2D signal $F_D$ [29, 30].

### *102.2.3 Sampling*

The 2D signal $F_D$ is adaptively sampled into a regular signal $F_U$ under the user specified precision threshold.

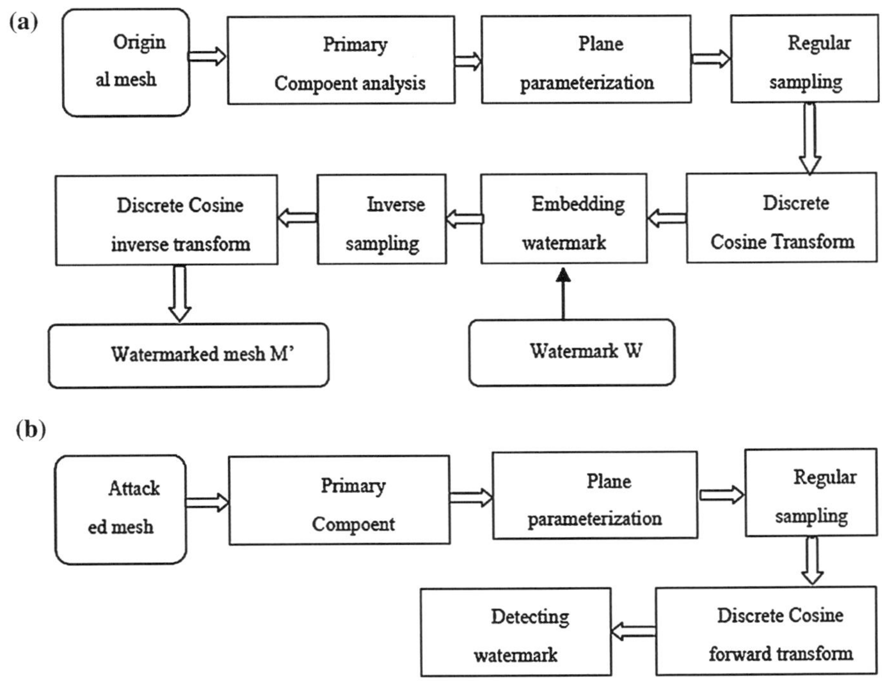


**Fig. 102.1** 3D mesh model watermarking scheme **a** Watermark insertion, **b** Watermark detection

### *102.2.4 Watermark Embedding and Detection*

A DCT-based watermark embedding scheme is proposed to transform the regular signal $F_U$ into frequency domain, the watermark can be adaptively embedded in some selected DCT coefficients under the user specified threshold, and then synthesized into a new watermarked signal $F'_U$ with the inverse DCT. In the watermark extraction process, first, the DCT is applied to both an original 2D regular signal and a watermarked 2D regular signal. Comparing their DCT coefficients, we detect the watermark.

### *102.2.5 Resampling and Inverse Mapping*

Together with the regular signal $F'_U$ and the parametric mesh D, a new signal $F'_M$ is reconstructed using linear interpolation and the inversed mapping of the parameterization.

## 102.3 DCT-Based Blind Watermarking Algorithm

Piva et al. [31] propose a frequency-domain DCT-based watermarking algorithm in, which was designed for DCT transform on the full image, then the watermark consisting of a pseudo-random sequence of real numbers will be embedded in some selected coefficients. Embedding watermark using the HVS properties of the process to ensure the invisibility of the watermark. Using the statistical properties of the watermark sequence, the watermark detection process does not require the original image data. We apply this algorithm to 3D mesh models.

### *102.3.1 Generation of Watermark*

We generate the watermark vector $w = (w_1, w_2, \ldots, w_m)^T$, where m is watermark length, m = 1,000. Its coefficients $w_i$ are real numbers sampled from a Gaussian distribution with zero mean and variance 1.

### *102.3.2 Watermark Insertion*

Watermark insertion process is composed of DCT of the 2D original regular signal, DCT coefficients select, watermark insertion, and IDCT of the coefficients with watermarks.

**DCT coefficients select:** The DCT coefficients selecting procedure can be summarized as follows.

The DCT coefficients of the 2D original regular signal are reordered into a zig–zag scan.

Select the coefficients from the (L + 1)th to the (L + m)th according to the zig–zag ordering of the DCT spectrum, where the first L coefficients are skipped to achieve the perceptual invisibility of the watermark.

**Insertion:** The watermark is inserted as follows:

$$\begin{aligned} v_{ix}^{'} &= v_{ix} + w_i \cdot \alpha \cdot |v_{ix}| \\ v_{iy}^{'} &= v_{iy} + w_i \cdot \alpha \cdot |v_{iy}| \\ v_{iz}^{'} &= v_{iz} + w_i \cdot \alpha \cdot |v_{iz}| \end{aligned} \tag{102.1}$$

where $v_{ix}'$, $v_{iy}'$, $v_{iz}'$ are the watermarked DCT coefficients for each of the three spatial coordinates X, Y, Z, $v_{ix}$, $v_{iy}$, $v_{iz}$ are the original DCT coefficients for each of the three spatial coordinates X, Y, Z, $\alpha$ is scaling factor, and $w_i$ is the $i$th watermark element in a watermark sequence of length m.

### 102.3.3 Watermark Detection

The watermark detection process is composed of DCT of watermarked 2D regular signal, DCT coefficients select and calculation of the correlation.

**Calculation of the correlation**: The correlation between the DCT coefficients are calculated as follows:

$$\begin{aligned} D_x &= \sum_{i=0}^{m} v_{ix}^* * w_i / .m \\ D_y &= \sum_{i=0}^{m} v_{iy}^* * w_i / .m \\ D_z &= \sum_{i=0}^{m} v_{iz}^* * w_i t / .m \end{aligned} \tag{102.2}$$

where $v_{ix}^*$, $v_{iy}^*$, $v_{iz}^*$ are the detected DCT coefficients for each of the three spatial coordinates X, Y, Z, $D_x$, $D_y$, $D_z$ are the correlation between the DCT coefficients of the three spatial coordinates X, Y, Z, and $w_i$ is the $i$th watermark element in a watermark sequence of length m.

The correlation $D_x$, $D_y$, $D_z$ can be used to determine whether a given watermark is exist or not with comparing to a predefined threshold $T_z$. When the correlation is greater than $T_z$, that is the given watermark is exist.

Because of the embedded watermark for each of the three spatial coordinates X, Y, Z is same, we can take the maximum of $D_x$, $D_y$, $D_z$ as the correlation.

## 102.4 Experimental Results

We implemented our algorithm in MSVC++ 6.0 on a P4 2.93 MHz machine. The test result in this section were obtained using a watermark length of m = 1,000 coefficients, the scaling factor $\alpha = 0.1$, L = 1,000.

### 102.4.1 Threshold Setting

The correlation threshold is calculated as follows. The paper [31] describes the technical details.

$$T_{\max} = \frac{\sum_{i=0}^{m} |v_i^*| * \alpha}{2 * m} \tag{102.3}$$

$$T_{\min} = 2 * T_{\max}/3 \tag{102.4}$$

where $T_{max}$ the maximum of the threshold is, $T_{min}$ is the minimum of the threshold, $\alpha$ is scaling factor, m is the watermark length.

### *102.4.2 Attack Analysis*

Table 102.1 shows the detection results for a host of attacks. We refer to the attack test method in the paper [32]. The watermark generating key is 123.
Entries with asterisks correspond to pictures in Fig. 102.2.

The experimental results show that our watermarking algorithm is robust under a variety of signal processing, including noise addition, multiple watermarking, mesh simplification, multiresolution filtering and enhancement, cropping and similarity transform.

**Table 102.1** The detection results for various attacks

| Attack | Venus | | |
|---|---|---|---|
| | Detected correlation | $T_{max}$ | $T_{min}$ |
| A: No attack | 2.246247 | 0.685582 | 0.457055 |
| B: Vertex reorder | 2.243400 | 0.685542 | 0.457028 |
| C: Noise 0.2 % | 2.242669 | 0.685874 | 0.457249 |
| D: Noise 0.45 % | 2.122979 | 0.687043 | 0.458029 |
| E: Noise 0.7 % | 2.090259 | 0.688402 | 0.458935 |
| F: Smoothing | 2.185404 | 0.680448 | 0.453632 |
| F1: Enhancing | 2.204550 | 0.698894 | 0.465929 |
| G: Similarty transform | 2.246247 | 0.685582 | 0.457055 |
| H: Simplify 50 % | 1.882965 | 0.684867 | 0.456578 |
| I: Simplify 80 % | 1.608073 | 0.684236 | 0.456157 |
| J: 2nd watermark | 2.187989 | 0.689851 | 0.459901 |
| K: Crop 1 | 2.050948 | 0.688203 | 0.458802 |
| L: Crop 2 | 2.143567 | 0.686389 | 0.457593 |
| M: Crop 3 | 2.149071 | 0.685979 | 0.457319 |
| N: B + C | 2.235771 | 0.684320 | 0.456214 |
| O: B + G | 2.243400 | 0.685542 | 0.457028 |
| P: C + G | 2.252669 | 0.685874 | 0.457249 |
| Q: G + N | 2.208022 | 0.687800 | 0.458533 |
| R: B + C + G | 2.261669 | 0.684420 | 0.456280 |
| S: B + G + H | 1.882965 | 0.684867 | 0.456578 |
| T: ALL (B,C,F,G,H,J,K) | 1.728268 | 0.694384 | 0.462922 |

## 102.5 Conclusions

In this paper a watermarking algorithm for 3D mesh models in the frequency domain is presented: 3D mesh models will be mapped to a 2D parametric mesh using a kind of planar parameterization method, geometric signals are then

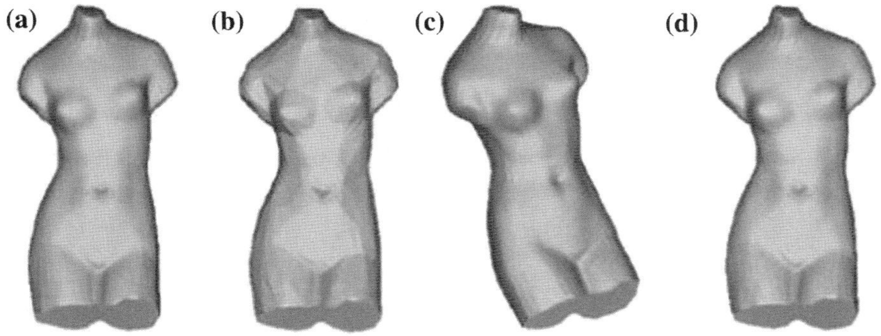


**Fig. 102.2** Watermarked models and various attacks. **a** Watermark model, **b** 80 % faces, **c** similarity transform, **d** 2nd watermark

transformed into 2D signals. The 2D signals are adaptively sampled into a regular signal under the user specified precision threshold. A pseudo-random sequence of real numbers having normal distribution with zero mean and unity variance is embedded in a selected set of DCT coefficients of 2D regual signals. The watermark can be detected without the original 3D models. Experimental results show that the embedded watermark is robust against various geometry signal processing.

## References

1. Anderson RJ (1996) Information hiding: first international workshop, vol 2. Springer, Berlin, pp 30–38
2. Zhao J, Koch E, Luo C (1998) In business today and tomorrow. Commun ACM 41(7):67–72
3. Memon N, Wong PW (1998) Protecting digital media content. Commun ACM 41(7):35–43
4. Bender W, Gruhl D, Morimoto N (1996) Techniques for data hiding. IBM Syst J 35(3–4):72–79
5. Jin LH, Li H, Song EM, Xu XY (2010) Impulsive noise removal using switching scheme and adaptive weighted median filters. Opt Eng 49(1):1–7
6. Plataniotis KN, Androutsos D, Venetsanopoulos ANN et al (1996) An adaptive nearest neighbor multichannel filter. IEEE Trans Circuits Syst Video Technol 6(6):699–703
7. Cox IJ, Kilian J, Leighton T, Shamoon T (1997) Secure spread spectrum watermarking for multimedia. IEEE Trans Image Process 6(12):1673–1687
8. Hartung F, Kutter M (1999) Multimedia watermarking techniques. Proc IEEE 87(7):1079–1107
9. Lukac R (2003) Adaptive vector median filtering. Pattern Recogn Lett 24(12):1889–1899
10. Jin LH, Li DH (2007) An efficient color impulse detector and its application to color images. IEEE Signal Process Lett 14(6):397–400
11. Jin LH, Li H, Xu XY, Song EM (2010) Quaternion-based color image filtering for impulsive noise suppression. J Electron Imaging 19(4):1–12
12. Pan J-S, Huang H-C, Jain LC, Fang W-C (eds) (2007) Intelligent multimedia data hiding: new directions, vol 9(8). Springer, Heidelberg, pp 467–475
13. Andreadis I, Louverdis G, Chatzianagnostou S (2004) New fuzzy color median filter. J Intell Rob Syst 41:315–330

14. Sangwine SJ (2000) Colour image filters based on hypercomplex convolution. IEEE Proc Vision Image Signal Process 147(2):89–93
15. Sangwine SJ (1998) Coulour image edge detector based on quaternion convolution. Electornics Lett 34(10):969–971
16. Jelali M, Kroll A (2003) Hydraulic servo-system modeling, identification and control. Springer, New York
17. Kim MY, Lee C-O (2006) An experimental study on the optimization of controller gains for an electro-hydraulic servo system using evolution strategies. Control Eng Pract 14(2):127–147
18. Sun H, Chiu GT-C (2002) Motion synchronization for dual-cylinder electro hydraulic lift systems. IEEE/ASME Trans Mechatron 7(2):171–181
19. Chen CY, Liu LQ, Cheng CC, Chiu GTC (2008) Fuzzy controller designs for synchronous motion in a dual-cylinder electro-hydraulic system. Control Eng Pract 16(6):658–673
20. Berger M (1996) Self-tuning of a PI controller using fuzzy logic for a construction unit testing apparatus. Control Eng Pract 4(6):785–790
21. Cheng CC, Chen C-Y (1998) A PID approach to suppressing stick-slip in the positioning of transmission mechanisms. Control Eng Pract 6(4):471–479
22. Franklin GF, Powell JD, Emanmi-Naeini A (1991) Feedback control of dynamic system. Addison-Wesley publishing company, New York
23. Nathans J, Thomas D, Hogness DS (1986) Molecular genetics of human color vision: the genes encoding blue, green, and red pigments. Science 232:193–202
24. Feng T, Xiang W, Jingao L (2008) Research and realization of innovative LED illumination system for DLP projector. In: Audio, language and image processing. International conference on ICALIP 2008, vol 1, pp 194–199
25. Hunt RWG, Pointer MR (1985) A color-apperarance transform for the CIE 1931 standard colorimetric observer. Color Res Appl 10:165–179
26. Cai S, Shen X-K (2011) Octree-based robust watermarking for 3D model. J Multimedia 6(1):83–90
27. Van Kessel PF (2001) Electronics for DLP TM technology based projection systems. In: Symposium on VLSI technology digest of technical papers, pp 91–94
28. Kalivas A, Tefas A, Pitas I (2003) Watermarking of 3D models using principal component analysis. In: Proceeding of acoustics, speech and signal processing(ICASSP'03), vol 4. ACM Press, NY, pp 676–679
29. Zhou K, Bao FJ, Shi JY (2002) A unified framework for digital geometry processing. Chin J Comput 25(9):904–909
30. Zhou K, Bao FJ, Shi JY (2003) 3D surface filtering using spherical harmonics. Comput Aided Design 3(3):478–487
31. Barini M, Bartolini F, Cappellini V, Piva A (1998) A DCT-domain system for robust image watermarking. Signal Process 66:357–372
32. Praun E, Hoppe H, Finkelstein A (1999) Robust mesh watermarking. SIGGRAPH Proc 4:69–76

Printed by Printforce, the Netherlands